W0111969

# Topics in Applied Physics
## Volume 106

Available **online** at
## Springer Link.com

Topics in Applied Physics is part of the SpringerLink service. For all customers with standing orders for Topics in Applied Physics we offer the full text in electronic form via SpringerLink free of charge. Please contact your librarian who can receive a password for free access to the full articles by registration at:

springerlink.com → Orders

If you do not have a standing order you can nevertheless browse through the table of contents of the volumes and the abstracts of each article at:

springerlink.com → Browse Publications

# Topics in Applied Physics

Topics in Applied Physics is a well-established series of review books, each of which presents a comprehensive survey of a selected topic within the broad area of applied physics. Edited and written by leading research scientists in the field concerned, each volume contains review contributions covering the various aspects of the topic. Together these provide an overview of the state of the art in the respective field, extending from an introduction to the subject right up to the frontiers of contemporary research.

Topics in Applied Physics is addressed to all scientists at universities and in industry who wish to obtain an overview and to keep abreast of advances in applied physics. The series also provides easy but comprehensive access to the fields for newcomers starting research.

Contributions are specially commissioned. The Managing Editors are open to any suggestions for topics coming from the community of applied physicists no matter what the field and encourage prospective editors to approach them with ideas.

## Managing Editor

Dr. Claus E. Ascheron

Springer-Verlag GmbH
Tiergartenstr. 17
69121 Heidelberg
Germany
Email: claus.ascheron@springer.com

## Assistant Editor

Adelheid H. Duhm

Springer-Verlag GmbH
Tiergartenstr. 17
69121 Heidelberg
Germany
Email: adelheid.duhm@springer.com

Marco Fanciulli    Giovanna Scarel  (Eds.)

# Rare Earth Oxide Thin Films

## Growth, Characterization, and Applications

With 210 Figures and 25 Tables

 Springer

Prof. Marco Fanciulli
Laboratorio Nazionale MDM CNR-INFM
Via C. Olivetti 2
20041 Agrate
Brianza (MI), Italy


Dr. Giovanna Scarel
Laboratorio Nazionale MDM CNR-INFM
Via C. Olivetti 2
20041 Agrate
Brianza (MI), Italy

Library of Congress Control Number: 2006935383

Physics and Astronomy Classification Scheme (PACS):
71.55.-i; 72.80.Sk; 73.20.At; 75.47.Lx; 77.55.+f


ISSN print edition: 0303-4216
ISSN electronic edition: 1437-0859
ISBN-10  3-540-35796-3  Springer Berlin Heidelberg New York
ISBN-13  978-3-540-35796-4  Springer Berlin Heidelberg New York

This work is subject to copyright. All rights are reserved, whether the whole or part of the material is concerned, specifically the rights of translation, reprinting, reuse of illustrations, recitation, broadcasting, reproduction on microfilm or in any other way, and storage in data banks. Duplication of this publication or parts thereof is permitted only under the provisions of the German Copyright Law of September 9, 1965, in its current version, and permission for use must always be obtained from Springer. Violations are liable for prosecution under the German Copyright Law.

Springer is a part of Springer Science+Business Media

springer.com

© Springer-Verlag Berlin Heidelberg 2007

The use of general descriptive names, registered names, trademarks, etc. in this publication does not imply, even in the absence of a specific statement, that such names are exempt from the relevant protective laws and regulations and therefore free for general use.

Typesetting: DA-TEX · Gerd Blumenstein · www.da-tex.de
Production: LE-TEX Jelonek, Schmidt & Vöckler GbR, Leipzig
Cover design: WMXDesign GmbH, Heidelberg

Printed on acid-free paper      57/3100/YL      5 4 3 2 1 0

# Preface

The technology for complementary metal-oxide-semiconductor (CMOS) is intensely searching for candidates to substitute $SiO_2$ as gate dielectric enabling the fabrication of small devices with low power consumption and high speed. The suitable candidates must have a high dielectric constant ($\kappa$ between 15-40), be thermodynamically stable in contact with the semiconductor substrate (Si, Ge, etc.) and with the metal gate, exhibit high conduction band offset with the semiconductor (to minimize leakage currents), and have low defect density both in the bulk and at the interfaces. Rare earth (RE, from La to Lu)-based oxides are among the candidates to replace $SiO_2$ as gate dielectrics first of all because of their predicted thermodynamical stability on silicon. In addition, high-$\kappa$ dielectrics, and RE oxides in particular, show potentials also for applications in nano-electronics, opto-electronics, and spintronics. Advantages and disadvantages of RE-based oxides are often related to the intrinsic properties of the RE elements:

1. the increasing $4f$ shell filling level (completely empty in La atoms, and totally filled in Lu ones);
2. the number and the value of the oxidation states (3 for La, Nd, Pm, Gd, Dy, Ho, Er, and Lu; 3 and 4 for Ce, Pr, and Tb; 3 and 2 for Sm, Eu, and Yb);
3. the decreasing ionic radius with increasing atomic number (from 0.123 nm for La to 0.092 nm for Lu), correlated with the decreasing Pauling electronegativity with increasing atomic number (from 2.34 for La to about 2.15 for Lu).

These intrinsic properties of the RE elements affect those of the corresponding oxides, especially in the form of nano-scaled films, and might influence their eligibility as high-$\kappa$ dielectrics. The main objective of this volume is to address the various properties of RE elements and to understand how to exploit them to obtain proper functionalities of their oxides.

The increasing $4f$ shell filling level could affect the trap density in the RE oxides, as the different energy and occupancy of the $d$ shells do in transition metal oxides. Moreover, it could determine the different reactivity and moisture sensitivity of the RE oxides. It is noteworthy that only four RE elements (i.e. La, Ce, Gd, and Lu, with respectively empty, mono-electron filled, half-filled and totally filled $4f$ shell) have also one electron in the outer $5d$ shell.

The existence of two oxidation states induces various stoichiometries in half of the RE elements (Ce, Pr, Sm, Eu, Tb, Tm, and Yb). This fact complicates the oxide chemistry and the electronic structure. The consequences of having pure $RE_2O_3$ oxides or a mixture of $RE_2O_3$ with REO or $REO_2$ oxides on the film functional properties might be remarkable and must be fully understood.

The mixed stoichiometries are expected to affect the band alignment. The reason is that the higher the stoichiometric ratio between the RE element and the oxygen in the oxide, the lower the height of the charge neutrality level (CNL) in the band gap and, finally, the higher also the conduction band offset (CBO) with Si, or another semiconductor substrate, e.g. Ge. Therefore, oxidation states of 2 are more favourable than 4. It is not clear, however, what are the consequences when thin films with mixed stoichiometries are produced.

In addition to the band alignment, the mixed stoichiometries might also affect the $\kappa$ value which is predicted to decrease as the atomic number of the RE decreases. In reality, however, the measured values are scattered, and sometimes anomalously high. Almost all RE elements with two oxidation states generate oxides with low band gaps ($\leqslant 4.0\,eV$) and very high $\kappa$ values. On the other hand, most RE elements with oxidation states 3 and 2 (Sm, Tm, and Yb) or with just oxidation state 3 (La, Nd, Gd, Dy, Ho, Er, and Lu) have a band gap around $5\,eV$, and a medium $\kappa$ value (about 13).

The decrease of the ionic radius with increasing atomic number of the RE elements accompanies the decrease of their Pauling electronegativity in a range that determines a constant oxygen coordination of 4 on all RE oxides. Such an oxygen coordination correlates with the disruption of the covalent network bonding, and leads to structures that readily crystallize at temperatures of interest in microelectronic processes ($500$–$1000\,^{\circ}C$ ). Ternary compounds based on RE atoms usually should exhibit lower oxygen coordination than the corresponding oxides, thus higher crystallization temperatures. They are therefore suitable alternatives to RE binary oxides. The electronegativity differences in the RE elements might also explain the different hygroscopic behaviour of the corresponding oxides.

The systematic investigation of the previously outlined issues is challenging because in first place good-quality thin films are needed. Several deposition techniques are currently being considered for nano-scale film fabrication, mainly on semiconductors: atomic layer deposition (ALD), electron beam evaporation, molecular beam epitaxy (MBE), metal-organic chemical vapour deposition (MOCVD), and pulsed laser deposition (PLD). Strength and weakness of these techniques are discussed in this volume, with a careful consideration of the growth mechanisms involved. Special attention is devoted to the synthesis and the properties of RE complexes used as precursors (e.g. for ALD and MOCVD). Methods to handle hygroscopicity in RE oxide films are considered.

In this volume, the structural and compositional properties of both the interface layer and the film are considered in detail, together with the crystalline or amorphous nature of the thin films, their roughness and homogeneity, and the interfacial layer between thin RE oxide layers and the substrate. These factors must be evaluated in order to proceed to a reliable electrical characterization and assess the potential of RE oxides for the various applications. The relationship between micro-structural and electrical properties is also considered. Finally, the real effectiveness of RE oxides in applications as high-$\kappa$ dielectrics for logic and memory devices, as active materials in laser technology, and in spintronics is discussed.

A significant investigation on RE oxide thin films is only at the beginning and requires expertise in many fields: growth methods, growth modeling and chemistry, physical-chemical characterization, and device technology. The effort naturally calls upon the convergence of many research groups on the same topic, whose attention we hope to catalyze through this volume.

Agrate Brianza (MI)                                                         *Marco Fanciulli*
November 2005                                                              *Giovanna Scarel*


## Acknowledgements

This volume collects most of the contributions given at the European Exploratory Workshop entitled *"Rare earth oxide thin films: growth, characterization, and applications"*, which took place at the Villa Nobel in Sanremo (IM), Italy, from May 11 to May 13, 2005. It is a pleasure to acknowledge and gratefully thank all the people and the institutions that supported the Workshop, and the preparation of this volume. The Workshop was funded by the European Science Foundation (ESF) through the Physics and Engineering Sciences Committee. The Province of Imperia made available for the workshop the beautiful and suggestive Villa Nobel. The staffs of Villa Nobel and of the Sanremo Promotion Agency have been very helpful in handling the logistics related to the Workshop. Sanremo Promotion also financially supported the Workshop. The National Institute for the Physics of Matter (INFM) contributed also to cover some of the expenses related to the Workshop. We also thank the administrative staff of the MDM National Laboratory (in particular Ms. Mara Lanati) for their help in the workshop organization, in the editorial work related to the collection and organization of the contributions for this volume, and in the handling of the financial part.

Agrate Brianza (MI)                                                         *Marco Fanciulli*
November 2005                                                              *Giovanna Scarel*

# Contents

## Molecular Beam Epitaxy of Rare-Earth Oxides
H. Jörg Osten, Eberhard Bugiel, Malte Czernohorsky, Zeyard Elassar,
Olaf Kirfel, Andreas Fissel .......................................  101

## Fabrication and Characterization of Rare Earth Scandate Thin Films Prepared by Pulsed Laser Deposition
Jürgen Schubert, Tassilo Heeg, Martin Wagner ..................... 115

## Film and Interface Layer Composition of Rare Earth (Lu, Yb) Oxides Deposited by ALD
Yuri Lebedinskii, Andrei Zenkevich, Giovanna Scarel, Marco Fanciulli . 127

## Electrical Characterization of Rare Earth Oxides Grown by Atomic Layer Deposition

## Dielectric Properties of Rare-Earth Oxides: General Trends from Theory

## Charge Traps in High-$k$ Dielectrics: Ab Initio Study of Defects in Pr-Based Materials

# Scientific and Technological Issues Related to Rare Earth Oxides: An Introduction

Giovanna Scarel[1], Axel Svane[2], and Marco Fanciulli[1]

[1] CNR-INFM MDM National Laboratory, Via C. Olivetti 2, 20041 Agrate Brianza (MI), Italy
giovanna.scarel@mdm.infm.it
[2] Department of Physics and Astronomy, University of Ååarhus, Ny Munkegade, DK 8000, Ååarhus, Denmark

**Abstract.** Significant research effort is currently being devoted to study deposition, dielectric, and electronic properties of binary rare earth oxides, as well as of complex oxides based on rare earth elements. Most of the motivations justifying this effort are found in the field of microelectronics – especially in the search of high dielectric constant oxides as candidates to substitute $SiO_2$ – and these will be mostly discussed. Open problems and issues from the scientific and technological point of view are discussed, and applications in fields other than microelectronics (such as spintronics) are also mentioned.

## 1 Introduction

The rare earth (RE) elements are the 15 elements of the Periodic Table (La, Ce, Pr, Nd, Pm, Sm, Eu, Gd, Tb, Dy, Ho, Er, Tm, Yb, and Lu) with atomic numbers from 57 through 71. Among them, Pm is radioactive and does not occur naturally, but might be prepared synthetically. In the outer electronic configuration of the RE element row, the $6s^2$ shell is always occupied, the $5d^1$ configuration appears in La, Ce, Gd and Lu, and finally the $4f$ shell is progressively filled as the atomic number increases. The degree of filling of the $4f$ shell is therefore the distinctive characteristic of the RE elements. In particular, the half-filled (Gd, with 7 electrons in the $4f$ shell) [1] and the totally filled (Lu, with 14 electrons in the $4f$ shell) configurations seem particularly stable. In the solid state, all 15 RE elements have the oxidation state +3, but some are stable also in the oxidation state +4 (Ce, Pr, and Tb), and others in the oxidation state +2 (Sm, Eu, Tm, and Yb). The classification of di-, tri-, and tetravalent RE elements and the RE–O bond lengths are summarized in Table 1. It is noteworthy that the +4 oxidation state appears in elements that follow one with a stable configuration, whereas the +2 oxidation state appears in elements that precede one with a stable configuration. These observations suggest that there might be "periodic" properties in the RE oxides. The stability of the tetravalent dioxides, trivalent sesquioxides, and divalent EuO and YbO, as well as the intermediate valent character of SmO are supported by self-interaction corrected total energy calculations [1–4] (see also the Chapter by *Petit* et al. in this volume).

M. Fanciulli, G. Scarel (Eds.): Rare Earth Oxide Thin Films,
Topics Appl. Physics **106**, 1–14 (2007)
© Springer-Verlag Berlin Heidelberg 2007

**Table 1.** RE–O bond lengths (in nm) of all known di-, tri-, and tetra-valent RE oxides [5]. "*" trivalent monoxides [6], "+" intermediate valent oxides [6]

| | La | Ce | Pr | Nd | Pm | Sm | Eu | Gd | Tb | Dy | Ho | Er | Tm | Yb | Lu |
|---|---|---|---|---|---|---|---|---|---|---|---|---|---|---|---|
| REO | 0.257* | 0.254* | 0.252* | 0.250* | – | 0.249+ | 0.257 | – | – | – | – | – | – | 0.244 | – |
| RE$_2$O$_3$ | 0.246 | 0.244 | 0.238 | 0.238 | 0.236 | 0.235 | 0.233 | 0.232 | 0.230 | 0.229 | 0.228 | 0.227 | 0.225 | 0.224 | 0.223 |
| REO$_2$ | – | 0.234 | 0.233 | – | – | – | – | – | 0.226 | – | – | – | – | – | – |

Other properties change monotonously, such as those depending on the ionic radii of the RE elements summarized in Table 2 for the oxides of the dominant trivalent RE elements. The RE atomic radii indeed decrease as the atomic number increases – this is the so-called "lanthanide contraction" – from 0.123 nm in La to 0.092 nm in Lu [7]. The lanthanide contraction is due to the increase in localization of the $4f$ shell as the atomic number increases [1].

**Table 2.** Values of the ionic radii (in nm) of the trivalent RE ions [7]

| La$^{3+}$ | Ce$^{3+}$ | Pr$^{3+}$ | Nd$^{3+}$ | Sm$^{3+}$ | Eu$^{3+}$ | Gd$^{3+}$ | Tb$^{3+}$ | Dy$^{3+}$ | Ho$^{3+}$ | Er$^{3+}$ | Tm$^{3+}$ | Yb$^{3+}$ | Lu$^{3+}$ |
|---|---|---|---|---|---|---|---|---|---|---|---|---|---|
| 0.123 | 0.115 | 0.114 | 0.112 | 0.106 | 0.106 | 0.104 | 0.100 | 0.099 | 0.098 | 0.096 | 0.094 | 0.093 | 0.092 |

One of the issues of major concern in microelectronics is how the periodic or monotonously changing characteristics of the RE elements could influence the properties required for making these oxides eligible, e.g., as high dielectric constant ($\kappa$) oxide candidates to substitute SiO$_2$ in complementary metal-oxide-semiconductor (CMOS) devices [8]. These properties are: thermodynamical stability on semiconductors (Si [9], Ge, GaAs, etc.), oxide band structure, band gap, and conduction band offsets with the semiconductors, $\kappa$ value of the oxides, and chemical stability. In the following text, the general requirements for high-$\kappa$ materials candidates in microelectronics will be discussed in the specific case concerning RE-based oxides, while other possible applications will be briefly addressed too.

## 2 Thermodynamical Stability

Current research accepts the group IV B-based oxides, in particular HfO$_2$ and other Hf-based materials, and partly also ZrO$_2$, as the main high-$\kappa$ materials candidates to substitute SiO$_2$ in advanced devices. All other transition metal oxides, which may even possess higher $\kappa$-values than the previously mentioned group IV B oxides, are ruled out because calculations [9] show that they are thermodynamically unstable on Si (i.e., the reaction between oxide and Si leading to the formation of a layer of SiO$_x$, with $x \leq 2$ , at the interface between the two, is favourable). Also Sc and Y oxides could be considered, but these have a significantly lower $\kappa$ than HfO$_2$ and ZrO$_2$. On the other hand, the oxides of all the RE elements (except Pm) are considered as eligible from the thermodynamical point of view [3], which leaves the choice on the oxides of 14 different elements.

The absence of a low-$\kappa$ interfacial layer of $SiO_x$, or Si-based compounds, between Si, or in general, between the semiconductor and the oxide, is essential in order not to lower the $\kappa$ of the entire insulating stack in CMOS and memory devices. Indeed, experimentally it was demonstrated that some of the RE oxides can be grown epitaxially on Si or GaAs, with sharp interfaces, for example, in the case of $Pr_2O_3$ on Si(111) deposited in a temperature range between 500–700 °C by molecular beam epitaxy (MBE) [10]. Another notable example is $Gd_2O_3$ on GaAs(100) deposited also by MBE at growth temperatures between 200–550 °C [11]. There is a factor, however, which the simple thermodynamics of RE oxides in contact with Si can not take into account. This factor is the possible catalytic role that the RE oxides play toward $O_2$, supplied either in post-deposition annealing environments or simply in air: the RE oxides easily dissociate $O_2$ into atomic oxygen [12]. The latter phenomenon promotes the formation of $SiO_x$, or Si-based compounds, between Si and the RE oxide. In turn, both experiment and theory indicate that the reaction between $SiO_2$ and $RE_2O_3$ readily forms RE silicates (free energy of formation $\Delta G \ll -100 \, kJ/ \, mol$), much more than in the case of transition metal oxides ($\Delta G \sim -1\text{–}10 \, kJ/ \, mol$) [13, 14]. X-ray reflectivity (XRR) measurements on as-grown and annealed $Lu_2O_3$ [15] and $Yb_2O_3$ [16] films deposited by atomic layer deposition (ALD) on Si(100) support the latter conclusions. The presence or absence of Si in the used precursors ($\{[\eta^5 - C_5H_4SiMe_3]_2LuCl\}_2$ for Lu [17] and $Yb(C_5H_5)_3$ for Yb) seem not to affect the final result.

On semiconductor substrates other than Si, the situation is different. X-ray photoelectron (XPS) data on ultra-thin $Lu_2O_3$ films deposited by ALD on Ge(100) and GaAs(100) indicate no interfacial $GeO_2$ or $Ga_2O_3$ layers (arsenic usually does not oxidize under standard cleaning procedures)(see the Chapter by *Zenkevich* et al. in this volume). Semiconductors other than Si, some of them offering high electron mobility, could therefore be a solution against interfacial layer formation, at least in as-grown layers. Nevertheless, other paths are sought in order to avoid interfacial layer formation also upon thermal treatments of the RE oxides/semiconductor stacks. If, however, an interfacial layer can not be completely avoided, it is necessary to single out the best possible substrate preparation method, deposition parameters, and deposition protocols to achieve a good-quality interfacial layer, which in real devices could limit mobility reduction due to remote phonon scattering [18].

# 3 Band Structure and Band Offsets

The +3 oxidation state is generally the stable one for the RE elements in the solid state. This is advantageous because, when one element has more than one stable oxidation state, more than one stoichiometry is possible which, in turn, could lead to a complicated band structure. In the lighter RE elements,

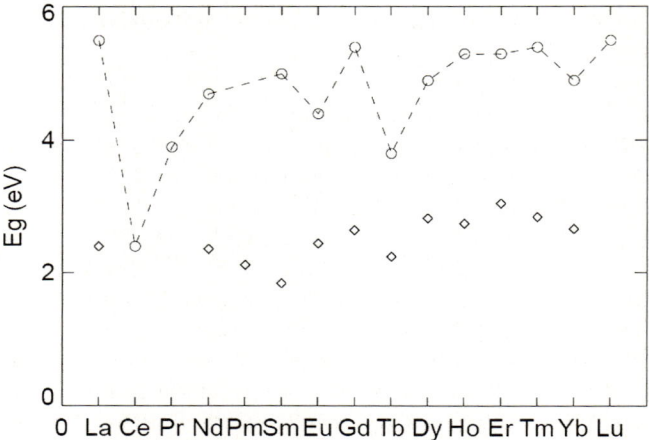

**Fig. 1.** Fundamental energy gap of the RE trivalent sesquioxides. Optical gaps [21] are shown as *circles* connected by a *dashed line*, while the $E_g$ values deduced from high-temperature conductivity experiments [22] are shown with *diamonds*. The origin of the smaller values of the conductivity gaps is not known

however, i.e., in Ce and Pr, where the $f$-electrons are less tightly bound, and in those with one extra $f$-electron on top of the half-filled $f$ shell, i.e., in Tb, a large oxygen coordination number is displayed and an oxidation state larger than 3 is favoured [1]. For the Ce, Pr, and Tb oxides, therefore, both the $REO_2$ and the $RE_2O_3$ stoichiometries can be found. If potentially a complication for the band structure, the ability of an oxide to easily transform from one stoichiometry to another might be usefully exploited. Indeed, the Ce oxide system was proposed as a potential switching device based on a reversible oxygen storage–release mechanism whereby the creation of an oxygen vacancy in a $CeO_2$ lattice would transfer an electron to Ce atoms, changing its oxidation state to +3. In reverse, supply of oxygen to the $Ce_2O_3$ lattice would increase the oxygen coordination of the cation and oxidize it to a Ce +4 state [19].

The oxides of the RE elements immediately before the very stable ones, and with a possible +2 oxidation state (Sm, Eu, Tm, and Yb), are instead oxygen deficient. Not much is known about them, apart from EuO, in which Eu has a half-filled $f$ shell. This compound is known as one of the few binary ferromagnetic oxides, with an insulator to metal transition below 69 K, and a nearly 100 % spin polarization of the electrons close to the conduction band edge, which makes it interesting for spintronic applications [20].

The high energy and the low occupancy of the outer $d$ shell in RE oxides is a beneficial property because it seems to avoid the occurrence of configuration instabilities of the Jahn–Teller type [23]. The latter, which may show up in Ti-based compounds (with a low-energy $d$ shell), indeed are supposed to lead to transport states for leakage current [24]. However, the absence of a bond strain reduction-driven self-organization at internal RE oxide/$SiO_2$ interfaces might increase the amount of fixed charges [25], and mitigate the previously mentioned advantages of the RE oxides.

The band gap ($E_g$) of the RE oxides vary in an almost "periodic" way with increasing atomic number. The optically measured $E_g$ values presented in Fig. 1 clearly show this trend. The $E_g$ values measured from high-temperature conductivity experiments [22] are also shown. For single crystals, it was found [21] that the oxides of the RE elements with the most stable configurations (La, Gd, and Lu) have the largest $E_g$ ($\sim$ 5.5 eV). These large values should correspond to transitions from the valence to the conduction bands in the optical spectra [1, 22]. Indeed, also high-temperature conductivity experiments [22] reveal that in $La_2O_3$, $Gd_2O_3$, $Lu_2O_3$, as well as in $Dy_2O_3$, $Ho_2O_3$, and $Er_2O_3$ the $f$-levels are not situated in the band gap, but appear as resonances within the bands. As a consequence, these oxides belong to class A according to the classification of Fig. 2. The oxides of the RE elements immediately following in atomic number those with the most stable configurations have significantly lower $E_g$ values (2.3 eV for Ce-, 3.9 eV for Pr-, and 3.8 eV for Tb-oxides [21]). These values likely correspond to transitions from the occupied $f$ levels above the valence band to the conduction edge [1, 22]. Thus, $Ce_2O_3$, $Pr_2O_3$, $Nd_2O_3$, and $Tb_2O_3$, in which the lowest energy excitations involve an $f$-electron being transferred into the conduction band, are considered to belong to class B (Fig. 2) [22]. Also the oxides of the RE elements located in the Periodic Table immediately before those with the most stable configurations have low $E_g$ values (4.2 eV for Eu-, and 4.4 eV for Yb-oxides). These values should correspond to transitions from the valence band to the empty $f$ levels below the conduction band [1, 22]. Therefore, $Sm_2O_3$, $Eu_2O_3$, $Tm_2O_3$, and $Yb_2O_3$, in which the lowest energy excitations involve a hole in the $p$-band and an extra $f$-electron captured on a RE site, belong to class C (Fig. 2) [22].

For thin films, experimentally the $E_g$ of $Lu_2O_3$ was measured to be 5.8 eV [26], whereas the $E_g$ values of $PrO_2$ and $Pr_2O_3$ were calculated to be respectively 1.1 and 2.6 eV [1], and 2–3 eV [27]. There is an urgent need for both accurate measurements and calculations of the $E_g$ and the band structure of RE oxides. In particular, calculations based on the GW approximation [27, 28] should be able to provide quantitatively reliable $E_g$ values, without the usual underestimation affecting the density-functional theory-based methods, which also includes the band gaps calculated with the self-interaction correction approach in [1](see also the Chapter by *Petit* et al. in this volume).

Given the above-described behaviour of the $E_g$ values in RE oxides, seeming to be "periodic" with the $f$ shell filling level throughout the RE element row, it is legitimate to ask whether the conduction band offset (CBO) with, e.g., Si, would also obey this "periodicity". This issue is important to clarify because the higher the CBO, the higher the probability to limit leakage currents through the insulating RE oxide layers. Therefore, CBO values are a

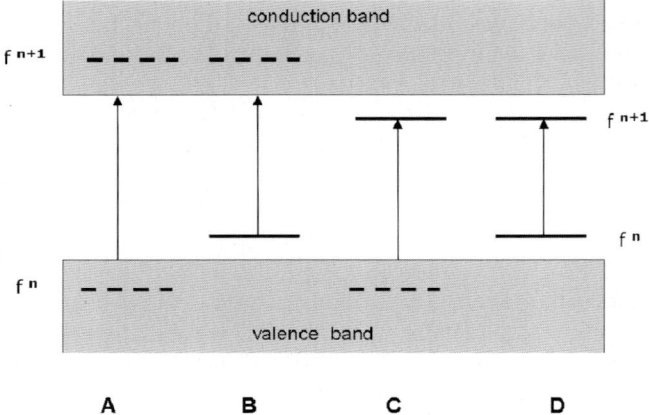

**Fig. 2.** The four different possible band gap structures of the RE oxides. The valence band refers to the occupied O $p$-states, while the conduction band is comprised of the unoccupied non-$f$ electron states, which are mainly of RE $d$ character. The localized occupied and unoccupied $f$-levels may fall above or below the band edges, giving rise to the four different possibilities: A: $v \rightarrow c$, B: $f \rightarrow c$, C: $v \rightarrow f$, D: $f \rightarrow f$. The classification is adapted from [22]

key factor for the choice of high-$\kappa$ dielectrics, candidates for microelectronic applications.

A systematic collection of experimental data on CBO values of RE oxides on Si is not available, but *Robertson* [29] calculated a CBO value of 1.5 eV for $HfO_2$ on Si and a value of 2.3 eV for $La_2O_3$ on Si. The result was explained in terms of the 2 : 3 metal:oxygen stoichiometry ratio in $La_2O_3$ films on Si being more favourable than the 1 : 2 one in $HfO_2$ films on Si in promoting high CBO values. The finding boosted the attention toward RE oxides as potential high-$\kappa$ dielectric candidates for microelectronics, certainly, however, not at the expense of $HfO_2$, which is currently still considered among the most promising. Experimental results do not support completely the described theoretical view on the CBO values of oxides on Si. Indeed, the CBO of $HfO_2$ on Si was measured to be 2.0 eV [30] using internal photoemission (IPE) spectroscopy and also XPS (see Chapter by *Seguini* et al. in this volume). The CBO of $Lu_2O_3$ was measured to be 2.1 eV using IPE [26]. The CBO values of the two small gap RE oxides $Pr_2O_3$ and $CeO_2$ (see Fig. 1) have recently been investigated: The CBO of $Pr_2O_3$ was measured using high field tunneling and determined to be 0.5–1.5 eV [31]. The CBO of $CeO_2$ was estimated to be 0.1 eV [32]. The CBO of $Yb_2O_3$ films on Si is under investigation, but literature data on energy barriers at the metal–$Yb_2O_3$–Si interfaces suggest its value also to be 2.1 eV [33]. This result is surprising, given the slightly lower $E_g$ of $Yb_2O_3$ [21] compared with that of $Lu_2O_3$ films. The CBO of RE oxides with a significantly lower $E_g$ than $La_2O_3$, $Gd_2O_3$, and $Lu_2O_3$ should be more carefully investigated to clarify if the $f$ shell filling level indeed affects the CBO value. $Tb_2O_3$ could be an oxide of choice for the test. For this investigation, the use of techniques for the direct measurement of the CBO, such as IPE, is highly recommended, because they do not need to rely on $E_g$ values. $La_2O_3$, $Gd_2O_3$, and $Lu_2O_3$ films in turn have a reported XPS-measured CBO on Si of, respectively, 2.3 eV, 3.1 eV, and 1.9 eV [34]. Again, further theoretical efforts, not just the experimental ones, are called for.

# 4 Dielectric Constant – $\kappa$

The $\kappa$ value of a compound is intimately related to frequencies of its dominant infrared optical modes [35], which in turn are related to the compound crystalline structure. The early RE oxides (from $La_2O_3$ to $Pr_2O_3$) have a stable hexagonal structure [1, 36]. $Nd_2O_3$ exhibits both the hexagonal and the cubic structure. Finally, in the oxides from $Sm_2O_3$ to $Lu_2O_3$, the cubic structure – also known as the bixbyite structure – is the most stable one, but a monoclinic distortion of the cubic structure might appear as well [1, 36]. The infrared vibrational spectra differ for the various structures and especially, so does the frequency of the most intense absorption band [37]. The $\kappa$ value is higher for crystalline films with structures having the most intense absorption band at lower frequencies. As an example, $La_2O_3$ crystallized in the hexagonal structure exhibits the most intense transverse optical mode around $200\,cm^{-1}$, and has an expected dielectric constant of 17–20 [35, 37]. On the other hand, $Lu_2O_3$ crystallized in the cubic bixbyite structure has the most intense transverse optical mode around $300\,cm^{-1}$, and both its calculated and measured $\kappa$ values are 12 [15, 35].

These observations suggest that the early RE oxides crystallized in the hexagonal structure should exhibit higher $\kappa$ values than the oxides crystallized in the cubic and monoclinic ones. Indeed, the limited available results on thin films confirm that this is the case, but establish also that the $\kappa$ values vary significantly depending on film deposition method, thickness, purity, and other factors, as the following examples clearly show. For $La_2O_3$ on Si(100), measurements for films grown using electron beam deposition in a thickness range between 2 and 7.8 nm resulted in $\kappa$ values from 8 to 23 (the latter value was found in the thicker film) [38]. On the other hand, for La oxide films deposited using metal-organic chemical vapor deposition (MOCVD), the maximum $\kappa$ was measured to be 19 [39] on films in a 5 to about 30 nm thickness range on a nitridated interfacial layer (the nitridated layer limited the formation of Si-rich both interfacial and La oxide layers). For a 11.4 nm thick $Pr_2O_3$ layer deposited on Si(001) using MBE, a $\kappa$ of 30±3 was measured [40]. On the other hand, even 80 nm to 100 nm thick films of Pr oxide deposited using ALD from $Pr[N(SiMe_3)_2]_3$ and $H_2O$ yielded a $\kappa$ of only half the value of the films deposited using MBE [41]. The high Si content in the latter films was considered to limit $\kappa$ so dramatically. Finally, for $CeO_2$ films deposited on Si(111) by MBE, for example, a large $\kappa$ around 50 was reported [32], but another research group needed to assume a value of 26 to correctly simulate the electrical characteristics of Si(111)/SiO$_2$/CeO$_2$ stacks [42]. The very high $\kappa$ value in the first case was explained to be a consequence of oxygen defects decreasing the Coulomb interactions in the oxide crystal, and consequently expanding its lattice spacing [32]. A summary of $\kappa$ values of RE oxides appears in Table 3.

## 5 Effects of the Monotonously Changing Properties: The Ionic Radii

As opposed to the RE oxide electronic properties, which are "periodic", other properties change monotonously through the RE element row. These properties are mainly related to the "lanthanide contraction" [1, 7], associated with a wealth of consequences for the properties of RE oxides. Two examples will be reported here. The first one is related to the hygroscopicity of the RE oxides, which affects their electrical characteristics, and is therefore a major issue for their applications.

**Table 3.** Summary of $\kappa$ values and $E_g$ of the corresponding RE oxides

| Oxide | $\kappa$ (bulk) | $\kappa$ (thin films) | $E_g$ (eV) |
|---|---|---|---|
| La$_2$O$_3$(h) | – | 8–23 [38]–19 [39] | 5.5 |
| CeO$_2$(h) | 16.6 | 52 [32]–26 [42] | 3.78 [43] |
| Pr$_2$O$_3$(h) | 14.9 | 30 [40]–15 [41] | 3.8 |
| Nd$_2$O$_3$ | 14.3–16 | 11.7 [44] | 4.6 |
| Sm$_2$O$_3$(c) | – | 10 [45]–43 [46] | 5.0 |
| Eu$_2$O$_3$(c) | 13.7 | – | 4.3 |
| EuO(c) | 23.9 | – | 1.12 [47] |
| Gd$_2$O$_3$(c) | 13.6 | 14 [48]–16 [49]–23 [50] | 5.4 |
| Tb$_2$O$_3$(c) | 13.3 | – | 3.8 |
| Dy$_2$O$_3$(c) | 13.1 | – | 4.9 |
| Ho$_2$O$_3$(c) | 13.1 | – | 5.3 |
| Er$_2$O$_3$(c) | 13.0 | 7–14 [51] | 5.3 |
| Tm$_2$O$_3$(c) | 12.6 | 7–22 [52] | 5.4 |
| Yb$_2$O$_3$(c) | 12–13.4 | 14 [38], $12 \pm 1$ [1] | 4.9 |
| Lu$_2$O$_3$(c) | 12.5 | $12 \pm 1$ [15, 35]–11 [53] | 5.5 |

$\kappa$ values for the bulk from [54]. $E_g$ values from [21].
Code: h = hexagonal structure, m = monoclinic structure
c = cubic structure.

---

[1] see the Chapter by *Fanciulli* et al. in this volume

Recently, *Jeon* et al. [55] showed that the reactivity with water decreases in RE oxides as the ionic radius of the RE element decreases. The conclusion was established from the analysis of the XPS O $1s$ peak area corresponding to the hydroxide for the considered RE oxides. Furthermore, the result was related to the higher reactivity of H atoms with cations having low electronegativity, such as the early RE elements as opposed to the late ones [55]. The lattice energy was also called into play to explain the phenomenon [53]. Another example refers to ternary RE-based manganites of interest as magnetoelectric materials, such as the $REMnO_3$ compounds [56]. At least two RE-oxygen types of bonds were singled out in the hexagonal $REMnO_3$ compounds, and both were found to increase within 2.3 %–3.6 % with the increase of the RE element ionic radius. This change in bond length was related to a change of the local dipole moment in the $REO_7$ polyhedra. This result is relevant, but the antiparallel coupling between the two local dipole moments at the two RE sites prevents a clear effect on the total dipole moment [56]. Nevertheless, in other complex RE-based manganites the final effect on the dipole moment could be significant, given that also the whole bond geometry is likely to be very different.

# 6 Summary

Several issues are identified to be relevant in both the scientific and the technological consideration of RE oxides, especially in microelectronic applications. 1. The thermodynamical stability of RE oxides on various semiconductors (e.g., Si, Ge, GaAs) differs significantly depending upon the semiconductor, and efforts are needed to define substrate preparation, deposition parameters, and deposition protocols to achieve the sharpest possible interface between RE oxide and semiconductor, or, at least, the best possible quality of the interfacial oxide capable of limiting mobility reduction due to remote phonon scattering [18]. 2. The electronic properties of the RE oxides are acknowledged to vary almost periodically throughout the RE element row in the Periodic Table, according to the filling level of the $f$ shell. The empty, half-filled, and completely filled $f$ shells offer the most stable configurations, and also those with the largest band gap [21]. It is not yet clear whether also the conduction band offsets of the RE oxides with respect to Si or other semiconductors exhibit a "periodic" behaviour. More investigation is required to clarify the issue. 3. $\kappa$ is expected to be higher for the early RE oxides than for the late ones. The former crystallize preferentially in the hexagonal phase, whose most intense absorption band in the infrared vibrational spectra appears at lower frequencies than the corresponding ones of the late RE oxides crystallizing in the cubic bixbyite structure. Notable deviations from this predicted behaviour in thin RE oxide films are ascribed to film deposition method, thickness, purity, and other factors (some examples are described in the text). 4. The properties of ternary compounds based on RE elements are not discussed in this Introduction, but given the large possible number of element and stoichiometry combinations, it can be argued that these can tune the properties of the RE-based compounds. This topic is an interesting and open chapter in Materials Science.

Various fields of applications are mentioned in this Introduction for the RE oxides, such as CMOS devices on Si and high-mobility semiconductors, memory devices, new switching mechanisms for logic and memory devices, ferromagnetic and magnetoelectric materials for spintronic applications. Other applications of RE oxides, and RE-based compounds, e.g., in optoelectronics, will be discussed throughout the book.

## Acknowledgements

The authors thank S. Spiga, C. Wiemer, E. Bonera, G. Tallarida, A. Debernardi, G. Seguini, S. Baldovino, M. Perego, and O. Costa for actively and fruitfully contributing to the reserach on RE oxides at the CNR-INFM MDM National Laboratory. The authors are indebted also to A. Zenkevich and Yu. Lebedinskii of the Moscow Engineering Physics Institute, and to I. L. Fedushkin of the G. A. Razuvaev Institute of Organometallic Chemistry in Nizhny Novgorod, Russia, for excellent collaboration. Prof. G. Lucovsky (North Carolina State University) is acknowledged for fruitful discussions. The Italian Ministry of Foreign Affairs, through Joint Projects between Italy and Russia, and the National Institute for the Physics of Matter (INFM), through the PAIS-REOHK project, are acknowledged for funding.

# References

[1] L. Petit, A. Svane, Z. Szotek, W. M. Temmerman: First principles study of rare-earth oxides, Phys. Rev. B
[2] W. M. Temmerman, Z. Szotek, A. Svane, P. Strange, H. Winter, A. Delin, B. Johansson, O. Eriksson, L. Fast, J. M. Wills: Electronic configuration of ytterbium compounds, Phys. Rev. Lett. **83**, 3900 (1999)
[3] M. Horne, P. Strange, W. M.Temmerman, Z. Szotek, A. Svane, H. Winter: The electronic structure of europium chalcogenides and pnictides, J. Phys. Condens. Matter **16**, 5061 (2004)
[4] A. Svane, V. Kanchana, G. Vaitheeswaran, G. Santi, W. M. Temmerman, Z. Szotek, P. Strange, L. Petit: Electronic structure of samarium monopnictides and monochalcogenides, Phys. Rev. B **71**, 45119 (2005)
[5] P. Villars, L. D. Calvert: *Pearson's handbook of crystallographic data for intermetallic phases*, 2nd ed. (ASM International, Ohio 1991)
[6] J. M. Leger, N. Yacoubi, J. Loriers: Synthesis of rare earth monoxides, J. Sol. State Chem. **36**, 1981 (1981)
[7] Y. Sakabe, Y. Hamaji, H. Sano, N. Wada: Effects of rare-earth oxides on the reliability of X7R dielectrics, Jpn. J. Appl. Phys. **41**, 5668 (2002)

[8] G. D. Wilk, R. M. Wallace, J. M. Anthony: High-$\kappa$ gate dielectrics: current status and materials properties, J. Appl. Phys. **89**, 5243 (2001)

[9] D. G. Schlom, J. H. Haeni: A thermodynamic approach to selecting alternative gate deielctrics, MRS Bull. **27**, 198 (2002)

[10] J. P. Liu, P. Zaumseil, E. Bugiel, H. J. Osten: Epitaxial growth of $Pr_2O_3$ on Si(111) and the observation of a hexagonal to cubic phase transition during post-growth $N_2$ annealing, Appl. Phys. Lett. **79**, 671 (2001)

[11] M. Hong, J. Kwo, A. R. Kortan, J. P. Mannaerts, A. M. Sergent: Epitaxial cubic gadolinium oxide as a dielectric for gallium arsenide passivation, Science **283**, 1897 (1999)

[12] V. Narayanan, S. Guha, M. Copel, N. A. Bojarczuk, P. L. Flaitz, M. Gribelyuk: Interfacial oxide formation and oxygen diffusion in rare earth oxide – silicon epitaxial heterostructures, Appl. Phys. Lett. **81**, 4183 (2002)

[13] S. Stemmer: Thermodynamic considerations in the stability of binary oxides for alternative gate dielectrics in complementary metal-oxide-semiconductors, J. Vac. Sci. Technol. B **22**, 791 (2004)

[14] L. Marsella, V. Fiorentini: Structure and stability of rare-earth and transition-metal oxides, Phys. Rev. B **69**, 172103 (2004)

[15] G. Scarel, E. Bonera, C. Wiemer, G. Tallarida, S. Spiga, M. Fanciulli, I. Fedushkin, H. Schumann, Y. Lebedinskii, A. Zenkevich: Atomic layer deposition of $Lu_2O_3$, Appl. Phys. Lett. **85**, 630 (2004)

[16] M. Malvestuto, G. Scarel, C. Wiemer, M. Fanciulli, F. D'Acapito, F. Boscherini: X-ray absorption spectroscopy study of $Yb_2O_3$ and $Lu_2O_3$ thin films deposited on Si(100) by atomic layer deposition, Nuc. Instr. Method. B **246**, 90 (2006)

[17] H. Schumann, I. Fedushkin, M. Hummert, G. Scarel, E. Bonera, M. Fanciulli: Crystal and molecular structure of $[(\eta^5$-$c_5h_4sime_3)_2lucl]_2$ – suitable precursor for $lu_2o_3$ films, Z. Naturforsch. **59b**, 1035 (2004)

[18] M. F. Fischetti, D. A. Neumayer, E. A. Cartier: Effective electron mobility in Si inversion layers in metal-oxide-semiconductor systems with a high-$\kappa$ insulator: the role of remote phonon scattering, J. Appl. Phys. **90**, 4587 (2001)

[19] N. V. Skorodumova, S. I. Simak, B. I. Lundqvist, I. A. Abrikosov, B. Johansson: Quantum origin of the oxygen storage capability of ceria, Phys. Rev. Lett. **89**, 166601 (2002)

[20] J. Lettieri, V. Vaithyanathan, S. K. Eah, J. Stephens, V. Sih, D. D. Awschalom, J. Levy, D. G. Schlom: Epitaxial growth and magnetic properties of EuO on (001) Si by molecular-beam epitaxy, Appl. Phys. Lett. **83**, 975 (2003)

[21] A. V. Prokofiev, A. I. Shelyakh, B. T. Melekh: Periodicity in the band gap variation of $Ln_2X_3$ (X = O, S, Se) in the lanthanide series, J. All. Comp. **242**, 41 (1996)

[22] H. B. Lal, K. Gaur: Electrical conduction in non-metallic rare-earth solids, J. Mater. Sci. **23**, 919 (1988)

[23] I. A. Bersuker: *Electronic Structure and Properties of Transition Metal Compounds* (Wiley 1996) Chap. 7

[24] G. Lucovsky, Y. Zhang, G. B. Rayner, G. Appel, H. Ade, J. L. Whitten: Electronic structure of high-$k$ transition metal oxides and their silicate and aluminate alloys, J. Vac. Sci. Technol. B **20**, 1739 (2002)

[25] G. Lucovsky, J. P. Maria, J. C. Phillips: Interfacial strain-induced self-organization in semiconductor dielectric gate stacks. II. Strain-relief at internal dielectric interfaces between $SiO_2$ and alternative gate dielectrics, J. Vac. Sci. Technol. B **22**, 2097 (2002)

[26] G. Seguini, E. Bonera, S. Spiga, G. Scarel, M. Fanciulli: Energy-band diagram of metal/$Lu_2O_3$/silicon structures, Appl. Phys. Lett. **85**, 5316 (2004)

[27] J. Dąbrowski, V. Zavodinsky, A. Fleszar: Pseudopotential study of $PrO_2$ and $HfO_2$ in fluorite phase, Microelectron. Rel. **41**, 1093 (2001)

[28] L. Hedin: On the correlation effects in electron spectroscopies and the GW approximation, J. Phys. Condens. Matter **11**, R489 (1999)

[29] J. Robertson: Electronic structure and band-offsets of high-dielectric-constant gate oxides, MRS Bull. **27**, 217 (2002)

[30] V. V. Afanas'ev, A. Stesmans, F. Chen, X. Shi, S. A. Campbell: Internal photoemission of electrons and holes from (100)Si into $HfO_2$, Appl. Phys. Lett. **81**, 1053 (2002)

[31] H. J. Osten, J. P. Liu, H. J. Müssig: Band gap and band discontinuities at crystalline $Pr_2O_3$/Si(001) heterojunctions, Appl. Phys. Lett. **80**, 297 (2002)

[32] Y. Nishikawa, T. Yamaguchi, M. Yoshiki, H. Satake, N. Fukushima: Interfacial properties of single crystalline $CeO_2$ high-$k$ gate dielectrics directly grown on Si(111), Appl. Phys. Lett. **81**, 4386 (2002)

[33] V. A. Rozhkov, A. Y. Trusova: Energy barriers at the interfaces in the MIS system Me–$Yb_2O_3$–Si, Tech. Phys. **44**, 404 (1999)

[34] T. Hattori, T. Yoshida, T. Shiraishi, K. Takahashi, H. Nohira, S. Joumori, K. Nakajima, M. Suzuki, K. Kimura, I. Kashiwagi, C. Ohshima, S. Ohmi, H. Iwai: Composition, chemical structure, and electronic band structure of rare earth oxide/Si(100) interfacial transition layer, Microel. Eng. **72**, 283 (2004)

[35] E. Bonera, G. Scarel, M. Fanciulli, P. Delugas, V. Fiorentini: Dielectric properties of high-$\kappa$ oxides: Theory and Experiment for $Lu_2O_3$, Phys. Rev. Lett. **94**, 27602 (2005)

[36] G. V. Samsonov, I. Y. Gil'man: Electronic structure and physical properties of the oxides of the lanthanides, Soviet Powder Metallurgy and Metal Ceramics **13**, 925 (1974)

[37] P. Delugas, V. Fiorentini: Dielectric properties of the two phases of crystalline lutetium oxide, Microelectron. Rel. **45**, 831 (2005)

[38] S. Ohmi, C. Kobayashi, I. Kashiwagi, C. Ohshima, H. Ishiwara, H. Iwai: Characterization of $La_2O_3$ and $Yb_2O_3$ thin films for high-k gate insulator application, J. Electrochem. Soc. **150**, F134 (2003)

[39] H. Yamada, T. Shimizu, A. Kurokawa, K. Ishii, E. Suzuki: MOCVD of high-dielectric-constant lanthanum oxide thin films, J. Electrochem. Soc. **150**, G429 (2003)

[40] A. Fissel, H. J. Osten, E. Bugiel: Towards understanding epitaxial growth of alternative high-$K$ dielectrics on Si(001): Application to praseodymium oxide, J. Vac. Sci. Technol. B **21**, 1765 (2003)

[41] K. Kukli, M. Ritala, T. Pilvi, T. Sajavaara, M. Leskelä, A. C. Jones, H. C. Aspinall, D. G. Gilmer, P. J. Tobin: Evaluation of a praseodymium precursor for atomic layer deposition of oxide dielectric films, Chem. Mater. **16**, 5162 (2004)

[42] L. Tye, N. A. El-Masry, T. Chikyow, P. McLarty, S. M. Bedair: Electrical characteristics of epitaxial $CeO_2$ on Si(111), Appl. Phys. Lett. **65**, 3081 (1994)

[43] T. R. Griffiths, M. J. Davies, H. V. S. A. Hubbard: Spectroscopic studies on single crystals having the fluorite lattics, J. Chem. Soc. Faraday Trans. II **4**, 765 (1976)

[44] A. A. Dakhel: Characterisation of $Nd_2O_3$ thick gate dielectric for silicon, Phys. Stat. Sol. (a) **201**, 745 (2004)

[45] J. Päiväsaari, M. Putkonen, L. Niinistö: A comparative study on lanthanide oxide thin films grown by atomic layer deposition, Thin Solid Films **472**, 275 (2005)

[46] A. A. Dakhel: Dielectric and optical properties of samarium oxide thin films, J. All. Comp. **365**, 233 (2004)

[47] P. Watcher: Europium Chalcogenides: EuO, EuS, EuSe, and EuTe, in K. A. Gschneider, Jr., L. Eyring (Eds.): *Handbook on the Physics and Chemistry of Rare Earths*

[48] J. Kwo, M. Hong, B. Busch, D. A. Muller, Y. J. Chabal, A. R. Kortan, J. P. Mannaerts, B. Yang, P. Ye, H. Grossmann, A. M. Sergent, K. K. Ng, J. Bude, W. H. Schulte, E. Garfunkel, T. Gustafsson: International conference on molecular beam epitaxy (cat. no. 02ex607), in (2002) p. 47

[49] D. Landheer, J. A. Gupta, G. I. Sproule, J. P. McCaffrey, M. J. Graham, K.-C. Yang, Z.-H. Lu, W. N. Lennard: Characterization of $Gd_2O_3$ films deposited on Si(100) by electron-beam evaporation, J. Electrochem. Soc. **148**, G29 (2001)

[50] M. P. Singh, C. S. Thakur, K. Shalini, S. Banerjee, N. Bhat, S. A. Shivashankar: J. Appl. Phys. **96**, 5631 (2004)

[51] V. Mikhelashvili, G. Eisenstein, F. Edelman, R. Brener. N. Zakharov, P. Werner: Structural and electrical properties of electron beam gun evaporated $Er_2O_3$ insulator thin films, J. Appl. Phys. **95**, 613 (2004)

[52] T. Zdanovicz, L. Zdanowicz: Preparation and some electrical properties of thulium oxide films, Thin Solid Films **58**, 390 (1979)

[53] S. Ohmi, M. Takeda, H. Ishiwara, H. Iwai: Electrical characteristics for $Lu_2O_3$ thin films fabricated by e-beam deposition method, J. Electrochem. Soc. **151**, G279 (2004)

[54] R. D. Shannon: Dielectric polarizabilities of ions in oxides and fluorides, J. Appl. Phys. **73**, 348 (1993)

[55] S. Jeon, H. Hwang: Effect of hygroscopic nature on the electrical characteristics of lanthanide oxides ($Pr_2O_3$, $Sm_2O_3$, $Gd_2O_3$, and $Dy_2O_3$), J Appl. Phys. **93**, 6393 (2003)

[56] B. B. Van Aken, T. T. M. Palstra: Influence of magnetic on ferroelectric ordering in $LuMnO_3$, Phys. Rev. B **69**, 134113 (2004)

# Index

# Atomic Layer Deposition of Rare Earth Oxides

Jani Päivässari[1], Jaakko Niinistö[1], Pia Myllymäki[1], Chuck Dezelah IV[2],
Charles H. Winter[2], Matti Putkonen[1], Minna Nieminen[1],
and Lauri Niinistö[1]

[1] Laboratory of Inorganic and Analytical Chemistry, Helsinki University of
Technology, P.O. Box 6100, FIN-02015 Espoo,
Finland
lauri.niinisto@hut.fi
[2] Department of Chemistry, Wayne State University, 5101 Cass Avenue, Detroit,
Michigan 48202
USA

**Abstract.** The principles of Atomic Layer Deposition (ALD) for thin film growth
are briefly introduced, emphasizing the aspects of a self-limiting mechanism. Binary
rare earth oxide (REO) thin films have been grown by ALD using various precursor
approaches, which are discussed starting with the $\beta$-diketonates (thd-complexes)
which require ozone to form the oxide. The focus of this review is on the most
recent developments in the precursor chemistry, viz. the use of precursors which
coordinate to the trivalent RE ions through carbon or nitrogen and react with water at reasonable temperatures, generally below 350 °C, to produce the oxide. The
growth rates obtained with RE-cyclopentadienyl complexes together with water as
oxygen source are significantly higher than those observed in the conventional $\beta$-
diketonate/ozone system, e.g., for $Y_2O_3$ using $(CpMe)_3Y/H_2O$ and $Y(thd)_3/O_3$ the
observed growth rates were 1.2–1.3 and 0.23 Å cycle$^{-1}$, respectively. Furthermore,
the resulting REO films from the organometallics and nitrogen-coordinated precursors have lower carbon and hydrogen impurity levels and good electrical characteristics. However, poor thermal stability of the novel carbon- or nitrogen-coordinated
precursors may restrict their use as ALD precursors.

In addition to the desired ALD-mode reactions, other reactions may occur causing some deleterious effects on the REO films. In particular, the large and basic
$La^{3+}$ ion tends to adsorb and react with environmental water and carbon dioxide to
form hydroxide and carbonate phases, respectively. Generally, dielectric properties
may be improved by annealing, e.g., the permittivity for as-deposited $Gd_2O_3$ film
was 10.4 but annealing in oxygen at 700 °C raised it to 15.4.

The ALD processes for the binary REOs form the basis for depositing multi-
component RE-containing films. Besides ternary oxides with the perovskite structure, e.g., $LaAlO_3$ and $LaGaO_3$, phases with two REOs can be processed as solid
solutions. An example thereof is $YScO_3$ having permittivity above 15 and remaining amorphous up to 800 °C to 1000 °C, depending on the ALD process selected,
thus offering a very attractive alternative to the existing high-$\kappa$ gate dielectrics.

M. Fanciulli, G. Scarel (Eds.): Rare Earth Oxide Thin Films,
Topics Appl. Physics **106**, 15–32 (2007)
© Springer-Verlag Berlin Heidelberg 2007

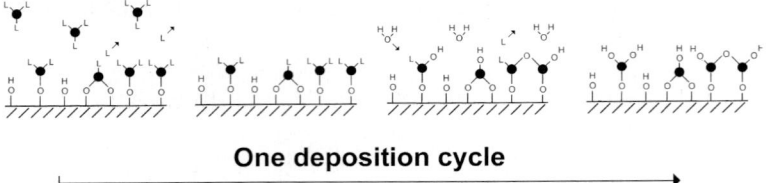

**One deposition cycle**

**Fig. 1.** Schematic illustration describing an ALD deposition cycle leading to the formation of a binary oxide film consisting of metal (o) and oxygen (•) atoms. $L$ refers to the precursor ligand

# 1 Introduction

The binary rare earth oxide (REO) thin films and overlayers have found a wide range of applications, including use as protective coatings, chemical sensors, catalysts and electrolytes in intermediate temperature solid oxide fuel cells (IT-SOFC). Additionally, the REOs are desired materials for microelectronics; most notably they have potential as highly insulating (high-$\kappa$) gate oxides [1]. The first ALD processes involving the rare earth precursors were designed during thin film electroluminescent (TFEL) display development in the 1980s in order to dope ZnS and SrS thin films with trivalent terbium and cerium, and to produce green and blue-green luminescence, respectively [2]. Subsequent to these initial studies, the ALD of REO thin films has grown to incorporate new precursors and processes for many of the rare earth elements.

# 2 Principles of ALD

Atomic Layer Deposition (ALD), also known as Atomic Layer Epitaxy (ALE) or Atomic Layer Chemical Vapor Deposition (ALCVD), can be considered as an advanced variant of the well-known Chemical Vapor Deposition (CVD) technique. ALD was developed in Finland some 30 years ago by *T. Suntola* and coworkers to meet the industrial needs for producing high-quality and long-life TFEL displays [3].

In ALD, the growth of thin films takes place by surface-controlled growth cycles [4]. An ideal growth cycle consists of 1. exposure of the substrate surface to the pulse of the first gaseous precursor and its chemisorption onto the surface, 2. inert gas purge to remove the unreacted precursor, 3. introduction of the second precursor followed by surface reaction between the precursors to produce the desired film material, and finally 4. inert gas purge to expel the excess of precursor and reaction by-products (Fig. 1).

The surface-controlled growth mode produces some obvious advantages, among them the fact that the precursor pulse length, i.e., dose, has no effect on the growth rate provided that the surface is saturated, meaning that all available surface sites are occupied by the precursor molecules. Furthermore,

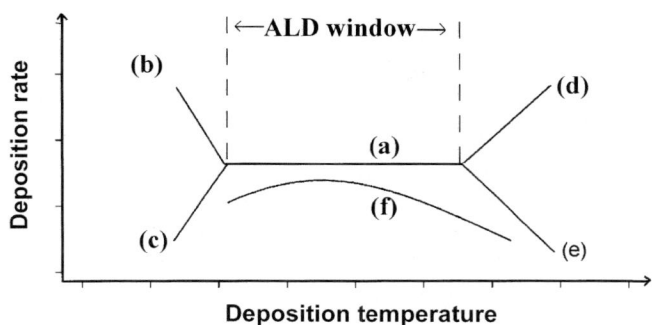

**Fig. 2.** Scheme of (**a**) ALD processing window limited by (**b**) precursor condensation, (**c**) insufficient reactivity, (**d**) precursor decomposition and (**e**) precursor desorption. If deposition rate is dependent on the number of available reactive sites as in (**f**), no actual ALD window is observed

because the growth is by cycles, the thickness control is facile and is achieved by monitoring the number of ALD cycles. Similarly, uniform doping is easy to accomplish by replacing, at a desired interval, the growth cycle by a doping cycle.

Another advantageous and inherent feature of ALD originates from its surface-controlled nature which allows substrates of various sizes and geometries to be uniformly coated. However, a main disadvantage of ALD in certain applications is the fact that it is a relatively slow technique when thicker films, measuring hundreds of nanometers and more, need to be deposited. Usually ALD does not produce a full monolayer of a film in a one-deposition cycle, but only a fraction thereof [5]. This can be attributed to steric hindrance of the precursor molecules bearing bulky ligands. However, in some special cases such as silica nanolaminates, a self-limiting mechanism with a deposition rate well over a one monolayer per cycle has been observed [6].

The ALD growth temperature is dictated by precursor chemistry, to a regime where the surface-limited reactions occur. Reactivity of the metal precursor mainly determines the temperature range where ALD growth occurs. In addition to self-liming growth, a region with a constant deposition rate, a so-called ALD window, is often observed [4, 7] (Fig. 2). The ALD window can extend over hundreds of degrees but it can be in some cases very narrow, as small as a few tens of degrees. However, many recently studied processes for oxide films do not exhibit a distinct ALD window, but nevertheless they can still be used for self-limiting ALD growth [8, 9].

## 3 Binary RE-Oxide Thin Films by ALD

The availability of suitable ALD precursors for rare earth oxides is limited compared to the main group and early transition metal oxide processes. The

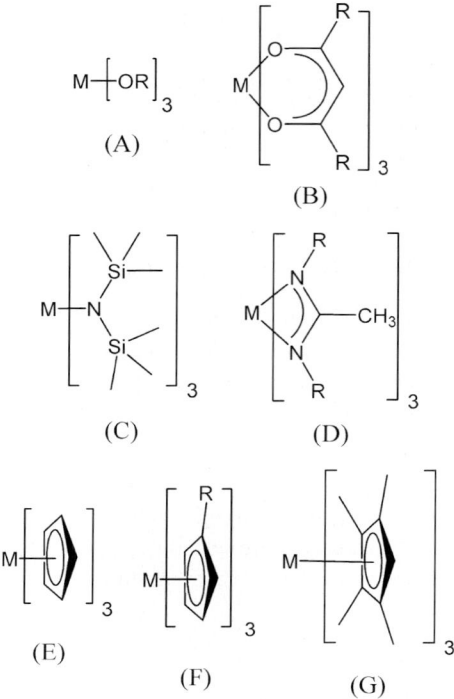

**Fig. 3.** Different precursors used for the RE depositions by ALD. Volatile (**a**) alkoxides, (**b**) β-diketonates, (**c**) amides, (**d**) amidinates and (**e**)–(**g**) organometallic cyclopentadienyl-type compounds with different substitutions have been exploited

first ALD processes for $Y_2O_3$ and $CeO_2$ were studied in the 1990s and were based on the β-diketonate complexes and ozone [10, 11]. Since then, a variety of new precursors and processes have been developed, but nevertheless oxygen-coordinated precursors continue to receive frequent attention. Among the oxygen-coordinated precursors, β-diketonate precursors are thermally very stable, whereas the corresponding metal alkoxides usually suffer from insufficient thermal stability for ALD-type processes. Known processes based on the β-diketonates and ozone cover almost all rare earth elements (Table 1) [12]. In microelectronics, however, it is difficult to obtain equivalent oxide thicknesses (EOT) of 10 Å or below by ozone-based processes and therefore new precursor classes are constantly studied. For example, various rare earth cyclopentadienyl (Cp, $C_5H_5$) and nitrogen-coordinated compounds are currently being investigated as precursors (Fig. 3). The reactivity of these novel compounds is much higher towards mild oxygen sources, such as water, than that of the β-diketonates but sometimes they suffer from limited thermal stability.

**Table 1.** The ALD processes for RE oxides. Growth rate value is obtained at the preferred deposition temperature

| Film material | Metal precursor | Oxygen source | Studied dep. temp.($^\circ$C) | Growth rate/Å cycle$^{-1}$ | Ref. |
|---|---|---|---|---|---|
| $Sc_2O_3$ | $Sc(thd)_3$ | $O_3$ | 175–500 | 0.13 | [13] |
| | $Cp_3Sc$ | $H_2O$ | 175–500 | 0.75 | [13] |
| $Y_2O_3$ | $Y(thd)_3$ | $O_3$ | 200–600 | 0.23 | [10, 14, 15] |
| | $Y(thd)_3$ | $O_2$ Plasma | 200–300 | Not reported | [16] |
| | $Y(thd)_3$phen | $O_3$ | 200–425 | 0.22 | [14] |
| | $Y(thd)_3$bipy | $O_3$ | 200–425 | 0.23 | [14] |
| | $Cp_3Y$ | $H_2O$ | 175–450 | 1.4–1.7 | [17] |
| | $(CpMe)_3Y$ | $H_2O$ | 175–500 | 1.2–1.3 | [17] |
| $La_2O_3$ | $La(thd)_3$ | $O_3$ | 180–425 | 0.36 | [18] |
| | $La[^iPrNC(CH_3)N^iPr]_3$ | $H_2O$ | 300 | 0.9 | [19] |
| | $La[N(SiMe_3)_2]_3$ | $H_2O$ | 200–300 | 0.5–0.3 | [20–23] |
| $CeO_2$ | $Ce(thd)_4$ | $O_3$ | 175–500 | 0.32 | [11, 24] |
| | $Ce(thd)_3$phen | $O_3$ | 225–350 | 0.42 | [24] |
| $Nd_2O_3$ | $Nd(thd)_3$ | $O_3$ | 200–450 | 0.44 | [12, 25] |
| $PrO_x$ | $Pr(OCMe_2CH_2OMe)_3$ | $H_2O$ | 150–350 | 0.1–0.7 | [26] |
| | $Pr[N(SiMe_3)_2]_3$ | $H_2O$ | 200–400 | 3.0–0.15 | [27–29] |
| $Sm_2O_3$ | $Sm(thd)_3$ | $O_3$ | 300 | 0.36 | [12] |
| $Eu_2O_3$ | $Eu(thd)_3$ | $O_3$ | 300 | 0.32 | [12] |
| $Gd_2O_3$ | $Gd(thd)_3$ | $O_3$ | 225–400 | 0.31 | [12, 30] |
| | $Gd(OCMe_2CH_2OMe)_3$ | $H_2O$ | 200–300 | 0.2–1.0 | [26] |
| | $(CpMe)_3Gd$ | $H_2O$ | 150–350 | 2.1–2.9 | [30] |
| | $Gd[N(SiMe_3)_2]_3$ | $H_2O$ | 150–300 | 0.5–2.2 | [28, 29] |
| $Dy_2O_3$ | $Dy(thd)_3$ | $O_3$ | 300 | 0.31 | [12] |
| $Ho_2O_3$ | $Ho(thd)_3$ | $O_3$ | 300 | 0.25 | [12] |
| $Er_2O_3$ | $Er(thd)_3$ | $O_3$ | 200–450 | 0.25 | [12, 31] |
| | $Er(thd)_3$ | $O_2$ Plasma | 200–300 | 0.5 | [16] |
| | $(CpMe)_3Er$ | $H_2O$ | 175–450 | 1.5 | [32] |
| | $Er(^tBuNC(CH_3)N^tBu)$ | $O_3$ | 225–300 | 0.39–0 55 | [33] |
| $Tm_2O_3$ | $Tm(thd)_3$ | $O_3$ | 300 | 0.22 | [12] |
| $Lu_2O_3$ | $\{[Cp(SiMe_3)]_2LuCl\}_2$ | $H_2O$ | 360 | 0.5 | [34] |

## 4 Oxygen-Coordinated Precursors

$\beta$-Diketonates are the most widely studied precursor group for the ALD of rare earth oxides [35]. In fact, all rare earth oxide thin films, apart from a few exceptions (Pr, Pm and Tb), have been deposited using the $\beta$-diketonate-type thd-complexes (thd = 2,2,6,6-tetramethyl-3,5-heptanedione) as metal precursors and ozone as the oxygen source (cf. Table 1) [10–15, 18, 24, 25, 30, 31].

Rare earth thd-compounds are volatile, thermally stable solids. Additionally, they are easy to synthesize [36] and easy to handle and store, being relatively stable even in contact with air.

Typical optimized deposition temperatures in all $RE(thd)_x/O_3$ ALD processes have been around $300\,°C$, with the metal precursor evaporation temperatures in the range of $115\,°C$ to $175\,°C$ ($\approx 2\,mbar$), depending on the size of the RE element. Processes have shown an ALD-type self-limiting growth in a wide temperature range, e.g., between $250\,°C$ and $375\,°C$ for the $Y_2O_3$ process [14]. Elemental analyses show that REO films deposited by $RE(thd)_x/O_3$ processes at optimized temperatures contain typically 1 at% to 5 at% of carbon and approximately 1 at% to 2 at% of hydrogen [12, 14, 15, 18, 24, 25, 30, 31]. The carbon concentrations in the $Sc_2O_3$ and $La_2O_3$ films markedly deviated from these limits, being only $< 0.1$ at% for $Sc_2O_3$ but up to 10 at% to 12 at% for $La_2O_3$ [13, 18]. Large deviations in the carbon content are most probably due to the difference in basicity of rare earth oxides causing a range of carbonate impurities in the films. Electrical characterizations for the $Al/Ln_2O_3/native\text{-}SiO_2/Si(100)/Al$ capacitor structures (Ln = Nd, Sm, Eu, Gd, Dy, Ho, Er and Tm) have proven that the ALD-grown films are insulating, with relative permittivies of 8.4–11.1 [12].

Despite the fact that relatively pure REO thin films can be grown by ALD using $\beta$-diketonate-type precursors, these processes are considered to have some drawbacks. Above all, water cannot be used as the oxygen source with the stable thd-complexes, and more reactive ozone is needed instead. Unfortunately, ozone has been reported to react with silicon substrate forming an interfacial $SiO_x$ layer [15, 37, 38], increasing the EOT value and potentially limiting the use of $RE(thd)_x/O_3$ ALD processes for high-$\kappa$ applications. However, the formation of the interfacial oxide layer may also depend on the deposited material itself and not on the oxygen source alone, because $H_2O$ and $O_3$ have been reported to form an equally thick $SiO_x$ layer between $HfO_2$ and the silicon substrate [39]. Also, the growth rates in $RE(thd)_x/O_3$ processes are rather low due to the steric profile of the large thd-ligand and low reactivity of $RE(thd)_x$ compounds. Typical growth rate values obtained are in the order of $0.2\,\text{Å cycle}^{-1}$ to $0.4\,\text{Å cycle}^{-1}$ [12, 14, 25, 30, 31].

$Y_2O_3$ and $CeO_2$ films have also been grown using neutral N-donor adducts of $Y(thd)_3$ or $Ce(thd)_3$ complexes as precursors [14, 24]. Comparative ALD studies using $Y(thd)_3$, $Y(thd)_3(1,10\text{-phenantroline})$ and $Y(thd)_3(bipyridine)$ as yttrium precursors as well as those on $Ce(thd)_4$ and $Ce(thd)_3(phen)$ as cerium precursors have shown, however, that the adducts do not bring any significant advantages. In $Y_2O_3$ depositions the growth rate, deposition temperature range and impurity levels of carbon and hydrogen were very similar in every case, but $Y_2O_3$ films deposited from $Y(thd)_3(phen)$ or $Y(thd)_3(bipy)$ were noticed to contain 0.1 at% to 0.3 at% of nitrogen [14]. $Ce(thd)_3(phen)$ may be useful for doping, however, particularly when cerium is needed in the oxidation state of $+3$, e.g., in the ALD of electroluminescent SrS:Ce thin films [40].

Er(thd)$_3$ or Y(thd)$_3$ precursors have also been used in radical-enhanced ALD, where metal precursors reacted with oxygen radicals from oxygen plasma [16]. The growth rate of 0.5 Å cycle$^{-1}$ obtained in the radical-enhanced ALD of Er$_2$O$_3$ was higher than in the conventional Er(thd)$_3$/O$_3$ ALD process [31]. At the same time, however, the carbon impurity level significantly increased to $\approx$ 26 at%. Furthermore, pulse times of several minutes were needed in the radical-enhanced ALD process for Er$_2$O$_3$, which are considerably longer than in conventional ALD.

The alkoxides are another commonly used precursor group in ALD and CVD. There is only one report, however, on the ALD of RE oxides from metal alkoxides [26]. PrO$_x$ and Gd$_2$O$_3$ films were deposited by so-called liquid injection ALD using RE(OCMe$_2$CH$_2$OMe)$_3$, i.e., RE(mmp)$_3$ (mmp = 1-methoxy-2-methyl-2-propanolate), and water as precursors. RE(mmp)$_3$ was dissolved in toluene, and tetraglyme [CH$_3$O(CH$_2$CH$_2$O)$_4$CH$_3$] was added to inhibit the condensation and bridging reactions of the RE precursors. The precursor solution was then injected at room temperature into a vaporizer, from where the precursor vapor was further directed into the reaction chamber. This technique did not prevent thermal decomposition of the precursor, however, and the growth rate of PrO$_x$ and Gd$_2$O$_3$ films was found to increase with increasing precursor dose, thus indicating lack of an ALD-type growth mode.

# 5 Carbon-Coordinated Precursors

Another interesting group of rare earth compounds that are potentially suitable as ALD precursors are organometallic compounds, which are complexes possessing direct metal to carbon bonds. Suitable metal alkyls do not exist for rare earth elements, but several Cp-compounds have been recently used for ALD (Table 1) [41]. Cp compounds may have a range of bonding modes, but must have at least one direct metal-carbon bond to the cyclopentadienyl ligand. They are more reactive towards water than the thd-complexes and thus aggressive oxidants like ozone are not necessary. However, the thermal stability of unsubstituted Cp compounds is generally lower than that of the $\beta$-diketonate compounds, restricting their utility for larger rare earths, e.g., La, Ce and Pr. Recently, ALD processes based on the Cp compounds for binary rare earth oxides have been developed for Sc$_2$O$_3$ [13], Y$_2$O$_3$ [17], Er$_2$O$_3$ [32], Gd$_2$O$_3$ [30] and Lu$_2$O$_3$ [34].

Because of their high reactivity towards water, Cp compounds offer the possibility of higher growth rates than those obtained with the thd-compounds, often even at lower temperatures. For example, the ALD growth rate of Sc$_2$O$_3$ obtained with Cp$_3$Sc/H$_2$O (0.75 Å cycle$^{-1}$) is about six times higher than that found for the Sc(thd)$_3$/O$_3$ process (0.13 Å cycle$^{-1}$) (Fig. 4) [13]. A similar trend has been observed with the Y$_2$O$_3$ processes [14, 17].

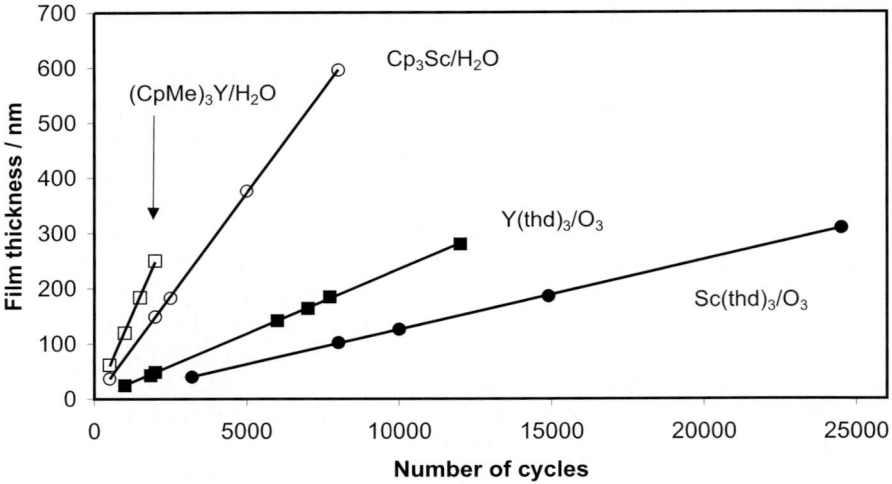

**Fig. 4.** Growth rate of $Sc_2O_3$ and $Y_2O_3$ thin film as a function of deposition temperature from $Sc(thd)_3/O_3$, $Y(thd)_3/O_3$, $(Cp)_3Sc/H_2O$ or $(CpMe)_3Y/H_2O$

With the Cp processes, the stoichiometry of the oxide film is often considerably closer to the ideal value, e.g., $Y_2O_3$ films deposited using $(CpMe)_3Y$ or $Cp_3Y$, as the metal source contained a lower amount of impurities than the thd-processed films [17]. The analogous Er compound, $(CpMe)_3Er$, functioned well as an ALD precursor, producing pure $Er_2O_3$ films at temperatures as low as 200 °C to 300 °C [32]. The permittivity for the $Er_2O_3$ layer was measured to be about 14, viz. higher than in the thd-processed $Er_2O_3$ films (10) [12]. On the other hand, the applicability of $(CpMe)_3Gd$ suffered from its partial thermal decomposition but uniform $Gd_2O_3$ films with almost ideal stoichiometry were still obtained [30]. Again, the capacitance–voltage characteristics were improved by the use of Cp-based precursor. However, higher degree of crystallinity and consequently development of leaky grain boundaries in many Cp-processed films as compared to thd-processed films increase the leakage current. *Scarel* et al. applied a bis-Cp compound, namely $\{[Cp(SiCH_3)]_2LuCl\}_2$, together with water to deposit $Lu_2O_3$ films [34]. Promising electrical properties were reported, despite the fact that a true ALD-type growth was not achieved due to thermal decomposition of the precursor above 250 °C, resulting in chlorine contamination.

Obviously, Cp compounds offer a wide range of possibilities in ALD of rare earth oxide films, but more synthetic work is needed to improve the thermal stability. To improve stability, more bulky Cp-compounds could be the solution to deposit high-quality films of some important binary rare earth oxides containing large rare earths, such as $La_2O_3$ or $PrO_x$.

# 6 Nitrogen-Coordinated Precursors

The ALD of rare earth oxide thin films from nitrogen-coordinated metalorganic complexes has been an area of much recent interest (Table 1) [19–23, 27–29, 33, 42, 43]. Two classes of such complexes have been employed for rare earth oxide ALD: amides and amidinates (Fig. 3). In general, amide-based precursors have been investigated in a wide range of ALD processes, with particular attention given to early transition metal oxides and nitrides. Unfortunately, the utility of amide complexes of rare earth metals as precursors for ALD is limited, due to either lack of volatility or insufficient thermal stability [20]. A notable exception, however, has been the complexes of bis(trimethylsilyl)amide ligand, for which volatile lanthanum, praseodymium, and gadolinium complexes have been prepared.

The deposition of $La_2O_3$ by ALD from $La[N(SiMe_3)_2]_3$ and water vapor was first described by *Gordon* et al., but only a few details were given [20]. Subsequently, *Triyoso* et al. investigated the $La_2O_3$ film growth behavior in the same process using a variety of substrate surfaces and also examining the physical properties of the deposited films [21, 22]. Surface saturative growth of uniform films was found at deposition temperatures between 225 °C and 275 °C for films deposited on silicon substrates having a chemically grown oxide layer, with growth rates between 0.3 and 0.5 Å cycle$^{-1}$, depending on the temperature. Deposition outside this temperature range resulted in non-uniform growth and sub-saturative behavior. The film growth rate on other surface types varied only modestly with the nature of the surface, with growth rates between 0.26 and 0.35 Å cycle$^{-1}$ at 275 °C. The films were generally amorphous and were comprised of stoichiometric $La_2O_3$, although a considerable amount of silicon was present, viz. $\approx$ 8 at% and $\approx$ 10 at% for films deposited at 225 °C and 275 °C, respectively. The carbon content was < 1 at% for a film deposited at 275 °C. The electrical characteristics were described as being reasonably well-behaved for both as-deposited and annealed (up to 800 °C) $La_2O_3$ layers. *He* and coworkers recently reported a growth rate of 0.43 nm cycle$^{-1}$ from $La[N(SiMe_3)_2]_3$ and $H_2O$ precursors [23], while *Triyoso* et al. presented a growth rate of $\approx$ 0.5 Å cycle$^{-1}$ using the same precursors and deposition temperature. Detailed study of the relationship between growth rate and temperature or precursor dose was not given in the report by *He* et al. The films were nearly amorphous for both the as-deposited films and annealed samples, with only a single very broad reflection from cubic $La_2O_3$ observed. Electrical characterization was described in the study, but it was not ideal. The calculated permittivity was 21.

Praseodymium oxide and gadolinium oxide have also been deposited by ALD using bis(trimethylsilyl)amido complexes as precursors [27–29]. $PrO_x$ films deposited at substrate temperatures of 200 °C and 300 °C from $Pr[N(SiMe_3)_2]_3$ and $H_2O$, with a growth rate of 0.30 nm cycle$^{-1}$ and 0.034 nm cycle$^{-1}$, respectively [27]. Growth was described as not entirely self-limiting, indicated by an increase in growth rate with the precursor pulse length. Deposition at temperatures $\geq 350\,°C$ afforded poor films due to precursor decomposition. All as-deposited films were amorphous, as revealed by X-ray diffraction. Annealing at $1000\,°C$ in $N_2$ resulted in the appearance of reflections due to monoclinic $Pr_2O_3$ and $Pr_{9.33}(SiO_4)_6O_2$. The as-deposited films were found to have high silicon and hydrogen impurity contents due to decomposition of the precursor. The calculated effective permittivity was around 15. Gadolinium oxide films grown by ALD from $Gd[N(SiMe_3)_2]_3$ and $H_2O$ have been recently described by *Jones* et al. [28, 29]. Films were deposited by the ALD method between $150\,°C$ and $300\,°C$, but the growth rate was found to significantly increase with pulse length at all temperatures, indicating that the growth was not self-limiting. The effective permittivity of an as-deposited $Gd_2O_3$ film was 10.4, but increased to 15.4 upon annealing.

Homoleptic amidinate complexes of the formula $M[RNC(R')NR]_x$, where R and $R'$ are alkyl groups, have been recommended as ALD precursors for a number of metals and their oxides, including $La_2O_3$ [19, 42, 43]. *Gordon* and coworkers have deposited $La_2O_3$ films grown in ALD-mode from $La[^iPrNC(CH_3)N^iPr]_3$ and $H_2O$ [19]. A growth rate of $0.9\,Å\,cycle^{-1}$ at a deposition temperature of $300\,°C$ was reported, but very few details on the growth behavior, film composition, or properties were given. Subsequently, *Gordon* reported that water-based $La_2O_3$ processes lead to lanthanum hydroxide phases which continuously desorb water under deposition conditions to give a non-ALD growth pathway and poor-quality films [43]. The atomic layer deposition of $Er_2O_3$ films was demonstrated using $Er(^tBuNC(CH_3)N^tBu)_3$ and ozone at substrate temperatures between $225\,°C$ and $300\,°C$ [33]. Growth rates increased linearly with temperature from 0.37 to $0.55\,Å\,cycle^{-1}$. The film growth in this study was not entirely self-limiting, as shown by a small increase in growth rate with precursor dose. The resulting films were slightly oxygen-rich $Er_2O_3$, with carbon, hydrogen, and fluorine levels of 1.0 at% to 1.9 at%, 1.7 at% to 1.9 at%, and 0.3 at% to 1.3 at%, respectively, at substrate temperatures of $250\,°C$ and $300\,°C$. Carbonate was suggested as the origin of both the carbon impurity and the slight excess of oxygen in the films. The as-deposited films were amorphous below $300\,°C$, but showed reflections due to cubic $Er_2O_3$ at $300\,°C$.

# 7 Multi-Component RE-Containing Oxide Thin Films

Owing to its cyclic nature, ALD is an ideal method for depositing multicomponent oxide films. In general, the deposition of a ternary oxide, for instance, is performed by alternately depositing two binary oxides. Precise composition control of the film is possible by varying the metal precursor pulsing ratio, i.e., by altering the numbers of binary oxide cycles relative to each other. However, the lack of suitable precursor combination often sets limitations. In an ideal case, the constituent binary oxides have overlapping ALD windows. Although this in practice is rarely possible, multi-component films can be deposited as long as the growth of constituent binary oxides is surface-saturated ALD-type growth. Indeed, several multi-component oxide thin films containing at least one RE metal have been deposited by ALD (Table 2). Most of them are ternary oxides where the goal has been to grow oxides with defined crystal structure and stoichiometry. Oxides with a wide range of metal ratios and thoroughly mixed binary oxides (solid solutions) are usually called mixed oxides whereas, in the case of nanolaminates, the film consists of two or more distinct alternate layers of binary phases.

**Table 2.** The reported ALD processes for RE-containing multi-component oxides

| Film material | Precursors | Oxygen source | Studied dep. temp./($^\circ$C) | Ref. |
|---|---|---|---|---|
| *Ternary oxides* | | | | |
| $LaNiO_3$ | $La(thd)_3$ - $Ni(thd)_2$ | $O_3$ | 150–450 | [44] |
| $LaCoO_3$ | $La(thd)_3$ - $Co(thd)_2$ | $O_3$ | 200–400 | [45] |
| $LaMnO_3$ | $La(thd)_3$ - $Mn(thd)_3$ | $O_3$ | 200–400 | [46] |
| $LaAlO_3$ | $La(thd)_3$ - $Al(acac)_3$ | $O_3$ | 325–400 | [47] |
| $LaGaO_3$ | $La(thd)_3$ - $Ga(acac)_3$ | $O_3$ | 325–425 | [48] |
| $NdAlO_3$ | $Nd(thd)_3$ - $Al(Me)_3$ | $O_3 + H_2O$ | 300 | [25] |
| *Pseudo-quaternary oxides* | | | | |
| $(La,Ca)MnO_3$ | $La(thd)_3$ - $Mn(thd)_3$ - $Ca(thd)_2$ | $O_3$ | 200–330 | [49] |
| *Mixed oxides* | | | | |
| $Y_2O_3$-$Sc_2O_3$ | $Y(thd)_3$ - $Sc(thd)_3$ | $O_3$ | 335–350 | [50] |
| | $(CpMe)_3Y$ - $Cp_3Sc$ | $H_2O$ | 300 | [50] |
| $Y_2O_3$-$ZrO_2$ | $Y(thd)_3$ - $Zr(thd)_4$ | $O_3$ | 375 | [51] |
| | $Y(thd)_3$ - $Cp_2Zr(Me)_2$ | $O_3$ | 310–365 | [51] |
| | $Y(thd)_3$ - $Cp_2ZrCl_2$ | $O_3$ | 275–350 | [51] |
| $Gd_2O_3$-$CeO_2$ | $Gd(thd)_3$ - $Ce(thd)_4$ | $O_3$ | 250 | [52, 53] |
| $LaAl_xO_y$ | $La[N(SiMe_3)_2]_3$ - $Al(Me)_3$ | $H_2O$ | 200–300 | [22] |
| $La_2O_3$-$(SiO_2)_x$ | $La[N(SiMe_3)_2]_3$ - $(^tBuO)_3SiOH$ | $H_2O$ | Not given | [20] |
| *Nanolaminates* | | | | |
| $La_2O_3/Al_2O_3$ | $La[^iPrNC(CH_3)N^iPr]_3$- $Al(CH_3)_3$ | $H_2O$ | 300–330 | [43] |

# 8 Multi-Component Oxides Containing One RE Metal

In the pioneering studies on the ALD growth of multi-component RE oxides, e.g., lanthanum-based $LaNiO_3$ [44] and $LaCoO_3$ [45], the approach was to grow one monolayer of each binary oxide alternately over the top of the other. It was expected that the monolayers of binary oxides would intermix, forming the desired ternary compound. However, XPS analysis indicated that the as-deposited $LaNiO_3$ films consisted of separate oxide layers that reacted only during annealing. In the as-deposited $LaCoO_3$ films no discrete layers were present, but the films contained excess of cobalt and a mixture of two-phase $LaCoO_3$ and $Co_3O_4$ was detected after annealing.

Subsequent studies, e.g., for $LaMnO_3$ [46], $LaAlO_3$ [47] and $LaGaO_3$ [48], showed that a better approach to grow ternary oxides is to sequentially grow submonolayers of the two constituent oxides so that the oxide structure is better mixed. A stoichiometric ratio of the two metals in the films is achieved by applying the appropriate precursor pulsing ratio. This method was also applied in the deposition of $NdAlO_3$ thin films using $Nd(thd)_3$ and $Al(CH_3)_3$ as metal precursors [25]. All the as-deposited ternary oxides were amorphous in the temperature range where the growth was ALD-type, and crystalline phases were obtained only after annealing. In the recent study of *Nilsen* et al., it was shown that the ALD method can be applied to grow even more complex oxides, such as pseudo-quaternary oxide $La_{1-x}Ca_xMnO_3$ [49].

The first ALD-deposited mixed oxide thin film containing one RE metal was yttria-stabilized zirconia (YSZ). Several different Zr precursors were used, e.g., $Zr(thd)_4$, $Cp_2Zr(Me)_2$ and $Cp_2ZrCl_2$, together with $Y(thd)_3$ as Y precursor [51]. The as-deposited films were identified to have polycrystalline cubic structure. Although the thd-complexes are still the most widely used precursors for RE metals, other alternatives have been studied for the deposition of multi-component oxides. $LaAl_xO_y$ [22] containing significant amounts of Si has been deposited using $La[N(SiMe_3)_2]_3$, $Al(CH_3)_3$ and water. Aluminum oxide was incorporated into the $La_2O_3$ in order to improve the dielectric properties of the material. The film stoichiometry was not reported. Also amorphous $La_2O_3(SiO_2)_x$ [54] has been deposited by ALD using $La[N(SiMe_3)_2]_3$ as La precursor. However, instead of water, tris(*tert*-butoxy)silanol $(t\text{-BuO})_3SiOH$ was used as combined silicon and oxygen source. Few details of the deposition conditions or the film characteristics were given.

*Lim* et al. have deposited $La_2O_3$–$Al_2O_3$ nanolaminates by ALD using $La(^iPrNC(CH_3)N^iPr)_3$, $Al(CH_3)_3$ and water as precursors [43]. Thin layers ($<$ a few nm thick) of binary oxides were deposited on top of each other. Amorphous, 9.8 nm thick film with a composition $La_{0.9\pm0.1}Al_{1.1\pm0.1}O_3$ had EOT 2.9 nm and permittivity 13. The as-deposited films had low leakage current, small flatband shifts and small hysteresis.

# 9 Multi-Component Oxides Containing Two REs

Only a few multi-component oxide thin films containing more than one rare earth have been deposited by ALD. The first example was ALD-deposited gadolinium doped ceria (CGO), which has been studied as a candidate material for IT-SOFC [52, 53]. Owing to the application, relatively thick films (200 nm to 1 μm) were deposited on several different substrate materials. $\beta$-diketonates were used as the metal precursors. The as-deposited films were identified as cubic $(Gd_{0.6}Ce_{0.4})O_{3.2}$ by XRD [52]. Recently $Y_xSc_zO_3$ thin

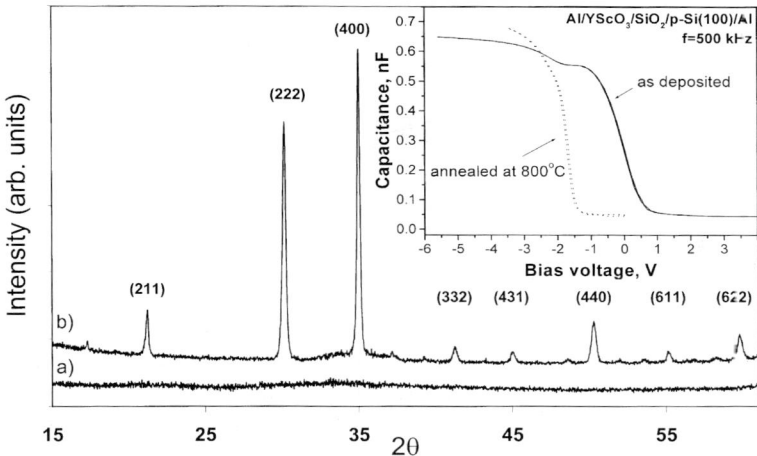

**Fig. 5.** XRD patterns of annealed films with composition $Y_{1.04}Sc_{0.96}O_3$ grown using thd-complexes as metal precursors. Annealing temperatures and film thicknesses were 800 °C and 140 nm (**a**) and 1000 °C and 126 nm (**b**), respectively. Miller indices for solid solution of cubic $Y_2O_3$ and $Sc_2O_3$ are given (JCPDS 5-629 and 25-1200, respectively). Inset shows capacitance-voltage behavior of 40 nm thick as-deposited and annealed film

films have been deposited by ALD using $\beta$-diketonates $M(thd)_3$ (M = Y, Sc) and cyclopentadienyl compounds $(CpMe)_3Y$ and $Cp_3Sc$ as metal precursors [50]. Both processes are ideal in a sense that they combine two binary processes with overlapping ALD windows. The composition of the films can be easily controlled by varying the precursor deposition cycles. A considerably higher growth rate ($1.07\,\text{Å cycle}^{-1}$) was achieved with the Cp compounds than with the thd-complex/ozone-process ($0.18\,\text{Å cycle}^{-1}$). The choice of the precursor type has also an effect on the crystallization behavior of the films. The as-deposited $YScO_3$ films were amorphous, regardless of the metal precursors used. Films deposited using $\beta$-diketonates remained amorphous after annealing at 800 °C, but crystallized as cubic solid solution of $Y_2O_3$ and $Sc_2O_3$ at 1000 °C. However, films deposited using Cp compounds showed signs of crystallization already at 800 °C. ALD-deposited $YScO_3$ films exhibited promising electrical properties, e.g., low hysteresis and low leakage current density. Relatively high effective permittivity ($\approx 15$) was measured for the $Al/YScO_3/$native $SiO_2/Si/Al$ capacitor structures.

## 10 Concluding Remarks

Atomic Layer Deposition, being a surface-controlled gas-phase deposition process, requires volatile and sufficiently stable but reactive precursors. In contrast to several other elements, most of the inorganic rare earth compounds are non-volatile and the oxides are highly refractive. Nevertheless, several processes and precursor systems have now been developed to deposit by ALD thin films and overlayers of REOs.

While the first processes were based on $\beta$-diketonates containing RE-oxygen bonds, several volatile organometallics with RE-carbon or RE-nitrogen bonds have currently become available to deposit oxide materials at reasonable temperatures. $\beta$-Diketonate processes of rare earth oxides have two drawbacks which may limit their applicability. First, $\beta$-diketonate complexes are quite unreactive and they require the use of a strong oxidizer such as ozone. Ozone can be unfavorable regarding the high-$\kappa$ applications because it leads to the formation of an interfacial $SiO_x$ layer, thus increasing the EOT value. In general, it has been observed that controlled oxidation of the interface leads to improved electrical properties as compared to the ALD deposition onto HF-treated silicon surfaces. Secondly, relatively low growth rates of $\beta$-diketonate processes may limit their usage for the deposition of thicker films.

An advantage in the use of the novel reactive and thermally stable precursors is the purity of films obtained under optimised ALD reaction conditions and the fact that ozone is not needed but water is sufficiently reactive to oxidize the precursor into the oxide on the substrate surface. For some rare earths, poor thermal stability of carbon- and nitrogen-coordinated precursors may restrict their use. ALD produces REO thin films and structures based thereon which have good electrical characteristics and interfaces.

It is also feasible to process RE-containing multi-component oxides by ALD. Several processes have recently been developed, e.g., for $LaAlO_3$ and $YScO_3$, by combining the binary oxide ALD processes.

## Acknowledgements

Financial support from the Academy of Finland (Projects 204742 and 205777) is gratefully acknowledged. The authors also wish to thank the Wihuri Foundation and Finnish Foundation for Technology for their valuable support.

# References

[1]  M. Leskelä, M. Ritala: Rare-earth oxide thin films as gate oxides in MOSFET transistors, J. Solid State Chem. **171**, 170 (2003)
[2]  M. Leskelä, M. Tammenmaa: Materials for electroluminescent thin films, Mater. Chem. Phys. **16**, 349 (1987)
[3]  T. Suntola, J. Antson: US Patent 4 058 430 (1977)
[4]  M. Ritala, M. Leskelä: Atomic layer deposition, in *Handbook of Thin Film Materials* (Academic Press, San Diego 2002) p. 103

[5] M. de Ridder, P. C. van de Ven, R. G. van Welzenis, H. H. Brongersma, S. Helfensteyn, C. Creemers, P. van der Voort, M. Baltes, M. Mathieu, E. F. Vansant: Growth of iron oxide on yttria-stabilized zirconia by atomic layer deposition, J. Phys. Chem. B **106**, 13146 (2002)

[6] D. Hausmann, J. Becker, S. Wang, R. G. Gordon: Rapid vapor deposition of highly conformal silica nanolaminates, Science **298**, 402 (2002)

[7] T. Suntola: Atomic layer epitaxy, Mat. Sci. Rep. **4**, 261 (1989)

[8] M. Putkonen, J. Niinistö, K. Kukli, T. Sajavaara, M. Karppinen, H. Yamauchi, L. Niinistö: $ZrO_2$ thin films grown onto silicon substrates by atomic layer deposition with $Cp_2Zr(CH_3)_2$ and water as precursors, Chem. Vap. Dep. **9**, 207 (2003)

[9] L. Niinistö, J. Päiväsaari, J. Niinistö, M. Putkonen, M. Nieminen: Advanced electronic and optoelectronic materials by atomic layer deposition: An overview with special emphasis on recent progress in processing of high-k dielectrics and other oxide materials, Phys. Stat. Sol. A **201**, 1443 (2004)

[10] H. Mölsä, L. Niinistö, M. Utriainen: Growth of yttrium oxide thin films from β-diketonate precursor, Adv. Mater. Opt. Electron. **4**, 389 (1994)

[11] H. Mölsä, L. Niinistö: Deposition of cerium dioxide thin films on silicon substrates by atomic layer epitaxy, Mater. Res. Soc. Symp. Proc. **355**, 341 (1994)

[12] J. Päiväsaari, M. Putkonen, L. Niinistö: A comparative study on the properties of lanthanide oxide thin films processed by atomic layer deposition, Thin Solid Films **472**, 275 (2005)

[13] M. Putkonen, M. Nieminen, J. Niinistö, L. Niinistö, T. Sajavaara: Surface-controlled deposition of $Sc_2O_3$ thin films by atomic layer epitaxy using β-diketonate and organometallic precursors, Chem. Mater. **13**, 4701 (2001)

[14] M. Putkonen, T. Sajavaara, L.-S. Johansson, L. Niinistö: Low-temperature ALE deposition of $Y_2O_3$ thin films from β-diketonate precursors, Chem. Vap. Dep. **7**, 44 (2001)

[15] E. P. Gusev, E. Cartier, D. A. Buchanan, M. Gribelyuk, M. Copel, H. Okorn-Schmidt, C. D'Emic: Ultrathin high-κ metal oxides on silicon: Processing and characterization and integration issues, Microelectron. Eng. **59**, 341 (2001)

[16] T. T. Van, J. P. Chang: Surface reaction kinetics of metal β-diketonate precursors with O radicals in radical-enhanced atomic layer deposition of metal oxides, Appl. Surf. Sci. **246**, 250 (2005)

[17] J. Niinistö, M. Putkonen, L. Niinistö: Processing of $Y_2O_3$ thin films by atomic layer deposition from cyclopentadienyl-type compounds and water as precursors, Chem. Mater. **16**, 2953–2958 (2004)

[18] M. Nieminen, M. Putkonen, L. Niinistö: Formation and stability of lanthanum oxide thin films deposited from β-diketonate precursor, Appl. Surf. Sci. **174**, 155 (2001)

[19] B. S. Lim, A. Rahtu, R. G. Gordon: Atomic layer deposition of transition metals, Nature Mater. **2**, 729 (2003)

[20] R. G. Gordon, J. Becker, D. Hausmann, S. Shu: Vapor deposition of metal oxides and silicates: possible gate insulators for future microelectronics, Chem. Mater. **13**, 2463 (2001)

[21] D. H. Triyoso, R. I. Hegde, J. Grant, P. Fejes, R. Liu, D. Roan, M. Ramon, D. Wherho, R. Rai, L. B. La, J. Baker, C. Garza, T. Guenher, B. E. White, P. J. Tobin: Film properties of ALD $HfO_2$ and $La_2O_3$ gate dielectrics grown

30    Jani Päiväsaari et al.

on Si with various pre-deposition treatments, J. Vac. Sci. Technol. B **22**, 2121 (2004)

[22]  D. H. Triyoso, R. I. Hegde, J. M. Grant, J. K. Schaeffer, D. Roan, B. E. White, P. J. Tobin: Evaluation of lanthanum based gate dielectrics deposited by atomic layer deposition, J. Vac. Sci. Technol. B **23**, 288 (2005)

[23]  W. He, S. Schuetz, R. Solanki, J. Belot, J. McAndrew: Atomic layer deposition of lanthanum oxide films for high-k gate dielectrics, Electrochem. Solid-State Lett. **7**, G131 (2004)

[24]  J. Päiväsaari, M. Putkonen, L. Niinistö: Deposition of cerium dioxide buffer layers by atomic layer epitaxy, J. Mater. Chem. **12**, 1828 (2002)

[25]  A. Kosola, J. Päiväsaari, M. Putkonen, L. Niinistö: ALD growth of neodymium oxide and neodymium aluminate thin films, Thin Solid Films **479**, 159 (2005)

[26]  R. J. Potter, P. R. Chalker, T. D. Manning, H. C. Aspinall, Y. F. Loo, A. C. Jones, L. M. Smith, G. W. Critchlow, M. Schumacher: Deposition of $HfO_2$, $Gd_2O_3$ and $PrO_x$ by liquid injection ALD techniques, Chem. Vap. Dep. **11**, 159 (2005)

[27]  K. Kukli, M. Ritala, T. Pilvi, T. Sajavaara, M. Leskelä, A. C. Jones, H. C. Aspinall, D. C. Gilmer, P. J. Tobin: Evaluation of a praseodymium precursor for atomic layer deposition of oxide dielectric films, Chem. Mater. **16**, 5162 (2004)

[28]  A. C. Jones, H. C. Aspinall, P. R. Chalker, R. J. Potter, K. Kukli, A. Rahtu, M. Ritala, M. Leskelä: Some recent developments in the MOCVD and ALD of high-k dielectric oxides, J. Mater. Chem. **14**, 3101 (2004)

[29]  A. C. Jones, H. C. Aspinall, P. R. Chalker, R. J. Potter, K. Kukli, A. Rahtu, M. Ritala, M. Leskelä: Recent developments in the MOCVD and ALD of rare earth oxides and silicates, Mater. Sci. Eng. B **118**, 97 (2005)

[30]  J. Niinistö, N. Petrova, M. Putkonen, L. Niinistö, K. Arstila, T. Sajavaara: Gadolinium oxide thin films by atomic layer deposition, J. Cryst. Growth **285**, 191 (2005)

[31]  J. Päiväsaari, M. Putkonen, T. Sajavaara, L. Niinistö: Atomic layer deposition of rare earth oxides: Erbium oxide films from $\beta$-diketonate and ozone precursors, J. Alloys Compd. **374**, 124 (2004)

[32]  J. Päiväsaari, J. Niinistö, K. Arstila, K. Kukli, M. Putkonen, L. Niinistö: High growth rate of erbium oxide thin films in atomic layer deposition from $(CpMe)_3Er$ and water precursors, Chem. Vap. Dep. **11**, 915 (2005)

[33]  J. Päiväsaari, C. L. Dezelah, D. Back, H. M. El-Kadri, M. J. Heeg, M. Putkonen, L. Niinistö, C. H. Winter: Synthesis, structure, and properties of volatile lanthanide complexes containing amidinate ligands: Application for $Er_2O_3$ thin film growth by atomic layer deposition, J. Mater Chem. **15**, 4224 (2005)

[34]  G. Scarel, E. Bonera, C. Wiemer, G. Tallarida, S. Spiga, M. Fanciulli, I. L. Fedushkin, H. Schumann, Y. Lebedinskii, A. Zenkevich: Atomic-layer deposition of $Lu_2O_3$, Appl. Phys. Lett. **85**, 630 (2004)

[35]  M. Tiitta, L. Niinistö: Volatile metal $\beta$-diketonates: ALE and CVD precursors for electroluminescent device thin films, Chem. Vap. Dep. **3**, 167 (1997)

[36]  K. J. Eisentraut, R. E. Sievers: Volatile rare earth chelates, J. Am. Chem. Soc. **87**, 5254 (1965)

[37]  S.-C. Ha, E. Choi, S.-H. Kim, J. S. Roh: Influence of oxidant source on the property of atomic layer deposited $Al_2O_3$ on hydrogen-terminated Si substrate, Thin Solid Films **476**, 252 (2005)

[38] J. Niinistö, M. Putkonen, L. Niinistö, K. Kukli, M. Ritala, M. Leskelä: Structural and dielectric properties of thin $ZrO_2$ films on silicon grown by atomic layer deposition from cyclopentadienyl precursor, J. Appl. Phys **95**, 84 (2004)

[39] H. B. Park, M. Cho, J. Park, S. W. Lee, C. S. Hwang, J.-P. Kim, J.-H Lee, N.-I. Lee, H.-K. Kang, J.-C. Lee, S.-J. Oh: Comparison of $HfO_2$ films grown by atomic layer deposition using $HfCl_4$ and $H_2O$ or $O_3$ as the oxidant, J. Appl. Phys. **94**, 3641 (2003)

[40] T. Leskelä, K. Vasama, G. Härkönen, P. Sarkio, M. Lounasmaa: Potential cerium precursors for blue colour in thin film electroluminescent devices grown by atomic layer epitaxy, Adv. Mater. Opt. Electron. **6**, 169 (1996)

[41] M. Putkonen, L. Niinistö: Organometallic precursors for atomic layer deposition in CVD precursors, Topics Organomet. Chem. **9**, 125 (2005)

[42] B. S. Lim, A. Rahtu, J.-S. Park, R. G. Gordon: Synthesis and characterization of volatile thermally stable reactive transition metal amidinates, Inorg. Chem. **42**, 7951 (2003)

[43] B. S. Lim, A. Rahtu, P. deRouffignac, R. G. Gordon: Atomic layer deposition of lanthanum aluminum oxide nano-laminates for electrical applications, Appl. Phys. Lett. **84**, 3957 (2004)

[44] H. Seim, H. Mölsä, M. Nieminen, H. Fjellvåg, L. Niinistö: Deposition of $LaNiO_3$ thin films in an atomic layer epitaxy reactor, J. Mater. Chem. **7**, 449 (1997)

[45] H. Seim, M. Nieminen, L. Niinistö, H. Fjellvåg, L.-S. Johansson: Growth of $LaCoO_3$ thin films from $\beta$-diketonate precursors, Appl. Surf. Sci. **112**, 243 (1997)

[46] O. Nielsen, M. Peussa, H. Fjellvåg, L. Niinistö, A. Kjekshus: Thin film deposition of lanthanum manganite perovskite by ALE process, J. Mater. Chem. **9**, 1781 (1999)

[47] M. Nieminen, T. Sajavaara, E. Rauhala, M. Putkonen, L. Niinistö: Surface-controlled growth of $LaAlO_3$ thin films by atomic layer epitaxy, J. Mater. Chem. **11**, 2340 (2001)

[48] M. Nieminen, S. Lehto, L. Niinistö: Atomic layer epitaxy growth of $LaGaO_3$ thin films, J. Mater. Chem. **11**, 3148 (2001)

[49] O. Nilsen, H. Fjellvåg, A. Kjekshus: The AVS topical conference on atomic layer deposition (ALD 2004), in *Growth of $La_{1-x}Ca_x MnO_3$ by ALD* (2004)

[50] P. Myllymäki, M. Nieminen, J. Niinistö, M. Putkonen, K. Kukli, L. Niinistö: High permittivity $YScO_3$ thin films by atomic layer deposition using two precursor approaches, J. Mater. Chem. **16**, 563 (2006)

[51] M. Putkonen, T. Sajavaara, L. Niinistö, L.-S. Johansson, L. Niinistö: Deposition of yttria-stabilized zirconia thin films by atomic layer epitaxy from $\beta$-diketonate and organometallic precursors, J. Mater. Chem. **12**, 1 (2002)

[52] E. Gourba, A. Ringuedé, M. Cassir, J. Päiväsaari, J. Niinistö, M. Putkonen, L. Niinistö: Microstructural and electrical properties of gadolinium doped ceria thin films prepared by atomic layer deposition (ALD), Proc. Electrochem. Soc. **2003-7**, 267 (2003)

[53] E. Gourba, A. Ringuedé, M. Cassir, A. Billard, J. Päiväsaari, J. Niinistö, M. Putkonen, L. Niinistö: Characterisation of thin films of ceria-based electrolytes for intermediate temperature-solid oxide fuel cells (IT-SOFC), Ionics **9**, 15 (2003)

[54] R. G. Gordon, J. Becker, D. Hausmann, S. Suh: Vapor deposition of metal oxides and silicates: Possible gate insulators for future microelectronics, Chem. Mater. **13**, 2463 (2001)

# Index

# MOCVD Growth of Rare Earth Oxides: The Case of the Praseodymium/Oxygen System

Raffaella Lo Nigro[1], Graziella Malandrino[2], Roberta G. Toro[2], and Ignazio L. Fragalà[2]

[1] CNR-IMM, Stradale Primosole 50, 95127 Catania, Italy
    raffaella.lonigro@imm.cnr.it
[2] Chemistry Department, University of Catania, Viale A. Doria 6, 95125 Catania, Italy

**Abstract.** Chemical Vapor Deposition (CVD) uses one or more gaseous species (precursors) to form on a substrate, solid phase materials through an activated process. While today a large variety of precursors is known and rather complex deposition routes are involved, a user-friendly classification of precursor compounds as well as a viable discussion of their physical and chemical characteristics can be useful to MOCVD practitioners.

In this Chapter, an overview of both the exploitation and challenges of MOCVD fabrication of praseodymium oxides will be highlighted from different points of view, including the more suited precursors, the synthesis of thin films and their stability on silicon substrates.

## 1 Introduction

In the last years, modern technologies are demanding innovative families of materials. Rare earth oxides represent a new, technologically useful class of materials targeted to a wide variety of potential applications, including optical coatings [1], catalysts used in chemical processing [2], gate insulators [3], and protective coatings [4]. In particular, the growth of rare earth oxide thin films on semiconductors, whether epitaxial, polycrystalline or amorphous, is a rapidly developing area fueled by a variety of potential microelectronics applications. These include considerations of novel candidates for gate dielectrics that may replace silicon dioxide in hyperscaled Si devices [5].

These new materials, in turn, demand new synthethic routes for the fabrication of thin solid films and new process integration strategies in order to achieve worldwide acceptance.

Chemical Vapor Deposition (CVD) is a suitable method for the formation of dense and conformal films or coatings on a substrate through the decomposition of relatively high vapor pressure precursors [6, 7]. Precursors are sublimed/evaporated (whatever liquid or gas), transported to the surface of a substrate, and finally decomposed via an activated (thermal, photo, plasma, etc.) reaction to form the required materials. Reaction by-products

M. Fanciulli, G. Scarel (Eds.): Rare Earth Oxide Thin Films,
Topics Appl. Physics **106**, 33–52 (2007)
© Springer-Verlag Berlin Heidelberg 2007

are vented out of the system. In addition to capabilities of penetrating porous films, blind holes, and high-aspect ratio vias, the advantages offered by CVD over other deposition processes (such as physical vapor deposition, sputtering or ion-metal plasma coating) include:

- Versatility: it is possible to deposit any element or compound
- Materials formation well below the melting point
- Conformal and near-net shape coatings
- Economical production, with simultaneous coating of several parts.

The deposited layers can fulfill many functions, and range from metals and alloys to semiconductors and insulators. In essence, CVD is a remarkably simple yet versatile technique and the various reaction types including thermal deposition, oxidation and reduction, hydrolysis and ammonolysis can be used in several device fabrications.

The various CVD processes include what is generally known as thermal CVD, which represents the original process, photo CVD, and plasma CVD. Differences, of course, rely on the method of applying the energy required for the precursor reaction to take place.

In this Chapter, the classical thermal CVD process for the fabrication of rare earth oxide thin films is addressed. It is founded on the use of metallo-organic precursors, and therefore, referred to as Metal Organic Chemical Vapor Deposition (MOCVD).

Among the rare earth oxides, the praseodymium/oxygen system represents the most complex case. In fact, praseodymium oxides may have different praseodymium/oxygen compositions depending on the praseodymium $+3/+4$ oxidation state and may form either stoichiometric compounds of general formula $Pr_nO_{2n-2}(n = 4, 7, 9, 10, 11, 12,$ and $\infty)$ or non-stoichiometric phases [8–10]. The $\vartheta$-$Pr_2O_3$ phase $(n = 4)$ has the sesquioxide structure $(c = 11.15\,\text{Å})$ of manganese oxide $(Mn_2O_3)$ and remains stable below $500\,°C$. The hexagonal phase $(h$-$Pr_2O_3\colon a = 3.85\,\text{Å}, c = 6.01\,\text{Å})$, related to the lanthanum oxide $(La_2O_3)$, is stable at temperatures higher than $500\,°C$. Phases with $O/Pr$ ratios greater than $1.5$ have a relaxed fluorite structure proportional to the extent of oxygen vacancies. The stoichiometric oxide involving the $+4$ oxidation state is the $\alpha$-$PrO_2$ phase, which has a fluorite $(CaF_2)$-type structure and lattice parameter of $c = 5.39\,\text{Å}$. The intermediate phases, including $\beta$-$Pr_6O_{11}$ $(n = 12)$, $\delta$-$Pr_{11}O_{20}$ $(n = 11)$, $\varepsilon$-$Pr_5O_9$ $(n = 10)$, and non-stoichiometric compounds, all contain $Pr$ cations with both the $+3$ and $+4$ oxidation states. Selective formation of these phases depends upon the oxygen partial pressure and reaction temperature.

The MOCVD fabrication of praseodymium oxide thin films will be discussed as the most representative case of issues involved in the synthesis of rare earth oxide films. In fact, the complexity of the praseodymium/oxygen system finds counterpart in several practical applications, including varistor ceramic materials [11, 12], photocatalytically active materials [13], anodes for organic light-emitting diodes [14] and microelectronic applications. In the last

Pr(tmhd)$_3$ indicates that this complex sublimes in a single step, in the 200 °C to 300 °C temperature range, with a 3 % to 4 % residue left at 450 °C. Isothermal gravimetric measurements show that the vaporization rate of Pr(tmhd)$_3$ remains constant throughout the investigated time (60 min) in the entire 140 °C to 180 °C temperature range. Nevertheless, the mass loss increases significantly at temperatures higher than 160 °C. Concluding, the Pr(tmhd)$_3$ shows clean vaporization with low residue, as required for MOCVD applications.

Based on previous observations on "M(hfa)$_n$•glyme" adducts [26–30], the synthesis of the low-melting, fluorinated $\beta$-diketonate Pr(III) with ancillary polyether has been attempted as a promising strategy to achieve better thermal and mass transport properties than those of the first-generation Pr(tmhd)$_3$ precursor [39].

The adopted synthetic route to this second-generation precursor has proven well suited for many metal ions including alkaline- and rare earth metals [23–30]. The procedure uses solution/suspension of the required metal oxide/salt, hexafluoroacetylacetonate and of the neutral glyme ligand in a one-pot synthesis. Advantages are:

 – very high yields (> 80 %)
 – short time reaction (1–2 h)
 – normal open-bench laboratory conditions using commercially available reactants
 – almost pure products that do not require further purification.

Thus, the Pr(hfa)$_3$•diglyme has been readily prepared by reaction of praseodymium nitrate with the appropriate $\beta$-diketonate molecules and diglyme polyether in dichloromethane solvent. The polyether length has been tuned to the Pr$^{3+}$ ionic radius. The reaction has been carried out under the most usual and simple laboratory conditions in a single step:

$$Pr(NO_3)_3 \cdot 6H_2O + 3H\text{-hfa} + \text{diglyme} + 3NaOH \rightarrow$$
$$Pr(hfa)_3 \cdot \text{diglyme} + 3NaNO_3 + 9H_2O$$

Evaporation of the solvent leaves behind the product in form of green crystals. Pr(hfa)$_3$•diglyme adduct is stable in air and melts at 73 °C. The single-crystal X-ray data indicate that the adduct is a mononuclear complex with the Pr$^{3+}$ ion lying in a 9-coordinated environment of the six oxygen atoms of the hfa framework and of the three oxygen atoms of the diglyme. Details of synthetic procedure, X-ray single-crystal structure and complete physico-chemical characterization have been reported elsewhere [39]. The Pr(hfa)$_3$•diglyme adduct is thermally stable and evaporates in a single step in the 150 °C to 240 °C range. Sizeable evaporation of this second-generation precursor is achieved at lower temperature than in the case of the first-generation thmd precursor. Isothermogravimetric experiments, in the 100 °C to 140 °C range, show that the vaporization rate of Pr(hfa)$_3$•diglyme remains

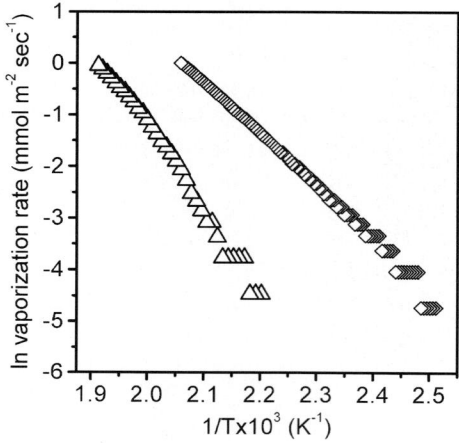

**Fig. 1.** Thermogravimetric vaporization rate data of Pr(hfa)$_3$•diglyme (*open triangle*) and Pr(tmhd)$_3$ (*open diamond*) as a function of temperature

constant at every temperature investigated and that this parameter increases linearly upon increasing the temperature.

Properties of Pr(hfa)$_3$•diglyme and Pr(tmhd)$_3$ precursors are compared in Fig. 1. It becomes evident that evaporation from the molten Pr(hfa)$_3$•diglyme adduct is more efficient than sublimation of the "first-generation" Pr(tmhd)$_3$ powder. The lower vaporization rate can be confidently related to surface effects during sublimation of surface and near-surface layers. Evaporation from melt, by contrast, involves thin surface layers continuously renewed. Both complexes show a linear behaviour of the vaporization rate, a clear indication of no partial decomposition during vaporization.

# 3 Thermal MOCVD

The above-mentioned Pr(III) precursors have been investigated as possible candidates for MOCVD of praseodymium oxide thin films. Operational conditions have been carefully optimized to selectively grow single phases of each, among the large variety of possible Pr oxide phases.

## 3.1 Film Growth from Pr(tmhd)$_3$

Different oxide phases have been stabilized over Si(001) substrates, depending on the oxygen partial pressure ($P_{O2} = 2 \times 10^{-3}$ Torr) adopted in the deposition process. In all the experiments, the Pr(tmhd)$_3$ precursor has been sublimed at 170 °C and the deposition temperature has been maintained at 750 °C [39–41]. The deposition temperature value has been chosen on the basis of the data previously reported on Molecular Beam Epitaxy (MBE) growth processes [17].

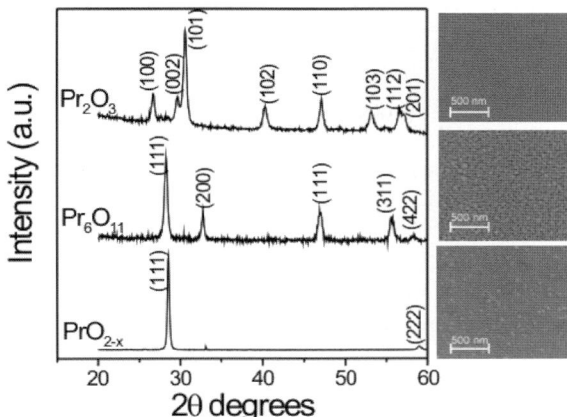

**Fig. 2.** XRD patterns of films deposited on Si (001) substrate from Pr(ɔmhd)$_3$ precursor and their SEM image surface morphologies

Of course, highly oxidized phases are expected under greater $P_{O_2}$ values. X-ray diffraction (XRD) data (Fig. 2) of thin films grown at 750 °C and $P_{O_2} = 2$ Torr show the (111) and (222) reflections of the PrO$_2$ phase, thus pointing to <111> oriented crystallites. The full width at half maximum (FWHM) of the rocking curve of about 2° indicates a good out-of-plane orientation. This value lies close to the typical values of pulsed laser deposition film growth under similar conditions [42]. XRD reflections suffer from a slight shift relative to values expected for the stoichiometric PrO$_2$ phase. Some oxygen vacancies in the fcc PrO$_2$ system can account for this observation.

Under lower $P_{O_2}$ values (0.7 Torr), the XRD pattern (Fig. 2) points to the formation of polycrystalline Pr$_6$O$_{11}$ films, since XRD peaks can be safely associated with (111), (200), (220), and (311) Pr$_6$O$_{11}$ reflections, respectively. In this case, the identification of the grown phase is unambiguous and no texture is evident.

The hexagonal polycrystalline Pr$_2$O$_3$, which is the dominant phase under oxygen-deficient conditions, has been obtained at $P_{O_2}$ values as low as $10^{-3}$ Torr. The XRD peaks of the deposited films are related to characteristic reflections of the hexagonal Pr$_2$O$_3$ polycrystalline phase. No preferential texturing is evident.

In all cases, the praseodymium oxide films show smooth surfaces, uniform distribution grain size and no cracking. The scanning electronic microscopy (SEM) images (Fig. 2) give evidence of spherical grains with mean size decreasing parallel to the oxidation state. Thus, smoother surfaces are obtained for Pr$_2$O$_3$ films than for PrO$_2$.

To conclude, the structure of praseodymium oxide films strongly depends on the oxygen partial pressure ($P_{O_2}$), and Pr$_2$O$_3$ films can be reproducibly grown for $P_{O_2} = 10^{-3}$ Torr.

## 3.2 Study of the Praseodymium Oxide Growth Process Through the Thermal MOCVD Process

It is well known that quality and reproducible synthesis of films critically depend upon the accurate control of deposition conditions, and therefore every MOCVD process needs the optimization of several parameters.

Effects associated with variations of deposition temperatures have been fully investigated in the 500 °C to 850 °C range, while maintaining constant all the other parameters. The XRD patterns have shown that films grown below 550 °C appear amorphous, while those deposited at 600 °C are poorly crystalline. Finally, those obtained beyond 650 °C consist of highly crystalline hexagonal $Pr_2O_3$ phase.

Growth rates vs. deposition temperatures have been evaluated by measuring the film thickness using TEM cross-section images. Rates vary in the 1.0 nm/min to 4.5 nm/min range in the 500 °C to 850 °C interval for a deposition time of 60 min. The slope of the Arrhenius plot points to an apparent activation energy of $34 \pm 2$ kJ · mol, and therefore, the present process occurs under kinetic regime. Depending upon deposition time (15 min to 90 min), the growth rate values at 750 °C point to different growth regimes. In the initial deposition stage, (0 min to 30 min), the growth rate increases linearly with a marked slope. In the 30 min to 60 min range, a less steep slope is observed up to a 2.7 nm/min value after 60 min. Longer times do not affect the growth rate, which remains almost constant with a flat plateau. This intriguing dependence is a clear indication of phenomena occurring in the initial growth stage at the film/substrate interface.

### 3.3 Characterization of the Film/Substrate Interface

Microstructure and composition of the film/substrate interface have been carefully investigated by high-resolution (HR) TEM imaging and X-ray photoelectron spectroscopy (XPS) [41]. In the cross-section HR-TEM image (Fig. 3), three layers are clearly visible: 1. a $SiO_2$ layer in direct contact with the Si substrate, 2. a bottom 8 nm Pr oxide-based layer having a different contrast than the $SiO_2$ layer and no long range order, and 3. well-defined $Pr_2O_3$ crystalline grains.

The composition of the bottom praseodymium layer has been highlighted by XPS analyses of samples consisting only of the amorphous layer.

XPS spectra of the Si2p, Pr3d and O1s core level spectral regions are displayed in Fig. 4. The Pr3d XPS signals cannot be simply related to a particular chemical environment, since $3d^{-1}$ ionizations represent several possible final states, reached upon ionizing open-shell structures, and are affected by intervening differential electron relaxation energies in the ion states [43]. This behavior, typical of open shell lanthanide systems, precludes any reliable detection of different Pr environments [43–45]. Detailed compositional

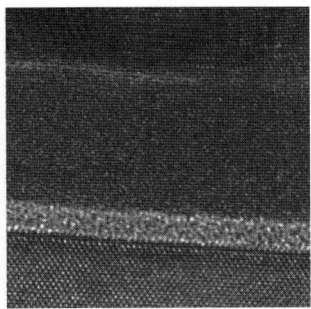

**Fig. 3.** High-resolution TEM image of the film/substrate interface showing a $SiO_2$ layer, a praseodymium-containing intermediate layer and crystalline $Pr_2O_3$ grains on the *top*

information about the amorphous layer has been, however, obtained from the Si2p and O1s signals.

Thus, the Si $2p$ XPS spectrum shows the Si $2p_{3/2\text{-}1/2}$ spin-orbit doublet (0.6 eV) typical of the elemental silicon, due to the substrate contribution [46]. However, the Si $2p$ peak showing the highest intensity is centred at higher binding energies and can be considered the superposition of two different components: the main contribution (101.7 eV) can be associated with rare earth silicate phases [43, 47, 48] while the peak at 102.9 eV is attributed to the $SiO_2$. There is no evidence of features representative of Pr–Si bonding, thus ruling out any formation of silicides [44], in accordance with data of other high-$\kappa$ dielectrics on Si.

Further evidence of silicate formation has been found in the O1s signals of which the shape appears rather complex due to the overlap of different contributions. The lowest binding energy (529.6 eV) feature can be safely associated with the Pr–O bonding in the $Pr_2O_3$ crystals [49] while the peak at 531.0 eV can be attributed to the O1s ionization of the silicate phase. Finally, the peak lying at the highest binding energies ($\approx 532$ eV) consists of a broad signal due to the overlap of different components associated with $SiO_2$ and carbonates and/or hydroxides that can form on the film surface. The presence of carbonates is further corroborated by the peak at 289.8 eV in the C1s spectrum [48].

This observation is not surprising, since it is well known that rare earth oxides readily absorb water vapor and carbon dioxide. Thus, for instance, it is well known that hexagonal $La_2O_3$ rapidly reacts with $CO_2/H_2O$ and forms $La(OH)_3$ and several carbonate phases [50]. $Y_2O_3$ films, also, show a similar reactivity towards atmospheric components, and films sputter-deposited on Si have OH groups and/or $H_2O$ molecules on their surfaces [51]. Similarly, it has been reported that rare earth oxides, depending on their ionic radius and electronegativity, have high affinities for $H_2O$ [52].

The influence of the deposition temperature on the formation of the praseodymium silicate phase has been highlighted through further XPS analyses. The Si 2p features of films deposited at 350 °C, 450 °C and 750 °C show in all cases evidence of the silicate phase. All these observations clearly point

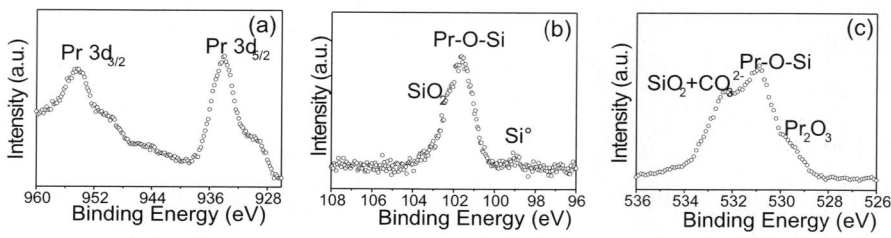

**Fig. 4.** XPS spectra of a praseodymium silicate film: praseodymium (**a**), silicon (**b**) and oxygen (**c**) regions

to some interaction between the growing $Pr_2O_3$ and the silicon substrate, even at the lower deposition temperatures. It is worth noting that the silicate formation seems to be independent of the growing technique, since data of MBE-grown $Pr_2O_3$ films also show the formation of a thin silicate layer at the film/Si interface [53].

It is, therefore, clearly shown that in the initial stage of the MOCVD process, the praseodymium silicate formation competes with the growth of $Pr_2O_3$ grains. Self-limitation, however, occurs in about 30 min and the silicate thickness remains almost constant. Under these conditions, a steady state is reached with an almost negligible diffusion rate of Si in the growing praseodymium oxide film.

## 3.4 Electrical Characterization

Electrical characteristics of polycrystalline $Pr_2O_3$ films have been investigated by capacitance–voltage (C–V) measurements. Au-gate MOS capacitors with an active area ranging from $7.85 \times 10^{-5}\,cm^2$ to $9.07 \times 10^{-4}\,cm^2$ were fabricated on n-type (001) Si substrates [54, 55]. Au was sputter-deposited, patterned and wet-etched to form the top electrode. In particular, a $I_2/2I^-$ solution was used to selectively remove the gold layer after resist patterning. The complete gold removal was optically evaluated, and the $Pr_2O_3$ surface roughness and morphology, analyzed after etching, was found to be identical to that of the as-deposited film, i.e., no surface damage was detected by atomic force microscopy after etching. The capacitance value in accumulation capacitance was used to determine the effective dielectric constant. The results obtained from 100 kHz C–V measurements of a typical MOS device with a polycrystalline $Pr_2O_3$ dielectric layer are plotted in Fig. 5. Evaluation of the dielectric constant ($\varepsilon$) provides values for polycrystalline $Pr_2O_3$ films in the 26–15 interval for thickness ranging from 44 nm to 15 nm. It turns out that thinner layers have smaller $\varepsilon$ values due to the greater contribution of the bottom praseodymium silicate layer having an $\varepsilon$ of about 8 [56]. Therefore, the multilayer $Pr–Si–O/SiO_2$ structure of the entire deposited stack causes, no doubt, a sizeable decrease of the measured dielectric constant.

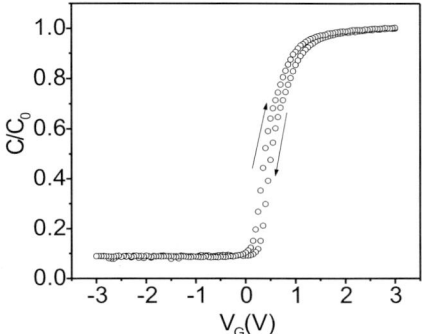

**Fig. 5.** C–V curve of Au/dielectric/Si capacitors, where the dielectric layer is a polycrystalline $Pr_2O_3$ film

The C–V characteristics, using metal gates, are directly related to the charge defects in the insulator and at the insulator/semiconductor interface. In fact, the unbalanced charges arising from defects cause a C–V curve shift moving the flat band from the ideal condition. The shift value can be used to quantify the number of fixed charges ($N_f$ $cm^{-2}$) in the dielectric layer. The observed $V_{fb}$ shift has been used to calculate the quantity of fixed charges ($N_f = 4.77 \times 10^{11}$ $cm^{-2}$).

A further phenomenon which can be commonly observed in the C–V curves of high-$\kappa$ materials is the hysteresis, the minimization of which is required for their use in memory applications. The curves show small hysteresis, indicating the poor trapping of a net negative charge (electrons) in the insulator during the measurements. A positive $\Delta V_{fb}$ is observed, due to a negative charge attributed to electrons injected from the gate and trapped at the dielectric/semiconductors interface. The number of trapped negative charges ($N_{ot} = 1.6 \times 10^{11}$ $cm^{-2}$) has been evaluated from the negligible hysteresis observed.

The study of the slope of the C–V curve provides information on the interfacial state density ($D_{it}$ $cm^{-2} \cdot eV^{-1}$), which represents a crucial issue to evaluate the device quality.

In present cases, the $D_{it}$ values have been obtained by mathemathical simulation assuming a "U" distribution of the state density inside the dielectric. The $D_{it}$ value is about $2 \times 10^{12}$ $eV^{-1}$ $cm^{-2}$.

The electrical measurements presently reported represent a preliminary step to the understanding of the relationship between dielectric properties and chemical effects associated with depositions.

### 3.5 Film Growth from Pr(hfa)$_3$•Diglyme

MOCVD experiments using the "second-generation" Pr(hfa)$_3$•diglyme adduct have been carried out maintaining parameters identical to those selected in the case of deposition from the Pr(thmd)$_3$ [39]. The source has been evaporated from melt at 120 °C. Attempts to scavenge fluorine by-products adopt-

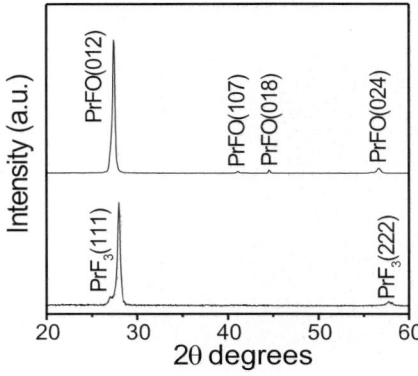

**Fig. 6.** XRD patterns of films deposited on Si (001) substrate using the Pr(hfa)$_3\bullet$diglyme precursor: PrF$_3$ (dry O$_2$ reaction gas) and PrOF (H$_2$O saturated O$_2$ reaction gas)

ing dry oxygen ($P_{O_2} = 0.4$ Torr), as reaction gas, failed and no oxide phases have been obtained under these conditions. XRD data (Fig. 6) point, by contrast, to (111) oriented PrF$_3$ films. This behavior is not surprising when fluoride precursors have been used, since pure O$_2$ ambient and high deposition temperature are not the best suited conditions to avoid fluoride phase formation. Other experiments have been carried out using water-saturated O$_2$ ($P_{O_2} = 0.4$ Torr). The introduction of water vapor into the oxygen stream is a well-known in-situ method for hydrolysis of fluorine-containing phases [57]. The typical XRD pattern of the praseodymium films deposited under these conditions provides evidence of the formation of polycrystalline PrOF phase, since peaks at $2\theta = 27.4°, 41.2°, 44.8°, 56.5°$ can be safely attributed to (012), (107), (018) and (024) reflections, respectively. The observed intensities are different from those expected for a polycrystalline powder and suggest a (012) preferential texturing. Systematic studies to preclude PrOF formation by varying the process parameters, including the deposition temperature (in the 450 °C to 750 °C range) and the water-saturated oxygen partial pressure (0.4 Torr to 1 Torr), proved unsuccessful. There is, therefore, evidence of a greater thermodynamic stability of the praseodymium oxyfluoride phase compared to oxides, and this is an indication of the need of an alternative fluorine-free praseodymium precursor.

To conclude, the novel Pr precursor possesses higher volatility and thermal stability than the first-generation Pr(thmd)$_3$ precursor, but the fluorine backbone favors the formation of fluoride and oxyfluoride films, which can not be used as dielectric materials. They, however, possess intriguing luminescence properties suitable for optical applications [58, 59].

# 4 MOCVD Growth of Mixed Pr/Al Oxide Films

In the last years, much attention has been devoted to amorphous silicate and aluminate systems as alternative to the pure oxide high-$\kappa$ dielectrics.

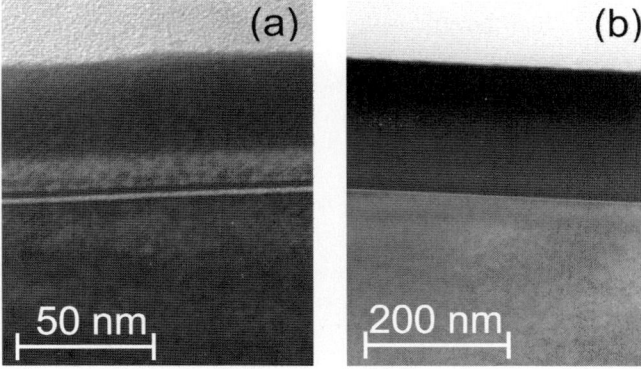

**Fig. 7.** Cross-section TEM images of praseodymium alluminate PrAlOx films grown from Pr(tmhd)$_3$/Al(acac)$_3$ mixture (**a**) and from Pr(hfa)$_3$•diglyme/Al(acac)$_3$ mixture (**b**)

Here two different approaches will be shown for the MOCVD fabrication of praseodymium aluminate films. Thus, a mixture of two "first-generation" precursors, Pr(tmhd)$_3$ and Al(acac)$_3$, can be expedient for this purpose even though a multi-element single source consisting of the low-melting second-generation Pr(hfa)$_3$•diglyme in mixture with the Al(acac)$_3$ precursor presents, no doubt, great advantages.

Atmospheric pressure TG curves of both mixtures point to singular sublimation steps in the 200 °C to 300 °C range in the case of the Pr(tmhd)$_3$/Al(acac)$_3$ mixture, while the molten Pr(hfa)$_3$•diglyme/Al(acac)$_3$ mixture shows a weight loss in the 150 °C to 200 °C range. These results point to a higher volatility of the Pr(hfa)$_3$•diglyme/Al(acac)$_3$ mixture than in the case of Pr(tmhd)$_3$/Al(acac)$_3$.

Deposition experiments of praseodymium aluminate layers have been performed using similar conditions to those discussed for praseodymium oxides. The deposition temperature, however, has been changed to 450 °C to obtain amorphous layers, as requested by the microelectronics applications.

Depositions carried out from the solid Pr(tmhd)$_3$/Al(acac)$_3$ mixture lead to the formation of two almost separate oxide layers. TEM cross-section images as well as EDX analysis in TEM mode (Fig. 7a) show that the bright layer lying over the silicon substrate consists of an aluminium oxide-rich layer, while the dark layer on its top is a praseodymium oxide-rich layer.

The use of the molten single source, namely the Pr(hfa)$_3$•diglyme/Al(acac)$_3$ mixture, beside some experimental advantages due to the constant evaporation rate of both precursors, results in the formation of films consisting of single PrAlO$_x$ layers (Fig. 7b).

In this case, the XPS analysis appears consistent with no F contamination, in contrast with previous data of the Pr(hfa)$_3$diglyme precursor. It is, therefore, evident that the aluminate phase is more stable than the fluoride phase.

The molten multi-element mixture represents a user-friendly source for an easy and simultaneous control of evaporation rates of both the praseodymium and aluminium precursors, it rules out all problems related to the crystallite size, and it results in a constant evaporation rate even for very long deposition times. Moreover, it is compatible with simpler reactor designs and requires simpler control of deposition parameters. In fact, there is no more need of controlling and optimising different sublimation-evaporation temperatures and carrier gas flows, but just one internal source can ensure an improvement of the deposition process.

These data lend themselves to more general considerations. The "second-generation" $M(hfa)_n \bullet$glyme adducts represent really good MOCVD precursors for the deposition of simple and mixed oxide films over a wide range of substrates. Either unwanted fluorinated phases or fluorine contamination should be addressed time by time depending on the nature of the metal ions and on the thermodynamic stability of the desired phase. Typical examples involve using the Ce or Pb adducts that form high-quality oxide thin films [31, 60] whilst the Ag homologues adduct favors the formation of metallic nanoparticles [61]. In several other cases, different mixed oxide films can be obtained [30, 33].

## 5 Summary

Herein, the authors report on the fabrication of praseodymium oxide thin films grown on silicon substrates by a widely used technique in ultra large scale integration (ULSI) processing, namely the Metal-Organic Chemical Vapor Deposition.

The first prerequisite for a successful MOCVD process is the availability of a volatile and thermally stable precursor that must possess high vapor pressures and constant mass transport properties.

In this general context, the properties of two different Pr $\beta$-diketonates have been compared. It has been demonstrated that the $Pr(tmhd)_3$ precursor possesses the appropriate mass transport properties and it can be easily used for the selective fabrication of different pure praseodymium phases ($PrO_2$, $Pr_6O_{11}$, $Pr_2O_3$). A novel precursor, namely the $Pr(hfa)_3 \bullet$diglyme adduct, has been also synthesized and tested. It possesses better thermal properties than the first-generation precursors. Nevertheless, fluoride and oxyfluoride phases compete favorably with the oxide formation.

Finally, the same second-generation $Pr(hfa)_3 \bullet$diglyme precursor has been tested within a well-assessed methodology which uses the intrinsic capacity of the low-melting precursor to act as solvent for solid precursors. Thus, the molten single $Pr(hfa)_3 \bullet$diglyme/$Al(acac)_3$ source has shown better performance for growing mixed Pr/Al oxide phase than the conventional

Pr(tmhd)$_3$/Al(acac)$_3$ solid mixture. The liquid mixture, in fact, can be easily and cleanly evaporated from melt, thus providing constant mass transport rate. On the contrary, solid precursor mixtures present several drawbacks associated with thermal instability, sintering and surface passivation upon sublimation, which in turn may render transport processes largely irreproducible.

The MOCVD routes to the synthesis of praseodymium-containing films represent a real, great challenge for large-scale process development which might be superior or complementary to other film growth methods.

# References

[1] S. Kimura, F. Arai, M. Ikezawa: Optical study on electronic structure of rare-earth sesquioxides, J. Phys. Soc. Jpn. **69**, 3451 (2000)

[2] H. L. Wan, X. P. Zhou, W. Z. Weng, R. Q. Long, Z. S. Chao, W. D. Zhang, M. S. Chem, J. Z. Luo, S. Q. Zhou: Catalytic performance, structure, surface properties and active oxygen species of the fluoride-containing rare earth (alkaline earth)-based catalysts for the oxidative coupling of methane and oxidative dehydrogenation of light alkanes, Catal. Today **51**, 161 (1999)

[3] J. Robertson: Band offsets of wide-band-gap oxides and implications for future electronic devices, J. Vac. Sci. Technol. B **18**, 1785 (2000)

[4] S. Chevalier, G. Bonnet, J. P. Larpin: Metal-organic chemical vapor deposition of Cr$_2$O$_3$ and Nd$_2$O$_3$ coatings. Oxide growth kinetics and characterization, Appl. Surf. Sci. **167**, 125 (2000)

[5] S. Jeon, K. Im, H. Yang, H. Lee, H. Sim, S. Choi, T. Jang, H. Hwang: Excellent electrical characteristics of lanthanide (Pr, Nd, Sm, Gd and Dy) oxide and lanthanide-doped oxide for MOS gate dielectric applications, Proc. IEDM Tech. Dig. p. 471 (2001)

[6] H. O. Pierson: *Handbook of Chemical Vapor Deposition* (Noesy, New York 1992)

[7] M. L. Hitchman, J. F. Jensen: *Chemical Vapor Deposition: Principles and Applications* (Academic Press, London 1993)

[8] R. G. Haier, L. Eyring: *Handbook on the Physics and Chemistry of the Rare Earths* (North-Holland, Amsterdam 1994)

[9] Z. C. Kang, L. Eyring: Fluorite structural principles: Disordered $\alpha$-phase to ordered intermediate phases in praseodymia, J. Alloy and Comp. **275–277**, 721 (1998)

[10] L. Eyring, N. C. Baenzige: On the structure and related properties of the oxides of praseodymium, J. Appl. Phys. **33**, 428 (1962)

[11] N. Horio, M. Hiramatsu, M. Nawata, K. Imaeda, T. Torii: Preparation of zinc oxide metal oxide multilayered thin films for low-voltage varistors, Vacuum **51**, 719 (1998)

[12] C. W. Nahm: Electrical properties and stability of praseodymium oxide-based ZnO varistor ceramics doped with Er$_2$O$_3$, J. Eur. Ceram. Soc. **23**, 1345 (2003)

[13] D. W. Hwang, J. S. Lee, W. Li, S. H. Oh: Electronic band structure and photocatalytic activity of Ln(2)Ti(2)O(7) (Ln = La, Pr, Nd), J. Phys. Chem. B **107**, 4963 (2003)

[14] C. Qiu, H. Chen, Z. Xie, M. Wong, H. S. Kwok: Praseodymium oxide coated anode for organic light-emitting diode, Appl. Phys. Lett. **80**, 3485 (2002)

[15] H. J. Osten, J. P. Liu, P. Gaworzewski, E. Bugiel, P. Zaumseil: High k gate dielectric with ultra-low leakage current based on praseodymium oxide, Proc. IEDM Tech. Dig. p. 653 (2000)

[16] G. Adachi, N. Imanaka: The binary rare earth oxides, Chem. Rev. **98**, 1479 (1998)

[17] H. J. Osten, J. P. Liu, H. J. Mussig: Band gap and band discontinuities at crystalline $Pr_2O_3$/Si(001) heterojunctions, Appl. Phys. Lett. **80**, 297 (2002)

[18] L. R. Morss: Thermochemical properties of yttrium, lanthanum and lanthanide elements and ions, Chem. Rev. **76**, 827 (1976)

[19] T. J. Marks: Coordination chemistry ruotes to films for superconducting electronics, Pure Appl. Chem. **67**, 313 (1995)

[20] M. Tiitta, L. Niinisto: Volatile metal beta-diketonates: ALE and CVD precursors for electroluminescent device thin films, Chem. Vap. Dep. **3**, 167 (1997)

[21] J. A. Belot, D. A. Neumayer, C. J. Reedy, D. B. Studebaker, B. J. Hinds, C. L. Stern, T. J. Marks: Volatility by design. Synthesis and characterization of polyether adducts of bis(1,1,1,5,5,5-hexafluoro-2,4-pentanedionato)barium and their implementation as metal-organic chemical vapor deposition precursors, Chem. Mater. **9**, 1638 (1997)

[22] N. L. Edleman, A. C. Wang, J. A. Belot, A. W. Metz, J. R. Babcock, A. M. Kawaoka, J. Ni, M. V. Metz, C. J. Flaschenriem, C. L. Stern, L. M. Liable-Sands, A. L. Rheingold, P. R. Markworth, R. P. H. Chang, M. P. Chudzik, C. R. Kannewurf, T. J. Marks: Synthesis and characterization of volatile, fluorine-free beta-ketoiminate lanthanide MOCVD precursors and their implementation in low-temperature growth of epitaxial $CeO_2$ buffer layers for superconducting electronics, Inorg. Chem. **41**, 5005 (2002)

[23] G. Malandrino, F. Castelli, I. L. Fragalà: A novel route to the $2^{nd}$ generation alkaline-earth metal precursors for metal organic chemical vapor deposition: One step synthesis of M(hfa)$_2$tetraglyme (M= Ba, Sr, Ca, and hfa = 1,1,1,5,5,5-hexafluoro-2,4-pentadione), Inorg. Chim. Acta **224**, 203 (1994)

[24] S. B. Turnipseed, R. M. Barkley, R. E. Siever: Synthesis and characterization of alkaline-earth-metal beta-diketonate complexes used as precursors for chemical vapor deposition of thin films superconductors, Inorg. Chem. **30**, 1164 (1991)

[25] G. Rossetto, A. Polo, F. Bentollo, M. Porchia, P. Zanella: Studies on molecular barium precursors for MOCVD: Synthesis and characterization of barium 2,2,6,6-Tetramethyl-3,5- heptanedionate – X-ray crystal structure of $[BA(THD)_2.ET_2O]_2$, Polyhedron **11**, 979 (1992)

[26] G. Malandrino, C. Benelli, F. Castelli, I. L. Fragalà: Synthesis, characterizations, crystal structure and mass transport properties of lanthanum $\beta$-diketonate glyme complexes, volatile precursors for metal-organic chemical vapor deposition applications, Chem. Mater. **10**, 3434 (1998)

[27] G. Malandrino, R. L. Nigro, F. Castelli, I. L. Fragalà, C. Benelli: Volatile Ce(III)hexafluoroacetylacetonate glyme adducts as promising precursors for MOCVD deposition of $CeO_2$ thin films, Chem. Vap. Dep. **6**, 233 (2000)

[28] K. D. Pollard, H. A. Jenkins, R. J. Puddephat: Chemical vapor deposition of cerium oxide using the precursors [Ce(hfac)$_3$(glyme)], Chem. Mater. **12**, 701 (2000)

[29] G. Malandrino, O. Incontro, F. Castelli, I. L. Fragalà, C. Benelli: Synthesis, characterization and mass transport properties of two novel gadolinium(III) hexafluoroacetylacetonate polyether adducts: Promising precursors for MOCVD of GdF$_3$ films, Chem. Mater. **8**, 1292 (1996)

[30] G. Malandrino, M. Bettinelli, A. Speghini, I. L. Fragalà: Europium "second generation" precursors for metal-organic chemical vapor deposition: Characterization and optical spectroscopy, Eur. J. Inorg. Chem. p. 1039 (2001)

[31] R. Lo Nigro, G. Malandrino, I. L. Fragalà: MOCVD of cerium dioxide (100) oriented films on random Hastelloy C 276, Chem. Mater. **13**, 4402 (2001)

[32] R. Lo Nigro, R. Toro, G. Malandrino, I. L. Fragalà: Heteroepitaxial growth of nanostructured cerium dioxide thin films by MOCVD on a (001) TiO$_2$ substrate, Chem. Mater. **15**, 1434 (2003)

[33] G. Malandrino, I. L. Fragalà, P. Scardi: Heteroepitaxy of LaAlO$_3$ (100) on SrTiO$_3$ (100): *In situ* growth of LaAlO$_3$ thin films by metal-organic chemical vapor deposition from a liquid single-source, Chem. Mater. **10**, 3765 (1998)

[34] J. M. Zhang, F. DiMeo Jr, B. W. Wessels, D. L. Schultz, T. J. Marks, J. L. Schindler, C. R. Kanerwurf: A new route to high-$T_c$ superconducting Bi–Sr–Ca–Cu–O thin films: improved deposition efficiency and film morphology using ammonia–argon mixtures as the carrier gas, J. Appl. Phys. **71**, 2769 (1992)

[35] S. B. Turnipseed, R. M. Barkley, R. E. Sievers: Synthesis and characterization of alkaline-earth-metal beta-diketonate complexes used as precursors for chemical vapor deposition of thin film superconductors, Inorg. Chem. **30**, 1164 (1991)

[36] S. Liang, C. S. Chern, Z. Q. Shi, P. Lu: Control of CeO$_2$ growth by metalorganic chemical vapor deposition with a special source evaporator, J. Cryst. Growth **151**, 359 (1995)

[37] H. A. Luten, W. S. Rees Jr., V. L. Goedkend: Preparation and structural characterization, and chemical vapor deposition studies with, certain yttrium tris(beta-diketonate) compounds, Chem. Vap. Dep. **2**, 149 (1996)

[38] K. J. Eisentraut, R. E. Siever: Volatile rare earth chelates, J. Am. Chem. Soc. **87**, 5254 (1965)

[39] R. Lo Nigro, R. G. Toro, G. Malandrino, I. L. Fragalà: Study of the thermal properties of Pr(III) precursors and their implementation in the MOCVD growth of praseodymium oxide films, J. Electrochem. Soc. **151**, F206 (2004)

[40] R. Lo Nigro, R. G. Toro, G. Malandrino, V. Raineri, I. L. Fragalà: A simple route to the synthesis of Pr$_2$O$_3$ high-k films, Adv. Mater. **15**, 1071 (2003)

[41] R. Lo Nigro, R. G. Toro, G. Malandrino, G. G. Condorelli, V. Raineri, I. L. Fragalà: Praseodymium silicate as a high k dielectric candidate: An insight on the Pr$_2$O$_3$ film/Si substrate interface fabricated through an MOCVD process, Adv. Funct. Mater. **15**, 838 (2005)

[42] D. K. Fork, D. B. Fenner, T. H. Geballe: Growth of epitaxial PrO$_2$ thin films on hydrogen terminated Si (111) by pulsed lased deposition, J. Appl. Phys. **68**, 4316 (1990)

[43] H. Ogasawara, A. Kotani, R. Potze, G. A. Sawatzky, B. T. Thole: Praseodymium 3D-core and 4D-core photoemission spectra of Pr$_2$O$_3$, Phys. Rew. B **44**, 5465 (1991)

[44] J. X. Wu, Z. M. Wang, F. Q. Li, M. S. Ma: Photoemission study of the oxidation and the post-annealing behaviors of a Pr-covered Si(100) surface, Appl.
Surf. Sci. **225**, 229 (2004)
[45] S. Lütkehoff, M. Neumann, A. Ślebarski: $3d$ and $4d$ X-ray-photoelectron spectra of Pr under gradual oxidation, Phys. Rev. B **52**, 13808 (1995)
[46] G. F. Cerefolini, C. Galati, S. Lorenti, L. Renna, O. Viscuso, C. Bongiorno,
V. Raineri, C. Spinella, G. G. Condorelli, I. L. Fragalà, A. Terrasi: The early
oxynitridation stages of hydrogen-terminated (100) silicon after exposure to
$N_2$:$N_2O$. III. Initial conditions, Appl. Phys. A **77**, 403 (2003)
[47] D. Schmeisser, H. J. Mussig: The $Pr_2O_3$/Si(001) interface studied by synchrotron radiation photo-electron spectroscopy, Solid State Electron. **47**, 1607
(2003)
[48] J. F. Malder, W. F. Stickel, P. E. Sobol, K. D. Bomben: *Handbook of X-ray
Photoelectron Spectroscopy* (Perkin-Elmer, Eden Proucie 1992)
[49] D. D. Sarma, C. N. R. Rao: XPES studies of oxides of 2nd-row and 3nd-row
transition-metals including rare earths, J. Electron. Spectrosc. Relat. Phenom.
**20**, 25 (1980)
[50] T. Gouguosi, D. Niu, R. W. Ashcraft, G. N. Parson: Carbonate formation
during post-deposition ambient exposure of high-$k$ dielectrics, Appl. Phys.
Lett. **83**, 3543 (2003)
[51] J. J. Chambers, G. N. Parson: Physical and electrical characterization of ultrathin yttrium silicate insulators on silicon, J. Appl. Phys. **90**, 918 (2001)
[52] S. Jeon, H. Hwang: Electrical and physical characteristics of $PrTi_xO_y$ for
metal-oxide-semiconductor gate dielectric applications, Appl. Phys. Lett. **81**,
4856 (2002)
[53] A. Fissel, J. Dabrowski, H. J. Osten: Photoemission and *ab initio* theoretical
study of interface and film formation during epitaxial growth and annealing of
praseodymium oxide on Si(001), J. Appl. Phys. **91**, 8986 (2002)
[54] R. Lo Nigro, V. Raineri, C. Bongiorno, R. Toro, G. Malandrino, I. L. Fragalà:
Dieletric properties of $Pr_2O_3$ high-k films grown by metalorganic chemical
vapor deposition on silicon, Appl. Phys. Lett. **83**, 129 (2003)
[55] R. Lo Nigro, R. Toro, G. Malandrino, P. Fiorenza, V. Raineri, I. L. Fragalà:
Effects of deposition temperature on the microstructural and electrical properties of praseodymium oxide based films, Mater. Sci. Eng. B **118**, 117 (2005)
[56] R. Lo Nigro, R. Toro, G. Malandrino, V. Raineri, I. L. Fragalà: Electrical
properties of MOCVD praseodymium oxide based MOS structures, Proc. ESS-
DERC p. 375 (2003)
[57] D. Chadwick, J. McAleese, K. Senliw, B. C. H. Steele: On the application of
XPS to ceria films grown by MOCVD using a fluorinated precursors, Appl.
Surf. Sci. **99**, 417 (1996)
[58] Y. J. Cho, M. Noma, Y. Hamakawa: Filtered full-color thin-film electroluminescent device with ZnS:TbOF/ZnS:PrOF phosphor layers, Sensors Mater. **9**,
25 (1997)
[59] S. Kuck, I. Sokolska: Room temperature emission from the $Pr^{3+1}S_0$-level in
$PrF_3$, Appl. Phys. A: Mater. Sci. Process. **77**, 469 (2003)
[60] G. Malandrino, R. Lo Nigro, P. Rossi, P. Dapporto, I. L. Fragalà: A volatile
Pb(II) $\beta$-diketonate diglyme complex as a promising precursor of MOCVD of
lead oxide films, Inorg. Chim. Acta **357**, 3927 (2004)

[61] M. E. Fragala, G. Compagnini, G. Malandrino, C. Spinella, O. Puglisi: Silver
     nanoparticles dispersed in polyimide thin film matrix, Eur. Phys J. D **9**, 631
     (1999)

# Index

# Requirements of Precursors for MOCVD and ALD of Rare Earth Oxides

Helen C. Aspinall

Department of Chemistry, Donnan and Robert Robinson Laboratories, University of Liverpool, Crown Street, Liverpool, L69 7ZD, United Kingdom
hca@liv.ac.uk

**Abstract.** This Chapter gives an account of the coordination chemistry of precursors for MOCVD and ALD of rare earth oxides. It opens with a brief introduction to the coordination chemistry of the rare earth elements, and then describes the strategies that have been employed to design and synthesize rare earth complexes with reasonable volatility. The chemical requirements of precursors for MOCVD and ALD are outlined, and it is shown that these requirements may be mutually exclusive. Important classes of precursors, including amides, alkoxides, diketonates and organometallics, are surveyed.

## 1 Introduction to Rare Earth Chemistry

In order to understand the requirements for rare earth precursors for MOCVD and ALD, it is first important to understand some of the basic principles of the chemistry of the rare earth elements. Use of the term 'rare earth' (which is historical in origin) for the elements Sc, Y and La–Lu gives these elements an undeserved reputation for scarcity and exoticism. The elements in general are relatively abundant: even Tm, the least abundant, is more abundant than either I or Hg, and the most abundant, Ce, Nd and La, have crustal abundances of 68 ppm, 38 ppm and 32 ppm respectively, which are comparable to those of Zn (75 ppm) and much higher than those of Pb (14 ppm) or Sn (2.2 ppm). In general, the early lanthanides (La to Gd) are more abundant than the later lanthanides (Dy to Lu). Y is found in ores with the later Ln (Ln will be used as a general symbol for the elements La–Lu); Sc is reasonably abundant, but much more widely distributed and found in lower concentrations. The prices of the elements show an approximate inverse dependance on abundance and, apart from Tm and Lu, the prices of rare earths are relatively low. China has the major share of the world's rare earth resources (43 %) and is by far the largest producer. Separation of the rare earth elements from mixed ores was the main obstacle in the development of rare earth chemistry. However, an extremely efficient and environmentally friendly solvent extraction process is now used for the separation, and high-purity rare earth elements and compounds are now available in large quantities.

The rare earth elements are all highly electropositive and are most stable in the +3 oxidation state. In some cases, the +4 and +2 oxidation states are

M. Fanciulli, G. Scarel (Eds.): Rare Earth Oxide Thin Films,
Topics Appl. Physics **106**, 53–72 (2007)
© Springer-Verlag Berlin Heidelberg 2007

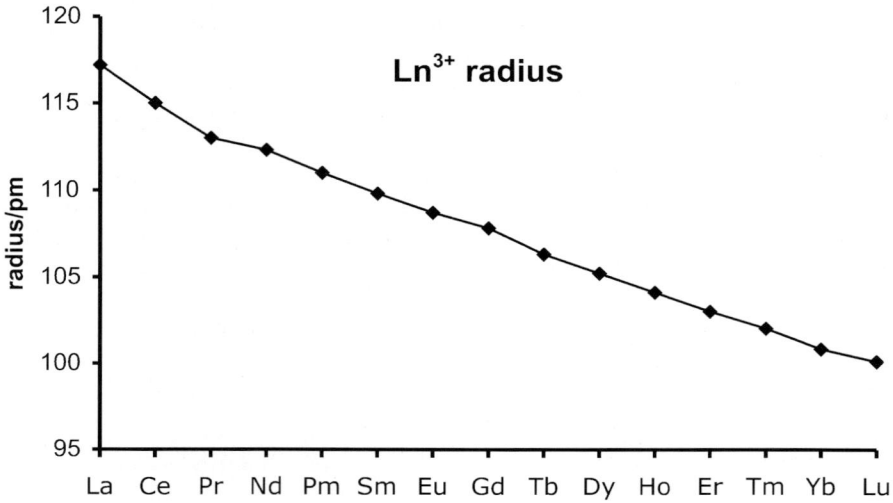

**Fig. 1.** Six-coordinate radii (Shannon) of $Ln^{3+}$ ions

available, particularly where this results in a $4f^0$, $4f^7$ or $4f^{14}$ configuration. For example, $Ce^{4+}$ ($4f^0$), $Pr^{4+}$ ($4f^1$) and $Tb^{4+}$ ($4f^7$) are all stable in the solid state, and $Ce^{4+}$ is also stable in solution. Ce is the only rare earth element for which molecular precursors in the +4 oxidation state are available. $Eu^{2+}$ ($4f^7$) and $Yb^{2+}$ ($4f^{14}$) are particularly stable examples of the +2 oxidation state.

The six-coordinate radii of the $Ln^{3+}$ ions [1] show a steady decrease from $La^{3+}$ to $Lu^{3+}$ as shown in Fig. 1, and this decrease in radius (the *lanthanide contraction*) is reflected in the structures of complexes: there is a corresponding decrease in Ln-to-ligand bond distances, and in many cases there are changes in the structures and reactivities of complexes as the series is traversed. The radius of $Y^{3+}$ (104 pm) falls between those of $Ho^{3+}$ and $Er^{3+}$ and its chemistry closely resembles that of the late lanthanides; $Sc^{3+}$ has a six-coordinate radius of 88.5 pm, significantly smaller than that of the other rare earth elements, and its chemistry is in many ways unique. The $Ln^{3+}$ ions are large compared with most transition metal ions, e.g., ($Hf^{4+}$ = 85 pm, $Ti^{4+}$ = 74.5 pm), and this results in higher coordination numbers (usually greater than 6 and often as high as 9–11) in their complexes, compared with those of transition metal complexes where octahedral coordination is most commonly observed. The steady decrease in $Ln^{3+}$ radius often results in structural changes as the series is traversed, with complexes of the early $Ln^{3+}$ having higher coordination numbers than those of the later $Ln^{3+}$.

The 4f orbitals are essentially core orbitals and are not involved in bonding; indeed, there is rarely any significant covalent contribution to bonding in rare earth complexes, although some contribution of 5d orbitals in some cyclopentadienyl complexes has been proposed. This general lack of covalent contribution means that electrostatic interactions dominate in the bonding of rare earth complexes. As a result, the coordination geometry of rare earth complexes is often very irregular, depending only on the size of the ligands, and not on the relative orientation of the ligands with respect to metal-centred orbitals. This is a significant difference between the chemistry of the rare earths and that of transition metals, where covalent contributions to bonding are of great importance. The electrostatic nature of the bonding also results in many of the complexes being very labile (i.e., the Ln-to-ligand bond is readily exchanged).

The highly electrostatic nature of Ln-to-ligand bonds means that alkoxides, alkylamides and organometallics of the lanthanides are usually significantly more reactive towards air and moisture than the corresponding complexes of transition metals. Their chemistry is much more closely related to that of the Group 2 (alkaline earth) elements than to that of the transition metals.

## 2 Requirements for MOCVD and ALD Precursors

Requirements for MOCVD and ALD precursors are concisely summarized as:

- adequate volatility
- appropriate reactivity

These two aspects will be dealt with separately.

### 2.1 Making Rare Earth Complexes Volatile

Volatility is achieved by reduction of intermolecular forces and has been the main goal of rare earth precursor chemists, although increased use of liquid injection MOCVD [2] means that high volatility is not quite so essential as it once was. However, rare earth chemists still have a considerable problem to address: the large radius of the $Ln^{3+}$ ions, and consequent requirement for high coordination numbers to satisfy steric demands, means that many lanthanide complexes with simple ligands such as methoxide, $CH_3O^-$ are oligomeric or even polymeric, and thus completely involatile. Neutral $Ln^{3+}$ complexes can accommodate only three negatively charged ligands and so steric saturation cannnot be achieved as easily as in more highly charged transition metal ions such as $Hf^{4+}$, which also have the advantage of a smaller ionic radius. The formation of polymeric species arises due to ligands bridging between Ln centres in order to increase the coordination number and hence achieve steric saturation at the Ln. This tendency to form polymeric species can be prevented by:

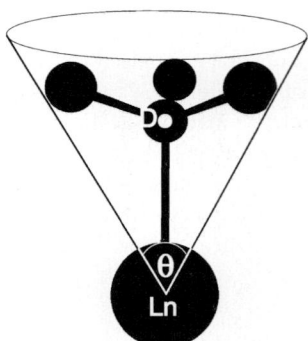

**Fig. 2.** The definition of ligand cone angle

- use of very large (sterically demanding) ligands such as the 'silylamide' ligand $(Me_3Si)_2N^-$
- use of multidentate ligands which bind to the Ln atom via two or more donor atoms

As well as ideally being monomeric, complexes should also be non-polar in order further to reduce intermolecular forces; the presence of large alkyl groups on the periphery of the complex helps to achieve this. Trialkylsilyl substituents are excellent for increasing volatility, but as will be seen later, the presence of Si in precursors often results in the incorporation of Si in deposited films. Low-symmetry complexes which pack poorly in the solid state will have reduced intermolecular forces and hence low melting points and enhanced volatility.

*Sterically Demanding Ligands*

The cone angle of a ligand, defined as the angle subtended at the metal by the ligand, is a useful concept in coordination chemistry, and is illustrated in Fig. 2.

For a monodentate ligand (i.e., one with a single donor atom), a larger cone angle can be achieved where the donor atom D is trivalent (e.g., N) compared with a divalent donor such as O, and complexes with bulky silylamide ligands were the first three-coordinate lanthanide complexes to be prepared and characterized [3]. These complexes have occupied a key position in lanthanide coordination chemistry since they were first reported in 1973. The large size of the silylamide ligand is effective in filling the coordination sphere of the $Ln^{3+}$ ion; the non-polar nature of the exterior of the complex results in weak intermolecular forces, and $[Ln\{N(SiMe_3)_2\}_3]$ are all very soluble in hydrocarbon solvents and have reasonable volatility, subliming without decomposition at ca. $120\,°C$ and $10^{-3}$ Torr. The structure of $[Ln\{N(SiMe_3)_2\}_3]$ is shown in Fig. 3. Less sterically demanding amide ligands such as $(Me_2HSi)_2N^-$ [4] or di-isopropylamide, $Pr_2^iN^-$, $(Pr^i = (CH_3)_2CH)$

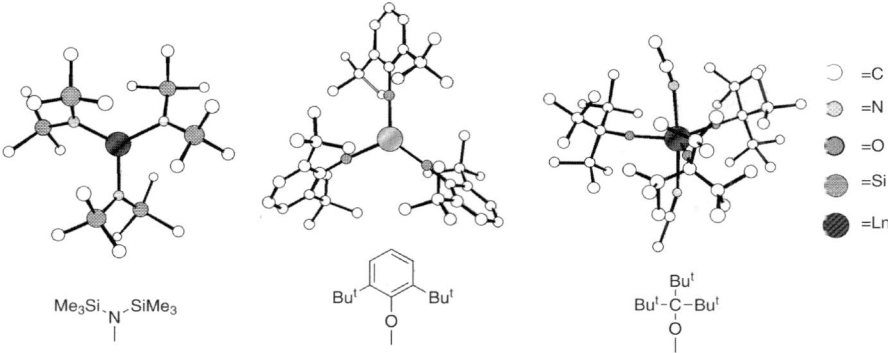

Fig. 3. Structures of monomeric lanthanide complexes with bulky monodentate ligands

also give monomeric complexes with Ln, but these complexes contain additional neutral ligands (usually THF; THF = tetrahydrofuran) and cannot be sublimed without decomposition, e.g., $[Ln(NPr_2^i)_3]$ decompose at $80\,°C$ and $10^{-4}$ Torr [5]. Complexes with the $Me_2N^-$ (dimethylamide) ligand can be stabilized only in the presence of $MMe_3$ (M=Ga or Al) to form $[Ln(NMe_2)_3(MMe_3)_3]$ [6].

The first example of a monodentate O-donor ligand to give three-coordinate monomeric complexes with the lanthanides was 2,6-di-tert-butylphenolate [7]. $[Ln(O-2,6-Bu_2^t-C_6H_3)_3]$ ($Bu^t = (CH_3)_3C$) are less volatile than the tris-silylamides: they sublime at $180–190\,°C$ and $10^{-5}$ Torr. The structure of $[Ce(O-2,6-Bu_2^t-C_6H_3)_3]$ is shown in Fig. 3. The aliphatic alkoxide $Bu_3^tCO^-$ ('tritox') and its Si analogue $Bu_3^tSiO^-$ ('silox') are even more sterically demanding than $2,6-Bu_2^t-C_6H_3O^-$ and give monomeric complexes with Ln [8,9]. The fluorinated analogue $(CF_3)_2MeCO^-$ gives monomeric solvated complexes [10].

The bulky alkyl ligand $^-CH(SiMe_3)_2$, which is a C analogue of the silylamide ligand, also gives monomeric three-coordinate complexes, with analogous structures to the silylamides [11], but sublimation of these complexes has not been reported.

The higher charge and smaller radius of $Ce^{4+}$ compared with $Ce^{3+}$ mean that it is easier to obtain monomeric complexes. However, it should be noted that the oxidizing power of $Ce^{4+}$ means that it is incompatible with some ligands, e.g., aryloxides and organic ligands. There are no known cerium(IV) alkylamides. Monomeric $[Ce(OBu^t)_4(THF)_2]$ has been crystallographically characterized, but condenses on standing in solution at room temperature to give the oxo-bridged cluster $[Ce_3(OBu^t)_{10}O]$ [12]. The methoxide complex $[Ce(OMe)_4]$ and $n$-octyloxide complex $[Ce\{O(CH_2)_7CH_3\}_4]$ have been reported; they have not yet been characterized crystallographically but they are unlikely to be monomeric [13].

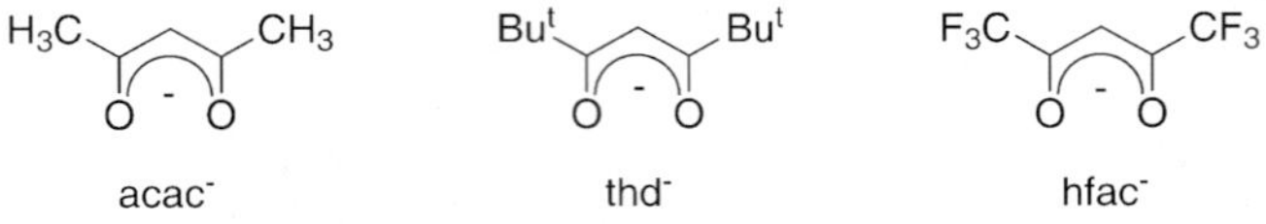

Fig. 4. Structures of $\beta$-diketonate ligands

## Complexes with Chelating Ligands

There is a wealth of chemistry of the lanthanides with chelating ligands, and a selection of those most relevant as MOCVD or ALD precursors is given below.

### $\beta$-Diketonates

Lanthanide $\beta$-diketonates have an important place in the history of lanthanide coordination chemistry: they are simple to prepare, stable to oxygen, and apart from reversible adduct formation, they are unreactive with $H_2O$. Examples of diketonate ligands are shown in Fig. 4.

Complexes with acetylacetonate (acac$^-$) have been known since the early years of the 20th century, but the simple [Ln(acac)$_3$] complexes for all but the smallest Ln are oligomeric and involatile. This is due to the inability of three small acac$^-$ ligands to satisfy the steric demands of Ln$^{3+}$. Monomeric complexes can be prepared by the addition of donor ligands such as bipyridyl or phenanthrolein, and these complexes are sufficiently volatile to be used as precursors for CVD processes [14, 15].

Complexes with the more sterically demanding thd ligand were reported in 1965 [16] and were investigated initially because of their volatility: they are sufficiently volatile to be subjected to gas chromatography and were considered as possible candidate compounds to allow separation of the rare earth elements. The complexes of the early lanthanides La to Gd form dimers in the solid state; those of the later lanthanides Dy to Lu are monomeric in the solid state, and [Tb(thd)$_3$] can exist as a monomer or a dimer, depending on the conditions employed for crystallization. All of the [Ln(thd)$_3$] complexes sublime as monomers, dimerizing only upon crystallization.

The fluorinated ligand hfac$^-$, which is a fluoro analogue of acac$^-$, has been investigated more recently. Fluorination is well-known to increase volatility, and complexes with hfac$^-$ are no exception. Sterically, hfac$^-$ is a little larger than acac$^-$, but smaller than thd$^-$. Monomeric complexes with hfac$^-$ can be formed by addition of further neutral ligands such as the tridentate polyether diglyme ($CH_3O(CH_2CH_2O)_2CH_3$) [17–19]. As well as increasing the volatility of the complex, the highly electronegative fluorine atoms increase the Lewis acidity of the Ln centre so that the neutral ligand is more strongly bound than in the analogous acac$^-$ or thd$^-$ complexes.

The effects of diketonate ligand structure on the volatility of [Sc(diket)$_3$] have been investigated. The vapour pressures at 155°C of [Sc(acac)$_3$] (3.22 Torr)

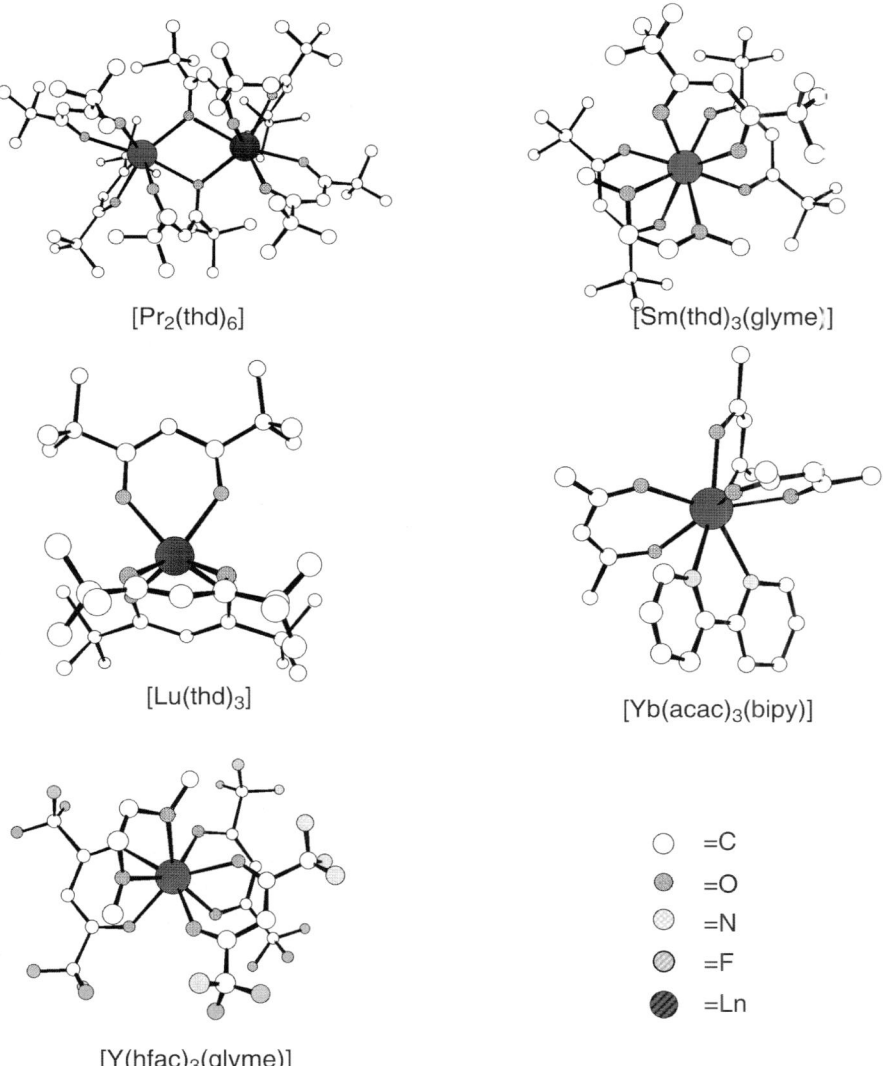

[Pr$_2$(thd)$_6$]

[Sm(thd)$_3$(glyme)]

[Lu(thd)$_3$]

[Yb(acac)$_3$(bipy)]

○ =C
◉ =O
◎ =N
◍ =F
⬤ =Ln

[Y(hfac)$_3$(glyme)]

**Fig. 5.** Structures of lanthanide tris-$\beta$-diketonate complexes

and [Sc(thd)$_3$] (3.31 Torr) are very similar, whereas [Sc(hfac)$_3$], despite its larger molecular mass, has a dramatically increased vapour pressure of 28.89 Torr [20]. There is no corresponding study of the vapour pressures of complexes with Ln. The structures of lanthanide $\beta$-diketonates are shown in Fig. 5.

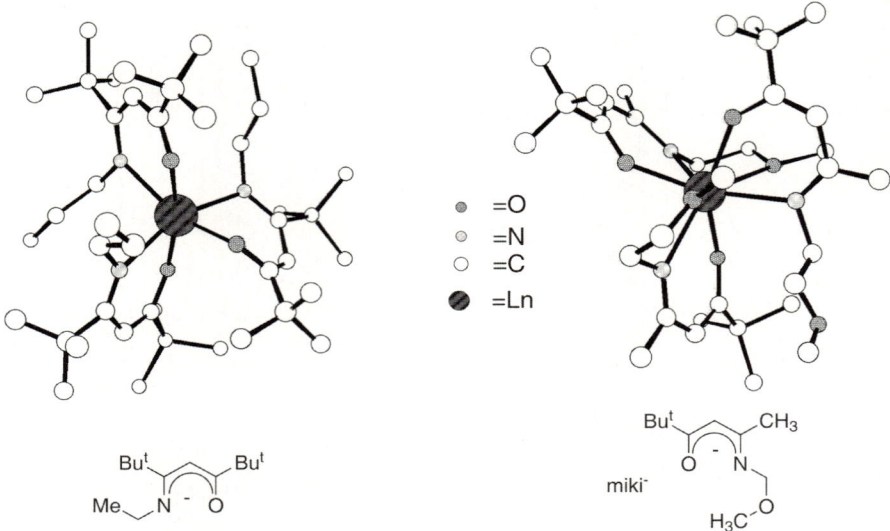

**Fig. 6.** Structures of Ln complexes with ketoiminate ligands

## Ketoiminates

Ketoiminate ligands are derived synthetically from the well-known $\beta$-diketonate ligands, but have one N-donor and one O-donor atom. Monomeric lanthanide tris (ketoiminates) have been prepared by *Rees* et al. [21]; the Yb complex is remarkably thermally robust and can be sublimed without decomposition at 600 °C and atmospheric pressure. This thermal robustness is a disadvantage for MOCVD processes but may be an advantage for ALD. *Marks* has prepared donor functionalized ketoiminates where the N atom carries an ether group to give a potentially tri-dentate ligand with two O-donors and one N-donor. The complexes of these ligands with Ln have been investigated by *Marks* et al. [22, 23]. The structures of the miki⁻ ligand and of its eight-coordinate complex with Nd are shown in Fig. 6. The Nd complex is eight coordinate with one of the ether groups 'dangling'. Smaller, late Ln give seven-coordinate complexes where two of the ether groups are 'dangling'. These complexes have low melting points and good volatility, subliming at 80–110 °C and $10^{-4}$ Torr.

## Donor-Functionalized Alkoxides

As described earlier in the section on "Sterically demanding ligands", the only monomeric three-coordinate lanthanide alkoxides are those with the very sterically demanding 2, 6-di-tert-butyl-phenolate or $Bu_3^tCO^-$ ligands. Otherwise, donor-functionalized alkoxides which have the potential to act as bidentate ligands must be used. A series of such donor-functionalized alcohols (shown

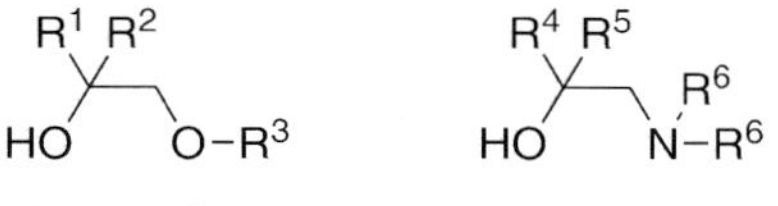

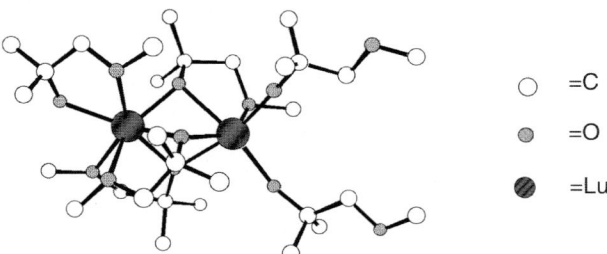

**Fig. 7.** Structures of donor-functionalized alcohols

**Fig. 8.** Structure of [Lu(mmp)₃]₂

in Fig. 7) and their lanthanide derivatives have been synthesized by *Anwander* and *Herrmann* [24, 25]. Of these alcohols, Hmmp (HOCMe₂CH₂OMe) has so far been the most important for MOCVD and ALD precursors.

Several of the lanthanide complexes with donor-functionalized alcoxides show good volatility, for example at $10^{-3}$ mbar. The Nd complexes with Et₂NCH₂C(H)(Buᵗ)O⁻, EtOCH₂CBu₂ᵗO⁻ and EtOCH₂CPr₂ⁱO⁻ sublime at 150 °C, 125 °C and 115 °C respectively. Unfortunately, most of these complexes have resisted crystallization and so their solid state structures have not been unambiguously characterized. One exception is [Lu(mmp)₃]₂, which has been characterized by X-ray diffraction: it is dimeric as shown in Fig. 8, with one six-coordinate and one seven-coordinate Lu. The OMe group of mmp is not a sufficiently strong donor always to chelate, and two of the mmp ligands are monodentate or 'dangling'.

Complexes of the early lanthanides with mmp are extremely susceptible to condensation reactions to form oxo-bridged species of reduced volatility; this can be prevented by the addition of the pentadentate polyether tetraglyme to [Ln(mmp)₃] solutions [26]. Tetraglyme does not form an inner-sphere complex with [Ln(mmp)₃] but is sufficiently closely associated to prevent the intermolecular interactions that are necessary for condensation reactions to occur. The alkyl substituents on the α-C atom of mmp are important in preventing the formation of involatile oligomers: if R¹ and R² are both H, then a cyclic decamer is formed with Ln [27].

A recent development in donor-functionalized alkoxide ligands is the use of the oxazolinyl substituent as a neutral donor, as in 2-(4, 4-dimethyl-4, 5-dihydrooxazol-2-yl)propan-2-ol (Hdmop) shown in Fig. 9. The imine nitrogen of dmop is expected to be a stronger donor than the OMe group of mmp, and, together with the added steric bulk of the Me substituents on the oxazoline

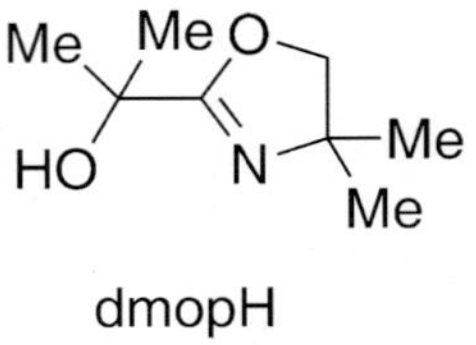

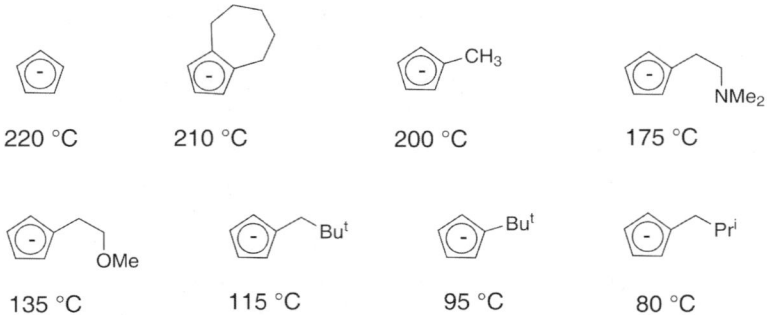

dmopH

**Fig. 9.** Stucture of 2-(4,4-dimethyl-4,5-dihydrooxazol-2-yl)propan-2-ol, Hdmop

220 °C          210 °C          200 °C          175 °C

135 °C          115 °C          95 °C          80 °C

**Fig. 10.** Cyclopentadiene ligands and the sublimation temperatures ($10^{-3}$ mbar) of their complexes with Nd

ring, this should favour the formation of monomeric complexes with Ln. Some Zr and Hf complexes with dmop have already been characterized by X-ray diffraction [28].

*Organolanthanide Complexes*

The applications of organolanthanides (compounds which involve a direct bond between C and Ln) in materials science have been reviewed [29]. The first organolanthanides to be prepared were the tris-cyclopentadienyls [LnCp$_3$] in 1956 [30]. These complexes adopt a range of structures depending on the Ln$^{3+}$ radius, and are monomeric only in the solid state for Ln around the middle of the series, such as Sm. Because the bonding is essentially ionic, these complexes, unlike transition metal cyclopentadienyls, are highly reactive with H$_2$O and O$_2$. They are, however, thermally robust and can be sublimed without decomposition at 200–250 °C and $10^{-4}$ Torr. Volatility of the simple [LnCp$_3$] complexes increases from La to Yb, i.e., with decreasing Ln$^{3+}$ radius. Substituents can be introduced into the Cp ring to increase volatility, as shown for complexes of Nd in Fig. 10 [31].

   Simple neutral alkyl complexes of the lanthanides are known only with extremely bulky alkyl groups such as (Me$_3$Si)$_2$CH$^-$, as discussed above.

*Volatile Lanthanide Complexes: Summary*

Since the late 1980s, major efforts have been directed towards the synthesis of volatile monomeric lanthanide complexes. The requirements for sterically demanding monodentate or chelating ligands are now well-understood. Fluorination of ligands results in enhanced volatility, although, as discussed below, the presence of fluorine in the ligand frequently leads to incorporation of F into the films. Similarly, use of $R_3Si$ groups to enhance volatility can result in Si incorporation. Although the functionalization of the periphery of complexes should enhance volatility, this is an area which is very much unexplored, except in the case of cyclopentadienyls. There have been few systematic studies of vapour pressures of lanthanide complexes.

## 2.2 Chemical Requirements of Precursors for MOCVD

It has already been seen that steric hindrance at the lanthanide centre is usually a requirement in order to achieve reasonable volatility, which is necessary for both MOCVD and ALD processes. Ideally, a precursor for MOCVD will also show clean thermal decomposition so that pure films uncontaminated by C can be deposited under reasonably mild conditions, and there should be a sufficient temperature window between evaporation and decomposition.

*β-Diketonate Precursors*

Lanthanide β-diketonates are easy to prepare, and because of their stability in air, they are easy to handle and store. Their volatility is well-documented, and many are commercially available. It is therefore not surprising that most of the MOCVD of rare earth oxides has been carried out with these compounds. However, from a chemical perspective, they are not the ideal precursors. Most films grown from $[Ln(thd)_3]$ require high growth temperatures, which are likely to be incompatible with device manufacture, and they are heavily contaminated with C. This occurs because there is no facile thermal decomposition pathway available for these complexes. Studies on $[Ln(thd)_3]$ show that these complexes are transported intact in the MOCVD reactor [32] and decompose thermally by a stepwise stripping of $Bu^t$ groups from the ligands by C-C bond homolysis, followed by dissociation of the remaining $C_3O_2$ fragment as shown in Fig. 11 [33]. In the absence of $O_2$, metallic Ln is deposited, rather than rare earth oxide.

In the case of $[Ln(hfac)_3]$ complexes, MOCVD results in formation of the very stable $LnF_3$, rather than the oxide [34]. $[La(hfac)_3(diglyme)]$ can, however, be used together with $[Al(acac)_3]$ for MOCVD of lanthanum aluminate, which contains no fluorine contamination [35].

Studies of MOCVD using Ba β-diketonates have shown that when the diketonate ligand has a β-hydrogen, the complex is much less thermally stable than the thd analogue, which has no β-hydrogen. The proposed mechanism for this decomposition is shown in Fig. 12 [36]. No such studies with Ln precursors have been reported.

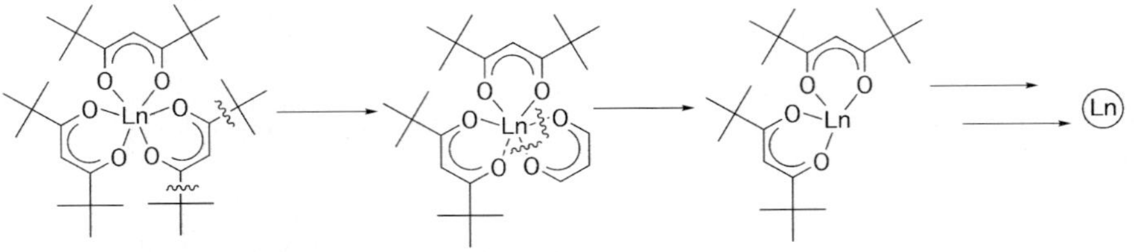

**Fig. 11.** Thermal decomposition of [Ln(thd)$_3$]

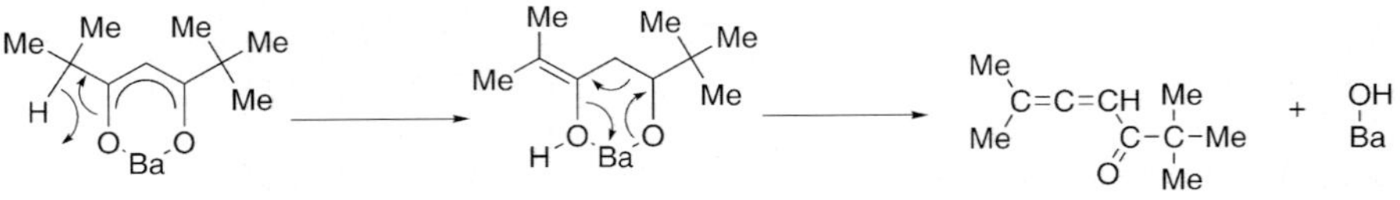

**Fig. 12.** Thermal decomposition of a Ba $\beta$-diketonate containing a $\beta$-hydrogen

*Alkoxide Precursors*

Despite the importance of metal alkoxides as precursors for MOCVD of oxide thin films, there are relatively few mechanistic studies of thermal decomposition of these compounds, and only one report of studies of a lanthanide alkoxide [8]. *Bradley* and *Faktor* showed in 1959 that Zr tert-amyloxide decomposes at temperatures between 208 and 247 °C according to the reaction schemes shown in Fig. 13. No detailed mechanism was proposed for the initial formation of alcohol. Dehydration of an alcohol to give one molecule of alkene and one molecule of H$_2$O is a well-known reaction in organic chemistry, and in Fig. 13, the dehydration of the alcohol is a rate-determining step. The resulting H$_2$O can then react with further molecules of Zr alkoxide to form ultimately ZrO$_2$ and alcohol. The decomposition is thus an autocatalytic process, and the rate of thermal decomposition of a metal alkoxide by this route should therefore be dependent upon the ease of dehydration of the parent alcohol.

Dehydration of silanols R$_3$SiOH yields one molecule of H$_2$O for every two molecules of silanol, as shown in the equation below and so the decomposition of siloxide complexes is not autocatalytic, and these complexes are more thermally stable than their alkoxide analogues.

$$2R_3SiOH \rightarrow 2R_3Si{-}O{-}SiR_3 + H_2O$$

Alkoxides of late transition metals are well-known to undergo $\beta$-hydride elimination reactions to yield a transition metal hydride; this is an unlikely pathway for lanthanide alkoxides because Ln$^{3+}$ ions are oxophilic hard Lewis acids, and formation of a bond to the soft H$^-$ ion is not a favourable process.

MOCVD using [Ln(mmp)$_3$] precursors forms pure Ln oxide films even in the absence of O$_2$, and it is proposed that the thermal decomposition proceeds via a $\beta$-hydrogen elimination process shown in Fig. 14. In this scheme,

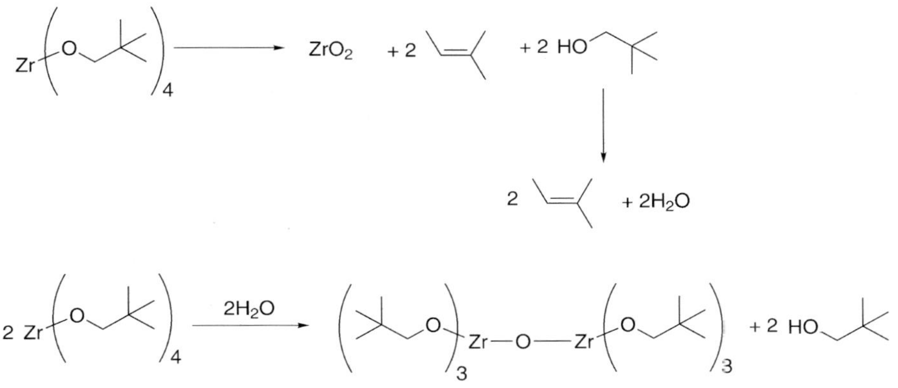

**Fig. 13.** Thermal decomposition of $[\text{Zr(OCH}_2\text{CMe}_3)_4]$

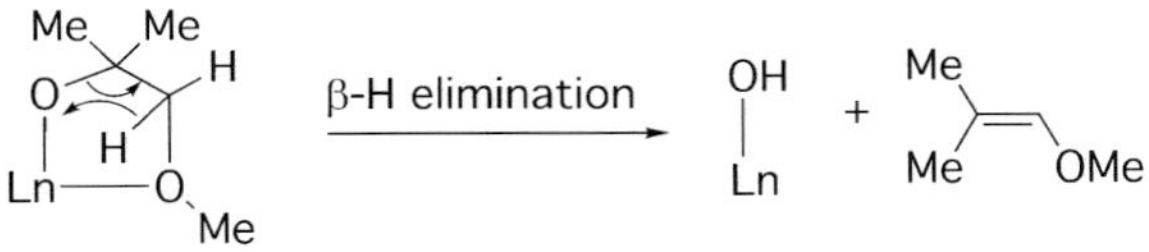

**Fig. 14.** $\beta$-hydrogen elimination reaction of a Ln mmp complex

a $\beta$-hydrogen migrates to the alkoxide O atom, the alkoxide O-to-C bond is cleaved and a molecule of alkene is eliminated in what is effectively an intramolecular dehydration. This is consistent with the mechanism proposed for the thermal decomposition of $[\text{Hf(mmp)}_4]$ in the MOCVD of $\text{HfO}_2$ [37].

*Silylamide Precursors*

Transition metal dialkylamides have been used for the liquid injection MO-CVD of oxides [38]. As there are no simple volatile dialkylamides of the lanthanides, the volatile lanthanide tris-silylamides have been investigated. Liquid injection MOCVD in the presence of $\text{O}_2$ gives Ln silicate films where the stoichiometry of the films is not well-defined and the Si content cannot be controlled [39, 40]. This demonstrates that, although $\text{SiMe}_3$ substituents are good for enhancing volatility, they generally result in incorporation of varying amounts of Si in deposited films and therefore should be avoided when pure oxide films are required.

## 2.3 Chemical Requirements of Precursors for ALD

The ALD process is represented simply in Fig. 15. The fundamental reactions are 1. the reaction of the precursor $\text{ML}_n$ with a surface-bound OH group followed by 2. the reaction of surface-bound substrate molecules with

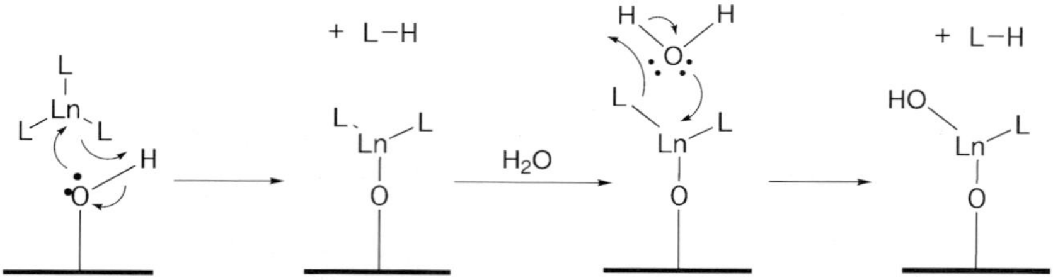

**Fig. 15.** General reaction scheme for the ALD process

an incoming acidic oxygen source to liberate HL and generate reactive OH sites. The initial step in the reaction with the oxygen source almost certainly involves the interaction of a lone pair of electrons on O with an electron-deficient (Lewis acidic) metal centre.

It is clear from the reacion scheme that in addition to volatility, an ALD precursor should also meet the following requirements:

– the complex should have reasonable thermal stability so that it does not thermally decompose on the substrate before reacting with surface OH groups or incoming $H_2O$
– the ligand $L^-$ should be a sufficiently strong Brønsted base to react readily with $H_2O$ or surface OH groups
– the complex should be sufficiently sterically unhindered to allow facile reaction with surface OH groups or incoming $H_2O$.

ALD of transition metal oxides such as those of Ti or Hf has used volatile chlorides $MCl_n$ as precursors; this option is not available for the rare earths as all their halides are involatile solids. The requirement in ALD for relatively high thermal stability, facile reaction with $H_2O$ and the minimum of steric hindrance at the metal centre means that, apart from volatility, the other requirements for precursors for MOCVD and ALD are mutually exclusive. These factors must be considered when choosing precursors for the two processes. ALD of rare earth oxides is a very promising technique, but it is still a very limited field; published examples of ALD studies using various classes of rare earth complexes are briefly summarized below.

*β-Diketonate Complexes*

The β-diketonates, which have so many advantages with regard to ease of preparation and handling, are chemically far from ideal as ALD precursors. The β-diketonate anion is a very weak Brønsted base (pK$_a$ for Hacac = 8.2 [41]) and is not protonated to any significant extent by $H_2O$. As a result, ALD processes using [Ln(diket)$_3$] precursors require the powerful oxidizing agent $O_3$ in order to form rare earth oxide, and contamination with C is still a problem [42, 43].

*Silylamide Complexes*

Transition metal dialkylamides have proved effective as precursors for ALD of oxides [44], but in rare earth chemistry the simple dialkylamides do not have sufficient volatility or thermal stability. Dialkylamide anions are strong Brønsted bases (pK$_a$ for Pr$^i_2$NH = 35.7; pK$_a$ for (Me$_3$Si)$_2$NH = 25.8 [45]) and so their complexes with the lanthanides react very readily with H$_2$O. The volatile lanthanide tris-silylamides have been investigated for ALD of oxides with H$_2$O as the oxygen source. Growth was not self-limiting and the layers had low levels of C contamination (2.5 to 2.8 at.%) and Si contents of between 4 and 12 %, depending on the growth conditions and annealing. It has not been established whether the source of the Si is diffusion from the substrate, or hydrolysis of the N–SiMe$_3$ linkage in the precursor [46]. However, liquid injection ALD of PrO$_x$ using a [Pr(mmp)$_3$] gives Si-free layers (see below), suggesting that Si diffusion from the substrate does not occur [47].

*Organometallic Complexes*

Organometallics of the rare earths are extremely reactive with H$_2$O, as the negatively charged organic ligands are very powerful Brønsted bases (pK$_a$ for CpH = 18.0 [48]), and these complexes are therefore very promising precursors for ALD. [Cp$_3$Y] and the substituted derivative [(C$_5$H$_4$CH$_3$)$_3$Y] have both been investigated for ALD of Y$_2$O$_3$ using H$_2$O as the oxygen source [49]. The Brønsted basicity of the cyclopentadienyl ligands and consequent facile reaction with H$_2$O resulted in growth rates that were at least five times greater than when [Y(thd)$_3$]/O$_3$ were used. [Cp$_3$Y] is less sterically hindered than [(C$_5$H$_4$CH$_3$)$_3$Y], allowing more facile reaction with H$_2$O or surface OH groups, and growth rates were correspondingly higher for [Cp$_3$Y]. Both precursors gave self-limiting growth, and the resulting films had very low C contamination. Similar results were obtained using [Cp$_3$Sc]/H$_2$O for growth of Sc$_2$O$_3$ [50].

*Alkoxide Complexes*

Alkoxide anions are strong Brønsted bases (e.g., pK$_a$ for Bu$^t$OH = 19.00 [51]) and so lanthanide alkoxides have the correct reactivity for use as ALD precursors. They have not yet been investigated by conventional ALD, but liquid injection ALD has been investigated with mmp complexes of Pr and Gd [47]. The Gd oxide films contained no carbon, and in the Pr oxide films carbon was either not detected or at levels of < 3.3 at.%. Self-limiting growth was, however, not achieved, and this was ascribed to facile β-H elimination from the precursor at the relatively low temperature of 225 °C, generating OH groups without reaction with H$_2$O. ALD growth rates were significantly lower than those achieved by MOCVD.

Liquid injection ALD of [Hf(mmp)$_4$] was investigated at the same time as that of the rare earth mmp complexes. Growth rates were found to be rather low and this was partially explained by steric hindrance at Hf due to the proximity of the two Me groups of each mmp ligand, increasing steric hindrance at Hf and reducing access of $H_2O$ or OH.

# 3 Conclusions

The factors influencing the volatility of rare earth complexes are now well-understood, and monomeric complexes can be prepared reliably. The effects of ligand symmetry and substitution patterns have been recognized, but except in the case of cyclopentadienyl complexes, these have not been systematically investigated. The use of fluorination or trimethylsilyl substituents, although enhancing volatility, usually leads to incorporation of F or Si into films and should be avoided. The $\beta$-hydrogen elimination reaction is important in the clean thermal decomposition of rare earth alkoxides. Although ALD of rare earth oxides is still in its infancy, the requirements of precursors have been identified: appropriate thermal stability (usually higher than that required for MOCVD), volatility, and facile reactivity with $H_2O$ or other oxygen sources. Apart from volatility, these requirements are almost mutually exclusive with those for MOCVD. $\beta$-diketonates have occupied a predominant role in the MOCVD and ALD of rare earth oxides, but apart from ease of synthesis and handling, these compounds are far from ideal and do not have the required reactivity for optimum performance. Future contributions from synthetic chemists must address the requirements of reactivity in addition to simply producing volatile complexes.

# References

[1] R. D. Shannon: Revised effective ionic-radii and systematic studies of interatomic distances in halides and chalcogenides, Acta Crystallogr. Sect. A **32**, 751 (1976)

[2] A. C. Jones: Molecular design of improved precursors for the MOCVD of electroceramic oxides, J. Mater. Chem. **12**, 2576 (2002)

[3] D. C. Bradley, J. S. Ghotra, F. A. Hart: Low coordination numbers in lanthanide and actinide compounds. 1. Preparation and characterization of tris[bis(trimethylsilyl)-amido]lanthanides, J. Chem. Soc. – Dalton Trans. p. 1021 (1973)

[4] R. Anwander, O. Runte, J. Eppinger, G. Gerstberger, E. Herdtweck, M. Spiegler: Synthesis and structural characterisation of rare-earth bis(dimethylsilyl)amides and their surface organometallic chemistry on mesoporous MCM-41, J. Chem. Soc. – Dalton Trans. p. 847 (1988)

[5] D. C. Bradley, J. S. Ghotra, F. A. Hart: Tris(di-2-propylamido)yttrium and related neodymium and ytterbium compounds, Inorg. Nucl. Chem. Lett. **12**, 735 (1976)

[6]  W. J. Evans, R. Anwander, R. J. Doedens, J. W. Ziller: The use of heterometal-
     lic bridging moieties to generate tractable lanthanide complexes of small lig-
     ands, Angew. Chem. – Int. Ed. Engl. **33**, 1641 (1994)

[7]  H. A. Stecher, A. Sen, A. L. Rheingold: Synthesis, structure, and reactiv-
     ity of tricoordinate cerium(III) aryloxides – the 1$^{st}$ structurally characterized
     monomeric Ln(OR)$_3$ complexes, Inorg. Chem. **27**, 1130 (1988)

[8]  H. A. Stecher, A. Sen, A. L. Rheingold: Synthesis, structure, and reactivity
     of cerium(iii) alkoxides. 2. Thermal decomposition of Ce(OC$^t$Bu$_3$)$_3$ and the
     structure of [Ce(OCH$^t$Bu$_2$)$_3$]$_2$, Inorg. Chem. **28**, 3280 (1989)

[9]  W. A. Herrmann, R. Anwander, M. Kleine, W. Scherer: Complexes of the
     lanthanides. 1. Solvent-free alkoxide complexes of neodymium and dyspro-
     sium – crystal and molecular structure of trans-bis(acetonitrile)tris(tri-tert-
     butylmethoxy)neodymium, Chem. Ber.-Recl. **125**, 1971 (1992)

[10] D. C. Bradley, H. Chudzynska, M. E. Hammond, M. B. Hursthouse, M. Mote-
     valli, W. Ruowen: The preparation and characterization of volatile deriva-
     tives of trivalent metals using fluorinated alkoxide ligands – X-ray structures
     of Sc(OCH(CF$_3$)$_2$)$_3$(NH$_3$)$_2$, 7pr(ocme(cf$_3$)$_2$)$_3$(nh$_3$)$_2$0, (OCMe(CF$_3$)$_2$)$_3$(THF)$_3$
     and Pr(OCMe$_2$(CF$_3$))$_3$, Polyhedron **11**, 375 (1992)

[11] P. B. Hitchcock, M. F. Lappert, R. G. Smith, R. A. Bartlett, P. P. Power: Syn-
     thesis and structural characterization of the 1$^{st}$ neutral homoleptic lanthanide
     metal(III) alkyls - [LnR$_3$] [Ln = La or Sm, R = Ch(SiMe$_3$)$_2$], J. Chem. Soc. –
     Chem. Commun. p. 1007 (1988)

[12] W. J. Evans, T. J. Deming, J. M. Olofson, J. W. Ziller: Synthetic and struc-
     tural studies of a series of soluble cerium(IV) alkoxide and alkoxide nitrate
     complexes, Inorg. Chem. **28**, 4027 (1989)

[13] P. S. Gradeff, F. G. Schreiber, K. C. Brooks, R. E. Sievers: Simplified method
     for the synthesis of ceric alkoxides from ceric ammonium-nitrate, Inorg. Chem.
     **24**, 1110 (1985)

[14] M. P. Singh, S. A. Shivashankar: Structural and optical properties of polycrys-
     talline thin films of rare earth oxides grown on fused quartz by low pressure
     MOCVD, J. Cryst. Growth **276**, 148 (2005)

[15] M. P. Singh, C. S. Thakur, K. Shalini, S. Banerjee, N. Bhat, S. A. Shivashankar:
     Structural, optical, and electrical characterization of gadolinium oxide films
     deposited by low-pressure metalorganic chemical vapor deposition, J. Appl.
     Phys. **96**, 5631 (2004)

[16] K. J. Eisentraut, R. E. Sievers: Volatile rare earth chelates, J. Am. Chem. Soc.
     **87**, 5254 (1965)

[17] W. J. Evans, D. G. Giarikos, M. A. Johnston, M. A. Greci, J. W. Ziller:
     Reactivity of the europium hexafluoroacetylacetonate (hfac) complex,
     Eu(hfac)$_3$(diglyme), and related analogs with potassium: Formation of the
     fluoride hfac 'ate' complexes, LnF(hfac)$_3$K(diglyme)$_2$, J. Chem. Soc. – Dalton
     Trans. p. 520 (2002)

[18] G. Malandrino, R. Licata, F. Castelli, I. L. Fragalà, C. Benelli: New thermally
     stable and highly volatile precursors for lanthanum MOCVD – Synthesis and
     characterization of lanthanum beta-diketonate glyme complexes, Inorg. Chem.
     **34**, 6233 (1995)

[19] K. D. Pollard, H. A. Jenkins, R. J. Puddephatt: Chemical vapor deposition
     of cerium oxide using the precursors [Ce(hfac)$_3$(glyme)], Chem. Mat. **12**, 701
     (2000)

[20] B. D. Fahlman, A. R. Barron: Substituent effects on the volatility of metal beta-diketonates, Adv. Mater. Opt. Electron. **10**, 223 (2000)

[21] W. S. Rees, O. Just, S. L. Castro, J. S. Matthews: Synthesis and magnetic and structural characterization of the first homoleptic lanthanide beta-ketoiminate, Inorg. Chem. **39**, 3736 (2000)

[22] J. A. Belot, A. C. Wang, R. J. McNeely, L. Liable-Sands, A. L. Rheingold, T. J. Marks: Highly volatile, low-melting, fluorine-free precursors for MOCVD of lanthanide oxide-containing thin films, Chem. Vap. Dep. **5**, 65 (1999)

[23] N. L. Edleman, A. C. Wang, J. A. Belot, A. W. Metz, J. R. Babcock, A. M. Kawaoka, J. Ni, M. V. Metz, C. J. Flaschenriem, C. L. Stern, L. M. Liable-Sands, A. L. Rheingold, P. R. Markworth, R. P. H. Chang, M. P. Chudzik, C. R. Kannewurf, T. J. Marks: Synthesis and characterization of volatile, fluorine-free beta-ketoiminate lanthanide MOCVD precursors and their implementation in low-temperature growth of epitaxial $CeO_2$ buffer layers for superconducting electronics, Inorg. Chem. **41**, 5005 (2002)

[24] W. A. Herrmann, R. Anwander, M. Denk: Complexes of the Lanthanides. 3. Volatile neodymium and yttrium alkoxides with new bulky chelating ligands, Chem. Ber. – Recl. **125**, 2399 (1992)

[25] R. Anwander, F. C. Munck, T. Priermeier, W. Scherer, O. Runte, W. A. Herrmann: Volatile donor-functionalized alkoxy derivatives of lutetium and their structural characterization, Inorg. Chem. **36**, 3545 (1997)

[26] H. C. Aspinall, J. Gaskell, P. A. Williams, A. C. Jones, P. R. Chalker, P. A. Marshall, J. F. Bickley, L. M. Smith, G. W. Critchlow: Growth of praseodymium oxide thin films by liquid injection MOCVD using a novel praseodymium alkoxide precursor, Chem. Vap. Dep. **9**, 235 (2003)

[27] O. Poncelet, L. G. Hubertpfalzgraf, J. C. Daran, R. Astier: Alkoxides with polydentate alcohols – synthesis and structure of $[Y(OC_2H_4OMe)_3]_{10}$, a hydrocarbon soluble cyclic decamer, J. Chem. Soc. – Chem. Commun. p. 1846 (1989)

[28] Y. F. Loo, R. O' Kane, A. C. Jones, H. C. Aspinall, R. J. Potter, P. R. Chalker, J. F. Bickley, S. Taylor, L. M. Smith: Deposition of $HfO_2$ and $ZrO_2$ films by liquid injection MOCVD using new monomeric alkoxide precursors, J. Mater. Chem. **15**, 1896 (2005)

[29] Y. K. Gunko, F. T. Edelmann: Organolanthanides in materials science, Comments Inorg. Chem. **19**, 153 (1997)

[30] J. M. Birmingham, G. Wilkinson: The cyclopentadienides of scandium, yttrium and some rare earth elements, J. Am. Chem. Soc. **78**, 42 (1956)

[31] W. A. Herrmann, R. Anwander, F. C. Munck, W. Scherer: Complexes of the lanthanides. 4. Alkyl-substituted and donor- substituted cyclopentadienyl complexes of neodymium, Chem. Ber.-Recl. **126**, 331 (1993)

[32] T. Nakamura, T. Nishimura, R. Tai, K. Tachibana: Reaction mechanism of a lanthanum precursor in liquid source metalorganic chemical vapor deposition, Mater. Sci. Eng. B – Solid State Mater. Adv. Technol. **118**, 253 (2005)

[33] H. A. Luten, W. S. Rees, V. L. Goedken: Preparation and structural characterization of, and chemical vapor deposition studies with, certain yttrium tris(beta-diketonate) compounds, Chem. Vap. Dep. **2**, 149 (1996)

[34] G. Malandrino, O. Incontro, F. Castelli, I. L. Fragalà, C. Benelli: Synthesis, characterization, and mass-transport properties of two novel gadolinium(III) hexafluoroacetylacetonate polyether adducts: Promising precursors for MOCVD of $GdF_3$ films, Chem. Mat. **8**, 1292 (1996)

[35] G. Malandrino, G. G. Condorelli, R. Lo Nigro: MOCVD of $LaAlO_3$ films from a molten precursor mixture: Characterization of liquid and gas and deposited phases, Chem. Vap. Dep. **10**, 171 (2004)

[36] R. G. Gordon, S. Barry, R. N. R. Broomhall-Dillard, D. J. Teff: Synthesis and solution decomposition kinetics of flash- vaporizable liquid barium beta-diketonates, Adv. Mater. Opt. Electron. **10**, 201 (2000)

[37] S. Horii, K. Yamamoto, M. Asai, H. Miya, M. Niwa: Metalorganic chemical vapor deposition of $HfO_2$ films through the alternating supply of tetrakis(1-methoxy-2-methyl-2-propoxy)-hafnium and remote-plasma oxygen, Jpn. J. Appl. Phys. Part 1 - Regul. Pap. Short Notes Rev. Pap. **42**, 5176 (2003)

[38] P. A. Williams, A. C. Jones, N. L. Tobin, P. R. Chalker, S. Taylor, P. A. Marshall, J. E. Bickley, L. M. Smith, H. O. Davies, G. W. Critchlow: Growth of hafnium dioxide thin films by liquid-injection MOCVD using alkylamide and hydroxylamide precursors, Chem. Vap. Dep. **9**, 309 (2003)

[39] H. C. Aspinall, P. A. Williams, J. Gaskell, A. C. Jones, J. L. Roberts, L. M. Smith, P. R. Chalker, G. W. Critchlow: Growth of lanthanum silicate thin films by liquid injection MOCVD using tris[bis(trimethylsilyl)amido]lanthanum, Chem. Vap. Dep. **9**, 7 (2003)

[40] H. C. Aspinall, J. Gaskell, P. A. Williams, A. C. Jones, P. R. Chalker, P. A. Marshall, L. M. Smith, G. W. Critchlow: Growth of praseodymium oxide and praseodymium silicate thin films by liquid injection MOCVD, Chem. Vap. Dep. **10**, 83 (2004)

[41] M. J. Hynes, M. T. Mooney, A. Moloney: The reactions of zinc(II) with 1,3-diketones in aqueous solution – Catalysis by cacodylic acid during complex formation, J. Chem. Soc.-Dalton Trans. p. 313 (1993)

[42] J. Paivasaari, M. Putkonen, L. Niinistö: A comparative study on lanthanide oxide thin films grown by atomic layer deposition, Thin Solid Films **472**, 275 (2005)

[43] J. Paivasaari, M. Putkonen, T. Sajavaara, L. Niinistö: Atomic layer deposition of rare earth oxides: Erbium oxide thin films from beta-diketonate and ozone precursors, J. Alloy. Compd. **374**, 124 (2004)

[44] D. M. Hausmann, E. Kim, J. Becker, R. G. Gordon: Atomic layer deposition of hafnium and zirconium oxides using metal amide precursors, Chem. Mat. **14**, 4350 (2002)

[45] R. R. Fraser, T. S. Mansour, S. Savard: Acidity measurements on pyridines in tetrahydrofuran using lithiated silylamines, J. Org. Chem. **50**, 3232 (1985)

[46] K. Kukli, M. Ritala, T. Pilvi, T. Sajavaara, M. Leskelä, A. C. Jones, H. C. Aspinall, D. C. Gilmer, P. J. Tobin: Evaluation of a praseodymium precursor for atomic layer deposition of oxide dielectric films, Chem. Mat. **16**, 5162 (2004)

[47] R. J. Potter, P. R. Chalker, T. D. Manning, H. C. Aspinall, Y. F. Loo, A. C. Jones, L. M. Smith, G. W. Critchlow, M. Schumacher: Deposition of $HfO_2$, $Gd_2O_3$ and $PrO_x$ by liquid injection ALD techniques, Chem. Vap. Dep. **11**, 159 (2005)

[48] F. G. Bordwell, J. P. Cheng, G. Z. Ji, A. V. Satish, X. M. Zhang: Bond-dissociation energies in DMSO related to the gas-phase, J. Am. Chem. Soc. **113**, 9790 (1991)

[49] J. Niinistö, M. Putkonen, L. Niinistö: Processing of $Y_2O_3$ thin films by atomic layer deposition from cyclopentadienyl-type compounds and water as precursors, Chem. Mat. **16**, 2953 (2004)

[50] M. Putkonen, M. Nieminen, J. Niinistö, L. Niinistö: Surface-controlled deposition of $Sc_2O_3$ thin films by atomic layer epitaxy using beta-diketonate and organometallic precursors, Chem. Mat. **13**, 4701 (2001)

[51] A. Gervasini, A. Auroux: Thermodynamics of adsorbed molecules for a new acid-base topochemistry of alumina, J. Phys. Chem. **97**, 2628 (1993)

# Index

# Models for ALD and MOCVD Growth
# of Rare Earth Oxides

Simon D. Elliott

Tyndall National Institute, University College, Lee Maltings, Prospect Row, Cork,
Ireland
simon.elliott@tyndall.ie

**Abstract.** Atomic layer deposition (ALD) and metal organic chemical vapour deposition (MOCVD) are suitable techniques for the controlled deposition of high-
quality oxide films. Increasingly, modelling is being used to complement deposition
experiments, and a brief overview of modelling approaches is presented here. The
main focus is on atomic-scale models using ab initio electronic structure theory
to investigate the reaction steps involved in growth, in particular precursor adsorption and elimination of by-products. The common water-based ALD process
is considered, using simulations of the ALD of alumina from trimethylaluminium
and water as a specific example. In addition, analytical models of film growth are
reviewed. Finally, models for gas transport within the reactor are presented, with
the possibility of incorporating feature-scale and atomic-scale descriptions as well.

## 1 Introduction

Process modelling is an increasingly important element of the design and optimisation of materials for technology. Deposition is the crucial first step in materials processing and this paper considers models that have been developed
for two thin film deposition processes: atomic layer deposition (ALD) (see
the Chapter by *Niinist* in this volume) and metal-organic chemical vapour
deposition (MOCVD) (see the Chapter by *Lo Nigro* et al. in this volume).
Concerning oxides, research to date has concentrated on oxides of the main
group and transition metals, and we will use these findings to point to future
directions for the modelling of rare earth oxides.

Consider, for example, the ALD reaction to deposit a metal oxide $MO_x$
using water as the oxygen source. This can in the first instance be described
by a balanced chemical equation:

$$ML_{2x} + xH_2O \rightarrow MO_x + 2xHL .\tag{1}$$

Ligands (L) are chosen for the M-precursor so as to ensure volatility and reactivity, as appropriate to the process (the criteria differ for ALD and MOCVD
(see the Chapter by *Aspinall* in this volume). As an example, one of the most
successful ALD precursor combinations is that of trimethylaluminium (TMA)

M. Fanciulli, G. Scarel (Eds.): Rare Earth Oxide Thin Films,
Topics Appl. Physics **106**, 73–86 (2007)
© Springer-Verlag Berlin Heidelberg 2007

and water, which yields an alumina film and the by-product methane. (Incidentally, most rare earth oxides are sesquioxides, like alumina.)

$$2\mathrm{Al(CH_3)}_{3(g)} + 3\mathrm{H_2O}_{(g)} \rightarrow \mathrm{Al_2O_3}_{(s)} + 6\mathrm{CH_4}_{(g)} \,. \tag{2}$$

However, the overall reaction does not reveal the discrete steps of the mechanism by which deposition occurs. Individual reaction steps are considered in Sect. 2, along with atomic-scale models of the surface reactions. Some analytic models of ALD and MOCVD growth are based on assumptions about the operation of the process (Sect. 3). A complete model must also consider the transport of precursor and product gases to and from the surface (Sect. 4). A substantial multi-scale modelling challenge is then posed by the integration of accurate surface chemistry into the transport models.

# 2 Modelling Deposition Reactions

## 2.1 Suitability of Electronic Structure Theory

There is a hierarchy of methods that provide approximate solutions of the Schrödinger equation and thus describe the electrons of the reacting system: semi-empirical methods (neglect of differential overlap, NDO), Hartree–Fock (HF), density functional theory (DFT) – especially gradient-corrected approximation (GGA) for transition metal oxides and organometallics – and a range of correlated methods, such as perturbation theory (MP2) or coupled cluster (CCSD) [1]. HF and post-HF methods are referred to as ab initio. In choosing a method, a balance must be struck between accuracy and computational cost, for a given size of system.

DFT has become popular in materials science, because of its outstanding performance: high accuracy is obtained at reasonable computational cost (routine calculations on $\sim$100 atoms). DFT can quantify the relative energies of structural minima (typically to $\pm 10\,\mathrm{kJ/mol} = 0.01\,\mathrm{eV}$) and is suited to $T = 0\,\mathrm{K}$ thermodynamics. It has therefore successfully been applied to describing the saturated surface in ideal ALD in thermodynamic equilibrium with gas-phase precursors [2]. Examples are given in Sect. 2.3. In addition, DFT can be applied to ideal MOCVD processes, since oxide clusters are the thermodynamic as well as kinetic product and reactions occur primarily at the surface [3, 4].

To simulate realistic reactor temperatures at this level of accuracy is more complicated. Rapid events ($< 10\,\mathrm{ps}$) can be examined with ab initio molecular dynamics (MD) [1]. Alternatively, entropic corrections $T\Delta S$ to ab initio energetics can be estimated from rotational and vibrational partition functions, but analysing the rotational and vibrational states is straightforward only for the gas phase (cluster model, Sect. 2.2) [5, 6].

Non-ideal deposition results from other effects, including insufficient precursor dose, desorption of active species, and competing non-growth sidereactions. To include these in an atomic-scale growth model requires a comprehensive knowledge of the reaction kinetics. Kinetic parameters can also be used in transport models (more detail in Sect. 4) or in other higher-level simulations such as kinetic Monte Carlo [7]. Full rotational and vibrational data are needed to evaluate Arrhenius pre-exponential factors for kinetics; instead, these are often estimated from handbook data for similar molecules [8]. For computing transition states and activation energies, HF and DFT are unreliable, especially in cases where the electronic structure changes strongly (e.g., redox reactions). However, there is evidence that reasonably accurate activation energies are obtained from DFT for the acid-base reactions of the $H_2O$ process (Sect. 2.3) [8]. In any case, DFT can be used to explore the potential energy hypersurface in a qualitative way and identify plausible reaction pathways.

## 2.2 Atomic-Scale Models

Because of the high accuracy of electronic structure methods, the greatest source of error in atomic-scale modelling is arguably the choice of model, i.e., what atoms are chosen in what arrangement. A model is necessarily a simplification of reality, and a balance must be found between system size and computational cost, at a given level of theory (Sect. 2.1). Examples of models for deposition chemistry are given here.

In the periodic calculations that are typical of condensed-phase simulations, the simulation cell is repeated in all directions. This is a good model of a crystalline surface [1]. In ALD, the conformality and sub-monolayer growth rate means that relatively small cells with flat surfaces are a good model – there is no "growth front" as in other crystal growth models. Spurious interactions between adsorbate images in adjacent cells can be a problem.

In cluster or molecular models, the atoms form an isolated cluster, surrounded by vacuum. Clearly, this is suitable for reproducing a precursor molecule or for MOCVD cluster reactions, but accounts poorly for steric effects or for surface diffusion on a solid film.

In some cases, periodic and cluster results agree. For instance, the computed energetics for the adsorption of TMA onto a hydroxylated substrate are $\Delta E = -0.6\,\mathrm{eV}$ from a cluster model [9] or $\Delta E = -0.7\,\mathrm{eV}$ from a periodic model [5] (Fig. 2, 1A→1B). This is because it is a Lewis acid-base reaction, where electron transfer and coordination changes dominate. By contrast, the steric arrangement of $H^+$ and $CH_3^-$ on the surface is more important for the subsequent elimination reaction of $CH_4$, and so periodic and cluster models yield different reaction pathways and different activation energies.

In both periodic and cluster models, some atoms may be omitted in order to make calculations tractable. It is common to replace large alkyl chains with smaller ones, unless the steric bulk of the chain is specifically of interest. More

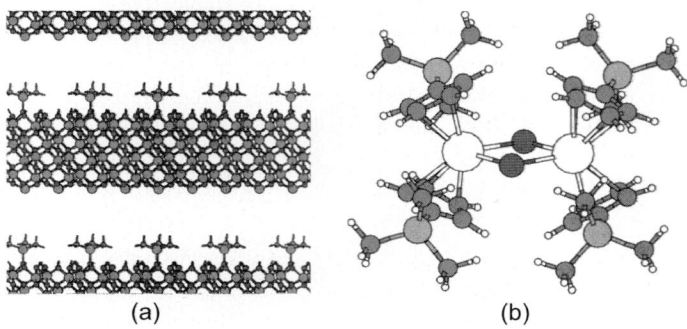

**Fig. 1.** Examples of atomic-scale models: (**a**) periodic model of $Al(CH_3)_3$ adsorption onto hydroxylated alumina, showing slabs with 3D-periodicity (*black* = O, *grey* = $Al_/C$, *white* = H). (**b**) cluster model of $[LuCl(C_5H_4Si(CH_3)_3)_2]_2$ precursor (*large white* = Lu, *black* = Cl, *mid grey* = C, *large grey* = Si, *small white* = H)

realistic is the embedding approach, where different parts of the system are treated at different levels of theory, allowing larger systems to be considered in tractable computational time [10].

For heavy atoms, such as the rare earths, most ab initio approaches explicitly compute wavefunctions only for the outer valence electrons. Relativistic effects, along with the faster inner electrons and the nucleus, are packed into an unchanging core, and this is termed a pseudopotential or effective core potential [1]. Some pseudopotential libraries do not yet include all of the rare earths, so that alternative atoms may have to be used in the model.

As an example, Y was used as a model for Lu in calculations of the precursor $[LuCl(C_5H_4Si(CH_3)_3)_2]_2$ [11], yielding a structure that closely matches the X-ray crystal structure (Y bond lengths < 3 % longer) [12]. The success of the Y model is due to the chemical similarity between Y and Lu (both $M^{3+}$, closed electronic shells, similar ionic radii), and such a close match is not to be expected for the other members of the lanthanide series that have larger ionic radii. With 92 atoms and a diameter of 1.5 nm, it is currently not possible to accommodate the $[LuCl(C_5H_4Si(CH_3)_3)_2]_2$ precursor as well as a substrate slab in a periodic simulation cell (sides typically ~1 nm). Our study therefore looked at the energetics of growth reactions using a cluster, where ligands were replaced by OH to simulate elimination and coordination to the oxide surface [12].

## 2.3 Reaction Steps in ALD

We identify the following steps in oxide ALD: chemisorption of a precursor molecule, dissociation into ions on the highly polar oxide surface, diffusion of ions along the surface and association of ions into a molecule, which desorbs. These reaction steps are listed in Table 1. Reactions that are expected to be highly exoergic (e.g., desorption of ions, cleavage of M−O) are not included in the list. Further extension of this list to include ligand reactions would be necessary in some cases, e.g., ozone-based ALD. Many MOCVD precursors likewise decompose via reductive elimination, and this has been modelled at ab initio level [13]. Radical reactions and gas-phase cluster formation may also play a role in MOCVD [3, 4].

**Table 1.** Products of elementary reaction steps (and of the reverse reactions) that occur at the surface for ALD with precursors $ML_{2x}$ and $H_2O$, as in reaction (1). The notation "surf-M", "surf-O" is not intended to show the proper coordination of surface species; to stress this, alternative coordination is illustrated in reactions A, E and F. Gas-phase reactions and physisorption are not shown

| | **Molecular chemisorption** | | **Molecular desorption** |
|---|---|---|---|
| A. | surf-O–$ML_{2x}$ | $\rightleftharpoons$ | surf-O + $ML_{2x(g)}$ |
| | or surf-OH–$ML_{2x}$ | | surf-OH + $ML_{2x(g)}$ |
| B. | surf-M–$OH_2$ | $\rightleftharpoons$ | surf-M + $H_2O_{(g)}$ |
| C. | $HL_{(surf)}$ | $\rightleftharpoons$ | $HL_{(g)}$ |
| | **Dissociation of molecule** | | **Diffusion of ions and** |
| | **and diffusion of ions** | | **association into molecule** |
| D. | surf-O–$ML_{2x-1}$ + surf-M-L | $\rightleftharpoons$ | surf-O–$ML_{2x}$ + surf-M |
| E. | surf-OH + surf-M–OH | $\rightleftharpoons$ | surf-O + surf-M–$OH_2$ |
| | or surf-OH + surf-O | | surf-O + surf-OH |
| F. | surf-OH + surf-ML | $\rightleftharpoons$ | surf-O–M-surf + $HL_{(surf)}$ |
| | or surf-$OH_2$ + surf-ML | | surf-OH–M-surf + $HL_{(surf)}$ |
| | **Densification** | | **Porosification** |
| G. | $-MO_{x-(bulk)}$ | $\rightleftharpoons$ | $-MO_x$ $-_{(surf)}$ |

Our experience with TMA+$H_2O$ (reaction (2)) has shown that the individual steps can be classified as either Lewis acid-base (reactions A, B, D, F) or Brønsted acid-base (reactions E, F, G). The elementary steps can also be grouped into likely reaction sequences. The reverse of reaction C would normally be followed by the reverse of F, together constituting elimination of a HL molecule. A+D and B+E are both examples of dissociative chemisorption. Many authors refer to "ligand exchange", which in the notation of Table 1 is A (or B) followed by the reverse reactions of F and C. *Puurunen* lists five possible mechanisms for the $ML_{2x}$ pulse, of which the first four are ligand exchange (A+F+C), whereas the fourth and fifth include dissociation/association (D) [14].

Depending on coverage and kinetics, forward and reverse reactions D–G (Table 1) are in principle possible throughout the ALD process. Adsorption/desorption (A–C) is dependent on the partial pressure of $ML_{2x}/H_2O/HL$ and so varies during the pulse-purge sequence. There has been little modelling of the coordination changes that must accompany formation of bulk-like ox-

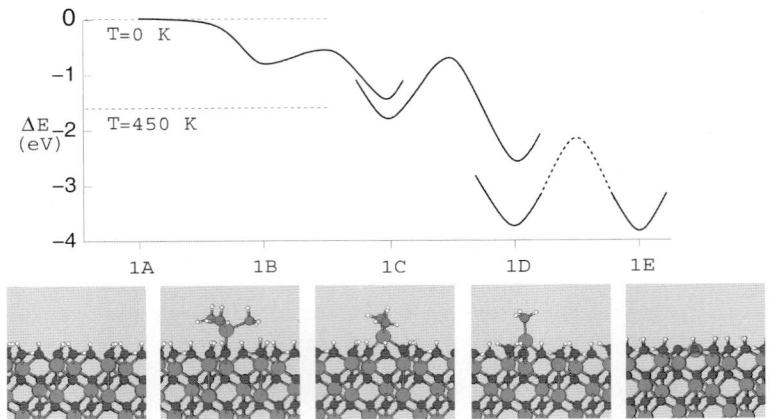

**Fig. 2.** Energetics and structures from periodic calculations for TMA adsorption and $CH_4$ elimination on a fully hydroxylated alumina substrate. *Dashed horizontal lines* show an upper boundary to the contribution of gas-phase $T\Delta S$ for the adsorption step at the temperatures given. *Ball-and-stick structures: large grey* = Al, *medium grey* = C, *medium black* = O, *small white* = H. Reprinted from [5]

ide (G). Most studies have focussed on adsorption (A, B) and elimination (F), as the following examples illustrate.

The structures and reaction energies for TMA adsorption and subsequent elimination reactions on a highly hydroxylated substrate, computed using a periodic model, are shown in Fig. 2. Real surfaces at 450–500 K have about 60 % of this OH coverage. It is found that the adsorbed TMA fragment shifts from bonding to terminal–O to bridging–$O_2$ to capping–$O_3$ as $CH_4$ is eliminated [5]. The entropy estimate shows that desorption of TMA is favoured at $T = 450$ K, except when elimination occurs before desorption. During the water pulse, the computed energetics of molecular chemisorption and $CH_4$ elimination are similar [5]. The difference lies in the energetics of dissociation into H + OH (E in Table 1), which show a strong dependence on OH-coverage, consistent with the experimental finding of temperature-dependent OH-coverage. Further mechanistic aspects of the TMA + $H_2O$ process, combining insights from theory and experiment, are discussed in a review by *Puurunen* [15].

All reaction steps in TMA + $H_2O$ that lead to $Al_2O_3$ growth are exoergic because of the weakness of Al–$CH_3$ and the strength of H–$CH_3$. The reverse reaction of $CH_4$ with Al–O units is highly endoergic. Experimentally, TMA+$H_2O$ is a near-ideal ALD process. By contrast, ab initio cluster calculations on the $AlCl_3$+$H_2O$ system show that this precursor behaves quite differently [6]. In this case, the adsorption complex surf-OH–$AlCl_3$ is more stable than elimination products such as surf-O–$AlCl_2$ + HCl. This is illustrated in Fig. 3 for TMA vs. $AlCl_3$, as well as for hafnium and zirconium

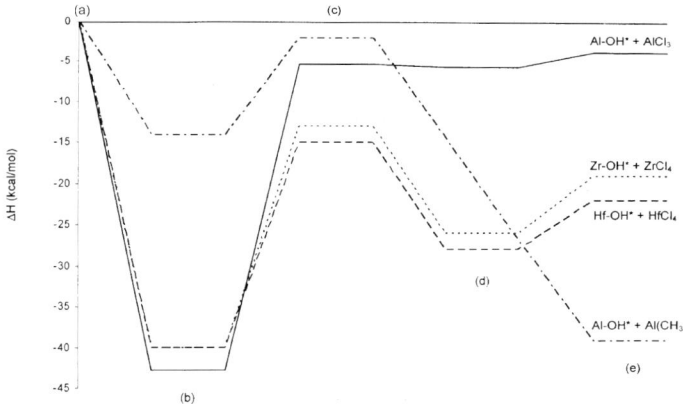

**Fig. 3.** Reaction paths and predicted energies for the first half-reaction for four different ALD precursors with $H_2O$: $Al_2O_3$ from $AlCl_3$ (*solid*), $Al_2O_3$ from TMA (*dash-dotted*), $HfO_2$ from $HfCl_4$ (*dashed*), and $ZrO_2$ from $ZrCl_4$ (*dotted*). (**a**) Reactants, (**b**) adsorbed complex, (**c**) transition state, (**d**) by-product physisorbed, (**e**) dissociated products. Reprinted from Fig. 6 of [6] with permission from the American Chemical Society

chlorides. Consistent with this, the $HfCl_4 + H_2O$ process suffers from the re-adsorption of HCl product into the growing film. The computed energetics agree with other ab initio work on $HfCl_4 + H_2O$ [10, 16, 17]; further validation comes from integrating these data into transport models and comparing with experiment (Sect. 4).

This work illustrates that when designing a precursor, it is important to take into account the relative energetics of adsorption and ligand elimination. For ALD processes that use $H_2O$ as oxygen source, the ligands (L) on the metal-containing precursor should have a high affinity for $H^+$ and then desorb readily as HL.

Finally, the prospective use of ALD for depositing oxide in transistor gate stacks in the electronics industry has prompted much work on the initial growth reactions on Si. As well as cluster models [10, 16–18], analytic models have been developed of nucleation and growth [19] and island formation [20]. The interface to Si has also been investigated using sophisticated Kinetic Monte Carlo schemes, which incorporate ab initio reaction parameters into an atomistic model spanning experimental length and time scales [7, 21].

## 3 Analytical Models for ALD Film Growth

In optimal ALD, the surface becomes completely saturated with precursor fragments at the end of a sufficiently long pulse, in a self-limiting half-reaction. This is followed by a sufficiently long purge so that there are no gas-phase reactions during the next pulse. In this ideal situation, the rate of deposition clearly depends on the saturating coverage, and so it is appealing to develop an analytical model of surface coverage and film growth.

If ALD by reaction (1) proceeds optimally, then L–L steric interaction between ligands limits the amount of M which can be adsorbed during the $ML_{2x}$ pulse. By modelling the bulk of precursor fragments and their packing efficiency and then comparing to experimental growth rates, it can be determined what surface intermediates are formed during ALD [22]. In general, the more L that can be eliminated during this pulse as HL (the "efficiency of ligand exchange" [23]), the more M that can be deposited. The equilibrium ratio M : L and the rate of deposition of $MO_x$ therefore depend both on the bulk of L and on the reactivity of L towards $H^+$ at the surface.

During the $H_2O$ pulse, similar individual reactions occur (see Sect. 2). Once more, the growth rate is dependent on $H_2O$ adsorption and elimination of HL. However, in this case, crowding between fragments H and OH is negligible and it is the desorption of $H_2O$ during the purge that impacts ALD rate. This is because the availability of surface $H^+$ limits the elimination of HL during the subsequent $ML_{2x}$ pulse.

We have proposed a graphical representation of the link between precursor pulses and the coverage of precursor fragments, and of the link between saturating coverages and growth rate. Such a "reaction portrait" is shown in Fig. 4 and examples of its application are given in [24]. It is useful to compare this detailed model of the individual pulses with in situ detection of mass changes and by-products during growth experiments [25].

Even an optimal ALD process is therefore subject to the following limitations, which can be addressed by precursor design:

– bulk of adsorbed ligands (e.g., L);
– inertness of ligands with respect to elimination (as HL);
– volatility of adsorbed ligands (e.g., of $H^+$ as $H_2O$).

In real ALD, many more factors come into play, including precursor condensation, gas-phase decomposition, incomplete elimination reactions, readsorption of elimination products, and deleterious side-reactions. These can lead to incomplete surface saturation, in which case a study of the rate as a function of available sites (Langmuir adsorption isotherm) may be relevant [26].

As noted in Sect. 2.3, five $ML_{2x}$ mechanisms are identified in [23] and these are used to generate quantitative growth mechanisms that can be compared with experiment [14]. For $Y(thd)_3+O_3$, the results are inconclusive. However, for $TiCl_4+H_2O$, there is a clear indication that OH coverage is not limiting (A, B, C, Fig. 5) and that Cl–Cl hindrance dominates instead, with some dissociation/association of $TiCl_4$ (D).

There has been little work to date on explaining the outstanding conformality of ALD films, even though sub-monolayer growth would suggest the formation of islands and an increase in roughness, which does not occur. One

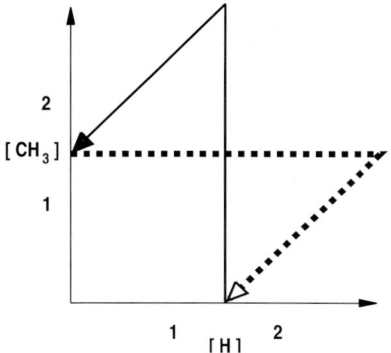

**Fig. 4.** Reaction portrait to illustrate ALD growth of $Al_2O_3$ from TMA + $H_2O$. The space is defined by surface concentrations of reactive intermediates ($CH_3$, H) in units of atoms per unit surface area. This means that chemisorption (*horizontal and vertical lines*) and elimination/desorption (*diagonal lines*) are shown separately. During the TMA pulse (*solid line*), one TMA molecule containing 3 $CH_3$ is adsorbed per surface unit and 1.5 $CH_4$ are eliminated, giving a final coverage of 1.5 $CH_3$/unit after the purge. The net adsorption of 3 H as 1.5 $H_2O$ occurs in the $H_2O$ pulse (*dotted line*), the remaining 1.5 $CH_4$ desorb and the final H coverage is 1.5/unit. Reprinted from Fig. 1 of [24]

study uses rather artificial reconstructions to show that intra-surface interaction ensures sub-monolayer deposition [27]. As dipole–dipole interaction between surf-OH would be an example of this effect, we suggest that this may provide a clue to the conformality of ALD.

## 4 Continuum Models for Gas Transport

Reactor geometry and gas-flow conditions are vital elements in film deposition by MOCVD and ALD. Variables such as precursor partial pressure, pulse/purge time, flow rate, reactor dimensions and reactor layout must be optimised. It is therefore useful to model the flow of gas from the inlet, through the reactor, across the reactive surface and to the exhaust, and this can be achieved by solving the Navier–Stokes equations for gas fluid dynamics. Fluid properties such as viscosity and turbulence must be known or estimated.

One of the most attractive aspects of ALD is the conformality of thickness and stoichiometry that can be achieved down high aspect ratio features. More advanced models therefore consider the coupled transport of gases through the macroscopic reactor and within a microscopic feature [28].

At the surface, gas-phase precursors are converted into a unit of solid oxide film and gas-phase products are produced, by the complex series of

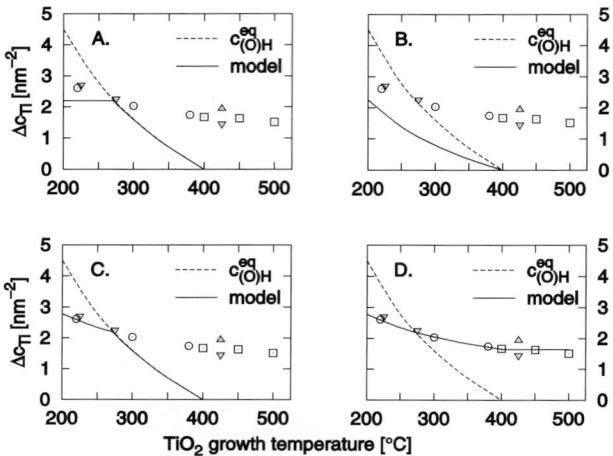

**Fig. 5.** Growth of $TiO_2$ from $TiCl_4 + H_2O$: Ti adsorbed per unit surface area per cycle ($\Delta c_{Ti}$) as a function of growth temperature (°C) based on predictions for mechanisms A–D (*solid lines*) at experimental OH coverages (*dashed line*), and compared with experimental growth rate (*points*). Reprinted from Fig. 4 of [14] with permission from Wiley-VCH

reactions sketched in Sect. 2.3. We present two approaches by which this gas-surface interaction may be incorporated into the transport model.

The simpler approach assigns an effective sticking coefficient $0 \leq s \leq 1$ to each precursor, quantifying in one parameter the total growth probability as a result of reaction steps at the surface. The effective sticking coefficient can be found by fitting simulated film profiles to the results of undersaturation experiments. For example, in $Al_2O_3$ ALD it is found that $s(\text{TMA}) = 0.1-0.9$, $s(\text{H}_2\text{O}) = 0.1-0.01$ but $s(\text{O}_3) = 0.001-0.01$ [28].

An alternative is to incorporate explicit chemical kinetics for the surface reactions into the fluid dynamics model. For each of the elementary reaction steps described in Sect. 2.3, an activation energy $E_a$ and Arrhenius pre-exponential factor $A$ must be computed (Sect. 2.1) or estimated from experiment, so as to obtain a rate constant for each step:

$$k = AT^n \exp(E_a/RT).$$  (3)

Surface species are then allowed to react with each other and with the gas phase at these rates, while gas-phase species are transported to/from the surface according to fluid dynamics. The accuracy and flexibility of such a model is appealing, but it is time-consuming to generate the input parameters. For example, 28 elementary reactions were needed to simulate $ZrCl_4+H_2O$ ALD [8]; as shown in Fig. 6, the results are in good agreement with experiment both in terms of mass evolution over one cycle and temperature dependence of the process. Because of the chemical detail contained in the

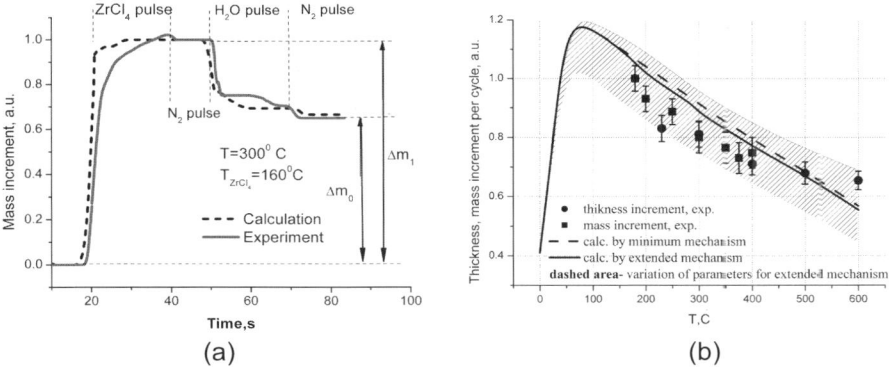

**Fig. 6.** ALD of $ZrO_2$ from $ZrCl_4 + H_2O$. (**a**) Film mass increment during one ALD cycle measured in experiment (*solid line*) and predicted by multi-scale model (*dashed*). (**b**) Predicted growth rate vs. process temperature (*solid* and *dashed lines; hatched area* reflects sensitivity analysis) compared with experiment (*points*). Reprinted from Figs. 5 and 7 of [8] with permission from Elsevier

model, *Deminsky* et al. were able to determine that insufficient surface-OH is responsible for the drop in rate with increasing temperature.

# 5 Conclusion

ALD and MOCVD are methods for the deposition of thin films that are largely chemistry-driven. Modelling has played an important part in understanding the underlying chemistry, whether by analytical models of growth per cycle or quantum mechanical treatments of the electronic and atomic structure during reactions. We discuss the applicability of various electronic structure methods and the most common models of reacting systems in ALD/MOCVD. In this way, insight has been obtained into the reactivity of precursors and the dependence on process temperature for a wide variety of oxide systems, even when deposition is "ideal". Applying these insights will simplify the design of new precursors and the optimisation of processes for rare earth oxide deposition.

Gas dynamics within the reactor and within narrow features can also be addressed with simulation, allowing such effects as precursor dose, flow rate and reactor geometry to be examined. In a convergence of these approaches, a multi-scale model has been demonstrated that successfully combines gas transport with ab initio derived reaction kinetics. Taken together, these approaches make up a "tool kit" for process simulation of ALD and MOCVD, with each approach suited to addressing a certain question at a certain level of sophistication.

**Acknowledgements**

This work was made possible by the support of the HIKE project, funded by the European Commission under the FP5 "Information Society Technologies" programme. Particular thanks to Alfred Kersch (Infineon A.G.), Jens Schmidt (Freescale Halbleiter GmbH), Alain Estève and Mehdi Djafari-Rouhani (CNRS-LAAS), Jacob Gavartin and Alex Shluger (University College London) for their contributions as project partners, and to Jim Greer (Tyndall National Institute) and Anatoli Korkin (formerly Motorola Inc.) for initiating the project. http://www.tyndall.ie/hike

# References

[1] R. M. Martin: *Electronic Structure: Basic Theory and Practical Methods* (Cambridge Univ. Press, Cambridge 2004)

[2] S. D. Elliott: First principles modelling of the deposition process for high-k dielectric films, Electrochem. Soc. Proc. **2003-14**, 231 (2003)

[3] A. Y. Timoshkin, H. F. Bettinger, H. F. Schaefer, III: The chemical vapor deposition of aluminium nitride: unusual cluster formation in the gas phase, J. Am. Chem. Soc. **119**, 5668–5678 (1997)

[4] A. Y. Timoshkin, H. F. Bettinger, H. F. Schaefer, III: DFT modeling of chemical vapor deposition of GaN from organogallium precursors. 1. Thermodynamics of elimination reactions, J. Phys. Chem. A **105**, 3240–3248 (2001)

[5] S. D. Elliott, J. C. Greer: Simulating the atomic layer deposition of alumina from first principles, J. Mater. Chem. **14**, 3246–3250 (2004)

[6] A. Heyman, C. B. Musgrave: A quantum chemical study of the atomic layer deposition of $Al_2O_3$ using $AlCl_3$ and $H_2O$ as precursors, J. Phys. Chem. B **108**, 5718–5725 (2004)

[7] G. Mazaleyrat, A. Estève, L. Jeloaica, M. Djafari-Rouhani: A methodology for the kinetic Monte Carlo simulation of alumina atomic layer deposition onto silicon, Comp. Mater. Sci. **33**, 74–82 (2005)

[8] M. Deminsky, A. Knizhnik, I. Belov, S. Umanskii, E. Rykova, A. Bagatur'yants, B. Potapkin, M. Stoker, A. Korkin: Mechanism and kinetics of thin zirconium and hafnium oxide film growth in an ALD reactor, Surf. Sci. **549**, 67–86 (2004)

[9] Y. Widjaja, C. B. Musgrave: Quantum chemical study of the mechanism of aluminium oxide atomic layer deposition, Appl. Phys. Lett. **80**, 3304–3306 (2002)

[10] M. D. Halls, K. Raghavachari: Importance of steric effects in cluster models of silicon surface chemistry: ONIOM studies of the atomic layer deposition of $Al_2O_3$ on H/Si(111), J. Phys. Chem. A **108**, 2982–2987 (2004)

[11] G. Scarel, E. Bonera, C. Wiemer, G. Tallarida, S. Spiga, M. Fanciulli, I. L. Fedushkin, H. Schumann, Y. Lebedinskii, A. Zenkevich: Atomic-layer deposition of $Lu_2O_3$, Appl. Phys. Lett. **85**, 630–632 (2004)

[12] S. D. Elliott, G. Scarel, C. Wiemer, M. Fanciulli, T. Lebedinskii, A. Zenkevich, I. L. Fedushkin: Precursor combinations for ALD of rare earth oxides and silicates - a quantum chemical and X-ray study, in (Proc. Electrochem. Soc. 2005)

[13] T. R. Cundari, S. O. Sommerer: Quantum modeling of the CVD of transition metal materials, Chem. Vap. Dep. **3**, 183–192 (1997)

[14] R. L. Puurunen: Growth per cycle in atomic layer deposition: real application examples of a theoretical model, Chem. Vap. Depos. **9**, 327–332 (2003)

[15] R. L. Puurunen: Surface chemistry of atomic layer deposition: a case study for the trimethylaluminium/water process, J. Appl. Phys. **97**, 121301 (2005)

[16] L. Jeloaica, A. Estève, M. D. Rouhani, D. Esteve: Density functional theory study of HfCl$_4$, ZrCl$_4$, and Al(CH$_3$)$_3$ decomposition on hydroxylated SiO$_2$: Initial stage of high-k ALD, Appl. Phys. Lett. **83**, 542–544 (2003)

[17] V. V. Brodskii, E. A. Rykova, A. A. Bagatur'yants, A. A. Korkin: Modelling of ZrO$_2$ deposition from ZrCl$_4$ and H$_2$O the Si(100) surface: initial reactions and surface structures, Comp. Mater. Sci. **24**, 278–283 (2002)

[18] J. H. Han, G. L. Gao, Y. Widjaja, E. Garfunkel, C. B. Musgrave: A quantum chemical study of ZrO$_2$ atomic layer deposition growth reactions on the SiO$_2$ surface, Surf. Sci. **550**, 199–212 (2004)

[19] M. L. G. M. A. Alam: Mathematical description of atomic layer deposition and its application to the nucleation and growth of HfO$_2$ gate dielectric layers, J. Appl. Phys. **94**, 3403–3413 (2003)

[20] R. L. Puurunen, et al.: Island growth in the atomic layer deposition of zirconium oxide and aluminium oxide on hydrogen-terminated silicon: Growth mode modeling and transmission electron microscopy, J. Appl. Phys. **96**, 4878–4889 (2004)

[21] A. A. Knizhnik, A. A. Bagaturyants, I. V. Belov, B. V. Potapkin, A. A. Korkin: An integrated kinetic Monte Carlo molecular dynamics approach for film growth modeling and simulation: ZrO$_2$ deposition on Si(100) surface, Comp. Mater. Sci. **24**, 128–132 (2002)

[22] M. Yliliammi: Monolayer thickness in atomic layer deposition, Thin Solid Films **279**, 124–130 (1996)

[23] R. L. Puurunen: Growth per cycle in atomic layer deposition: a theoretical model, Chem. Vap. Depos. **9**, 249–257 (2003)

[24] S. D. Elliott: Predictive process design: A theoretical model of atomic layer deposition, Comp. Mater. Sci. **33**, 20–25 (2005)

[25] A. Rahtu, T. Alaranta, M. Ritala: In situ quartz crystal microbalance and quadrupole mass spectrometry studies of ALD of aluminium oxide from TMA and water, Langmuir **17**, 6506–6509 (2001)

[26] H.-S. Park, J.-S. Min, J.-W. Lim, S.-W. Kang: Theoretical evaluation of film growth rate during atomic layer epitaxy, Appl. Surf. Sci. **158**, 81–91 (2000)

[27] M. Ahr, M. Biehl: Modelling sublimation and atomic layer epitaxy in the presence of competing surface reconstructions, Surf. Sci. **488**, L553–L560 (2001)

[28] G. Prechtl, A. Kersch, G. S. Icking-Konert, W. Jacobs, T. Hecht, H. Boubekeur, U. Schröder: A model for Al$_2$O$_3$ ALD conformity and deposition rate from oxygen precursor reactivity, in *2003 IEDM Techn. Digest* (2003) pp. 245–248

# Index

# Growth of Oxides with Complex Stoichiometry by the ALD Technique, Exemplified by Growth of $La_{1-x}Ca_xMnO_3$

Ola Nilsen, Martin Lie, Helmer F. Fjellvåg, and Arne Kjekshus

Department of Chemistry, University of Oslo, P.O. Box 1033 Blindern, N-0315
Oslo, Norway
ola.nilsen@kjemi.uio.no

**Abstract.** To grow films with a complex stoichiometry by the ALD approach has long been regarded as a complicated task and hitherto not pursued to a large extent. This review will cover some of the milestones on the way to controlled deposition of compounds with intricate stoichiometry with the ALD technique. The survey shows that some of the foreseen obstacles might not necessarily be that severe. The growth of films of $La_{1-x}Ca_xMnO_3$ will be used as a model to describe the process for growth of oxides with complex composition, and a mathematical model to predict the deposited stoichiometry is advanced. It is shown that the ALD window found for deposition of $MnO_2$ is extended to higher temperatures when Ca and/or La is present on the film surface.

## 1 Introduction

ALD has since its birth in the 1970s been used for synthesis of thin films of numerous types of compounds [1–3]. However, among these there are relatively few examples of compounds with complex stoichiometry. There has been a general opinion in the ALD community that controlled deposition of compounds with complex stoichiometry will be exposed to the sum of the limitations known from deposition experiments of each of the binary compounds of the constituents involved. Hence the progress in deposition of complex types of materials has been slow until now.

Reported (some yet unpublished) accounts for ternary compounds synthesized by the ALD technique cover a list of almost 30 examples (Table 1), whereas the examples of quaternary compounds are almost nil (listed at the bottom of Table 1). One reason for this situation is that there exists no well-documented procedure to control the transfer of pulsed composition to the substrate. This overview will try to elaborate on this aspect and discuss how deposition of such compounds with the ALD technique may be approached. The system Ca–La–Mn–O will be used to illustrate the accomplishments, since it was in fact used as a model for the development of the methodology which is discussed below.

We expect that as one becomes accustomed with procedures to deposit materials with complex stoichiometry, the applicability of the technique will

M. Fanciulli, G. Scarel (Eds.): Rare Earth Oxide Thin Films,
Topics Appl. Physics **106**, 87–100 (2007)
© Springer-Verlag Berlin Heidelberg 2007

**Table 1.** Ternary compounds reported synthesized by the ALD technique. The bottom of table gives available examples of quaternary compounds prepared correspondingly

| Compound | Reference | Compound | Reference |
|---|---|---|---|
| $BaTiO_3$ | [4–6] | $SrTa_2O_6$ | [7, 8] |
| Bi–Ti–O | [9, 10] | $SrNb_2O_6$ | [7] |
| $CaMnO_3$ | [11] | $Ti_xHf_yO_z$ | [12] |
| $LaAlO_3$ | [13] | $Ti_xZr_yO_z$ | [12] |
| $LaCoO_3$ | [14] | $TaO_xN_y$ | [15] |
| $LaFeO_3$ | [16] | $Ti_xSi_yN_z$ | [17, 18] |
| $LaGaO_3$ | [19] | $Zr_xSi_yO_z$ | [12] |
| $LaMnO_3$ | [20] | $Y_2O_2S$ | [21] |
| $LaNiO_3$ | [22] | $ZnO_{1-x}S_x$ | [23] |
| $SrTiO_3$ | [5, 6, 24–28] | $La_2O_2S$ | [29] |
| $CoFe_2O_4$ | [30] | $CuGaS_2$ | [31] |
| $Al_xCr_yO_z$ | [32] | $CuInS_2$ | [33] |
| $Al_xHf_yO_z$ | [12] | $SrS_{1-x}Se_x$ | [34, 35] |
| $Al_xTi_yO_z$ | [12] | $ZnS_{1-x}Se_x$ | [35–39] |
| $Al_xZr_yO_z$ | [12] | $GaP_{1-x}As_x$ | [40, 41] |
| — | — | — | — |
| $La_{1-x}Ca_xMnO_3$ | [11] | $Bi_{1-x-y}Ti_xSi_yO_z$ | [42, 43] |
| $La_{1-x}Sr_xFeO_3$ | [16] | $SrBi_2Ta_2O_9$ | [44] |

expand rapidly. It will then be possible to produce materials with even more exciting electric and magnetic properties of interest for pure science as well as for applied purposes. An additional feature is the possibility to tune desired properties. This is considered in some detail for materials in the chosen model system, $R_{1-x}A_xMnO_3$ ($R$ = rare-earth element, $A$ = alkali or alkaline earth element), where the combination of $R$ and $A$, and the value of $x$ appears to be the determining factor for the colossal magnetoresistant (CMR) properties.

## 2 Conceptional Basis for ALD Deposition of Complex Stoichiometries

There are several ways to handle multiple-element situations in ALD synthesis. Complex oxides have previously been grown by the use of multi-element single-source metallorganic precursors [7, 8]. The obvious advantage of this procedure is that the stoichiometry can be fixed at a high precision provided the compound under consideration is sufficiently thermally stable. This approach also simplifies the design of the deposition process since only one kind of metal precursor is needed for the metal components. However, the benefit of having a fixed stoichiometry between the metal components renders such an approach useless when tuning is needed. Also, the accessibility of suitable precursors for such exercises is today rather limited.

Another approach that maintains the benefit of a single type of precursor for all metal components is to switch to a well-defined mixture of precursors. This is commonly used today in MOCVD depositions by the aid of flash evaporators. However, when precursors are mixed, most commonly in solution, it is likely that the different compounds as well as the solvent will evaporate at different rates. The practice is therefore to continuously flash evaporate parts of the solution in order to obtain constant stoichiometry of the vapor phase. Indeed, this can be accomplished with the ALD technique in so-called liquid injection ALD systems [9, 44, 45], which has been demonstrated to deposit materials with rather complex stoichiometry [9,44]. According to such an approach, the surface chemistry may be rather complex even though self-limited growth has been demonstrated. It is not intuitively obvious that the composition of the precursor mixture will be transferred to the substrate, owing to spread in the sticking coefficients involved. These brief comments emphasize that the procedure requires precursors with similar chemistry and sticking coefficients, which in turn may not present a too severe obstacle. Nevertheless, the transferability of the stoichiometry from the mixture in the solution to the film, as well as the distribution of precursor mixture throughout the gas phase of the ALD apparatus have to be mapped.

The procedure which we would like to elaborate on in more detail is sequential use of multiple single-element precursors in a controlled manner. An advantage of this procedure is that it enables one to build on the already available experience for growth of binary compounds. However, the surface chemistry may be different since other species are present and this can, in turn, affect the growth with regards to the kind and amount of surface active sites, and result in alterations of the ALD windows.

## 2.1 The Model System

The model system for deposition of compounds with complex stoichiometries is, as already mentioned, $La_{1-x}Ca_xMnO_3$. This belongs to the perovskite type family where different members exhibit a large variety of properties (e.g., insulators – semiconductors – metals – superconductors, diamagnets – paramagnets – ferrimagnets – ferromagnets, pyroelectrics – piezoelectrics – ferroelectrics). Ca–La–Mn–O phases have previously been deposited by various techniques ranging from physical procedures (e.g., MBE, PVD, PLD, and sputtering) to more chemical alternatives (e.g., chemical bath deposition, spray-pyrolysis, and MOCVD), and now also ALD [11]. For more extensive reviews on depositions of this type of phases, reference is made to a recent paper of *Prellier* et al. [46] and an overview by *Sun* et al. [47]. Most of the mentioned techniques require deposition on relatively warm substrates, and some techniques also have limitations with regard to sizes and shape. This is where the ALD technique may provide a methodological improvement toward deposition at lower temperatures and on larger substrates with arbitrary shapes.

$R_{1-x}A_x\text{MnO}_3$ shows a variety of interesting properties, among which the CMR effect established around $x = 0.3$ is the most extensively studied. The CMR effect has been ascribed to the coexistence of ferromagnetic insulator and ferromagnetic metallic states, and it is possible to affect the relative amount of these states by an external magnetic field. The largest CMR effects are found close to the paramagnetic insulator – ferromagnetic metal transition temperature ($T_C$). There are also reports which connect large MR effects with the grain boundaries in the material, but these appear to have a different temperature dependence than the regular CMR effect. By and large, the CMR phenomena are greatly influenced by sample features such as: composition, texture, strain imposed by the substrate-to-film interface (for more information on CMR effects, see [46, 48–55]).

## 2.2 Deposition of Binary Compounds

It is relatively easy to control the composition of deposited films of binary compounds provided, of course, that the desired composition is attainable under the available physical and chemical conditions. With respect to growth of oxides, it is possible to control the oxygen stoichiometry to some extent by the use of different types of oxidizing agents during the growth. The applicability of the latter means requires that the metal constituent exhibits some flexibility with respect to oxidation state. This can be exemplified by the growth of both MnO and $\text{MnO}_2$ with the ALD technique using, respectively, the precursors $\text{Mn(hmds)}_2$ (hmds = hexamethyldisilasane) and water [56], and $\text{Mn(thd)}_3$ (Hthd = 2,2,6,6-tetramethylheptan-3,5-dione) and ozone [57] as oxidizing agents. However, depending on the chemical system, there may be challenges connected to the alternating oxidation environment occurring during such ALD growth.

## 2.3 Deposition of Ternary Compounds

Growth of ternary compounds indeed increases the complexity of the ALD procedure. Nevertheless, a decent number of such phases have been grown by the ALD technique (see Table 1). A key element in the procedure for preparation of ternary oxides is first to choose a deposition temperature that suits the ALD windows for both metal components involved. Then, the correct transfer of composition from the gas-phase pulses to the deposited layer must be carefully probed.

On closer examination of some of the phases listed in Table 1, it becomes evident that the ALD windows involved do not overlap to the full extent. This is nicely demonstrated for the systems $\text{MnO}_2$–CaO (actually, $\text{CaCO}_3$) and $\text{MnO}_2$–$\text{La}_2\text{O}_3$ [11] (see Fig. 1). Growth of $\text{MnO}_2$ by $\text{Mn(thd)}_3$ and ozone exhibits uncontrolled growth at temperatures above ca. 240 °C. However, when $\text{Mn(thd)}_3$ and ozone are used as precursors for deposition of film on a surface that contains Ca and/or La atoms, it is possible to extend the usability of

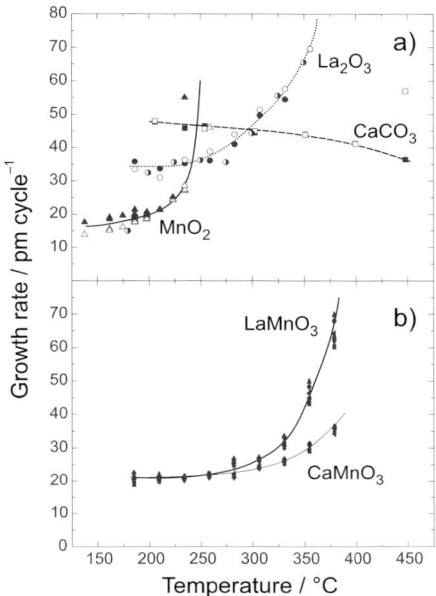

**Fig. 1.** Growth rate as function of temperature for deposition of (**a**) $MnO_2$, $La_2O_3$, and $CaCO_3$, (**b**) $CaMnO_3$ and $LaMnO_3$. Data from [11, 57, 58]

$Mn(thd)_3$ as a precursor up to deposition temperatures where $La(thd)_3$ or $Ca(thd)_2$ and ozone experience uncontrolled growth. Hence, an extension of ALD windows may be obtained (in other cases, it may be vice versa). In order to shed further light on such extension of ALD windows, one should tentatively focus attention on the limitations of the $Mn(thd)_3$ precursor. The authors believe that adsorbed $Mn(thd)_n$ ($n = 3-1$) partially decomposes on the growing film surface by liberation of thd-ligands. This opens up otherwise blocked surface sites for further deposition of $Mn(thd)_3$ and thus cancels the self-hindered growth. On a surface containing either Ca or La, these cations take over ligands from the $Mn(thd)_3$ precursor, block surface sites, and maintain a self-hindered growth. This may be the mechanism for extension of the ALD window to fit the surface chemistry of the involved Ca and/or La containing compounds.

Another previous concern in relation to deposition of materials with complex stoichiometry has been whether or to what extent the rules of traditional chemistry will be maintained. For example, when $Ca(thd)_2$ alone is used as a precursor for deposition of CaO, $CaCO_3$ is rather obtained [58] (Fig. 2). It would indeed be unfortunate if the same failure were to occur when $Ca(thd)_2$ is used as precursor for deposition of a calcium-containing ternary or quaternary oxide. However, as the compositional complexity of materials increases, the chemistry follows. In line with this, $La_{1-x}Ca_xMnO_3$ is "fortunately" more

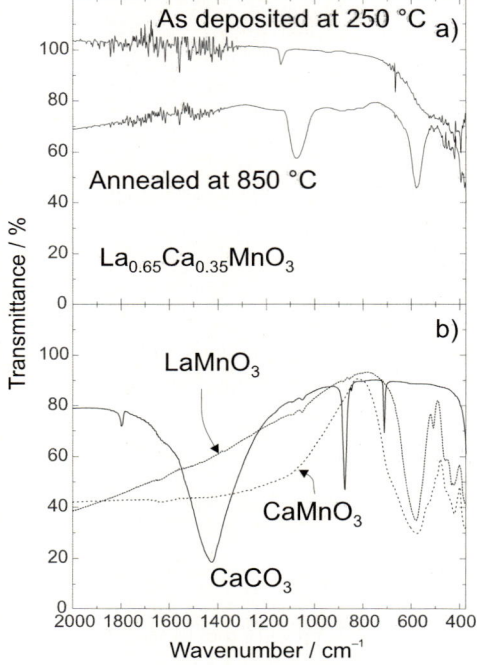

**Fig. 2.** FT-IR spectroscopy of film of (**a**) $La_{0.65}Ca_{0.35}MnO_3$, as deposited at $250\,°C$ and annealed at $850\,°C$ and (**b**) powders of $CaCO_3$ and the ternary end members of the $La_{1-x}Ca_xMnO_3$ phase. Note the lack of absorption around $1400\,cm^{-1}$ for the film sample. Quoted from [11]

stable than possible carbonates, and thus virtually carbonate-free films of this quaternary oxide are obtained (Fig. 2).

It is also interesting to note that the composition gradients in the films are greatly reduced as the composition of the deposited film approaches the ideal ($ABO_3$) stoichiometry for perovskites (Fig. 3). A proper explanation of this behavior is lacking at the moment.

Some information may be extracted by closer inspection of the transfer of stoichiometry from pulsed ratio to film composition. According to the model of *Ylilammi* [59], the growth rate of the binary end member's compounds should be inversely proportional to the area that the precursors of these metal components occupy on the film surface. On this basis, it is possible to construct a model that predicts the composition obtained with different pulsed ratios. The model relies on the physical size each kind of precursor occupies on the surface, but it is not straightforward to quantify this parame-

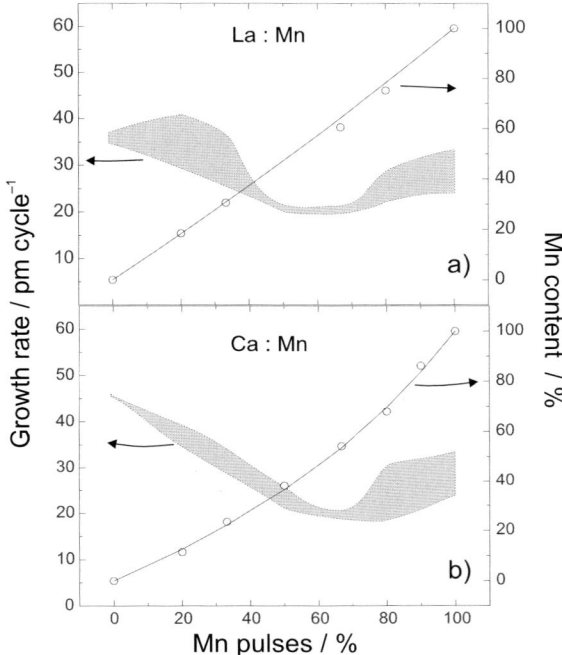

**Fig. 3.** Film stoichiometry versus growth rate (at a deposition temperature of 250 °C) for (**a**) LaMnO$_3$ and (**b**) CaMnO$_3$. Experimental scatter in growth rate is indicated by shading. The lines represent fitting of (1) to experimental data from [11]

ter. However, luckily enough, the absolute size of the chemiscrbed precursor fragment is not required, only the relative size:

$$D_{\mathrm{Mn}} = \frac{P_{\mathrm{Mn}}U_{\mathrm{Mn}}}{P_{\mathrm{Mn}}U_{\mathrm{Mn}} + P_{\mathrm{Ca}}U_{\mathrm{Ca}}}$$
$$D_{\mathrm{Ca}} = 1 - D_{\mathrm{Mn}}, \tag{1}$$

which can be solved to give the pulsing ratio necessary (2) for the desired stoichiometry:

$$P_{\mathrm{Mn}} = \frac{D_{\mathrm{Mn}}U_{\mathrm{Ca}}}{U_{\mathrm{Ca}}D_{\mathrm{Mn}} - U_{\mathrm{Mn}}D_{\mathrm{Mn}} + U_{\mathrm{Mn}}U_{\mathrm{Ca}}}$$
$$P_{\mathrm{Ca}} = 1 - P_{\mathrm{Mn}}. \tag{2}$$

The parameters $U_M$, $D_M$, and $P_M$ represent, respectively, relative growth rate (equivalent to a surface utilization coefficient), deposited stoichiometry, and pulsed stoichiometry for a metal component $M$.

This equation may be fitted to the experimental data in order to extract the relative growth rates (recalling that the growth rate is inversely proportional to surface occupation area). The fittings for the La–Mn and Ca–Mn systems result (Fig. 3) in relative growth rates of 1 : 0.891 for La : Mn and 1 : 0.573 for Ca : Mn. The combination of these growth rate ratios give relative growth rates of 1 : 0.643 : 0.573 for Ca : La : Mn.

## 2.4 Deposition of Quaternary Compounds

ALD deposition of quaternary phases (as well as deposition of multi-component phases) is merely an extension of the approach described for ternary phases. Equation (1) may be extended to take into account three metal components (3):

$$
\begin{aligned}
D_{\mathrm{Mn}} &= \frac{P_{\mathrm{Mn}}U_{\mathrm{Mn}}}{P_{\mathrm{Mn}}U_{\mathrm{Mn}} + P_{\mathrm{Ca}}U_{\mathrm{Ca}} + P_{\mathrm{La}}U_{\mathrm{La}}} \\
D_{\mathrm{La}} &= \frac{P_{\mathrm{La}}U_{\mathrm{La}}}{P_{\mathrm{Mn}}U_{\mathrm{Mn}} + P_{\mathrm{Ca}}U_{\mathrm{Ca}} + P_{\mathrm{La}}U_{\mathrm{La}}} \\
D_{\mathrm{Ca}} &= 1 - D_{\mathrm{Mn}} - D_{\mathrm{La}}
\end{aligned}
\tag{3}
$$

and may be solved to give the pulsing ratio necessary for the desired stoichiometry (4):

$$
\begin{aligned}
P_{\mathrm{La}} &= \frac{U_{\mathrm{Mn}}D_{\mathrm{La}}U_{\mathrm{Ca}}}{U_{\mathrm{Ca}}D_{\mathrm{Mn}}U_{\mathrm{La}} - U_{\mathrm{Mn}}D_{\mathrm{Mn}}U_{\mathrm{La}} + U_{\mathrm{Mn}}D_{\mathrm{La}}U_{\mathrm{Ca}} - U_{\mathrm{Mn}}M_{\mathrm{La}}U_{\mathrm{La}} + U_{\mathrm{Mn}}U_{\mathrm{La}}} \\
P_{\mathrm{Mn}} &= \frac{U_{\mathrm{La}}D_{\mathrm{Mn}}U_{\mathrm{Ca}}}{U_{\mathrm{Ca}}D_{\mathrm{Mn}}U_{\mathrm{La}} - U_{\mathrm{Mn}}D_{\mathrm{Mn}}U_{\mathrm{La}} + U_{\mathrm{Mn}}D_{\mathrm{La}}U_{\mathrm{Ca}} - U_{\mathrm{Mn}}M_{\mathrm{La}}U_{\mathrm{La}} + U_{\mathrm{Mn}}U_{\mathrm{La}}} \\
P_{\mathrm{Ca}} &= 1 - P_{\mathrm{La}} - P_{\mathrm{Mn}} \, .
\end{aligned}
\tag{4}
$$

With the help of (3) and (4), we have constructed a map visualizing the transfer of pulsed ratio to stoichiometry of the deposited film (Fig. 4). The illustration shows that the films quite generally obtain a higher Ca content than that based on the pulsed ratio. This is not surprising, given that the chemisorbed fragments of the $Ca(thd)_2$ precursor exhibit smaller steric hindrance than the corresponding fragments of $La(thd)_3$ and $Mn(thd)_3$. In turn, it is seen that the chemisorbed fragments of the $La(thd)_3$ precursor occupies slightly smaller areas than those of the $Mn(thd)_3$ precursor. This is, on the other hand, somewhat surprising given the fact that the $La^{III}$ is markedly larger than $Mn^{III}$. The reason for the behavior may originate from geometrical orbital constraints of $Mn(thd)_2$ fragments, rendering these more flat than the corresponding $La(thd)_2$ fragments.

The model has also been applied to simulate the growth of films in the Sr–La–Fe–O system (Fig. 5). The relative growth rates of Sr : La : Fe were found to be 1 : 0.528 : 0.338 for the precursors $Sr(thd)_2$, $La(thd)_3$, $Fe(thd)_3$ and ozone. This again reveals the expected dependence based on the composition of the chosen precursors [$Sr(thd)_2$, $La(thd)_3$, and $Fe(thd)_3$], and reveals

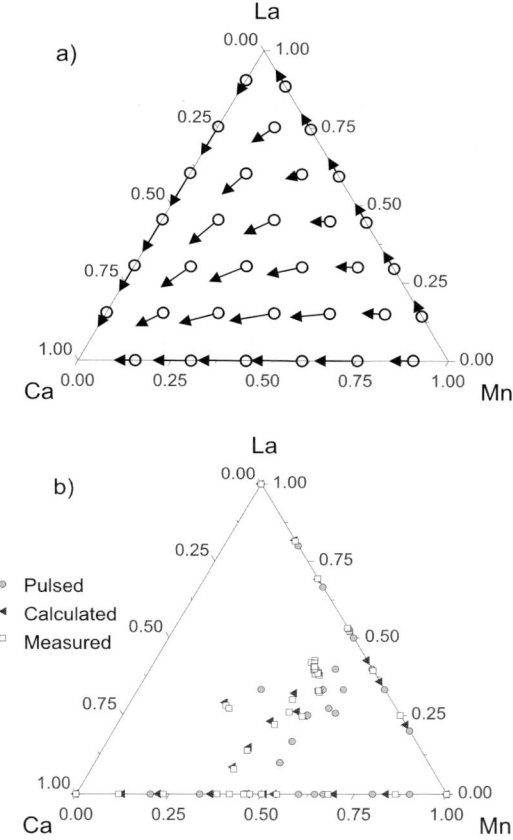

**Fig. 4.** (**a**) Transfer (indicated by *connecting lines*) of stoichiometry from the composition of the pulses to deposited films simulated for the Ca–La–Mn–O system. (**b**) Experimental data according to [11]

that $Fe(thd)_2$ fragments occupy much larger surface areas than $La(thd)_2$ fragments. The findings suggest that this simple model is applicable as guide in deposition experiments for multi-component systems.

## 3 Conclusions

The ALD technique is still not routinely used for deposition of materials with complex stoichiometry, but recent efforts have demonstrated its capabilities for such tasks. Previously foreseen obstacles for deposition of complex materials, such as limited ALD windows, have been demonstrated not to be a limiting factor in the described systems, and the deposited stoichiometry can be predicted by means of a few calibration runs.

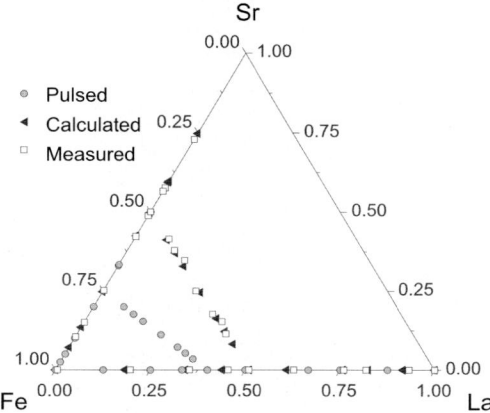

**Fig. 5.** (a) Transfer of stoichiometry from the composition of the pulses to the deposited films simulated for the Sr–La–Fe–O system. (b) Experimental data from [16]

# References

[1] M. Leskelä, M. Ritala: Atomic layer deposition (ALD): from precursors to thin film structures, Thin Solid Films **409**, 138 (2002)

[2] L. Niinistö, J. Päiväsaari, J. Niinistö, M. Putkonen, M. Nieminen: Advanced electronic and optoelectronic materials by atomic layer deposition: an overview with special emphasis on recent progress in processing of high-k dielectrics and other oxide materials, Phys. Stat. Solidi A **201**, 1443 (2004)

[3] M. Ritala, M. Leskelä: Atomic layer deposition, in H. S. Nalwa (Ed.): *Handbook of thin film materials*, vol. 1 (Academic Press, San Diego, CA 2002) p. 103

[4] T. Hatanpaeae, M. Vehkamaeki, I. Mutikainen, J. Kansikas, M. Ritala, M. Leskelä: Synthesis and characterisation of cyclopentadienyl complexes of barium: precursors for atomic layer deposition of $BaTiO_3$, Dalton Trans. **8**, 1181 (2004)

[5] M. Ritala, K. Kukli, M. Vehkamaki, T. Hanninen, T. Hatanpaa, P. I. Raisanen, M. Leskelä: Atomic layer deposition of high-k oxides, Proc. Electrochem. Soc. **2000-13**, 597 (2000)

[6] M. Vehkamaki, T. Hatanpää, T. Hänninen, M. Ritala, M. Leskelä: Growth of $SrTiO_3$ and $BaTiO_3$ thin films by atomic layer deposition, Electrochem. Solid-State Lett. **2**, 504 (1999)

[7] R. J. Potter, P. A. Marshall, J. L. Roberts, A. C. Jones, P. R. Chalker, M. Vehkamaeki, M. Ritala, M. Leskelä, P. A. Williams, H. O. Davies, N. L. Tobin, L. M. Smith: Liquid injection MOCVD and ALD studies of "single source" Sr–Nb and Sr–Ta precursors, Mater. Res. Soc. Symp. Proc. **784**, 97 (2004)

[8] M. Vehkamaki, M. Ritala, M. Leskelä, A. C. Jones, H. O. Davies, T. Sajavaara, E. Rauhala: Atomic layer deposition of strontium tantalate thin films from bimetallic precursors and water, J. Electrochem. Soc. **151**, F69 (2004)

[9] Y. J. Cho, Y.-S. Min, J.-H. Lee, B.-S. Seo, J. K. Lee, Y. S. Park, J.-H. Choi: Atomic layer deposition (ALD) of bismuth titanium oxide thin films using direct liquid injection (DLI) method, Integr. Ferroelectr. **59**, 1483 (2003)

[10] M. Schuisky, K. Kukli, M. Ritala, A. Harstå, M. Leskelä: Atomic layer CVD in the Bi–Ti–O system, Chem. Vap. Dep. **6**, 139 (2000)
[11] O. Nilsen, A. Kjekshus, H. Fjellvåg: to be published
[12] M. Ritala, K. Kukli, A. Rahtu, P. I. Raisanen, M. Leskelä, T. Sajavaara, J. Keinonen: Atomic layer deposition of oxide thin films with metal alkoxides as oxygen sources, Science **288**, 319 (2000)
[13] M. Nieminen, T. Sajavaara, E. Rauhala, M. Putkonen, L. Niinistö: Surface-controlled growth of $LaAlO_3$ thin films by atomic layer epitaxy, J. Mater. Chem. **11**, 2340 (2001)
[14] H. Seim, M. Nieminen, L. Niinistö, H. Fjellvåg, L.-S. Johansson: Growth of $LaCoO_3$ thin films from $\beta$-diketonate precursors, Appl. Surf. Sci. **112**, 243 (1997)
[15] M. Ritala, P. Kalsi, D. Riihelä, K. Kukli, M. Leskelä, J. Jokinen: Controlled growth of TaN, $Ta_3N_5$, and $TaO_xN_y$ thin films by atomic layer deposition, Chem. Mater. **11**, 1712 (1999)
[16] M. Lie, A. Kjekshus, H. Fjellvåg: Growth of $Pe_2O_3$ thin films by atomic layer deposition, Thin Solid Films **488**, 74 (2005)
[17] J.-S. Min, H.-S. Park, S.-W. Kang: Metalorganic atomic-layer deposition of titanium nitride silicide films, Appl. Phys. Lett. **75**, 1521 (1999)
[18] J.-S. Min, H.-S. Park, W. Koh, S.-W. Kang: Chemical vapor deposition of Ti–Si–N films with alternating source supply, Mater. Res. Soc. Symp. Proc. **564**, 207 (1999)
[19] M. Nieminen, S. Lehto, L. Niinistö: Atomic layer epitaxy growth of $LaGaO_3$ thin films, J. Mater. Chem. **11**, 3148 (2001)
[20] O. Nilsen, M. Peussa, H. Fjellvåg, L. Niinistö, A. Kjekshus: Thin film deposition of lanthanum manganite perovskite by the ALE process, J. Mater. Chem. **9**, 1781 (1999)
[21] K. Kukli, M. Peussa, L.-S. Johansson, E. Nykanen, L. Niinistö: Controlled growth of yttrium oxysulphide thin films by atomic layer deposition, Mater. Sci. Forum **315-317**, 216 (1999)
[22] H. Seim, H. Molsa, M. Nieminen, H. Fjellvåg, L. Niinistö: Deposition of $LaNiO_3$ thin films in an atomic layer epitaxy reactor, J. Mater. Chem. **7**, 449 (1997)
[23] B. W. Sanders, A. Kitai: Zinc oxysulfide thin films grown by atomic layer deposition, Chem. Mater. **4**, 1005 (1992)
[24] D.-S. Kil, J. M. Lee, J.-S. Roh: Low-temperature ALD growth of $SrTiO_3$ thin films from Sr $\beta$-diketonates and Ti alkoxide precursors using oxygen remote plasma as an oxidation source, Chem. Vap. Dep. **8**, 195 (2002)
[25] A. Kosola, M. Putkonen, L.-S. Johansson, L. Niinistö: Effect of annealing in processing of strontium titanate thin films by ALD, Appl. Surf. Sci. **211**, 102 (2003)
[26] O. S. Kwon, S. K. Kim, M. Cho, C. S. Hwang, J. Jeong: Chemically conformal ALD of $SrTiO_3$ thin films using conventional metallorganic precursors, J. Electrochem. Soc. **152**, C229 (2005)
[27] S. W. Lee, O. S. Kwon, C. S. Hwang: Chemically conformal deposition of $SrTiO_3$ thin films by atomic layer deposition using conventional metal organic precursors and remote-plasma activated $H_2O$, Microelectron. Eng. **80**, 158 (2005)

[28] M. Vehkamaki, T. Hanninen, M. Ritala, M. Leskelä, T. Sajavaara, E. Rauhala, J. Keinonen: Atomic layer deposition of $SrTiO_3$ thin films from a novel strontium precursor – strontium bis(tri-isopropylcyclopentadienyl), Chem. Vap. Dep. **7**, 75 (2001)

[29] K. Kukli, H. Heikkinen, E. Nykanen, L. Niinistö: Deposition of lanthanum sulfide thin films by atomic layer epitaxy, J. Alloys Comp. **275–277**, 10 (1998)

[30] M. Lie, A. Kjekshus, H. Fjellvåg: to be published

[31] N. Tsuboi, T. Isu, N. Kakuda, T. Terasako, S. Iida: Vapor-phase atomic layer epitaxy of $CuGaS_2$ at atmospheric pressure using metal chlorides and $H_2S$, Jpn. J. Appl. Phys. Part 2 **33**, L244 (1994)

[32] V. E. Drozd, A. A. Tulub, V. B. Aleskovski, Korol'kov: Synthesis of oxide superalloys by ML-ALE method, Appl. Surf. Sci. **82-83**, 587 (1994)

[33] M. Nanu, L. Reijnen, B. Meester, J. Schoonman, A. Goossens: $CuInS_2$ thin films deposited by ALD, Chem. Vap. Dep. **10**, 45 (2004)

[34] J. Ihanus, M. Ritala, M. Leskelä: ALE growth of $SrS_{1-x}Se_x$ thin films by substituting surface sulfur with elemental selenium, Proc. Electrochem. Soc. **25**, 1423 (1997)

[35] J. Ihanus, E. Lambers, P. H. Holloway, M. Ritala, M. Leskelä: XPS and electroluminescence studies on $SrS_{1-x}Se_x$ and $ZnS_{1-x}Se_x$ thin films deposited by atomic layer deposition technique, J. Cryst. Growth **260**, 440 (2004)

[36] C.-T. Hsu: Epitaxial growth of II–VI compound semiconductors by atomic layer epitaxy, Thin Solid Films **335**, 284 (1998)

[37] C.-T. Hsu: Variation with composition of the properties in $ZnS_xSe_{1-x}$, J. Cryst. Growth **193**, 33 (1998)

[38] C.-T. Hsu: Growth of $ZnS_xSe_{1-x}$ layers on Si substrates by atomic layer epitaxy, Mater. Chem. Phys. **58**, 6 (1999)

[39] J. Ihanus, M. Ritala, M. Leskelä, Rauhala: ALE growth of $ZnS_{1-x}Se_x$ thin films by substituting surface sulfur with elemental selenium, Appl. Surf. Sci. **112**, 154 (1997)

[40] H. Ikeda, Y. Miura, N. Takahashi, A. Koukitu, H. Seki: Substitution of surface-adsorbed As atoms to P atoms in atomic layer epitaxy, Appl. Surf. Sci. **82–83**, 257 (1994)

[41] T. Taki, T. Nakajima, A. Koukitu, H. Seki: Substitution reaction of surface adsorbed P atoms to As atoms in the GaP/GaAs atomic layer epitaxy, J. Cryst. Growth **183**, 75 (1998)

[42] Y.-S. Min, Y. J. Cho, C. S. Hwang: Amorphous high k dielectric $Bi_{1-x-y}Ti_xSi_yO_z$ thin films by ALD, Electrochem. Solid-State Lett. **7**, F85 (2004)

[43] Y.-S. Min, Y. J. Cho, I. P. Asanov, J. H. Han, W. D. Kim, C. S. Hwang: $Bi_{1-x-y}Ti_xSi_yO_z$ (BTSO) thin films for dynamic random access memory capacitor applications, Chem. Vap. Dep. **11**, 38 (2005)

[44] W. C. Shin, S. O. Ryu, I. K. You, S. M. Yoon, S. M. Cho, N. Y. Lee, K. D. Kim, B. G. Yu, W. J. Lee, K. J. Choi, S. G. Yoon: Low voltage switching characteristics of 60 nm thick $SrBi_2Ta_2O_9$ thin films deposited by plasma-enhanced ALD, Electrochem. Solid-State Lett. **7**, F31 (2004)

[45] R. J. Potter, P. R. Chalker, T. D. Manning, H. C. Aspinall, Y. F. Loo, A. C. Jones, L. M. Smith, G. W. Critchlow, M. Schumacher: Deposition of $HfO_2$, $Gd_2O_3$ and $PrO_x$ by liquid injection ALD techniques, Chem. Vap. Dep. **11**, 159 (2005)

[46] W. Prellier, P. Lecoeur, B. Mercey: Colossal magnetoresistive manganite thin films, J. Phys. Condens. Matter **13**, R915 (2001)

[47] J. Z. Sun, L. Krusin-Elbaum, A. Gupta, G. Xiao, P. R. Duncombe, S. S. P. Parkin: Magnetotransport in doped manganate perovskites, IBM J. Res. Dev. **42**, 89 (1998)

[48] J. M. D. Coey, M. Viret, S. Von Molnar: Mixed-valence manganites, Adv. Phys. **48**, 167 (1999)

[49] E. Dagotto, T. Hotta, A. Moreo: Colossal magnetoresistant materials: the key role of phase separation, Phys. Rep. **344**, 1 (2001)

[50] L. P. Gor'kov, V. Z. Kresin: Mixed-valence manganites: fundamentals and main properties, Phys. Rep. **400**, 149 (2004)

[51] A. M. Haghiri-Gosnet, J. P. Renard: CMR manganites: physics, thin films and devices, J. Phys. D Appl. Phys. **36**, R127 (2003)

[52] S. Jin, M. McCormack, T. H. Tiefel, R. Ramesh: Colossal magnetoresistance in La–Ca–Mn–O ferromagnetic thin films, J. Appl. Phys. **76**, 6929 (1994)

[53] E. L. Nagaev: Colossal-magnetoresistance materials: manganites and conventional ferromagnetic semiconductors, Phys. Rep. **346**, 387 (2001)

[54] L. Sheng, D. Y. Xing, D. N. Sheng, C. S. Ting: Theory of colossal magnetoresistance in $R_{1-x}A_xMnO_3$, Phys. Rev. Lett. **79**, 1710 (1997)

[55] R. von Helmolt, J. Wecker, K. Samwer, K. Baerner: Transport properties of manganates with giant magnetoresistance, J. Magn. Magn. Mater. **151**, 411 (1995)

[56] O. Nilsen, A. Kjekshus, H. Fjellvåg: to be published

[57] O. Nilsen, H. Fjellvåg, A. Kjekshus: Growth of manganese oxide thin films by atomic layer deposition, Thin Solid Films **444**, 44 (2003)

[58] O. Nilsen, H. Fjellvåg, A. Kjekshus: Growth of calcium carbonate by the atomic layer chemical vapour deposition technique, Thin Solid Films **450**, 240 (2004)

[59] M. Ylilammi: Monolayer thickness in atomic layer deposition, Thin Solid Films **279**, 124 (1996)

# Index

# Molecular Beam Epitaxy of Rare-Earth Oxides

H. Jörg Osten[1], Eberhard Bugiel[1], Malte Czernohorsky[1], Zeyard Elassar[2], Olaf Kirfel[2], and Andreas Fissel[2]

[1] Institute for Electronic Materials and Devices, University of Hannover, Appelstr. 11A, D-30167 Hannover, Germany
`osten@mbe.uni-hannover.de`
[2] Information Technology Laboratory, Schneiderberg 32, D-30167 Hannover, Germany

**Abstract.** We present results for crystalline lanthanide oxides on silicon with the $Ln_2O_3$ composition (Ln = Pr, Nd and Gd) in the cubic *bixbyite* structure grown by solid state molecular beam epitaxy (MBE). On Si(001)-oriented surfaces, crystalline $Ln_2O_3$ grows as (110)-oriented domains, with two orthogonal in-plane orientations. We obtain perfect epitaxial growth of cubic $Nd_2O_3$ on Si($\bar{1}11$) substrates. These layers can be overgrown epitaxially with silicon. The successfully demonstrated heteroepitaxy of such $Si/Ln_2O_3/Si(111)$ stacks opens the door to a wide range of novel tunneling devices. For all investigated lanthanide oxides grown under ultra-high vacuum conditions, we observed the formation of crystalline interfacial silicide inclusions. MBE in combination with real-time reflection high-energy electron diffraction and in vacuo X-ray photoelectron spectroscopy were used to gain a detailed understanding of the interface and film formation during epitaxial growth of $Nd_2O_3$ on silicon. Based on that understanding, the whole growth procedure had to be adapted accordingly. In particular, the partial oxygen pressure during the interface formation and during growth is a very critical parameter. Layers grown by an appropriately by modified MBE process display no silicide inclusions, and also no interfacial silicon oxide layers.

## 1 Introduction

Many materials systems are currently under consideration as potential replacements for $SiO_2$ as the gate dielectric material for sub-$0.1\,\mu m$ CMOS technology. A systematic consideration of the required properties of gate dielectrics indicates that the key guidelines for selecting an alternative gate dielectric are 1. permittivity, band gap, and band alignment to silicon, 2. thermodynamic stability, 3. film morphology, 4. interface quality, 5. compatibility with the current or expected materials to be used in processing for CMOS devices, 6. process compatibility, and 7. reliability [1]. Many dielectrics appear favorable in some of these fields, but very few materials are promising with respect to all of these guidelines [2]. The most promising of these are the simple binary metal oxides. Unfortunately, a number of these materials are not thermally stable on silicon [3]. The formation of $SiO_2$ and/or metal silicate interfacial layers often occurs when these materials are deposited upon silicon. Further growth of silicon dioxide or a silicate at the interface takes place

M. Fanciulli, G. Scarel (Eds.): Rare Earth Oxide Thin Films,
Topics Appl. Physics **106**, 101–114 (2007)
© Springer-Verlag Berlin Heidelberg 2007

during subsequent annealing steps. It is important to note, however, that the occurrence of an interfacial layer of $SiO_2$ or another low-permittivity material will limit the highest possible gate stack capacitance, or equivalently, the lowest achievable *Equivalent Oxide Thickness (EOT)* value.

The common approach for alternative high-$\kappa$ gate dielectrics has involved amorphous metal oxides. In general, attempts have been made to keep these materials amorphous, in particular, after post-deposition high-temperature processing in order to avoid increased surface roughness and additional leakage due to the formation of grain boundaries [2]. Generally, all known high-$\kappa$ materials are more ionic than $SiO_2$. There are some general trends related to the ionicity of an oxide: the crystallization temperature decreases and the dielectric constant increases with ionic character. Due to the relatively low re-crystallization temperature of highly ionic materials (like $ZrO_2$ or $HfO_2$), these materials are often not compatible with a CMOS process. It is possible to reduce the ionicity (and thereby increase the crystallization temperature) by alloying the metal oxides with Al or Si. These aluminates or silicates are thermally more stable on silicon but have lower dielectric constants. Consequently, for a given *EOT* the physical layer thickness has to be reduced, leading to an increase in leakage current.

Further, the dielectric/Si interface properties influence the device performance significantly. A good interface requires either that the oxide is amorphous, or that it is epitaxial and lattice-matched to the underlying silicon. Amorphous dielectrics are expected to be able to adjust the local bonding to minimize the number of Si dangling bonds at the interface.

The alternative is to use an epitaxial oxide. This involves more effort, but it has the advantage of enabling defined interfaces engineering. Molecular Beam Epitaxy (MBE), known for its superior capability in atomic-level engineering and interface control, has been used in the epitaxial growth of various high-$\kappa$ materials. For a better understanding of the interface and layer formation processes, well-defined experimental studies are needed which are performed under ultra-clean ultra-high vacuum conditions. MBE in combination with real-time reflection high-energy electron diffraction (RHEED), is here the most advantageous method because of its ability to monitor the formation of structures under ultra-high vacuum in situ within a wide range of different conditions. When combined with methods analyzing the bonding properties, like X-ray photoelectron spectroscopy (XPS), this approach results in a rather detailed insight into the interface formation and properties.

Generally, there are two groups of possible candidates for epitaxial growth on Si, namely 1. perovskite-type structures and 2. binary metal oxides [4]. Various perovskite-type metal oxides ($ABO_3$), in particular, have become a technologically important class of materials due to their unique dielectric, piezoelectric, ferroelectric, ferromagnetic, optical, electro-optic, and catalytic properties. These perovskite oxides cover a wide range of materials from insulators to conductors and superconductors, and everything in between.

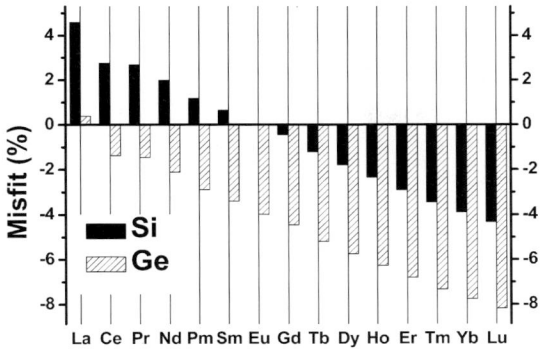

**Fig. 1.** Lattice mismatch of different lanthanide oxides relative to silicon and germanium (lattice constants from [5,6])

It is extremely important and desirable to integrate these highly functional metal oxides into mature semiconductor technologies.

## 2 Epitaxial Metal Oxides on Silicon

Here, we will focus on the epitaxial growth of binary metal oxides. Epitaxial growth on a clean surface requires matching in symmetry as well as in atomic spacing. Lanthanide oxides (LnOs) form the most interesting group for epitaxial growth on silicon. The LnOs can have different oxygen compositions $LnO_x$, with $x$ ranging from 1 to 2 due to the multiple oxidation states (+2, +3, and +4) of the rare-earth metals. This leads to oxides with different stoichiometries (LnO, $Ln_2O_3$, $LnO_2$). All known Ln(II) oxides, like EuO, are not insulating [5,6]. Therefore, they will not be considered here any further.

Different oxygen content can lead to different structural phases including two cubic phases, such as the calcium fluoride ($CaF_2$) structure for the Ln(IV) only, and the manganese oxide ($Mn_2O_3$) or *bixbyite* structure for Ln(III). The *bixbyite* structure is based on the calcium fluorite structure, where 1/4 of the oxygen atoms have been removed from specific lattice sites. Some LnOs also crystallize in the hexagonal lanthanum oxide structure, which is suitable for epitaxy only on Si(111). Also, monoclinic phases are known for various lanthanide oxides [5,6]. Both the calcium fluoride and the *bixbyite* structure have lattice symmetry suitable for epitaxial growth on Si(100) and Si(111). Commonly, the oxides are evaluated on the basis of lattice matching, with the misfit defined as the relative difference in the lattice constants ($a_{film}$ and $a_{Si}$). Figure 1 summarizes the lattice mismatch for all cubic lanthanide(III) oxides relative to silicon or germanium substrates.

Taken alone, this is misleading, because the lattice is made up of metal atoms occupying the positions of a face-centered cubic (*fcc*) lattice with a lattice constant $a_{film}$, where the tetrahedral holes are occupied by oxygen atoms.

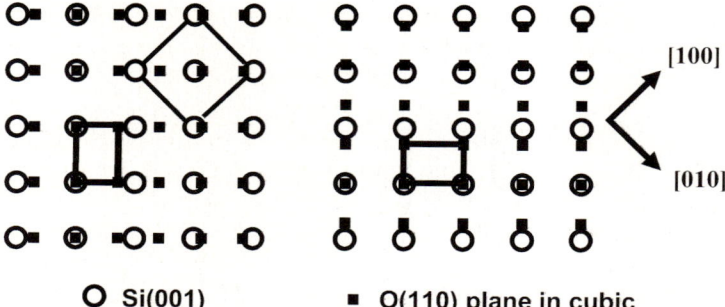

O  Si(001)          ■  O(110) plane in cubic

**Fig. 2.** $Ln_2O_3(110)//Si(001)$ alignment (schematically): 1 : 1 matching occurs along one direction. In the other direction, there is roughly a 3 : 2 matching relation. Two types of (110)-oriented domains are possible, with two orthogonal in-plane orientations

Therefore, the lattice mismatch would be identical for all three epitaxial relationships, i.e., (100)//(100), (110)//(110), and (111)//(111). All other combinations violate symmetry matching. However, the growth of LnOs is based on the deposition of metal oxide molecules. Due to the existence of highly ionic Ln–O bonds in combination with the high bonding strength of the covalent Si–O bonds, we can assume that the interface is predominantly formed by Si–O–Ln bonds. Therefore, the matching of the oxygen atoms is the important parameter.

The complete crystallographic calcium fluorite structure can also be described by two non-identical metal and oxygen lattices. The arrangement of the oxygen atoms forms a simple cubic lattice with a lattice constant of $a_{film}/2$. Thus, the interesting matching condition (Fig. 2) for epitaxial growth is the $LnO_2(110)[-110]//Si(001)[010]$ relation. In that case, nearly 1 : 1 matching should occur along one direction. In the other direction, there would be roughly a 3 : 2 matching relation. Layers grown in this orientation can exist with two orthogonal in-plane orientations (found experimentally for a large variety of binary metal oxides on Si(001)). Due to the 45° rotation of the (110) plane relative to the substrate, the mismatch for the 1 : 1 matching is identical to that shown in Fig. 2. For the calcium fluorite structure $(LnO_2)$, however, the 3 : 2 relation leads to unsaturated oxygen bonds that would create additional interface charges. Therefore, $Ln_2O_3(110)//Si(001)$ heterostructures (with the metal oxide having the *bixbyite* structure) are favorable due to the lower oxygen content in that structure.

For application in a device fabrication process, all lanthanide oxides exhibiting more than one valence state (+3 and +4) are not the best choice as epitaxial high-$\kappa$ materials because of the coexistence of phases with different oxygen content. For example, cerium(IV) oxide $(CeO_2)$ can release oxygen under reduction conditions forming a series of reduced oxides with stoichiometric cerium(III) oxide $(Ce_2O_3)$ as an end product, which in its turn easily takes up oxygen under oxidizing conditions, turning the cerium(III) oxide back into $CeO_2$. This reversible transition has been studied in a number of theoretical works. In [7], the authors show how the oxygen-vacancy formation process is facilitated in a most essential way by a simultaneous capture of two electrons into the localized $f$-level traps on two cerium atoms, which therefore change their valence from +4 to +3. In addition, stable mixed valence-state structures can occur for some LnOs. For example, $Pr_6O_{11}$ is the most stable phase for praseodymium oxide [5, 6].

All lanthanide oxides displaying only one valence state are easier to handle due to the absence of transitions between phases with different oxygen content. Based on that argument, we will focus our discussion mainly on lanthanide (III) oxides (occurring as $Ln_2O_3$).

# 3 Results for the Growth of Different Rare-Earth Oxides on Silicon

## 3.1 Experimental

In the following, we will compare results for the MBE growth of three LnOs, namely $Pr_2O_3$, $Nd_2O_3$, and $Gd_2O_3$. All experiments were performed in a multi-chamber ultra-high vacuum system (*DCA Instruments*) capable of handling 8 in wafers. This system includes a growth, an annealing, and an analysis chamber connected by an ultra-high vacuum (UHV) transfer system. The layers were grown on Si(001) and Si(111) substrates. Substrates were cleaned ex situ using as the last step diluted HF etch (HF : $H_2O$ = 1 : 10) followed by a dilution rinse, and then were immediately inserted into the vacuum system. Substrates were annealed in situ to transform the initial hydrogen-terminated $(1 \times 1)$ surface structure into the $(2 \times 1)$ superstructure, indicating a clean and well-ordered surface. Commercially available, granular $Ln_2O_3$ material was evaporated using an electron-beam evaporator. Growth temperatures were in the range 800 K to 1000 K. Typical growth rates were 0.05 nm/s to 0.01 nm/s. The surface and layer structure was evaluated during the growth by RHEED.

After growth, the wafers were transferred into the XPS analysis chamber without leaving the UHV. Non-monochromatized Al $K\alpha$ radiation ($h\nu$ = 1486.6 eV) was used for the excitation of photoelectrons. All measured wafers were electrically grounded to eliminate charging effects during long-time measurements. To minimize experimental uncertainties associated with energy variations caused by spectrometer instabilities and to improve the signal to noise ratio, the XPS data were collected by repeatedly scanning the silicon, the lanthanide, and the oxygen level. A multipeak Gaussian deconvolution procedure was used to extract the exact line position and intensities.

The layer thickness was measured ex vacuo by X-ray reflectivity (XRR) using a standard single-crystal diffractometer with graphite monochroma-

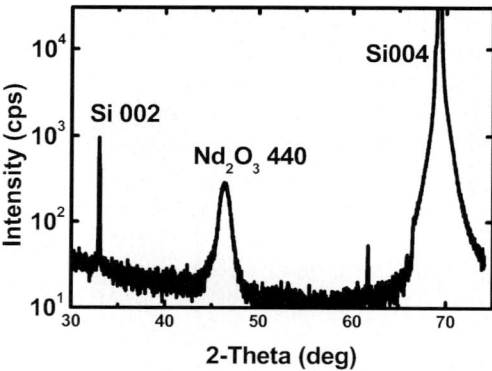

**Fig. 3.** Typical $\theta/2\theta$ X-ray scan of a 13.0 nm $Nd_2O_3$ film, grown on a Si(001) substrate. Besides the Si peaks, there is only one peak in a wide range scan

tor in front of the detector. Layers were also characterized by X-ray diffraction (XRD) ($\theta/2\theta$, $\omega$ and $\Phi$-scans) and transmission electron microscopy (TEM), i.e., high-resolution cross-section and plan-view images combined with selected area diffraction (SAD).

## 3.2 Epitaxial Growth on Si(100)

Recently, we reported results using MBE growth of crystalline praseodymium oxide (as $Pr_2O_3$ in the *bixbyite* structure) on Si(001) substrates [8, 9]. As predicted, the $Pr_2O_3$ was found to grow as (110)-single-crystalline domains, with two orthogonal in-plane orientations [8]. Also for the other two LnOs, we found the same crystalline orientation and domain formation. Figure 3 shows a typical $\theta/2\theta$ scan of a 13.0 nm thick $Nd_2O_3$ layer grown on a Si(001) substrate. Besides the Si peak, there is only one peak in a wide range scan. The intensity of this diffraction peak increases, and the peak width decreases, with increasing layer thickness, indicating that the peak originates from the overlayer. The layer spacing, determined from the peak position, agrees well with the (440) diffraction of $Nd_2O_3$ in *bixbyite* structure. The crystalline orthogonal domain structure of the layers is clearly visible in the plan-view TEM micrograph of $Nd_2O_3$ on Si(001) (Fig. 4). It is further confirmed by electron diffraction pattern on a plan-view sample of Fig. 4. All spots can be interpreted by a superposition of the Si(001) pattern, two $Nd_2O_3$(110) pattern rotated azimuthally by 90°, and multiple diffractions [10]. RHEED observations during growth indicate smooth layer-by-layer growth only to a limited thickness ($\approx$ 10 nm) for all three materials. For thicker layers, we observe a transition to three-dimensional growth.

Different investigations indicate that the occurrence of the two orthogonal domains is due to the nucleation on neighboring terraces with Si dimer rows ($2 \times 1$ reconstruction) perpendicular to each other [11]. Thus, the ori-

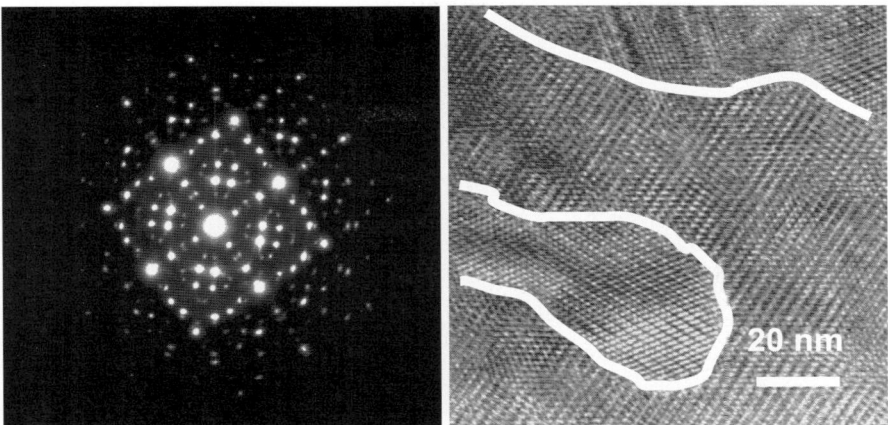

**Fig. 4.** Electron diffraction (*left*) and plan-view TEM micrograph (*right*) of an 11 nm thick $Nd_2O_3$ layer grown on Si(001). The domain boundaries are marked

entation of the domains correlates with monoatomic surface steps. Domains with different orientations can nearly be prevented with the use of vicinal substrates of $4°$ miscut. Such a miscut surface exposes surface steps of double atomic layers, thus giving Si terraces of $\sim 8$ nm spacing with the same dimer orientation. Single-domain $Gd_2O_3$ layers have been realized on vicinal substrates [12].

### 3.3 Epitaxial Growth on Si(111)

On Si(111), we obtained perfect epitaxial growth of hexagonal $Pr_2O_3$ [13]. We also found that a phase transition takes place during the annealing of the as-grown layers in $N_2$ below growth temperature. Annealed layers display a cubic structure isomorphic to manganese oxide, (111)-oriented but $180°$ rotated (orientation "B") about the Si(111) surface normal ("A"). For $Gd_2O_3$, no hexagonal structure is known. $Nd_2O_3$ and $Gd_2O_3$ already grow in the cubic structure. In addition, the $Ln_2O_3$ layer grown on Si(111) can be epitaxially overgrown with silicon. We found the typical "ABA" orientation for an epitaxial $Si/Nd_2O_3/Si(111)$ stack (Fig. 5).

The successfully demonstrated heteroepitaxy of such stacks opens the door to a wide range of novel device technologies with enhanced functionality and flexibility, like novel tunneling devices.

### 3.4 Interface Stability

Recently, we performed the following experiment to investigate the influence of air contact on $Pr_2O_3$ layers grown on Si(001).[13] Directly after the layer growth (without leaving the ultra-high vacuum), part of the wafer was sealed

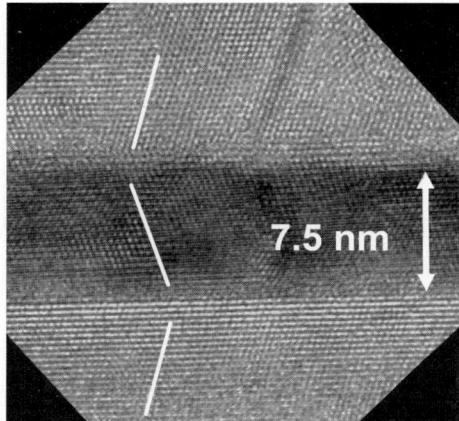

**Fig. 5.** Cross-sectional TEM image of a 7.5 nm thick $N_2O_3$ layer, overgrown epitaxially with silicon. The typical "ABA" orientation is marked

with 100 nm poly-Si using a special shadow mask technique. The wafer was then stored in a container connected to air at room temperature. One week after growth, we took two high-resolution TEM images of the same layer with and without the capping poly-Si. For the sample without the silicon cap, a bright amorphous layer was clearly visible between the substrate and the oxide layer. From the TEM contrast, we can conclude that mainly lighter elements are present. Capping the metal oxide layer with poly-Si completely suppresses the occurrence of this bright amorphous layer. It is well known that praseodymium oxide can be reduced and re-oxidized due to its different valence states in the oxides. Oxygen diffusion is fast in ionic oxide, even at room temperature. We postulated the formation of an interfacial silicon oxide layer due to oxygen supplied from the air and diffused through the praseodymium oxide layer. Now, we performed similar investigations as well as various annealing experiments with the single-valence lanthanide oxides $Nd_2O_3$ and $Gd_2O_2$. We found that the tendency to form interfacial silicon oxide layers due to air contact is significantly less pronounced for these oxides.

$Nd_2O_3$ layers grown under ultra-high vacuum conditions on Si(001) were investigated with TEM (Fig. 6). In the cross-sectional micrograph, we clearly find crystalline inclusions with the interface extending some nanometers into the Si substrates. Moreover, voids were found in such layers. In addition, the plan-view micrograph shows typical Moire fringes originating from these inclusions. Based on a detailed evaluation of the TEM results, we could identify the crystallographic structure of these inclusions to be identical to the known structure of tetragonal $NdSi_2$. Such silicide inclusions appear even at the lowest possible growth temperature (550 °C) which is needed for good epitaxial growth. After annealing in an oxygen-containing atmosphere, the

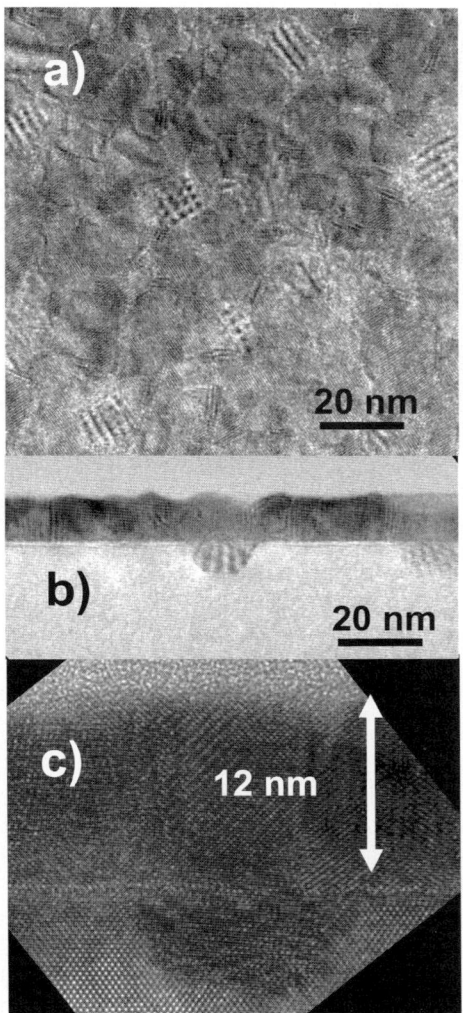

**Fig. 6.** TEM investigation of a 12 nm thick $Nd_2O_3$ layer grown on Si(001). (a) plane-view micrograph, showing pronounced Moire fringes; (b) and (c) cross-sectional micrographs with different magnification showing the formation of crystalline interfacial inclusions

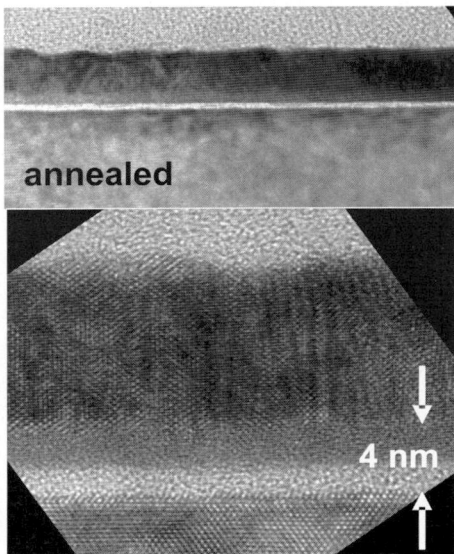

**Fig. 7.** Cross-sectional TEM micrograph of the same sample shown in Fig. 5, but after annealing in nitrogen ($2 \times 10^{-2}$ mbar, 600 °C/10 min). The formation of an amorphous interfacial layer is visible

silicide inclusions disappear, however, and an amorphous interfacial layer and regions within the layer appear (Fig. 7).

We found a similar formation of interfacial silicide inclusions also for the other investigated lanthanide oxides. These observations are similar to recent results for growth and annealing experiments of $HfO_2$ and $Al_2O_3$ layers on Si grown under UVH or oxygen-poor conditions [14–17].

We applied RHEED and XPS to investigate the $Ln_2O_3/Si(001)$ interface and layer formation during epitaxial growth. The growth process was interrupted several times to transfer the wafer into an XPS analysis chamber without leaving the UHV environment. Using such an approach, we investigated different $Ln_2O_3$ layers grown on Si(001) [18]. Here, we will discuss the typical behavior for $Nd_2O_3$; the main results hold in principle also for the other $Ln_2O_3$ layers. Figure 8 summarizes the XPS results for the Si-2p core-level spectrum for layers with different thicknesses. For layers with a thickness below 1 nm, we found that Si exhibits mainly a peak corresponding to elemental Si ($Si^0$ state). However, the spectra show significant broadening to the low-energy side (in particular visible in the second spectrum), indicating the appearance of a $Si^{1-}$ bonding state due to silicide-like bonding formation. Spectra for layers with thicknesses above 1 nm exhibit an additional peak at higher binding energies, but still lower than expected for $SiO_2$ formation. We attribute this peak to a silicate-like bonding [18].

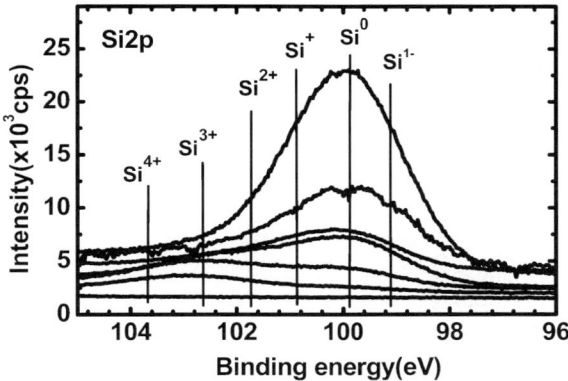

**Fig. 8.** Si-2p core-level XPS spectra for $Nd_2O_3$ layers of various thicknesses. The layer thickness increases stepwise from $\approx 0.5$ nm (*upper spectra*) to 3 nm (layer #6). The lowest spectrum was taken for a 6 nm thick layer. *Straight lines* indicate the different Si bonding states

From the deconvolution of the individual spectra combined with the appropriate O 1s and Nd $3d_{5/2}$ spectra, we extracted the following behavior during interface formation: in the initial stage of growth, silicon is mainly bound to neodymium (silicide-like bonding). During further growth, the number of silicate-like bonds (Si–O–Nd) increases initially, visible also by the shift of the O 1s signals to higher energies, before the peaks originating from Nd–O–Nd bindings in $Nd_2O_3$ become dominating. There was no indication of $SiO_2$ formation. For the second layer, we observe a shift of the Nd $3d_{5/2}$ to lower binding energies, as would be expected if the bonding becomes metal-like. That behavior together with the observed shifts in the Si and O spectra can be explained only by an exchange of oxygen bound to silicon by neodymium or by a movement of oxygen atoms from a Si–O–Nd interface site into the layer during the initial stage of growth. For thicker layers, we observe the appearance of a higher-energy (oxygen-rich environment) and a lower-energy (less oxygen) Si bonding state in a silicate-like bonding configuration. Thereby, the number of states in oxygen-rich environment increases with layer thickness. Such different bonding configurations can appear in different kinds of silicates. $Ln_2SiO_5$, for example, contains one oxygen atom in a Ln–O–Ln configuration, whereas $Ln_2Si_2O_7$ contains one oxygen atom in a Si–O–Si configuration, giving rise to different ratios in Si bonding states and, therefore, also to a shift in the measured 2p bonding energy [19, 20].

For the epitaxial growth of $Ln_2O_3$ oxides, a well-controlled interface and subsequent growth engineering are necessary to reach the required electrical properties. The appearance of silicide inclusions seems to be a serious drawback for future high-$\kappa$ applications. However, the formation and stability of the silicide-like phase depends on the oxygen chemical potential [21]. Considering the low oxygen partial pressure under UHV conditions (as used in conventional MBE growth), the chemical potential of oxygen can become negative. Thus, silicide formation will be favored over oxide formation. That is one of the most crucial points for the growth of dielectric layers, because the silicide growth can continue as long as the oxygen content remains low enough or the oxygen chemical potential remains strongly negative. This occurs even faster in a surface or interface region where the energetical equilibrium is out of order, for example, due to stress.

In conclusion, the partial oxygen pressure during the interface formation and during growth is a very critical parameter. Again, the single-valence LnO are favorable due to the absence of additional reduction/re-oxidation processes as observed, for example, for $Pr_2O_3$. Too low oxygen content will lead to the observed formation of silicide inclusions. By contrast, too high oxygen content might oxidize the Si surface, leading to a low-$\kappa$ interfacial $SiO_x$ layer. The available oxygen concentration can easily be controlled by using modified MBE growth processes. We found that MBE growth under defined oxygen partial pressures of $1 \times 10^{-7}$ mbar during the interface formation and/or during the subsequent growth can prevent any kind of silicide inclusions and void formation. On the other hand, it was still low enough to avoid the formation of interfacial $SiO_x$. This makes MBE-grown epitaxial $Ln_2O_3$ layers excellent candidates for application as very thin high-$\kappa$ materials.

# 4 Outlook

The material requirements for the alternative gate dielectric are very challenging to achieve performance comparable to $SiO_2$. A new approach is based on the development of epitaxial metal oxides grown directly on silicon surfaces. Here, we presented results for MBE-grown lanthanide oxides with the $Ln_2O_3$ composition in the cubic *bixbyite* structure. We showed experimental results for the growth of crystalline neodymium oxide on Si substrates. On Si(001)-oriented surfaces, crystalline $Nd_2O_3$ grows as (110)-oriented domains, with two orthogonal in-plane orientations. We obtain perfect epitaxial growth of cubic $Nd_2O_3$ on Si(111) substrates. These layers can be overgrown epitaxially with silicon.

MBE is known for its superior capability in atomic-level engineering and interface control, and is one of the techniques being investigated for the epitaxial growth of various high-$\kappa$ materials. The development of suitable epitaxial high-$\kappa$ materials requires a detailed understanding of all processes during interface formation and subsequent growth. Based on that understanding, the whole growth procedure has to be adapted accordingly. Layers grown by an appropriately modified MBE process display very promising electrical properties [22]. This makes epitaxial $Ln_2O_3$ layers excellent candidates for application as very thin high-$\kappa$ materials.

**Acknowledgements**

This work was partly founded by the German Federal Ministry of Education and Research (BMBF) under the KrisMOS project (01M3142D).

# References

[1] M. L. Green, E. P. Gusev, R. Degraeve, E. L. Garfunkel: Ultrathin ($< 4\,$nm) $SiO_2$ and Si–O–N gate dielectric layers for silicon microelectronics: understanding the processing, structure, and physical and electrical limits, J. Appl. Phys. **90**, 2057 (2001)

[2] G. D. Wilk, R. M. Wallace, J. M. Anthony: High-k gate dielectrics: current status and material properties consideration, J. Appl. Phys. **89**, 5243 (2001)

[3] K. J. Hubbard, D. G. Schlom: Thermodynamic stability of binary oxides in contact with silicon, J. Mater. Res. **11**, 2757 (1996)

[4] D. P. Norton: Synthesis and properties of epitaxial electronic oxide thin-film materials, Mat. Sci. Eng. **R 43**, 139 (2004)

[5] G. V. Samsonov (Ed.): *The Oxide Handbook*, 2nd ed. (IFI/Plenum, New York 1982)

[6] G. Y. Adachi, N. Imanaka: The binary rare earth oxides, Chem. Rev. **98**, 1479 (1998)

[7] N. V. Skorodumova, S. I. Simak, B. I. Lundqvist, I. A. Abrikosov, B. Johansson: Quantum origin of the oxygen storage capability of ceria, Phys. Rev. Lett. **89**, 166601 (2002)

[8] H. J. Osten, J. P. Liu, E. Bugiel, H. J. Muessig, P. Zaumseil: Growth of crystalline praseodymium oxide on silicon, J. Cryst. Growth **235**, 229 (2002)

[9] A. Fissel, H. J. Osten, E. Bugiel: Towards understanding epitaxial growth of alternative high-k dielectrics on Si(001): application to praseodymium oxide, J. Vac. Sci. Technol. **B 21**, 1765 (2003)

[10] E. Bugiel, J. P. Liu, H. J. Osten: TEM investigation of epitaxial praseodymium oxide on silicon, Inst. Phys. Conf. Ser. **169**, 411 (2001)

[11] H. J. Osten, J. P. Liu, E. Bugiel, H. J. Muessig, P. Zaumseil: Epitaxial growth of praseodymium oxide on silicon, Mat. Sci. Eng. **B 87**, 297 (2001)

[12] J. Kwo, M. Hong, A. R. Kortan, K. L. Queeny, Y. J. Chabal, R. L. Opila, D. A. Mueller, S. N. G. Chu: Properties of high-k gate dielectrics $Gd_2O_3$ and $Y_2O_3$ for Si, J. Appl. Phys. **89**, 3920 (2001)

[13] J. P. Liu, P. Zaumseil, E. Bugiel, H. J. Osten: Epitaxial growth of $Pr_2O_3$ on Si(111) and the observation of a hexagonal to cubic phase transition during postgrowth $N_2$ annealing, Appl. Phys. Lett. **79**, 671 (2001)

[14] N. Miyata, T. Nabatame, T. Horikawa, M. Ichikawa, A. Toriumi: Void nucleation in thin $HfO_2$ layer on Si, Appl. Phys. Lett. **82**, 3880 (2003)

[15] P. F. Lee, J. Y. Dai, H. L. Chan, C. L. Choy: Study of Hf–Al–O high-k gate dielectric thin films grown on Si, Integrated Ferroelectrics **59**, 1213 (2003)

[16] P. F. Lee, J. Y. Dai, H. L. Chan, C. L. Choy: Study of interfacial reaction and its impact on electric properties of Hf–Al–O high-k gate dielectric thin films grown on Si, Appl. Phys. Lett. **82**, 2419 (2003)

[17] N. Miyata, M. Ichikawa, T. Nabatame, T. Horikawa, A. Toriumi: Thermal stability of a thin $HfO_2$/ultrathin $SiO_2$/Si structure: interfacial Si oxidation and silicidation, Jpn. J. Appl. Phys. **42**, L138 (2003)
[18] A. Fissel, J. Dabrowski, H. J. Osten: Photoemission and *ab initio* theoretical study of interface and film formation during epitaxial growth and annealing of praseodymium oxide on Si(001), J. Appl. Phys. **91**, 8986 (2002)
[19] J. J. Chambers, G. N. Parsons: Physical and electrical characterization of ultrathin yttrium silicate insulators on silicon, J. Appl. Phys. **90**, 918 (2001) see also [20]
[20] Landolt-Börnstein: *Group III Condensed Matter*, 7 **D1A** (Springer, Berlin, Heidelberg 1985)
[21] D. Schmeisser, J. Dabrowski, H.-J. Muessig: $Pr_2O_3$/Si(001) interface reactions and stability, Mater. Sci. Eng. **B 109**, 30 (2004)
[22] H. J. Osten, M. Czernohorsky, O. Kirfel, A. Fissel: unpublished

# Index

# Fabrication and Characterization of Rare Earth Scandate Thin Films Prepared by Pulsed Laser Deposition

Jürgen Schubert, Tassilo Heeg, and Martin Wagner

Institute of Thin Films and Interfaces, ISG1-IT and Center of Nanoelectronic Systems for Information Technology, CNI, Research Center Jülich, D-52425 Jülich, Germany
j.schubert@fz-juelich.de

**Abstract.** The continuous structure size reduction in semiconductor technology is leading to a considerable attention for advanced high-$\kappa$ dielectrics. The rare earth scandates ($REScO_3$, where $RE$ is a rare earth element) were recently proposed as candidate materials for the replacement of $SiO_2$ in silicon MOSFETs in either amorphous or epitaxial form. Epitaxial rare earth scandate thin films have been prepared by pulsed laser deposition technique (PLD) on different substrates. Film stoichiometry and quality were investigated by means of Rutherford backscattering spectrometry (RBS), RBS channelling, high-resolution transmission electron microscopy (HRTEM) and X-ray diffraction (XRD). Electrical measurements on MIM-structures show a high dielectric constant of around 20 and low leakage currents.

## 1 Introduction

Rare earth-based oxides are an important candidate in the discussion to replace $SiO_2$ as the gate dielectric in highly scaled MOSFETs. The necessity of the replacement can be seen in the phrase "The early availability of manufacturing-worthy high-$\kappa$ dielectrics is one of the difficult challenges in process integration for highly scaled MOSFETs, and it is necessary to meet stringent gate leakage and performance requirements" [1] taken from the 2004 update of the International Technology Roadmap for Semiconductors. This reveals the urgency to find an appropriate high-$\kappa$ dielectric for the CMOS technology in order to keep up with Moore's Law in the near future.

Many candidates for high-$\kappa$ dielectrics have been investigated during the past years, like $Si_3N_4$, $Al_2O_3$, $ZrO_2$, $LaAlO_3$, $SrZrO_3$, or $HfO_2$ [2], but none of them seems to meet all the requirements (see Fig. 1). These are a sufficiently high permittivity, large bandgap and band offsets, thermodynamic stability in contact with silicon up to $1000\,^{\circ}C$, a good film morphology, a high Si/dielectric interface quality, compatibility with other materials used in the CMOS process and with the process itself, and the long-term reliability of the films [3]. The most frequently investigated material, $HfO_2$, which possibly will be introduced into the production of microelectronic devices, for example, has a low recrystallization temperature of around $700\,^{\circ}C$, which limits

M. Fanciulli, G. Scarel (Eds.): Rare Earth Oxide Thin Films,
Topics Appl. Physics **106**, 115–126 (2007)
© Springer-Verlag Berlin Heidelberg 2007

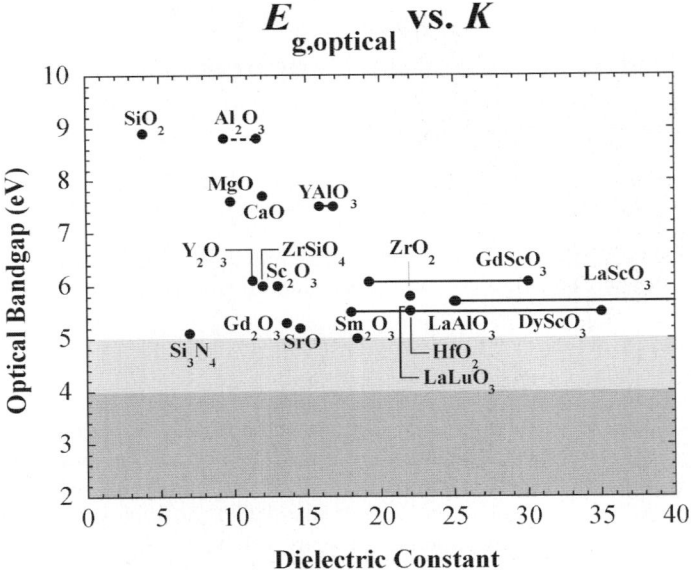

**Fig. 1.** Optical bandgap versus dielectric constant of several high-$\kappa$ dielectrics [2]. Three different scandate materials are added

the thermal budget in a CMOS process. Some other new high-$\kappa$ candidates are also seen in Fig. 1.

The $RE\mathrm{ScO_3}$ materials ($RE$ = Dy, Gd and La) possess a large optical bandgap of 5.5 to 6 eV and an anisotropic dielectric tensor according to their crystal orientation. For example, for $\mathrm{GdScO_3}$ the $\kappa$ in a- and b-direction of the orthorhombic unit cell is 19 and in the c-direction 30 [4]. Furthermore, these scandate materials are predicted to be stable in contact with silicon, as is $\mathrm{HfO_2}$, up to temperatures of at least 700 °C [2].

To investigate the physical properties of such rare earth-based oxide thin films, a suitable deposition technique is needed that can be used to prepare thin films from these materials. The MOCVD (metal organic chemical vapor deposition) technique typically used in microelectronic processes needs precursors for each element to be deposited, which must exhibit sufficient stability during thin film growth. However, stable precursors for the $RE\mathrm{ScO_3}$ materials are not known up to now, so at the moment, MOCVD can not be used. On the other hand, sputtering is a well-established deposition method for the growth of metal films and multilayers. Also binary oxide thin films are grown with high perfection by sputtering, but for multi-component materials the optimization of the growth conditions is time consuming. A fast screening is not possible. As a third example, molecular beam epitaxy (MBE) is a well-known technique which is used successfully for epitaxial oxide thin film growth, but it is very complex and therefore also time consuming to

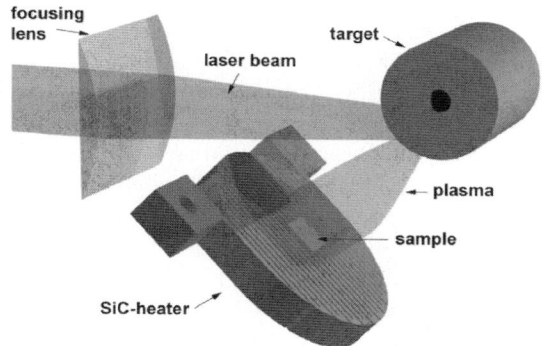

**focusing lens**

**laser beam**

**target**

**plasma**

**sample**

**SiC-heater**

**Fig. 2.** Schematic drawing of a typical pulsed laser deposition geometry

establish in a process [5–7]. A fast and feasible technique to perform material screening is the pulsed laser deposition technique (PLD) that has been used for this work. The boom of the PLD started with the epitaxial growth of high-temperature superconductor thin films [4, 8, 9].

## 2 Pulsed Laser Deposition

The sketch of a typical PLD setup is shown in Fig. 2. A high-energy laser beam is focused on a target to ablate material from it. Typically, a pulsed excimer laser beam with a wavelength of $\lambda = 248\,\text{nm}$ is used. The energy of one laser pulse can be varied in the range of 400 mJ up to 1.2 J with a pulse width of 25 ns. The laser beam is focused through a lens onto the target surface. Depending on the focus area, this results in an energy density in the range of 1.5 to 5 J cm$^{-2}$. The optical power of the laser beam ranges up to 0.2 GW cm$^{-2}$, calculated from the pulse width and the focus area of $0.1 \cdot 2\,\text{cm}^2$. Because of this enormous power and the short interaction period, the laser ablation is a non-equilibrium process. A plasma is generated which contains the species removed from the target. This process allows the stoichiometric transfer of the target material to the substrate, which is one main advantage of the pulsed laser deposition compared to other techniques.

Ceramic bulk material as well as single crystals can be used as target material. Typically, the substrate is placed on a heater to apply high temperatures during the thin film growth. These are needed to enable epitaxial growth on suitable substrate materials. $YBa_2Cu_3O_{7-x}$ grown on $SrTiO_3(100)$ is one example for such an epitaxial growth of an oxide film deposited from a complex stoichiometric target [4, 8, 9]. As the pulsed laser deposition system does not need ultra-high vacuum conditions for the growth of oxide thin films, a fast processing is possible. For more details on the pulsed laser deposition technique, see [10].

In our PLD system used for the growth of $RE$ScO$_3$ thin films, cylindrical targets are used. The irradiated surface of the target is not plane. We use the cylindrical target envelope in combination with a line focus to obtain a large laser-induced plasma plume. However, the area of uniform deposition in this setup is limited to $10 \cdot 10 \, \text{mm}^2$. The target is rotated around its axis to ensure a homogeneous removal of the target material over time. The scandate target material was made by mixing the 99.99 % pure stoichiometric oxide (Gd$_2$O$_3$, Dy$_2$O$_3$, La$_2$O$_3$ and Sc$_2$O$_3$) powders by ball milling. The resulting powder was pressed and sintered at 1500 °C in air for 12 h.[1] In order to heat the substrate to high temperatures during deposition, it is placed on a wire-EDM machined SiC-meander heater. This resistive heater reaches temperatures up to 1000 °C within 5 min in an oxygen ambient of 1 mbar. For the growth of thin films, this allows a short processing time. Typically, for the epitaxial growth of the $RE$ScO$_3$ thin films, SrTiO$_3$(100) is used as substrate material.

## 3 Epitaxial Films

The crystal structure of the $RE$ScO$_3$ materials is orthorhombic with the space group Pbnm(62). The lattice parameters, permittivity and optical bandgap values of the bulk material (single crystalline or polycrystalline) are listed in Table 1 [11,12]. In comparison to the lattice parameter of silicon (a = 5.43 Å), DyScO$_3$ fits the best to a Si(100) substrate surface, probably making it possible to grow epitaxial films on silicon. To compare the properties of $RE$ScO$_3$ thin films to those of single crystals, epitaxial films were grown on single-crystalline oxide substrates. The films were deposited at a substrate temperature ranging from room temperature up to 1000 °C. The typical oxygen pressure during the deposition was $0.2 \cdot 10^{-3}$ mbar. The target to substrate distance measured 5 cm and the energy density was typically 5 J cm$^{-2}$. The crystal structure of the thin films was investigated using Rutherford backscattering spectrometry in the random and the channelling mode (RBS/C) (1.4 MeV He$^+$), and 3- and 4-circle X-ray diffractometry (XRD). C–V-measurements were performed for electrical characterization.

**Table 1.** Properties of some $RE$ScO$_3$ single crystals. The lattice parameters, permittivity $\kappa$ and known optical bandgaps $E_\text{g}$ are listed

|          | a (Å) | b (Å) | C (Å) | $\kappa$ | $E_\text{g}$ (eV) |
|----------|-------|-------|-------|-------|---------|
| DyScO$_3$ | 5.440 | 5.714 | 7.887 | 19–35 | 5.5 |
| GdScO$_3$ | 5.488 | 5.746 | 7.934 | 19–30 | 5.9 |
| LaScO$_3$ | 5.678 | 5.787 | 8.098 | 25–68 | Unknown |

[1] The targets were prepared by Y. Jia and D. G. Schlom, Penn. State University, University Park, PA 16802, USA.

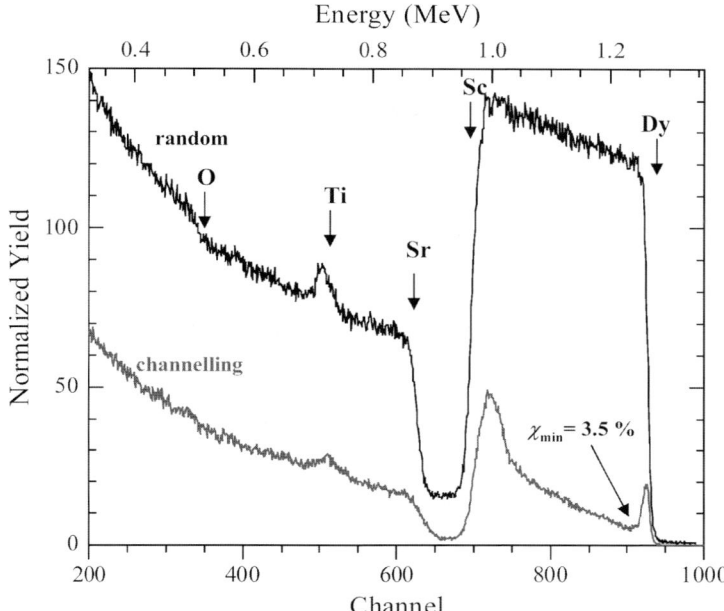

**Fig. 3.** RBS/channelling measurement of a 250 nm thick epitaxial DyScO$_3$ film grown on SrTiO$_3$(100). A minimum yield of 3.5 % reveals the good quality of the epitaxially grown film. The measurement was performed using 1.4 MeV He$^+$

In Fig. 3, the RBS/C spectra of an epitaxial DyScO$_3$ thin film grown at a substrate temperature of $T = 900\,°C$ is shown. Using the program RUMP[2], numerous characteristics can be derived from the measured spectrum and its simulation. The stoichiometric transfer of the target material to the thin film is verified. Within the accuracy of the measurement, the ratio Dy : Sc is 1 : 1. This stoichiometric transfer is valid also for the other two materials. The thickness of the film is 250 nm with a 5 % variation over the area of the sample. The minimum yield value $\chi_{min} = 3.5\,\%$ indicates the good crystal quality of the film grown on SrTiO$_3$(100). Measurements on a LaScO$_3$ thin film grown at $T = 900\,°C$ result in a $\chi_{min}$ of 2.5 %. $\theta$–$2\theta$ XRD measurements performed on LaScO$_3$ films grown at different temperatures are shown in Fig. 4. At 500 °C, only the 100 reflections of the SrTiO$_3$(100) substrate used (marked by asterisks) are observed. Starting from 600 °C, a second set of reflections appears and can be identified as the hk0 reflections of a twinned orthorhombic $RE$ScO$_3$ structure with an $a$, $b$ axis length of 5.84 Å. $\Phi$-scans of the 226 reflex verify the twinned structure, as four broad reflections instead of two are measured. Measurements performed on tilted samples give access to the $c$-axis with a length of $c = 8.06$ Å. Films grown at $T = 900\,°C$ show a

---

[2] RUMP was programmed by Larry Doolittle, Cornell University Ithaca, USA.

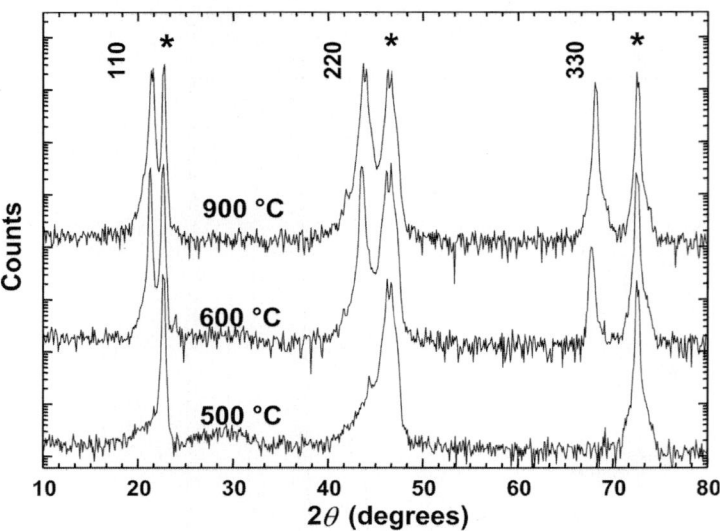

**Fig. 4.** XRD measurements of LaScO₃ films grown at different temperatures. Best epitaxial growth is achieved at a substrate temperature of 900 °C. This sample possesses lattice parameters of $a$, $b = 5.84$ Å and $c = 8.06$ Å. The Rocking curve width measured at the 220 reflex is $\Delta\omega = 0.1°$ FWHM

high degree of perfection, indicated by a rocking curve width of 0.1° FWHM measured at the 220 reflection.

The epitaxial growth of LaScO₃ on Si(100) could also be demonstrated. The direct deposition on silicon is not possible due to the native SiO₂ on the surface of the substrates, but using an epitaxial SrTiO₃ buffer-layer grown with MBE [5], epitaxial deposition is feasible. The RBS/C measurements of a film grown at 800 °C are shown in Fig. 5. The film has a $\chi_{min} = 3.5\%$ which is comparable to the films grown on SrTiO₃(100). The rocking curve width of 0.6° FWHM measured at the 220 reflex is slightly larger than the values obtained on oxide substrates, but the quality of the underlying SrTiO₃-buffer layer limits the thin film quality.

The electrical characterization was performed using an impedance analyzer (Hewlett Packard 4192 A). Metal-insulator-metal (MIM) capacitors were defined by using an epitaxial SrRuO₃ layer on the SrTiO₃(100) substrates as bottom electrode. The epitaxial SrRuO₃ electrode was also prepared by PLD at a temperature of 600 °C in 0.15 mbar oxygen ambient. The $RE$ScO₃ films were grown on this substrate under the same growth conditions as for the growth on pure SrTiO₃(100) substrates. An array of golden circular top electrodes for the capacitors was deposited by thermal evaporation of the metal through a shadow mask. The results of the electrical characterization for an epitaxial 105 nm thick LaScO₃ film are shown in Fig. 6. A small spread in the results is seen depending on the size of the circular electrodes. In the

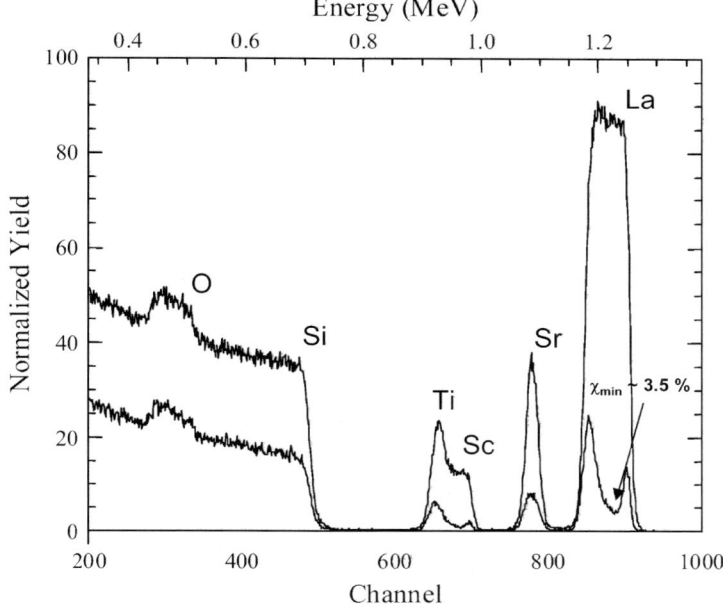

**Fig. 5.** RBS/channelling measurement of an 80 nm thick epitaxial LaScO₃ film grown on a 10 nm thick epitaxial SrTiO₃∥Si(100). The SrTiO₃ film was grown by MBE [5]. The measurement was performed using 1.4 MeV He⁺

electrical field range from $-2$ to $2 \cdot 10^5$ V cm$^{-1}$, a 0.05 mm² capacitor shows a capacitance of about 105 pF almost independent of frequency from 10 kHz to 1 MHz, corresponding to a dielectric constant $\kappa$ of 25 to 27. The observed losses are very low: $\tan \delta$ is in the range from 0.004 (at 10 kHz) to 0.03 (at 1 MHz). Similar results were attained for DyScO₃ ($\kappa = 20$–21, $\tan \delta = 0.03$–0.06) and GdScO₃ ($\kappa = 20$, $\tan \delta = 0.004$–0.03). These $\kappa$-values correspond to those measured on $RE$ScO₃ single crystals in the (110)-plane [11].

Leakage current measurements were performed using a semiconductor parameter analyzer (Hewlett Packard 4155B). Typical values of the films are summarized in Table 2. Leakage currents for a DyScO₃ film are in the range of 0.67 nA at an applied voltage of 2.6 V, corresponding to a current density of 1.4 μA cm$^{-2}$ at an electrical field of 250 kV cm$^{-1}$.

Summarizing the results, the epitaxial $RE$ScO₃ thin films show promising properties. The values of the dielectric constant are comparable to single-crystal data in a, b direction, and LaScO₃ has similar but slightly higher values than DyScO₃ and GdScO₃.

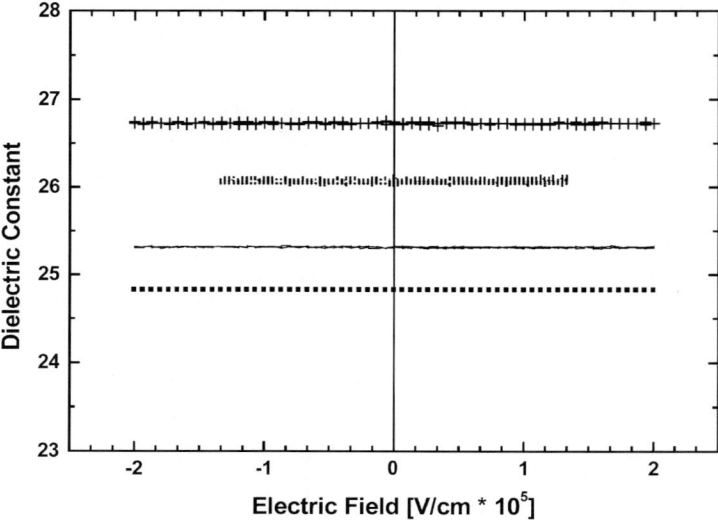

**Fig. 6.** Dielectric constants extracted from C–V measurements performed on capacitors of 0.1, 0.05 and 0.025 mm$^2$ in area

**Table 2.** Properties of epitaxial $RE\mathrm{ScO}_3$ thin films. The lattice parameters of the thin films as well as the leakage current $I_{\mathrm{leak}}$ and the breakdown field $E_{\mathrm{bd}}$ are listed

|  | a, b (Å) | c (Å) | $\kappa$ | $I_{\mathrm{leak}}$ (µA cm$^{-2}$) | $E_{\mathrm{bd}}$ (MV cm$^{-1}$) |
|---|---|---|---|---|---|
| DyScO$_3$ | 5.63 | Unknown | 20–21 | 1.4 | 0.25 |
| GdScO$_3$ | 5.70 | 7.88 | 20 | 0.85 | 1.2 |
| LaScO$_3$ | 5.84 | 8.06 | 25–27 | 6 | 0.6 |

## 4 Amorphous Films

After investigating epitaxial $RE\mathrm{ScO}_3$ thin films which have similar properties as single crystals, amorphous layers of the same materials were analyzed. The films were prepared in an oxygen ambient of $2 \cdot 10^{-3}$ mbar at a substrate temperature of approx. 500 °C either on Si(100) after a wet IMEC clean resulting in a chemical oxide of around 0.8 nm thickness, or on Si(100) wafers covered with a native SiO$_2$. RBS measurements on a 50 nm thick GdScO$_3$ film grown on a silicon substrate were performed. The comparison with the simulation reveals again a stoichiometric transfer of the target material to the substrate. Furthermore, no diffusion processes (Si into scandate or vice versa) are observed, as expected for these materials and growth conditions [2]. To analyze the thermal stability of the scandate thin films, samples were investigated by high-temperature XRD. The samples were heated successively up to 1100 °C and a complete XRD-scan was performed after each 100 °C.

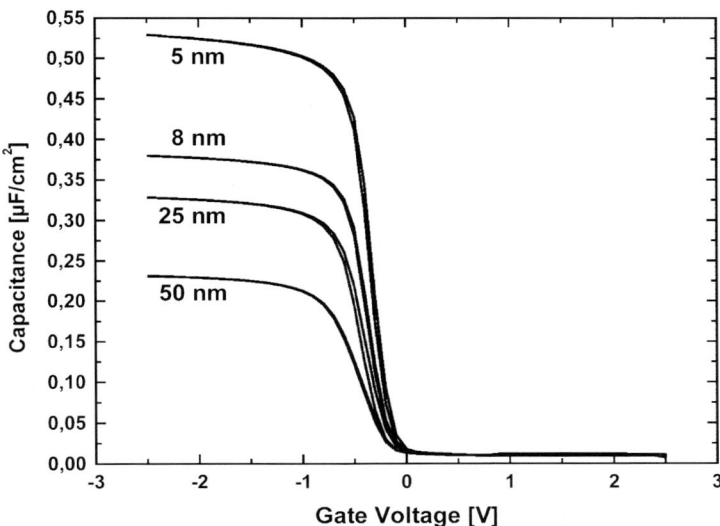

**Fig. 7.** C–V measurements on DyScO$_3$-films of various thickness. Only a small hysteresis is observed

LaScO$_3$ starts to recrystallize at 700 °C, which is a similar behavior to that of HfO$_2$. DyScO$_3$ and GdScO$_3$ remain amorphous up to 1000 °C. Starting at 1100 °C, a phase segregation is observed and the oxides of the rare earth elements start to crystallize [13]. Bandgap and band offset (conduction band and valence band) values were determined using internal photoemission and photoconductivity measurements. In the amorphous state, all three materials have a bandgap higher than 5.5 eV and band offsets to silicon of $\Delta E_c = 2$ eV (conduction band) and $\Delta E_v = 2.5$ eV (valence band), which is sufficient for the application as an alternative gate oxide [14]. As a drawback, additional states are observed which reduce the bandgap to 4.5 eV. Yet it is not clear whether these states are due to the amorphous phase or rather to intrinsic properties of the $RE$ScO$_3$.

AFM investigations on 20 nm thick films show a very low surface roughness of less then 1 nm peak to valley over a scanned area of 2 μm by 2 μm. To evaluate the dielectric properties of the amorphous films MIS-capacitors were manufactured similar to the MIM-stacks. An additional forming gas anneal (90 % N$_2$/10 % H$_2$) was performed for 10 min at 550 °C. In Fig. 7, C–V curves of DyScO$_3$ thin films are shown. Only a small hysteresis can be seen, suggesting a small number of defects in the oxide thin film. Also, a small flat band voltage shift of 0.2 V extracted from a Hauser fit is measured [15]. To extract the κ value of the studied insulators, C–V curves of the capacitors were analyzed. Figure 8 depicts the obtained EOT (equivalent oxide thickness) vs. physical thickness curve of DyScO$_3$ and GdScO$_3$ samples. The experimental data points distribute along a straight line with only small scatter. From the

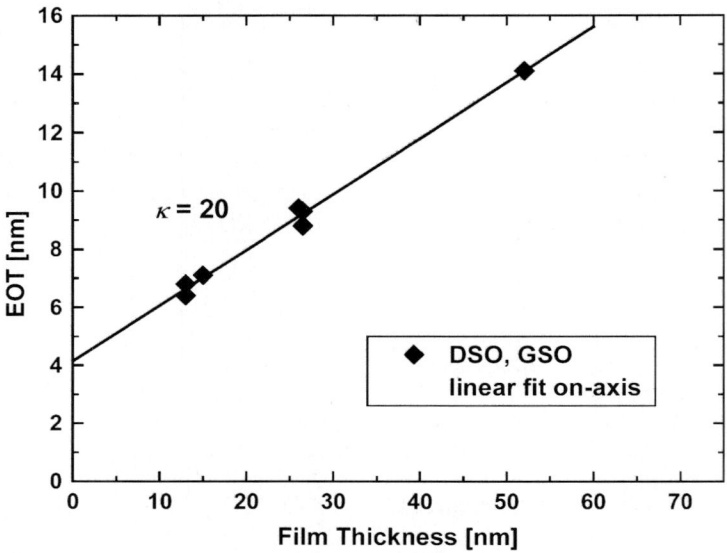

**Fig. 8.** EOT plot of MIS capacitors formed from GdScO$_3$ and DyScO$_3$ thin films. A $\kappa$ value of 20 can be extracted

slope of the line, the $\kappa$ value can be extracted. GdScO$_3$ and DyScO$_3$ show a $\kappa$ of 20. LaScO$_3$ has a similar value [13]. The measurements on the amorphous scandate films reveal that the scandates are a promising class of materials to replace SiO$_2$ as gate dielectric. Large bandgap and band offsets as well as high dielectric constants are comparable to these of HfO$_2$. The thermal stability of the scandates is even better.

# 5 Conclusions

The pulsed laser deposition method is a fast and reliable method to grow thin films from *RE* oxide-based materials. For an interesting class of materials, the rare earth scandates, the epitaxial growth on SrTiO$_3$(100) was demonstrated. The properties of the epitaxial thin films are comparable to those of single crystals. For LaScO$_3$, no single crystals of large size are available to measure the full dielectric tensor, but the dielectric properties of good crystalline thin films demonstrate high $\kappa$ values of 25–27 in $a$, $b$-direction of the crystal lattice. Amorphous films also reveal favourable properties. DyScO$_3$ and GdScO$_3$ stay amorphous up to an annealing temperature of 1000 °C and exhibit similar bandgap and band offsets to silicon as does HfO$_2$. The dielectric constant is around 20. In summary, the results obtained from these thin films demonstrate promising properties and turn the scandates into a possible replacement of SiO$_2$.

**Acknowledgements**

The authors thank M. Caymax and C. Zhao from IMEC, Belgium, as well as L. F. Edge, W. Tian, Y. Jia and D. G. Schlom, Penn State University, for stimulating discussions. TH gratefully acknowledges the financial support by the Deutsche Forschungsgemeinschaft (Graduiertenkolleg GRK 549 "Non-centrosymmetric Crystals"). MW gratefully acknowledges the financial support by the European network of excellence "SINANO".

# References

[1] International technology roadmap for semiconductors: (2004)
URL public.itrs.net/

[2] D. G. Schlom, J. H. Haeni: A thermodynamic approach to selecting alternative gate dielectrics, MRS Bull. **27**, 198 (2002)

[3] G. D. Wilk, R. M. Wallace, J. M. Anthony: High-$\kappa$ gate dielectrics: current status and materials properties considerations, J. Appl. Phys. **89**, 5243 (2001)

[4] J. Fröhlingsdorf, W. Zander, B. Stritzker: Direct preparation of high-$T_c$-superconducting films by laser ablation, Solid State Commun **67**, 965 (1988)

[5] J. Lettieri: *Critical Issues of Complex, Epitaxial Oxide Growth and Integration with Silicon by Molecular Beam Epitaxy*, Ph.D. thesis, Pennsylvania State University (2002)
URL    www.etda.libraries.psu.edu/theses/approved/WorldWideIndex/ETD-202/

[6] R. A. McKee, F. J. Walker, J. R. Conner, E. D. Specht, D. E. Zelmon: Molecular beam epitaxy growth of epitaxial barium silicide, barium oxide, and barium titanate on silicon, Appl. Phys. Lett. **59**, 782 (1991)

[7] H. J. Osten, E. Bugiel, O. Kirfel, M. Czernohorsky, , A. Fissel: MBE growth and properties of epitaxial metal oxides for high-$\kappa$ dielectrics, J. Cryst. Growth **278**, 18 (2005)

[8] D. Dijkkamp, T. Venkatesan, X. D. Wu, S. A. Shaheen, N. Jisrawi, Y. H. Min-Lee, W. L. McLean, M. Croft: Preparation of Y-Ba-Cu oxide superconductor thin films using pulsed laser evaporation from high $T_c$ bulk material, Appl. Phys. Lett. **51**, 619 (1987)

[9] B. Roas, L. Schultz, G. Endres: Epitaxial growth of $YBa_2Cu_3O_{7-x}$ thin films by a laser evaporation process, Appl. Phys. Lett. **53**, 1557 (1988)

[10] D. B. Chrisey, G. K. Hubler: *Pulsed Laser Deposition of Thin Films* (Wiley, New York 1984)

[11] J. H. Haeni: *Nanoengineering of Ruddlesden–Popper phases using molecular beam epitaxy*, Ph.D. thesis, Pennsylvania State University (2002)
URL    www.etda.libraries.psu.edu/theses/approved/WorldWideIndex/ETD-181/

[12] J. Schubert, O. Trithaveesak, A. Petraru, C. L. Jia, R. Uecker, P. Reiche, D. G. Schlom: Structural and optical properties of epitaxial $BaTiO_3$ thin films grown on $GdScO_3$(110), Appl. Phys. Lett. **82**, 3460 (2003)

[13] C. Zhao, T. Witters, B. Brijs, H. Bender, O. Richard, M. Caymax, T. Heeg, J. Schubert, V. V. Afanas'ev, A. Stesmans, D. G. Schlom: Ternary rare-earth metal oxide high-k layers on silicon oxide, Appl. Phys. Lett. **86**, 132903 (2005)

126    Jürgen Schubert et al.

[14] V. V. Afanas'ev, A. Stesmans, C. Zhao, M. Caymax, T. Heeg, J. Schubert, Y. Jia, D. G. Schlom, G. Lucovsky: Band alignment between (100)Si and complex rare earth/transition metal oxides, Appl. Phys. Lett. **85**, 5917 (2004)
[15] J. R. Hauser, K. Ahmed: Characterization of ultra-thin oxides using electrical C–V and I–V measurements, AIP Conf. Proc. **449**, 235 (1998)

# Index

# Film and Interface Layer Composition of Rare Earth (Lu, Yb) Oxides Deposited by ALD

Yuri Lebedinskii[1], Andrei Zenkevich[1], Giovanna Scarel[2], and
Marco Fanciulli[2]

[1] Moscow Engineering Physics Institute, Department 25 (Solid State
 Engineering), 31, Kashirskoe Chaussee, 115409 Moscow, Russian Federation
 `a.zenkevich@zmail.ru`
[2] CNR-INFM MDM National Laboratory MDM-INFM-CNR, Via C. Olivetti 2,
 20041 Agrate Brianza (MI), Italy

**Abstract.** Thin $Lu_2O_3$, $Yb_2O_3$, and Lu silicate layers were grown on Si using
Atomic Layer Deposition (ALD), and studied mainly with X-ray photoelectron
spectroscopy (XPS) and low-energy ion spectroscopy (LEIS). Experimental results
on elemental composition and chemical state both in thin oxide layers and at the interface with the Si substrate, and their evolution upon different heat treatments, are
presented. Besides oxygen states attributed to stoichiometric $RE_2O_3$ (RE = Lu, Yb)
and $OH^-$, an additional, loosely bound state is identified in a subsurface layer at
equilibrium conditions for both $Lu_2O_3$ and $Yb_2O_3$ layers. A variety of thin (amorphous) Lu silicate films grown by ALD with different compositions depending on
the precursor and/or growth conditions was unambiguously established by XPS,
the components' chemical shifts found to be a monotonous function of the composition. As grown continuous ultrathin ($\leq 5\,\mathrm{nm}$) Lu and Yb oxide films on Si appear
to consist of a hydroxide on top and of a silicate layer, $\sim 1$–$3\,\mathrm{nm}$ in thickness, at
the interface with the Si substrate, the stack is converted into amorphous silicates
upon post-deposition annealing in $N_2$ and appears further stable on Si at least up
to annealing at $T = 1100\,^\circ\mathrm{C}$ (in $N_2$).

## 1 Introduction

Rare earth (RE) oxides have recently attracted renewed interest as a class of
materials potentially suitable as a gate dielectric in advanced complementary
metal-oxide-semiconductor (CMOS) technology: they exhibit a high dielectric constant ($\kappa$), were predicted to be thermodynamically stable on silicon
and therefore may generally satisfy requirements to substitute $SiO_2$ as insulating layer in CMOS devices [1, 2]. In particular, $Lu_2O_3$ and $Yb_2O_3$ were
considered to be promising candidates since: 1. for both of these, relatively
wide band gaps were measured on single crystals [3]; 2. a large conduction
band offset (CBO) at the oxide/silicon interface is expected since the 2 : 3
metal : oxygen stoichiometry ratio promotes a low charge neutrality level [4];
3. besides, $Lu_2O_3$ has an unique oxidation state (equal to 3), which avoids
mixed metal oxide stoichiometries with different electronic structures; 4. for

M. Fanciulli, G. Scarel (Eds.): Rare Earth Oxide Thin Films,
Topics Appl. Physics **106**, 127–142 (2007)
© Springer-Verlag Berlin Heidelberg 2007

Lu, the intrinsic high energy of the $5d$ shell and its low occupancy (only one electron) should, in principle, limit the density of interfacial traps in $Lu_2O_3$ [5]. Recent experimental results on the electric properties of $Lu_2O_3$ on Si partially support optimistic predictions [6, 7].

In this work, thin layers of $Lu_2O_3$, $Yb_2O_3$, and Lu silicates grown on silicon using atomic layer deposition (ALD) are studied focusing on their elemental and chemical composition. Both as-grown samples and annealed ones in different atmospheres are monitored combining two analytical techniques, namely, X-ray photoelectron spectroscopy (XPS), and low-energy ion spectroscopy (LEIS). While XPS probes an about 5 nm subsurface layer and provides information on both the elemental and chemical composition in the layer, LEIS, which utilizes $He^+$ ions, is sensitive only to the uppermost surface layer and is used to determine surface elemental composition. The combination of the two techniques, exploiting either in situ vacuum or $O_2$ annealing, or ex situ rapid thermal annealing in $N_2$, allows a detailed characterization of the elemental versus chemical structure evolution of Lu and Yb oxides, and of Lu silicates on Si. The reaction paths found to occur in the different annealing environments provide the necessary experimental basis to better understand the thermodynamics of the $Lu_2O_3/Si$ and $Yb_2O_3/Si$ systems. In particular, the understanding of RE oxide/Si interfacial layer formation and evolution upon heat treatments in various atmospheres is of special importance, since the interfacial layer stability is one of the criteria that ultimately define the applicability of dielectric materials in devices. To directly assess the formation and evolution of RE oxide/Si interfacial layers with XPS, and to avoid detrimental effects of $Ar^+$ etching on interfacial chemical composition [8], ultrathin Lu and Yb oxide layers were grown using ALD on Si(100), and characterized with XPS (elemental/chemical composition in the layer) and LEIS (surface elemental composition).

Recently, Ge has re-emerged as a possible material of choice for high-mobility channels in future CMOS applications. Therefore, also $Lu_2O_3/Ge(100)$ interfacial layer formation was characterized and compared with that of the $Lu_2O_3/Si$ system.

The obtained experimental data allow to preliminarily conclude on the applicability of particular RE-based compounds as candidates for microelectronics.

## 2 Experimental

This work investigates Lu and Yb oxides, and Lu silicate films grown using ALD, a technique employed in microelectronics because of the smoothness, conformality, and good electrical characteristics of the deposited films. The organometallic compounds $\{[\eta^5\text{-}C_5H_4(SiMe_3)]_2LuCl\}_2$ [9] and $Lu(iPrO)_3$ were used as precursors for Lu, together with both $H_2O$ and $O_3$ as oxygen sources (Me = $CH_3$). On the other hand, $Yb(Cp)_3$ and $Yb(thd)_3$ were used

**Table 1.** ALD growth parameters for samples used in this work

| Precursor combination | Growth temperature (°C) | RE source temperature (°C) |
|---|---|---|
| $\{[\eta^5-C_5H_4(SiMe_3)]_2LuCl\}_2$ and $H_2O$ | 360 | 195 |
| $\{[\eta^5-C_5H_4(SiMe_3)]_2LuCl\}_2$ and $O_3$ | 360 | 195 |
| $Lu(iPrO)_3$ and $H_2O$ | 330 | 185 |
| $Lu(iPrO)_3$ and $O_3$ | 310 | 180 |
| $[(Me_3Si)_2N]_3Lu + H_2O$ | 380 | 150 |
| $Yb(Cp)_3$ and $H_2O$ | 360 | 100 |
| $Yb(Cp)_3$ and $O_3$ | 360 | 100 |
| $Yb(thd)_3$ and $O_3$ | 360 | 100 |

as precursors for Yb combined with respectively $H_2O$ and $O_3$ the former, and with only $O_3$ for the latter (Cp $= C_5H_5$, and thd $= C_{11}H_{19}O_2$). The films were deposited in an ALD F-120 ASM-Microchemistry reactor on Si(100) and Ge(100) wafers. Some of samples were also subjected to post-deposition rapid thermal processing (RTP) in $N_2$ for 60 s up to 1000 °C. The growth parameters of samples used in this work are listed in Table 1

For XPS analysis, an XSAM-800 (Kratos) electron spectrometer was employed utilizing an Mg $K_\alpha$ source ($E = 1253.6$ eV) to excite the photoelectrons. The energy resolution was measured to be $\Delta E = 0.9$ eV (Au $4\vec{j}$). The fitting of the spectra was performed by setting the line as a combination of Gaussian and Lorenzian shapes utilizing the Origin 6.0 software.

In situ annealing up to $T = 950$ °C was performed either in ultra-high vacuum (UHV) ($P = 10^{-9}$ Torr), or in $O_2$ atmosphere ($P_{O_2} = 10^{-6}$ to $10^{-2}$ Torr) in the preparation chamber of the spectrometer, using calibrated electric current transmission directly through the Si substrate mounted on the holder for heat control. Rapid thermal processing in $N_2$ was performed ex situ at $T = 950$ °C, for 60 s. For in situ XPS calibration, Au at few monolayer thickness was in some cases deposited on top of RE oxide layers using the pulsed laser deposition (PLD) module installed in the preparation chamber of the spectrometer ($E_{BAu4f7/2} = 84.0$ eV). To investigate possible effects of the particular growth technique (ALD) on the chemical transformations particularly in $Yb_2O_3$ films, similar oxide layers were grown on Si using PLD. In this case, metallic Yb layers were ablated in $O_2$ atmosphere ($P_{O_2} = 1 \cdot 10^{-2}$ Torr), and further analyzed in situ.

The $Ar^+$ ion beam at the energy $E = 2.5$ keV incident at 55° with respect to the sample normal was employed to etch the surface layer of oxide thin films.

For LEIS analysis, an $He^+$ beam was used instead, with an incident energy in the $E_0 = 100 - 1000$ eV range. The ions backscattered at $\theta = 125$° were analyzed with the same hemispheric analyzer of the electron spectrometer switched to the opposite polarity. Since it appeared important to accurately

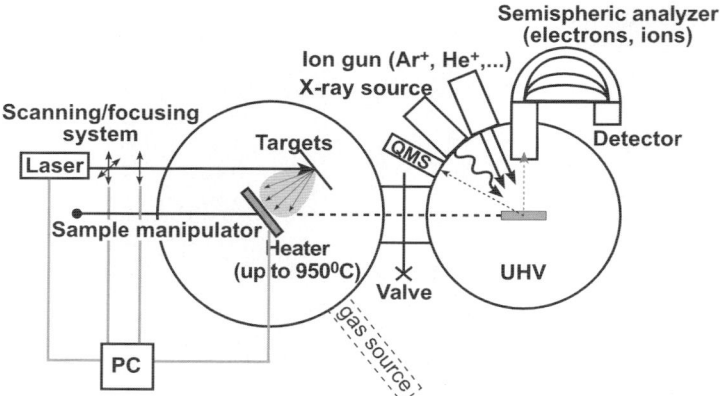

**Fig. 1.** The schematics of PLD-XPS-LEIS setup

calibrate the position of peaks in LEIS spectra, additional measurements were taken to set the incident ions energy within a 1 eV accuracy. A schematic representation of the PLD-XPS-LEIS setup used in this work is given in Fig. 1.

# 3 Results and Discussion

## 3.1 $Lu_2O_3$

Lu $4f$ and O $1s$ core level photoemission spectra taken from the surface of the as-grown 30 nm thick $Lu_2O_3$ layer deposited from $Lu(iPrO)_3$ and $O_3$ are displayed in Fig. 2a. Note that the film is partly amorphous (as revealed by X-ray diffraction measurements). Although for our experimental conditions the spin-orbit splitting of the components in the Lu $4f$ line are only partially resolved, it is clear that the two Lu $4f$ doublets are present in the spectrum. The existence of more than one component in the O $1s$ spectrum is also evident. The fitting of the spectra was performed with a pair of 3 : 4 double and single lines for Lu $4f$ and O $1s$, respectively. The position of lines attributed to pure $Lu_2O_3$ [6] and $Lu(OH)_3$, usually present in the subsurface layer of RE oxides [10], are given in Table 2.

Using the routine procedure of comparing peak areas from the ultrathin surface (hydroxide) layer and from the bulk oxide, the thickness of hydroxide layer is calculated to be $\approx 3$ nm. In fact, it is removed upon mild $Ar^+$ etching (Fig. 2b). However, an *additional* line in the O $1s$ spectrum at slightly lower binding energy (BE) is still visible. A much smaller hydroxide component is observed in the O $1s$ spectrum for the same $Lu_2O_3$ sample upon RTP (Fig. 3a), indicating that the hydroxide is removed from the surface layer

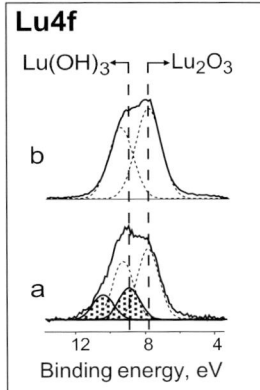
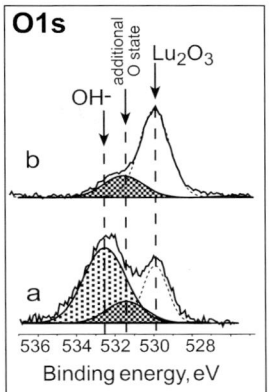

**Fig. 2.** XPS Lu $4f$ and O $1s$ lines taken from the surface of an as-grown $Lu_2O_3$ layer (**a**), and upon $Ar^+$ ion etching (**b**)

**Table 2.** Lu $4f$ and O $1s$ core level line position (BE) taken from the surface of an as-grown partly amorphous $Lu_2O_3$ film and upon $Ar^+$ ion etching

|  | BE(O $1s$) (eV) | BE(Lu $4f$) (eV) 5/2  7/2 |  |
|---|---|---|---|
| $Lu_2O_3$ | 529.8 (531.5) | 7.9 | 9.3 |
| $Lu(OH)_3$ | 532.3 | 9.0 | 10.4 |

during crystallization ("densification"). This phenomenon occurs presumably according to the reaction:

$$2\,Lu(OH)_3 \longrightarrow Lu_2O_3 + 3\,H_2O\,.$$

Further in situ annealing in vacuum ($T = 300\,°C$), and in $O_2$ ($P_{O_2} = 10^{-6}$ Torr, $T = 600\,°C$) results in the *additional* O $1s$ peak area diminishing and reemerging, respectively (Fig. 3b,c) in the two cases. The fact that the fraction of additional O state can be modified by the environment during annealing at moderate $T$ indicates that the chemical structure of $Lu_2O_3$ is not quite stable, at least in the subsurface layer. Several possibilities can be considered to explain the origin of the O chemical state giving rise to the additional peak in the O $1s$ core level spectrum at higher BE. In particular: 1. $OH^-$ groups, always present in the subsurface layer, but which appear at noticeably higher BE than those usually observed (see Table 2); 2. O bound to Si in the $SiO_2$ form, which is often observed on the surface of ALD-grown oxide layers; this hypothesis is weak, since the additional O state is found even when no Si is observed on the surface; 3. existence of two chemically inequivalent O states should also be ruled out for $Lu_2O_3$ crystallizing in

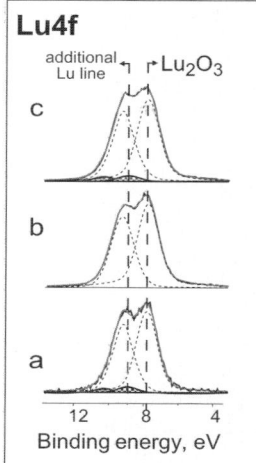 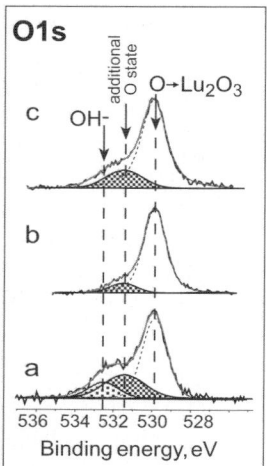

**Fig. 3.** XPS Lu $4f$ and O $1s$ lines taken from the surface of as-grown *polycrystalline* Lu$_2$O$_3$ layers (**a**), upon *in situ* vacuum annealing (**b**) and annealing in O$_2$ ($P_{O_2} = 10^{-6}$ Torr) (**c**)

bixbyite structure. From the analysis of the whole set of experimental data for both as-grown amorphous Lu$_2$O$_3$ and Yb$_2$O$_3$ layers, we conclude that the additional O state observed with XPS could be attributed to superstoichiometric oxygen, which is either trapped during ALD growth or absorbed upon exposure to air at least in the subsurface region at equilibrium conditions. Additional evidence for this conclusion is that ultrathin RE oxide layers on Si are fully converted into silicates upon annealing in vacuum, a process that possibly requires additional oxygen (see below).

### 3.2 Yb$_2$O$_3$

In Yb$_2$O$_3$ films grown using the Yb(Cp)$_3$ and H$_2$O precursor combination, the chemical bonding is qualitatively very similar to the characteristic one in Lu$_2$O$_3$ films (Fig. 4). To elucidate whether the additional O state is related to ALD growth process, Yb oxide layers were also grown for comparison with reactive PLD. The PLD films were deposited at room temperature with Yb laser ablation in O atmosphere at $P_{O_2} = 1 \cdot 10^{-2}$ Torr and then annealed at $T = 600\,°C$. It turns out that the oxygen chemical states in PLD- and ALD-grown Yb$_2$O$_3$ films, and their evolution upon different heat treatments, are remarkably similar, irrespective of the growth technique. The presence of the additional O state, at least in the subsurface layer of RE oxides exposed to atmospheric air, should therefore be considered as an intrinsic material property.

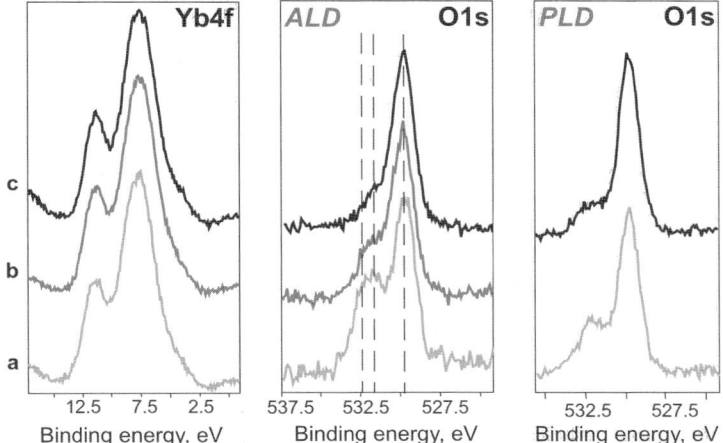

**Fig. 4.** Yb $4f$ and O $1s$ (ALD vs. PLD) core level lines taken from the surface of as-grown $Yb_2O_3$ layers (**a**), upon surface $Ar^+$ etching (**b**) and annealing up to $T = 300\,^\circ C$ in vacuum (**c**)

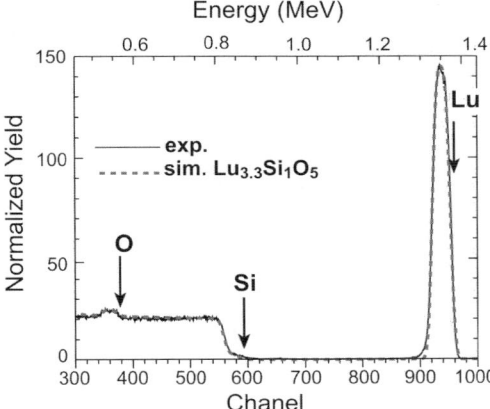

**Fig. 5.** Backscattering spectrum of a Lu silicate film with the composition $Lu_{3.3}Si_1O_5$

## 3.3 Lu Silicate Layers on Si

While using $\{[\eta^5\text{-}C_5H_4(SiMe_3)]_2LuCl\}_2$ and $O_3$, ALD produces Lu oxide layers with a large amount of silicon, homogeneously distributed in these. In Fig. 5a, the Rutherford backscattering spectrometry (RBS) spectrum measured for a $Lu_xSi_yO_z$ sample is displayed, where three components in the grown layer at the ratio $Lu : Si : O = 3.3 : 1 : 5$ are evident [11].

In Fig. 6, core level Lu $4f$, Si $2p$ and O $1s$ spectra of a silicate layer from $[(Me_3Si)_2N]_3Lu + H_2O$ are compared with those of pure $SiO_2$ and $Lu_2O_3$ (all the spectra are carefully calibrated with Au overlayer deposition). Several

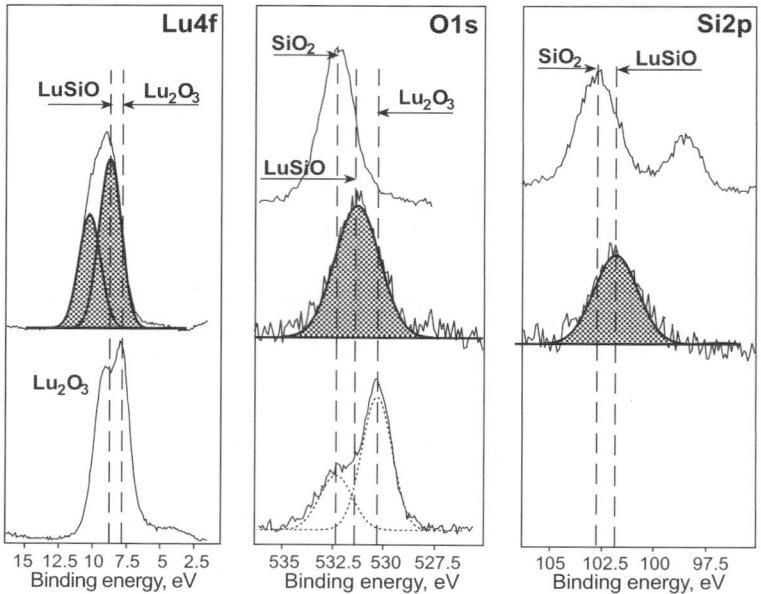

**Fig. 6.** XPS Lu $4f$, O $1s$, and Si $2p$ lines taken from the bulk of an ALD-grown Lu silicate layer obtained from $[(\text{Me}_3\text{Si})_2\text{N}]_3\text{Lu} + \text{H}_2\text{O}$, and compared with the corresponding ones from pure $\text{Lu}_2\text{O}_3$ and $\text{SiO}_2$

differences are evident in the BE of the typical lines in silicates and oxides: 1. the Lu $4f$ doublet in the former compound is found at higher BE than in $\text{Lu}_2\text{O}_3$ films, $\Delta\text{BE} \approx 1\,\text{eV}$; 2. the O $1s$ line position in silicates is between that in $\text{SiO}_2$ and in $\text{Lu}_2\text{O}_3$, and can be fitted with one, albeit broad, line; 3. the Si $2p$ line is found in a "suboxide" position with respect to $\text{SiO}_2$ (i.e., $\Delta\text{BE}_{\text{SiO}_2} \approx 3.1\,\text{eV}$ from $\text{Si}^0$ for the Lu silicate, and $\Delta\text{BE}_{\text{SiO}_2} \approx 3.8\,\text{eV}$ from $\text{Si}^0$ for $\text{SiO}_2$; it is worth noting that the position of $\text{Si}^0$ itself may vary depending on whether it is $n$- or $p$-type). These differences, i.e., the relative shifts of the XPS lines of all three constituents in the silicate with respect to the corresponding ones in the oxide, the single line in the O $1s$ spectrum, and the Si $2p$ line appearing at a lower BE than $\text{Si}^{4+}$, allow to unambiguously identify Lu silicates with XPS.

In Fig. 7, the BE shifts ($\Delta\text{BEs}$) of the Lu $4f$, Si $2p$, and O $1s$ lines in as-grown Lu silicates with respect to those in pure $\text{SiO}_2$ (grown by thermal oxidation of Si) and $\text{Lu}_2\text{O}_3$ are plotted as a function of the Lu/Si ratio measured by XPS in the silicate layers. It is evident that the BE shifts are a monotonous function of composition, which is not uniquely determined in as-grown silicate layers. It is indeed known that RE silicate layers are always amorphous, apparently because of the high barrier for the nucleation of the crystalline phase [12], which may prevent also the formation of a specific

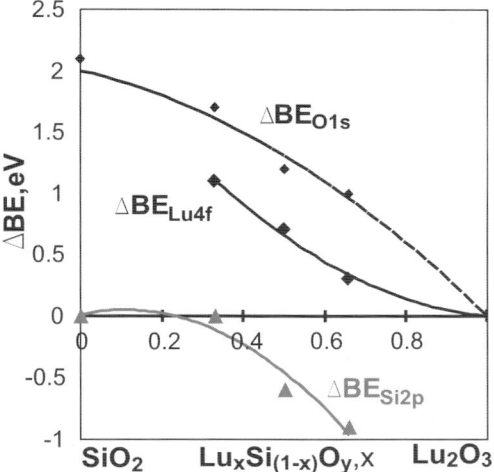

**Fig. 7.** BE shifts ($\Delta$BEs) of the Lu $4f$, Si $2p$, and O $1s$ lines in Lu silicates versus the Lu/Si ratio

phase stoichiometry. Interestingly, unlike RE oxide films, hydroxide formation is never observed in the surface layer of silicates.

To investigate the thermal stability of an amorphous Lu silicate layer, vacuum annealing of a thick $Lu_{0.75}Si_1O_3$ sample from $[(Me_3Si)_2N]_3Lu + H_2O$ was performed up to $900\,^{\circ}C$ in vacuum and monitored in situ with XPS (the heating was interrupted at each annealing step to allow the analysis).

The changes in composition as a function of annealing temperature $T$ in vacuum derived from XPS data are shown in Fig. 8. The composition of the subsurface layer in Lu silicate is quite sensitive to vacuum annealing. The Si/Lu ratio decreases with $T$, which suggests that Si desorbs from the surface. The available data are not sufficient to establish the true mechanism of Lu silicate composition evolution, but it seems reasonable that Si desorbs in the SiO form. Further annealing in $O_2$ ($P_{O_2} \sim 10^{-1}$ Torr) partially restores the initial stoichiometry. This phenomenon occurs only if we assume Si outdiffusion from the deeper layers (ultimately, from the Si substrate) to the surface, and its oxidation. These results show that Lu silicates are not as robust with respect to vacuum heat treatments as is generally expected from the thermodynamic point of view [2].

### 3.4 Ultrathin $Lu_2O_3$ and $Yb_2O_3$ Layers on Si(100)

Ultrathin oxide layers grown on single-crystalline Si substrate are of special interest since they are an ultimate test for the growth technique ability to fabricate suitable stacks for, e.g., CMOS technology. On the other hand, when the oxide layer thickness is $\leq\sim 5$ nm, the interfacial layer with the sub-

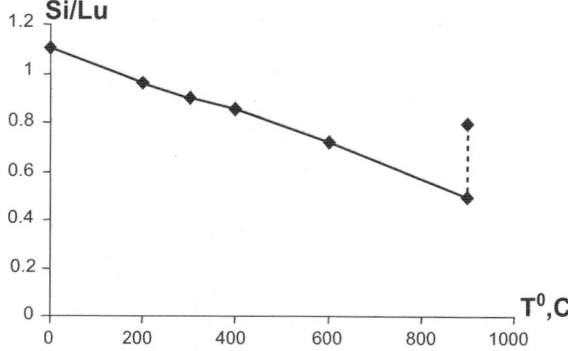

**Fig. 8.** Compositional changes in Lu silicate (originally $Lu_{0.75}Si_1O_3$) upon vacuum annealing

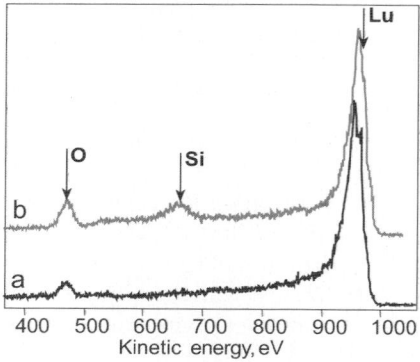

**Fig. 9.** LEIS spectrum ($E_0^{He^+} = 1\,keV$) taken from the surface of: (**a**) as-grown ultrathin $Lu_2O_3$ on $Si(100)$; (**b**) upon RTP in $N_2$ up to $950\,°C$

strate can be directly probed and characterized with XPS, for both as-grown and annealed films. $Lu_2O_3$ layers, 3 nm thick from RBS measurements, were grown from $\{[\eta^5\text{-}C_5H_4(SiMe_3)]_2LuCl\}_2$ and $H_2O$ on $Si(100)$. LEIS analysis was performed upon mild ($T = 150\,°C$) heating in vacuum to determine the surface elemental composition. The spectrum taken at $E_0^{He^+} = 1\,keV$ for an as-grown sample is shown in Fig. 9. The fact that the only observed peaks are those corresponding to Lu and O, while the one corresponding to Si is not visible, allows to conclude that the $Lu_2O_3$ layer is continuous [13]. Upon RTP in $N_2$ up to $950\,°C$, Si emerges in the LEIS spectrum, either because of the $Lu_2O_3$ crystalline grains coarsening on the surface and opening Si substrate surface, or because of Lu silicate formation.

XPS was employed to assess the $Lu_2O_3/Si(100)$ interfacial layer for an as-grown sample, and to monitor its evolution during heat treatments in $O_2$ and in vacuum. Core level Lu $4f$, Si $2p$, and O $1s$ spectra taken at different stages are presented in Fig. 10. The O $1s$ line for an as-grown ultrathin $Lu_2O_3$ layer

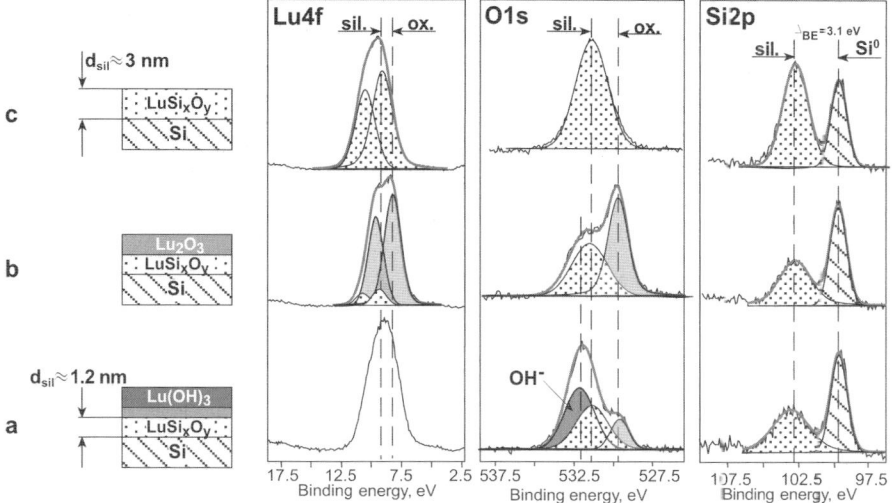

**Fig. 10.** Lu $4f$, O $1s$ and Si $2p$ core level lines taken from the surface of ALD-grown 3 nm thick $Lu_2O_3$ on Si(100) (MDM284): (**a**) as-grown, (**b**) upon vacuum annealing up to 300 °C; (**c**) upon RTP in $N_2$, $T = 950$ °C

on Si(100) can be fitted only with three lines, i.e. Lu hydroxide, pure $Lu_2O_3$, and Lu silicate (Si $2p$, Fig. 10a). The Si $2p$ line is a combination of a peak originating from the Si substrate, and of another one located in the "suboxide" region ($\Delta BE = 3.1$ eV from the $Si^0$ peak, compared to $\Delta BE = 3.8$ eV for $SiO_2$), indicative of a silicate present in the layer. (The Lu $4f$ line for the as-grown sample was not fitted since, at the employed resolution, it was difficult to obtain a unique fit with several doublets). It is therefore reasonable to conclude that 1. an about $2 - 3$ nm thick $Lu(OH)_3$ layer is present on the surface; 2. a silicate layer is forming at the $Lu_2O_3$/Si(100) interfacial layer during deposition or upon exposure to the atmosphere. Its thickness, calculated from the relative $Si^0$ versus $Si^{sil.+}$ peak areas, is $\sim 1.2$ nm.

Upon vacuum annealing at $T = 300$ °C, the $Lu(OH)_3$ layer disappears, in agreement with earlier observations on thicker $Lu_2O_3$ films, and the spectra consist of two components, one corresponding to $Lu_2O_3$ and the other to $Lu_xSi_yO_z$ (Fig. 10b). Upon RTP at $T = 950$ °C, the Lu $4f$, Si $2p$ and O $1s$ XPS lines consistently point toward a silicate as the only component in the spectra (Fig. 10c). The silicate composition, estimated by comparing the relative Si $2p$ and Lu $4f$ peak areas with those in pure $SiO_2$ and $Lu_2O_3$, and taking into account the sensitivity factors for the elements, is $Lu_2Si_xO_{3 + 2x}$, $1.2 < x < 1.5$.

For the $Yb_2O_3$/Si(100) system deposited from $Yb(thd)_3$ and $O_3$, the overall picture is qualitatively very similar to the one determined for the $Lu_2O_3$/Si(100) system (Fig. 11). The only difference is that an Yb silicate layer

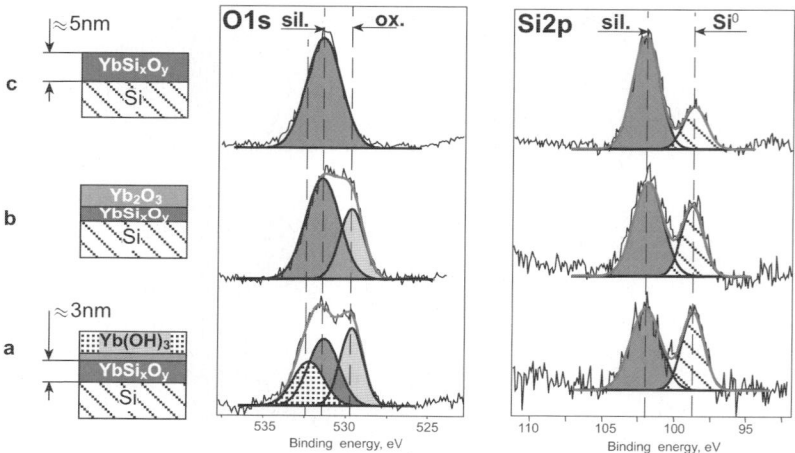

**Fig. 11.** Yb $4f$, O $1s$ and Si $2p$ core level lines taken from the surface of ALD-grown 3 nm thick $Yb_2O_3$ on Si(100) from $Yb(thd)_3$ and $O_3$: (**a**) as grown, (**b**) upon vacuum annealing up to 300 °C; (**c**) upon RTP in $N_2$ $T = 950$ °C

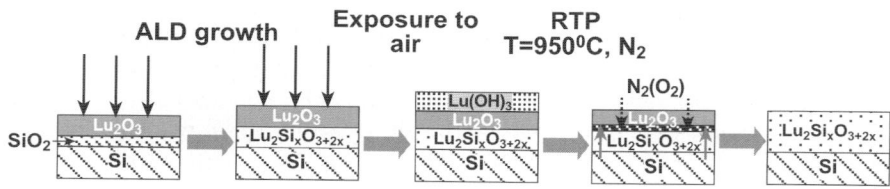

**Fig. 12.** Interface formation and evolution for ultrathin ALD-grown $Lu_2O_3$ layers on Si(100)

formed at the $Yb_2O_3$/Si interface is thicker ($d_{YbSiO} \approx 3$ nm) than the corresponding one at the $Lu_2O_3$/Si interface ($d_{LuSiO} \approx 1.2$ nm). This difference is mainly due to the use of $O_3$ as oxygen source in the case of the $Yb_2O_3$ film, as shown for other ALD-deposited oxide systems [14].

Therefore, according to our experimental observations, the $RE_2O_3$/Si(100) (RE = Lu, Yb) interfacial layer formation and evolution upon annealing in different environments can be summarized as follows (for clarity, we consider the $Lu_2O_3$/Si system, Fig. 12): 1. a few monolayer thick silicate interfacial layer presumably forms during the initial stage of $Lu_2O_3$ growth on a chemical oxide ($SiO_2$) terminated Si surface; 2. upon exposure to air, an ultrathin as-grown $Lu_2O_3$ layer is mostly converted into an $Lu(OH)_3$ layer because amorphous $Lu_2O_3$ is known to be highly hydroscopic; 3. the $Lu(OH)_3$ layer can be transformed back to oxide by vacuum annealing up to 300 °C; 4. annealing at higher temperatures converts the whole $Lu_2O_3$ layer into a Lu silicate layer (in case oxygen is present in the annealing enviroment).

The experimentally observed transformations are supported by thermodynamic considerations. In particular, a driving force for silicate formation exists if $Lu_2O_3$ is in contact with $SiO_2(Lu_2O_3 + SiO_2 \longrightarrow Lu_2SiO_5)$ [12]. In fact, using the bulk thermodynamic data of *Marsella* et al. [15], and assuming the $\Delta H$ value for Lu silicate to be the same as those for Y and La silicates, one obtains:

$$\Delta H_{Lu_2O_3}(-20.18\,\text{eV}) + \Delta H_{SiO_2}(-8.2\,\text{eV}) > \Delta H_{Lu_2SiO_5}(-35\,\text{eV}) ,$$

which is also consistent with the reaction path given by *Stemner* [12].

The silicate formation at the interfacial layer between RE oxide films and Si was experimentally shown [6]. It was also pointed out [12] that this phenomenon is likely due to excess oxygen incorporated into the oxide layer during growth, or to the oxygen in the annealing atmosphere. Indeed, the reaction:

$$Lu_2O_3 + Si + O_2 \longrightarrow Lu_2SiO_5 ,$$

where $O_2$ comes from the environment and goes through the oxide layer to react with the underlying Si, is also favorable from the thermodynamical point of view $[\Delta H_{Lu_2O_3}(-20.2\,\text{eV}) > \Delta H_{Lu_2SiO_5}(-35\,\text{eV})]$ [15]. Our observations of the additional O state, attributed to "superstoichiometric" oxygen in the as-grown oxide layer at equilibrium conditions, supports this model.

Finally, since hydroxide species are found to be the dominating ones for ultrathin "oxide" layers, the following reaction is suggested to take place:

$$2\,Lu(OH)_3 \longrightarrow Lu_2O_3 + 3\,H_2O \uparrow$$

The alternative pathway is [12]:

$$2\,Lu(OH)_3 + \tfrac{3}{2}Si \longrightarrow Lu_2O_3 + \tfrac{3}{2}SiO_2 + \tfrac{3}{2}H_2 \uparrow ,$$

and the products would eventually be converted into a silicate.

### 3.5 Ultrathin $Lu_2O_3$ on Ge(100)

LEIS spectra taken with $He^+$ ions at $E_0 = 1\,\text{keV}$ for ultrathin $Lu_2O_3$ films on Ge(100) from $\{[\eta^5\text{-}C_5H_4(SiMe_3)]_2LuCl\}_2$ and $H_2O$ are shown in Fig. 13, part 1. Both X-ray reflectivity (XRR) and RBS revealed a thickness of $\sim$ 3 nm. No Ge signal is detected in the LEIS spectra, indicating that the $Lu_2O_3$ layer is continuous [13]. Some Si, possibly released from the Lu precursor, is found on the film surface. The Lu $4f$, O $1s$ XPS, and the $GeL_3M_{45}M_{45}$ Auger lines for the as-grown and the annealed Lu oxide thin films are shown in Fig. 13, parts 2–4. The $GeL_3M_{45}M_{45}$ Auger line is the strongest one in the emission spectrum and is sensitive to the Ge chemical state. The Ge $L_3M_{45}M_{45}$ line for natural GeO is also shown for comparison.

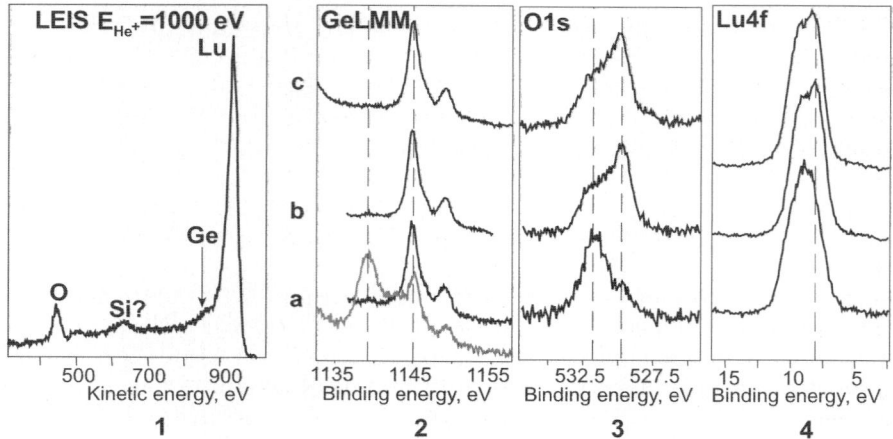

**Fig. 13.** LEIS/XPS data taken for ultrathin $Lu_2O_3$ thin films on $Ge(100)$ $\{[\eta^5\text{-}C_5H_4(SiMe_3)]_2LuCl\}_2$ and $H_2O$: (**1**) LEIS spectrum on the surface of as-grown sample; (**2**)-(**4**) Ge $L_3M_{45}M_{45}$, O $1s$, Lu $4f$ lines for a) as grown sample; b) upon vacuum annealing up to $500\,^\circ C$; c) upon annealing in $O_2$, $T = 500\,^\circ C$ (Ge LMM grey line is the spectrum for the oxidized Ge surface)

The XPS spectra reveal that: 1. as for the films on Si, the as-grown ultrathin "oxide" is mostly a hydroxide; 2. no oxidized Ge is found at the interface upon growth (compare the Ge $L_3M_{45}M_{45}$ line from the film and for oxidized Ge); 3. the $Lu_2O_3/Ge$ interface does not change upon annealing up to $500\,^\circ C$ both in vacuum and in $O_2$ ($P_{O_2} = 10^{-6}$ Torr).

The fact that no Ge oxide is formed at the interface is not surprising, since GeO is unstable. More importantly, unlike in the case of growth on Si, a continuous ultrathin pure $Lu_2O_3$ layer can be formed on Ge upon crystallization ("densification") during moderate annealing, with no sign of interfacial reactions. The formation of a sharp and stable interface makes the $Lu_2O_3/Ge$ system promising for further consideration in future CMOS applications.

## 4 Conclusions

Rare earth metal binary oxides are a class of materials considered for alternative gate dielectrics for Si-based CMOS. Although their electrical properties and interface with semiconductors were predicted to be suitable, thin $Lu_2O_3$ and $Yb_2O_3$ films until recently were rather poorly investigated representatives of the rare earth oxides. ALD-grown $Lu_2O_3$ and $Yb_2O_3$, and Lu silicates films were therefore studied with a combination of XPS/LEIS, which turned out to be a useful tool to characterize RE oxide films and their interfaces with Si and Ge. It was found that, as for all RE oxides, both $Lu_2O_3$ and $Yb_2O_3$ are highly hydroscopic, but the ability to absorb water depends on the crystallinity of the film surface. An additional O state is observed in the subsurface layer, the amount depending on the sample treatment, which led us to conclude it is loosely bound and may be attributed to superstoichiometric O in the oxide. A variety of amorphous Lu silicates can be grown using ALD. In this case, the stoichiometry depends on the growth parameters as well as on the post-deposition heat treatments. The stoichiometry in the subsurface region depends on annealing temperature and environment, and, particularly, vacuum annealing may affect the Si composition. As-grown ultrathin $Lu_2O_3$ and $Yb_2O_3$ layers on Si include $\sim$ 3 nm hydroxide on the surface, and 1.2 nm (3 nm) silicate at the interface with Si. Annealing of these ultrathin stacks removes the hydroxide, but converts the oxide into a homogeneous silicate layer during annealing in $O_2$. The latter observation allows us to conclude that, in agreement with theoretical predictions and with the experimentally observed behavior of other RE oxides, when there is no way to hinder oxygen penetration into the film, both $Lu_2O_3$ and $Yb_2O_3$ layers are not thermodynamically stable in contact with Si. However, the corresponding silicates formed after the annealing of the oxides in $O_2$ and/or $N_2$ atmosphere are stable. Therefore, Lu or Yb silicates could be a choice for microelectronics applications. Finally, interestingly no interfacial layer is revealed at the interface in $Lu_2O_3/Ge(100)$ systems.

## Acknowledgements

This work was partially supported by funds of the Italian Ministry of Foreign Affairs for Joint Projects between Italy and Russia.

# References

[1] G. D. Wilk, R. M. Wallace, J. M. Anthony: High-$\kappa$ gate dielectrics: Current status and materials properties, J. Appl. Phys. **89**, 5243 (2001)

[2] D. G. Schlom, J. H. Haeni: A thermodynamic approach to selecting alternative gate dielectrics, MRS Bull. **27**, 198 (2002)

[3] A. V. Prokofiev, A. I. Shelyakh, B. T. Melekh: Periodicity in the band gap variation of $Ln_2X_3$ (X = O, S, Se) in the lanthanide series, J. All. Comp. **242**, 41 (1996)

[4] J. Robertson: Band offsets of wide-band-gap oxides and implications for future electronic devices, J. Vac. Sci. Technol. B **18**, 1785 (2000)

[5] G. Lucovsky, Y. Zhang, G. B. Rayner, G. Appel, H. Ade, J. L. Whitten: Electronic structure of high-$k$ transition metal oxides and their silicate and aluminate alloys, J. Vac. Sci. Technol. B **20**, 1739 (2002)

[6] G. Scarel, E. Bonera, C. Wiemer, G. Tallarida, S. Spiga, M. Fanciulli, I. L. Fedushkin, H. Schumann, Y. Lebedinskii, A. Zenkevich: Atomic layer deposition of $Lu_2O_3$, Appl. Phys. Lett. **85**, 630 (2004)

[7] S. Ohmi, M. Takeda, H. Ishiwara, H. Iwai: Electrical characteristics for $Lu_2O_3$ thin films fabricated by e-beam deposition method, J. Electrochem. Soc. **151**, G279 (2004)

[8] Y. Lebedinskii, A. Zenkevich: Silicide formation at $HfO_2$/Si and $ZrO_2$/Si interfaces induced by $Ar^+$ ion bombardment, J. Vac. Sci. Technol. A **22**, 2261 (2004)
[9] H. Schumann, I. L. Fedushkin, M. Hummert, G. Scarel, E. Bonera, M. Fanciulli: Crystal and molecular structure of $[(\eta^5\text{-}C_5H_4SiMe_3)_2LuCl]_2$ – suitable precursor for $Lu_2O_3$ films, Z. Naturforsch. **59b**, 1035 (2004)
[10] J. P. Espinos, A. R. Gonzalez-Elipe, J. A. Odriozola: XPS study of lutetium oxide samples with different hydration / carbonation degrees as a function of the preparation method, Appl. Surf. Sci. **29**, 40 (1987)
[11] S. Elliott, G. Scarel, C. Wiemer, M. Fanciulli, Y. Lebedinskii, A. Zenkevich, I. L. Fedushkin: Precursor combinations for ALD of rare earth oxides and silicates – A quantum chemical and x-ray study, Electrochem. Soc. Proc. **2005-09**, 605 (2005)
[12] S. Stemmer: Thermodynamic considerations in the stability of binary oxides for alternative gate dielectrics in complementary metal-oxide-semiconductors, J. Vac. Sci. Technol. B **22**, 791 (2004)
[13] R. L. Puurunen, W. Vandervorst: Island growth as a growth mode in atomic layer deposition: A phenomenological model, J. Appl. Phys. **96**, 7686 (2004)
[14] S. Spiga, C. Wiemer, G. Tallarida, G. Scarel, S. Ferrari, G. Seguini, M. Fanciulli: Effects of the oxygen precursors on the electrical and structural properties of $HfO_2$ films grown by atomic layer deposition on Ge, Appl. Phys. Lett. **87**, 112904 (2005)
[15] L. Marsella, V. Fiorentini: Structure and stability of rare-earth and transition-metal oxides, Phys. Rev. B **69**, 172103 (2004)

# Index

# Local Atomic Environment of High-κ Oxides on Silicon Probed by X-Ray Absorption Spectroscopy

Marco Malvestuto and Federico Boscherini

Department of Physics and CNR-INFM, University of Bologna,
Viale C. Berti Pichat 6/2, 40127 Bologna, Italy
federico.boscherini@bo.infm.it

**Abstract.** We describe the use of X-ray absorption spectroscopy (XAS) with synchrotron radiation to study the local atomic structure of high-κ oxide thin and ultra-thin films deposited on silicon. A brief description of the advantages of XAS to probe local atomic arrangements in this context is given. We then describe two case studies: $Y_2O_3/Si(001)$ and $Lu_2O_3/Si(001)$.

## 1 X-Ray Absorption Spectroscopy in the Study of Rare Earth Oxides

A description and an understanding of the local atomic environment of thin and ultra-thin films and of their interfaces are of paramount importance since it is the local (i.e., first and second atomic shell) interactions which determine, to a significant extent, the electronic properties. The main advantages of X-ray Absorption Spectroscopy (XAS) are a rather direct structural interpretation, atomic selectivity, a high precision in the determination of interatomic distances and the possibility to study monolayer-thin systems ($\sim 10^{14}$ atoms/cm$^2$, if particular detection schemes are used; the possibility of applying XAS to these systems was recognized at the beginning of the modern development of the technique [1].

XAS [1–3] derives information on the local atomic structure from an analysis of the energy-dependent modulations of the X-ray absorption cross-section which occur in condensed matter; these modulations appear at energies immediately above the threshold for photoelectric absorption (the "absorption edge") in which the initial state is a deeply bound core orbital (most often $1s$ or $2p$) and may extend from a few hundred to over $1000\,\mathrm{eV}$. The energy of the absorption edge is a characteristic of each element, so that it is possible to selectively probe the local structure around each element in a compound.

In order to measure an XAS spectrum with adequate signal-to-noise ratio and energy resolution, a synchrotron radiation source is essential. Synchrotron radiation sources are relativistic electron storage rings which provide photon beams characterized by high brilliance and tunability over a wide energy range. Recently, a number of advanced "third-generation" synchrotron

M. Fanciulli, G. Scarel (Eds.): Rare Earth Oxide Thin Films,
Topics Appl. Physics **106**, 143–152 (2007)
© Springer-Verlag Berlin Heidelberg 2007

radiation sources have been constructed, one example of which is the European Synchrotron Radiation Facility (ESRF) in Grenoble, France. Third-generation sources provide high-quality photon beams which are essential to detect the weak signals typical of ultra-thin films. In order to increase the surface sensitivity of XAS, particular detection modes and experimental set-ups can be employed: total electron yield (TEY) or fluorescence yield (FY), possibly coupled to a grazing incidence geometry.

Traditionally, the interpretation of XAS spectra has been divided into the analysis of the region within a few tens of eV from the edge (XANES: X-ray Absorption Near Edge Structure, also known as NEXAFS: Near Edge X-ray Absorption Fine Structure) and the analysis of the fine structure present sufficiently far from the edge, say 50 eV above the edge and beyond (EXAFS: Extended X-ray Absorption Fine Structure). Scattering theory provides a unifying picture of XAS in these two energy regions [4]. Modulations in the X-ray-absorption cross-section arise from a modification of the final state of the photo-excited electron due to the scattering by the atoms surrounding the excited one. In the absence of neighboring atoms, the final state would be an outgoing spherical wave. By contrast, the presence of neighboring atoms scatters this wave and, depending on the wavelength of the photoelectron and on the distance between excited and scattering atom, gives rise to an interference effect leading to an increase or a decrease of the X-ray-absorption cross-section. What distinguishes the two energy regions mentioned above is that far from the edge (EXAFS), single scattering of the photoelectron is usually dominant (only two atoms are involved in the process), while in the near-edge region multiple scattering (more than two atoms involved) becomes increasingly more important. Each scattering path gives rise to a modulation of the X-ray-absorption cross-section which is a sinusoidal function of the photoelectron wave vector. XAS, whether in the extended or in the near-edge spectral region, is a technique with a local sensitivity, the physical origin of which is the limited lifetime of the core hole left by the photoabsorption process ($\sim 10^{-15}$ s) and short mean free path (5–10 Å) of the photoelectron; the combination of these effects guarantees that atomic correlations at distances greater than $\sim 10$ Å rarely contribute to an XAS spectrum.

As an illustrative example, in Fig. 1 we show raw XAS data at the Lu $L_{III}$ edge of a $Lu_2O_3$ polycrystalline powder while in the inset the corresponding extended fine structure oscillations (EXAFS, or $\chi(k)$) are shown. In Fig. 2 we report the magnitude of the Fourier Transform (FT) of the $\chi(k)$ data. This oxide, like many other rare earth oxides studied in the field of high-$\kappa$ materials, has the cubic bixbyite $Mn_2O_3$ structure with space group $T_h^7 - Ia3$. The bixbyite structure can be thought of as derived from a fluorite structure ($MO_2$ stoichiometry) in which one quarter of the anions is removed. There are two inequivalent cation sites, both of which are six-fold coordinated to oxygen atoms in a quasi-octahedral configuration. In site I, all oxygen atoms lie at the same distance (at 2.19 Å in $Lu_2O_3$). In site II, O atoms lie at different interatomic distances with average distances of 2.24 Å.

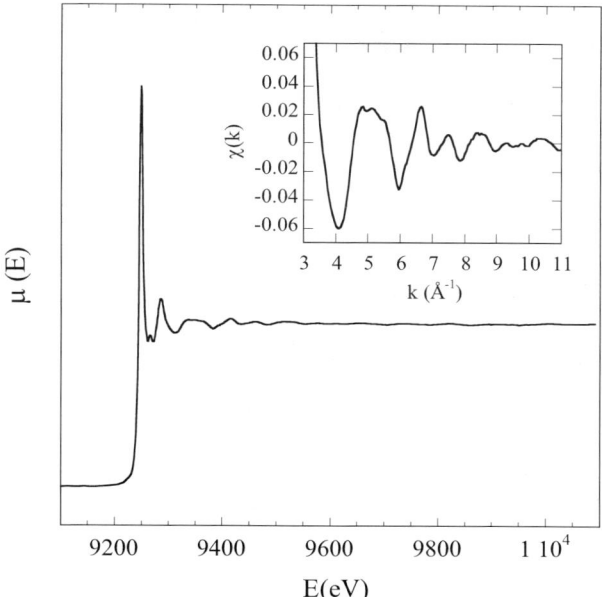

**Fig. 1.** The main figure reports a raw Lu $L_{III}$ edge XAS spectrum of $Lu_2O_3$ polycrystalline powder while in the inset the corresponding extracted fine structure oscillations $\chi(k)$ are shown

Second-shell cation distances are twofold and fourfold for site I and site II, with a characteristic splitting of 0.3–0.4 Å. The FT representation of EXAFS provides, at a qualitative level, a picture of the local bonding around the excited atom (Lu, in this case); in Fig. 2 the first peak is due to the first-shell Lu−O atomic correlations, while the second and third ones are due to the split Lu−Lu second shell. The presence of different, closely spaced, interatomic distances complicates quantitative analysis of EXAFS data. Despite this difficulty, good-quality fits (obtained as described below) are possible, as reported in Fig. 2. In general, from the analysis of an EXAFS spectrum, the following local structural parameters can be obtained: interatomic distances ($R$, typically ±0.01 Å for the first shell), identity of neighboring atoms and their coordination number ($CN$, typically ±10 %) and the variance of the Gaussian distance distribution function, also known as the "Debye–Waller factor" ($\sigma^2$, typically ±1 × 10$^{-3}$ Å$^2$).

An analysis of the near-edge spectrum (XANES) provides information on the site symmetry of the absorbing atom (e.g., substitutional or interstitial), on the oxidation state of cations and on the composition of the first few coordination shells. Analysis of the XANES from the structural point of view is greatly aided by the development of real-space multiple scattering theory [3–5]. An example of this kind of analysis is reported in Fig. 3, again

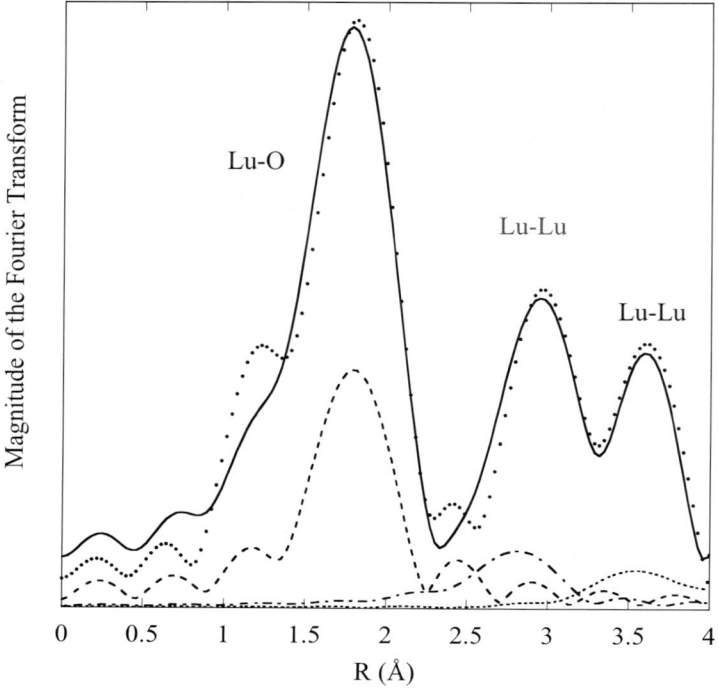

**Fig. 2.** Fourier Transform of the $\chi(k)$ data reported in Fig. 1 (*continuous line*), the fit based on the bixbyite structure (*dots*) and the separate contributions to the fit (*dashed lines*, rescaled for clarity)

for the $L_{III}$ edge in $Lu_2O_3$. We report the experimental spectrum and simulations performed with the FEFF 8.2 code as a function of the number of atoms included in the input cluster: up to the $1^{st}$, $2^{nd}$, $3^{rd}$ shell and finally for a cluster composed of 206 atoms (radius of 9 Å). This kind of analysis of XANES is particularly useful since it provides a direct link between the local structural environment and spectral features. An alternative approach is based on molecular orbital theory; for example, this has been used in the high-$\kappa$ field by *Lucovsky* et al. [6, 7].

## 2 Case Studies: $Y_2O_3$ and $Lu_2O_3$ on Si(001)

XAS has been applied by us to study the growth of $Y_2O_3$ on Si(001) [8]. The physical properties of $Y_2O_3$/Si which make it interesting include a relatively high $\kappa$ value (16–18) [9, 10], relatively high band offsets with respect to silicon [10, 11], a predicted thermodynamic stability [9, 12] and the lattice commensurability (the lattice parameter of cubic yttria being approximately

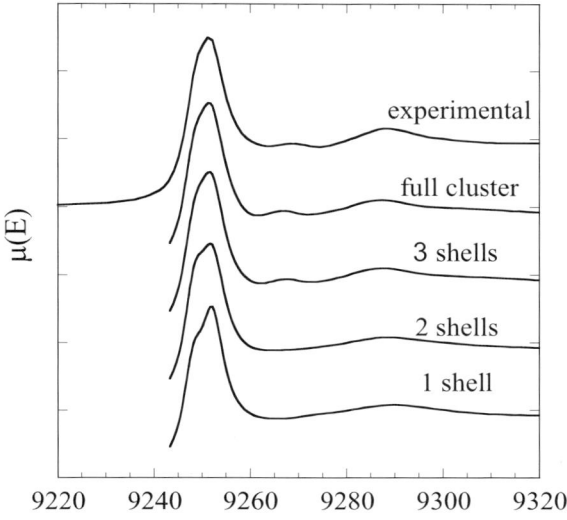

**Fig. 3.** Lu $L_{III}$ edge XANES spectrum of $Lu_2O_3$ polycrystalline powder: experiment and simulations performed as described in the text

twice that of silicon) which could, in principle, allow high-quality heteroepitaxial growth.

$Y_2O_3$ layers were grown by molecular beam epitaxy (MBE) in ultra-high vacuum (UHV) at a substrate temperature of 450 °C; $Y_2O_3$ was evaporated by means of an electron-gun from a sintered ceramic target. The oxide growth rate was $\sim 0.4$ Å/s for all films. Five samples with thicknesses ranging from 2–20 nm were deposited and studied; some samples were subsequently annealed in situ in UHV for 30 min at 500 °C.

In Fig. 4 we report FT of Y K-edge EXAFS data. These measurements were performed at the GILDA beam line (BM08) of the European Synchrotron Radiation Facility in Grenoble, France. Y absorption coefficients for all epilayers were collected in the fluorescence mode using a dynamically sagitally focusing Si(311) monochromator [13] and a 13-element hyperpure Ge detector. Due to the very limited thickness of this system, the samples were measured in the grazing incidence geometry [14] (Refle-XAFS). The advantage of the grazing incidence setup in the hard XAS study of a thin film deposited on a relatively light element substrate such as Si is the minimization of the (elastic and inelastic) scattering signal originating from the substrate due to the limited penetration of the beam in the sample; this leads to a smaller background signal and to the reduction of spectral distortions. The data shown in Fig. 4 indicate that the 20 nm sample has a local structure which bears a close resemblance to that of bulk $Y_2O_3$. The same is true for the annealed 4 nm sample, while the as-deposited one exhibits a less defined second-shell structure, indicating the presence of a greater degree of local

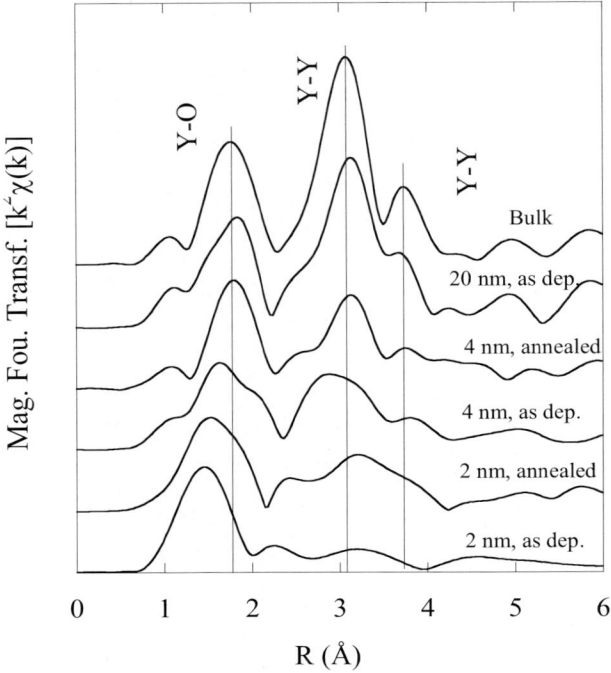

**Fig. 4.** Magnitude of the $k^2$-weighted Fourier Transform of the Y edge EXAFS for the five samples of $Y_2O_3$ epilayers on Si(001) and for bulk yttria; the *thin vertical lines* identify the positions of the main peaks in yttria

structural distortions. The 2 nm samples exhibit greater deviations with respect to the bulk. In fact, it can be clearly seen that the first-shell peak in the FT is significantly shifted to lower interatomic distance values. Quantitative analysis indicated that in the 2 nm samples a significant number of Y−Si bonds are present, at a distance which is close to that found in $YSi_2$. A long-range ordered silicide phase was observed by X-ray diffraction at the higher deposition temperature of 610 °C, but not at 450 °C; hence, it can be concluded that the Y−Si bonds observed are the precursors of the silicide phase which forms at higher deposition temperature. The possibility of detecting Y−Si bonds is a unique characteristic of XAS. It should be pointed out that no evidence for the formation of a silicate phase was found.

We now turn our attention to $Lu_2O_3$ thin and ultra-thin films deposited by atomic layer deposition (ALD) on Si(001) [15]. $Lu_2O_3$ has a large band gap, between 4.8–5.8 eV, and large conduction band offset with respect to silicon [16–18]. In the form of thin films, $Lu_2O_3$ has a moderate value of $\kappa$ equal to 12 [19]. The $Lu_2O_3$ films were deposited alternating injections of the newly synthesized complex $[(\eta^5\text{-}C_5H_4SiMe_3)_2LuCl]_2$ (11 s) as Lu source [20], and injections of $H_2O$ (11 s) as oxygen source, both carried in the reaction

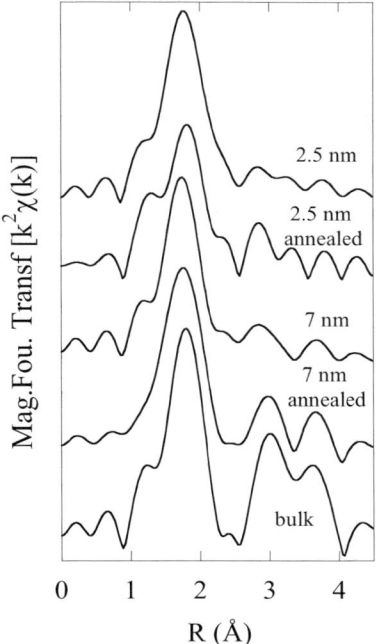

**Fig. 5.** Magnitude of the Fourier Transform of Lu $L_{III}$ edge EXAFS data for $Lu_2O_3/Si(001)$ thin films and bulk $Lu_2O_3$

chamber by an $N_2$ flux. The Lu and O precursors were kept respectively at 195 °C and 18 °C. An $N_2$ flux (8 s) purged away the reaction by-products after each step of the ALD cycle. In all, 82 and 18 cycles were applied for the films studied in this work. Four films were studied: two had a thickness of approximately 2.5 nm and the other two of approximately 7 nm; both as-deposited and annealed films (950 °C for 60 s in $N_2$) were studied. For the thicker films, in-house X-ray reflectivity showed the presence of an interface layer characterized by an electronic density intermediate between that of $Lu_2O_3$ and $SiO_2$ of thickness 0.9–1.5 nm; on the same films, in-house X-ray diffraction showed that the as-deposited films are poorly crystalline while annealing crystallizes these completely in the bixbyite structure. X-ray reflectivity and diffraction were not possible on the 2.5 nm films due to the limited thickness.

In Fig. 5 we report FT of Lu $L_{III}$ edge EXAFS data. The experiment was performed under conditions very similar to those described above for $Y_2O_3$. Inspection of the figure (and quantitative fitting) indicates that the local structure of the 7 nm films is very similar to bixbyite $Lu_2O_3$; however, the as-deposited film has weaker second- and third-shell peaks, indicating a greater degree of structural disorder. In the 2.5 nm films, atomic correlations above the first shell are absent, suggesting a high degree of structural disorder;

the first peak was found to be due to Lu−O bonds, as in the bulk, but with a significantly expanded ($\sim$ 0.05 Å) interatomic distance. Since a low-density interface layer was detected in these films, an important issue is its composition and structure. The EXAFS data show that this layer does not have a significant degree of local order; in particular, the presence of a well-ordered silicate phase can be ruled out in these samples. Again, we see that the local character of XAS is able to provide original insight in a high-$\kappa$ silicon system.

## 3 Conclusions

XAS is powerful tool to study local structure in thin and ultra-thin films. In the case of high-$\kappa$ on silicon, we have shown how it is possible to study and quantify the local structure as a function of thickness and sample treatment, detect local bonding arrangements even if they do not give rise to long-range ordered phases, and provide insight into the nature of interface phases.

### Acknowledgements

Our high-$\kappa$ work was performed in close collaboration with M. Fanciulli, G. Scarel, S. Spiga and G. Wiemer of the MDM-INFM group, and it is thus a pleasure to acknowledge their contribution to this research; we acknowledge F. D'Acapito for continuing collaboration in the field of XAS and stimulating discussions. Financial support was provided by project REOHK, funded by INFM. Measurements at ESRF and ELETTRA were supported by the Synchrotron Radiation Committee of INFM.

## References

[1] P. A. Lee, P. H. Citrin, P. Eisenberger, B. M. Kincaid: Extended X-ray absorption fine structure – its strengths and limitations as a structural tool, Rev. Mod. Phys. **53**, 769 (1981)
[2] D. C. Koningsberger, R. Prins (Eds.): *X-Ray Absorption: Principles, Applications, Techniques of EXAFS, SEXAFS and XANES* (Wiley, New York 1988)
[3] J. J. Rehr, R. C. Albers: Theoretical approaches to X-ray absorption fine structure, Rev. Mod. Phys. **72**, 621 (2000)
[4] C. R. Natoli, M. Benfatto: Beyond the pair distribution function in X-ray absorption spectra, J. Phys. Colloq. **47**, C–8 (1986)
[5] A. L. Ankudinov, B. Ravel, J. J. Rehr, S. D. Conradson: Real-space multiple-scattering calculation and interpretation of X-ray-absorption near-edge structure, Phys. Rev. B **58**, 7565 (1998)
[6] G. Lucovsky, J. L. Witten, Y. Zhang: A molecular orbital model for the electronic structure of transition metal atoms in silicate and aluminate alloys, Solid State Electron. **46**, 1687 (2002)

[7] G. Lucovsky, J. G. Hong, C. C. Fulton, Y. Zou, R. J. Nemanich, H. Ade: X-ray absorption spectra for transition metal high-dielectrics: Final state differences for intra- and inter-atomic transitions, J. Vac. Sci. Technol. B **22**, 2132 (2004)

[8] M. Malvestuto, R. Carboni, F. Boscherini, F. D'Acapito, S. Spiga, M. Farciulli, A. Dimoulas, G. Vellianitis, G. Mavrou: X-ray absorption study of the growth of $Y_2O_3$ on Si(001), Phys. Rev. B **71**, 075318 (2005)

[9] K. J. Hubbard, D. G. Schlom: Thermodynamic stability of binary oxides in contact with silicon, J. Mater. Res. **11**, 2757 (1996)

[10] P. W. Peacock, J. Robertson: Band offsets and Schottky barrier heights of high dielectric constant oxides, J. Appl. Phys. **92**, 4712 (2002)

[11] A. Ohta, M. Yamaoka, S. Miyazaki: Photoelectron spectroscopy of ultrathin yttrium oxide films on Si(100), Microelectron. Eng. **72**, 154 (2004)

[12] L. Marsella, V. Fiorentini: Structure and stability of rare-earth and transition-metal oxides, Phys. Rev. B **69**, 172103 (2004)

[13] S. Pascarelli, F. Boscherini, F. d'Acapito, J. Hardy, C. Meneghini, S. Mobilio: X-ray optics of a dynamical sagittal-focusing monochromator on the GILDA beamline at the ESRF, J. Synchrotron Rad. **3**, 147 (1996)

[14] F. D'Acapito, I. Davoli, P. Ghigna, S. Mobilio: The reflEXAFS station at the GILDA beamline (BM08) of ESRF, J. Synchrotron Rad. **10**, 260 (2003)

[15] M. Malvestuto, G. Scarel, C. Wiemer, M. Fanciulli, F. D'Acapito, F. Boscherini: X-ray absorption spectroscopy study of $Yb_2O_3$ and $Lu_2O_3$ thin films deposited on Si(100) by atomic layer deposition, Nucl. Instrum. Met. B (2006) accepted for publication

[16] G. V. Samsonov, I. Y. Gil'man: Electronic structure and physical properties of the oxides of the lanthanides – a review", Poroshkovaya Metallurgiya **11**, 73 (1974)

[17] A. V. Prokofiev, A. I. Shelykh, B. T. Melekh: Periodicity in the band gap variation of Ln,X3 (X = 0, S, Se) in the lanthanide series, J. All. Comp. **242**, 41 (1996)

[18] G. Seguini, E. Bonera, S. Spiga, G. Scarel, M. Fanciulli: Energy-band diagram of metal/$Lu_2O_3$/silicon structures, Appl. Phys. Lett. **85**, 5316 (2004)

[19] E. Bonera, G. Scarel, M. Fanciulli, P. Delugas, V. Fiorentini: Dielectric properties of high-oxides: theory and experiment for $Lu_2O_3$, Phys. Rev. Lett. **94**, 027602 (2005)

[20] H. Schumann, I. L. Fedushkin, M. Hummert, G. Scarel, E. Bonera, M. Fanciulli: Crystal and molecular structure of $[\eta^5\text{-}C_5H_4SiMe_3)_2LuCl]_2$: A precursor for the production of $Lu_2O_3$ films, Z. Naturforsch. **59b**, 1035 (2004)

# Index

# Local Structure, Composition and Electronic Properties of Rare Earth Oxide Thin Films Studied Using Advanced Transmission Electron Microscopy Techniques (TEM-EELS)

Sylvie Schamm[1], Giovanna Scarel[2], and Marco Fanciulli[3]

[1] Groupe NanoMatériaux CEMES/CNRS, BP 94347, 31055 Toulouse Cedex 04, France
   schamm@cemes.fr
[2] CNR-INFM MDM National Laboratory, Via C. Olivetti 2, 20041 Agrate Brianza (MI), Italy

**Abstract.** This contribution aims to demonstrate the ability of transmission electron microscopy (TEM) and its associated techniques, high-resolution transmission electron microscopy (HRTEM) and electron energy loss spectroscopy (EELS), to contribute to the understanding of the structural, chemical and electronic properties at the nanometre level of rare earth oxide (REO) thin films in the context of their potential use as alternative gate dielectrics to $SiO_2$. A review of the existing work on binary REO and a preliminary work on atomic layer deposited (ALD) $Lu_2O_3$/Si stack as-grown and annealed are proposed.

## 1 Introduction

Rare earth oxide (REO) thin films can be considered as useful materials for microelectronics, and for the fabrication of good-quality thin-film optical coatings [1]. REO films are transparent over a wide spectral range and have the refractive index in the range 1.8–1.92 at 550 nm. Moreover, very good dielectric and insulating properties classify them for applications in electronic microcircuits [2]. Since 5 years, search for high-permittivity (high-$\kappa$) dielectric materials to replace $SiO_2$ as a gate insulator in metal-oxide-semiconductor field effect transistors (MOSFETs) has prompted an intense scientific and technological activity on oxide compounds. The most studied materials are the metal oxides such as $HfO_2$, $ZrO_2$ and their silicates and aluminates [2,3]. However, they still have some problems such as interfacial layer (IL) and/or microcrystal formation during the post-deposition annealing process that lead to increase of equivalent oxide thickness and gate leakage current [4]. REO such as $La_2O_3$, $CeO_2$, $Pr_2O_3$, $Gd_2O_3$ and $Lu_2O_3$ are considered among the next generation of promising candidates either because of their large optical band gap, which will provide large enough energy barriers for electrons and holes in Si substrate, or because of their large dielectric constant (see the Chapter by *Scarel* et al. in this book). Moreover, for all REO, thermodynamical stability on Si is predicted, against both silicide and silicates formation [3].

M. Fanciulli, G. Scarel (Eds.): Rare Earth Oxide Thin Films,
Topics Appl. Physics **106**, 153–178 (2007)
© Springer-Verlag Berlin Heidelberg 2007

Nevertheless, this thermodynamic stability is controversial. A recent paper indicates some of the rare earth oxides, including $La_2O_3$, YbO and $Lu_2O_3$, to be stable only against silicide formation [5], and transmission electron microscopy analyses performed on films deposited using various methods seem to agree with these theoretical conclusions [6].

It is clear that one of the major issues concerning high-$\kappa$ oxide films is their thermodynamical stability on silicon or on the high-mobility semiconductor substrates on which they might be deposited, and with the metal gate [7]. The issue is of high importance because real devices undergo high-temperature annealing during their processing. Such a procedure might alter the composition and the thickness of the IL, a fact that could degrade the total capacitance of the device. As semiconductor process technology scales down to below the 65 nm node, the gate oxide thickness needs to be reduced to a few nanometres (see the Chapter by *Kakushima* et al. in this book). Therefore, transmission electron microscopy (TEM) is an invaluable technique to be investigated together with its associated analytical techniques in order to be able to visualise the devices at the scale of the structural and chemical modifications they suffer during their preparation.

Recent developments in analytical electron microscopy enable us to investigate the structural, chemical and electronic properties of materials on a sub-nanometre scale. The optical performance of modern transmission electron microscopes (TEM) equipped with a field emission gun (FEG) and with a post-column energy filter makes it possible to generate high-resolution electron microscopy (HREM) images of the material structure and, simultaneously, allows the determination of the chemical composition by electron energy loss spectroscopy (EELS) with a small electron probe of nanometre size. Moreover, with the energy resolutions better than 1 eV currently available with the last generation of spectrometers and energy filters, insight into the local electronic structure is also accessible at the same time with the same spatial resolution through a detailed study of the fine structures observed in the EELS spectrum, in the low energy-loss domain (VEELS) and near the core-loss edges (ELNES).

The information that can be brought by TEM-EELS to support the research on REO thin films as potential gate dielectric is discussed in the following. The specific methods, used nowadays to image the atomic structure and the chemical composition at the nanometre level, are presented together with recent results on investigations of REO thin films. Then, the TEM-EELS study of the particular case of ALD-$Lu_2O_3$ films will be illustrated.

## 2 State of the Art

TEM associated with EELS can address some important points in relation with the many requirements needed for alternative gate dielectrics for Si-based MOSFETs. Various studies have already been devoted to high-$\kappa$ materials by the community of microscopists with different equipments and different techniques, but all aim to solve the same problems related to high-$\kappa$ dielectric layers. They are: 1. *the precise thickness determination of the gate oxide film and of the IL.* Indeed, precision is important because even a 0.1 nm decrease in the oxide thickness can lead to an order of magnitude increase in leakage current; 2. *the crystallization state in the gate dielectric film* before and after post-deposition annealing treatments. Whether it is amorphous, polycrystalline or epitaxial has noticeable consequences. First, polycrystalline films, which present grain boundaries and interfacial roughness, are believed to be correlated to an undesired increase in leakage currents and reduction in transistor mobility [8]; second, a stable epitaxial structure could avoid a $SiO_2$ IL formation and, third, amorphous materials are preferred by the semiconductor industry; 3. *the quality of the interface*; indeed knowledge of the roughness of the atomic arrangement at the interface between Si and the REO will give an insight into the associated electronic structure in order to understand the electrical properties; 4. *the chemical nature of the different coexisting phases* in the stack; this point must be systematically addressed together with the previous three preceding ones. For example, the calculation of the equivalent oxide thickness (EOT) of the film as defined in [2] can be done if the dielectric properties of the layers are known (chemical nature known) together with their physical thickness; 5. *the band gap*; indeed this gap must be high enough.

In the following, examples from the literature will be used to illustrate how TEM-EELS can address these different points.

## 2.1 Thickness Measurement

Precise measurement of the oxide thickness is critical. Many examples of dielectric and IL thickness analyses are found in the literature. They support EOT determinations [9, 10], models for X-ray reflectivity (XRR) data fitting [11] and capacitance measurements or estimation of the dielectric constants of amorphous phases, as is the case for a $La_2Hf_2O_7$ film [12].

High-resolution lattice images are a reliable tool for precise thickness determination. The images can be obtained either with a coherent, phase-contrast imaging, i.e., with the high-resolution transmission electron microscopy (HRTEM) technique or with an incoherent, amplitude-contrast technique, i.e., high-angle anular dark field scanning transmission electron microscope (HAADF-STEM) imaging. In the former case, sample thickness (thickness travelled by the electron beam of the electron microscope) must be thin enough (10 nm–20 nm) to avoid any contrast artefact in the image of the interface of the film with the substrate due to roughness of this interface. In the latter case, thickness can be measured in samples as thick as 50 nm provided a large inner angle detector and a highly localized probe are used [13, 14]. A comparative and detailed investigation of dielectric film thickness determination by HRTEM and HAADF-STEM can be found in [14]. In general, thickness measured by HRTEM can be 0.5 nm thinner than the one determined by HAADF-STEM [13,15]. The interface roughness is the major source of error in ultra-thin gate dielectric thickness measurement. Averaging of the roughness over the sample thickness (travelled by the electrons) occurs. Both oxide thickness, 1.5 nm, and interface roughnesses, 0.175 nm and 0.3 nm, have been measured with HAADF images for the stack c-Si/SiO$_2$/poly-Si [13].

## 2.2 Lattice Images

In order to study the atomic structure of an interface, the TEM sample is prepared as a cross-section so that when observed in the TEM, it is oriented in such a way that the electron beam is parallel to the film/substrate interface, i.e., the (001) planes of Si in [110]-oriented Si. As a consequence of this procedure, the interface is edge-on. The electron microscopy modes are the same as these previously described for thickness measurements.

HRTEM is a parallel detection method where a nearly parallel 200 keV to 300 keV electron beam simultaneously illuminates the entire imaged area of the crystal. The primary and elastically scattered beams interfere and form the image by a coherent superposition. For a thin enough area, the image is related to the projected potential of the object (atomic positions) and therefore "represents" the crystalline structure of the specimen. For a thicker area, the image must be compared to computer simulations in order to be able to understand the more complicated contrasts (contrast reversals, for example).

HAADF is a serial detection method where a sub-0.25 nm diameter probe (atomic dimensions) of 100 keV to 300 keV electrons is focused and scanned across the imaged area of the crystal. On microscopes equipped with a high-angle annular dark field detector, large-angle elastic scattering can be detected and incoherently summed to obtain an imaging signal which depends strongly on the atomic number of the atoms illuminated by the probe (Z-contrast image). Moreover, since the nominal probe size is smaller than many inter-atomic distances, the scattered signal reveals atomic columns with good contrast [16]. The thicker the sample, the better is the contrast, with an upper limit of around 50 nm. Because the image is acquired serially pixel by pixel, the atomic position cannot be determined as accurately as with a parallel detection method. Due to its sensitivity to Z-contrast, an ADF image is also a first approach to chemical composition analyses. This is particularly useful in the case of REO deposited on Si with often SiO$_2$-based IL. In ADF images, the bright contrast corresponds to the REO and the dark contrast to Si and SiO$_2$, SiO$_2$ appearing darker than Si. For this reason, in the case of REO, HAADF is more often used for chemical investigation and precise probe positioning during EELS analysis than for atomic structure imaging [17].

HRTEM of films and interfaces in REO thin films/Si stacks are often investigated in the literature dealing with CMOS gate dielectrics. The most studied cases concern lanthanum-based films (lanthanum oxide, lanthanum silicates and lanthanum aluminates), gadolinium-based films and praseodymium oxide films. The structure and stability of the films are investigated before and after post-deposition annealing treatments.

In general, very thin $La_2O_3$ films are always amorphous or poorly crystalline. $La_2O_3$ layers, about 2 nm thick and deposited by molecular beam epitaxy (MBE) on thermal $SiO_2$ on Si, are amorphous after annealing at 600 °C for 30 s in $N_2$ and begin to crystallize when the temperature is raised to 800 °C [9]. $La_2O_3$ films, about 4 nm thick and deposited by ALD on a chemical oxide Si, are mostly amorphous with some crystalline regions after 900 °C anneal during 60 s in $N_2$ [18].

Circa 50 nm thick $Pr_2O_3$ films grown by metal-organic chemical vapour deposition (MOCVD) on a HF-last treated Si(001) are polycrystalline in the 650 °C to 850 °C temperature range [19]. A 10 nm $Pr_2O_3$ film deposited by MBE on hydrogen-terminated Si(001) grows with a cubic structure as (110) domains with two orthogonal in-plane orientations. On the other hand, on Si(111) and for a 6 nm thick film, a perfect epitaxial growth of the hexagonal form was obtained [20].

MOCVD 6.5 nm thick $Gd_2O_3$ films prepared by electron beam evaporation on HF-last treated Si(111) or with epitaxial $Si_3N_4$ IL on Si(111) are partially crystallized after in situ annealing at 700 °C for 5 min at $5 \times 10^{-8}$ Torr [21]. $Gd_2O_3$ films, with thicknesses between 3.7 nm and 5.4 nm, deposited by electron beam evaporation on a HF-last cleaned Si(001) at 500 °C and annealed for 10 min in $O_2$ at 500 °C, 700 °C or 780 °C, are polycrystalline with an increase of the crystalline grain size upon annealing [11].

The silicate forms of La and Gd crystallize at higher temperatures than their oxide counterparts. Lanthanum and gadolinium silicate films (respectively 23 nm and 30 nm thick) deposited by electron-beam evaporation on HF-last Si crystallize at temperatures between 900 °C and 950 °C, and 1000 °C and 1050 °C respectively [22]. Lanthanum aluminate films, a few nm thick and prepared by pulsed laser or MOCVD, remain amorphous up to temperatures of 860 °C and 900 °C respectively [23, 24].

Besides the crystallisation state of the film, the chemical stability of the interface between the Si substrate and the REO is of primary concern. Non-reactive interfaces appear sharp, with a direct transition between the (001) planes of Si and the amorphous or crystalline phase of the high-$\kappa$ observed with a dark contrast on HREM images. When reactivity occurs, often an IL of bright contrast is seen on the HRTEM images. Sometimes, for amorphous REO films, there is no particular change in the contrast of the image near the interface and it is difficult to conclude about the existence of a reaction [9, 18]. Moreover, it is seen in the literature that the same REO/Si stacks prepared in different ways can lead to interaction or no interaction, depending on the deposition method used (physical vapour deposition or chemical vapour deposition), and primarily depending on a critical parameter, the oxygen partial pressure. This point has been discussed for the $La_2O_3$ [18],

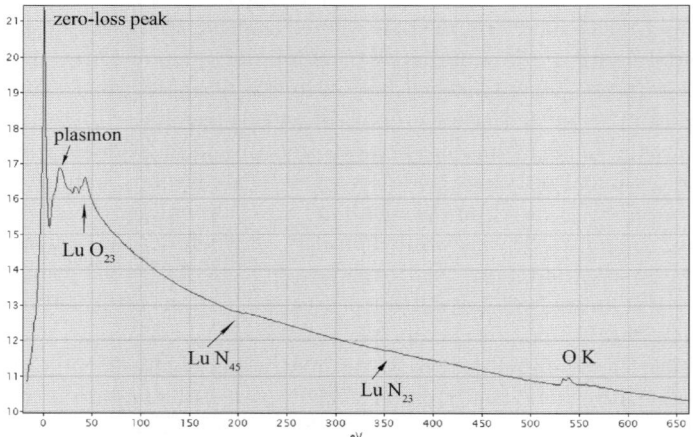

**Fig. 1.** Electron energy-loss spectrum of a reference $Lu_2O_3$ film shown with the electron intensity on a logarithmic scale, which is convenient to show simultaneously the zero-loss peak, the plasmon, and ionization edges due to each element

$Gd_2O_3$ [11] and $Pr_2O_3$ [25] cases. For these reasons, it is of great importance to go further in the investigation by using analytical techniques. X-ray photoelectron spectroscopy (XPS) is most systematically used [11, 18, 21, 24, 26, 27] but Medium Energy Ion Scattering Spectroscopy (MEIS) [28, 29] or Fourier transform infrared spectroscopy (FT-IR) [17, 25] have also been investigated. The existing TEM-EELS studies on REO will be presented in the following.

## 2.3 EELS Analysis

EELS has advantages over other techniques. It is performed on line with the TEM experiment. It adds a new dimension to the structural imaging with the electron microscope, which is high-resolution chemical analysis or imaging. The information provided by EELS results from electron–specimen interactions. These are elastic and inelastic scattering processes. The incident electrons have a high energy (typically between 100 keV and 300 keV). The basic signal is the EELS spectrum, which represents the scattered intensity as a function of the decrease in kinetic energy of the fast electrons. The EELS spectrum of $Lu_2O_3$ is shown as an example in Fig. 1.

In the low energy-loss domain, we find the zero-loss or "elastic" peak and the plasmon peak. The zero-loss peak represents electrons which are transmitted without suffering any measurable energy loss. The plasmon peak corresponds to the collective excitation of valence or conduction electrons. Then, edges are seen in the core-loss energy domain, above 100 eV, superimposed on a decreasing background, nearly as a power law of energy. They indicate the individual excitations of inner-shell electrons. The corresponding sharp rise in intensity occurs at the ionization threshold of which the energy is approximately the binding energy of the corresponding atomic shell. From the detailed study of the EELS spectrum, information on specimen thickness, dielectric response, gap (low energy-loss domain), elemental composition, chemical bonding and band structure (core-loss energy domain) is accessible, with an energy resolution depending essentially on the source of the TEM electron beam and a spatial resolution limited by the TEM optics and the uncertainty principle. Depending on the apparatus on which it is performed, EELS can be used in different ways. Conventional TEM-EELS (CTEM-EELS) probes elemental composition and electronic structure with energy resolution and spatial resolution that depend on the electron emission. The typical values are $0.7\,eV$ or $1\,eV$ and $1\,nm$ or $10\,nm$ in the case of field emission or thermionic emission respectively. Field Emission Gun-Scanning TEM-EELS (FEG-STEM EELS) does the same, with $0.3\,eV$ energy resolution and sub-nanometre spatial resolution. Energy Filtered TEM (EFTEM) maps elemental composition down to $1\,nm$ spatial resolution. All these modes have been performed for the study of REO thin films on Si [9, 19, 22, 30].

The classical way to perform EELS is to locate the electron probe at places of interest and acquire the EELS spectrum. In the particular case of Gd and La silicate films, HRTEM and EELS performed by this way on as-deposited and annealed films allow to see the modification of the topography of the interfaces with annealing temperature, together with the crystallization state of the film and the chemical nature of the IL [22]. Typically, the EELS signals are acquired in energy domains of the core-loss edges of interest: Si–$L_{23}$ near $100\,eV$, O–K near $530\,eV$, and RE edges (e.g., Gd–$N_{45}$ near $135\,eV$). The case of La is particular because the La–$N_{45}$ core-loss edge overlaps the Si–$L_{23}$ one. Thus, the La–$M_{45}$ edge near $850\,eV$ must be considered. Based on the quantitative elemental analysis and/or on the fine structure (ELNES) investigation and comparison of these edges to reference samples, important differences between La- and Gd-silicate films were found. After annealing thick films ($30\,nm$ for the Gd-silicate and $23\,nm$ for the La-silicate) at $900\,^{\circ}C$ in oxygen for $2\,min$, an interfacial $SiO_2$ layer $3.4\,nm$ thick is formed in the Gd-silicate, while this IL, $5.5\,nm$ thick, is a $SiO_2$-rich La-silicate layer in the La-silicate film [22]. For the Gd-silicate, some diffusion of Gd into the $SiO_2$ layer is observed after annealing at $1050\,^{\circ}C$.

The most commonly used technique for studying REO thin films as potential gate dielectric is the combination of HAADF imaging and EELS performed in the STEM mode (HAADF STEM-EELS). In this case, the probe size can be as small as $0.2\,nm$. A lattice resolution Z-contrast image is used to position the small probe at different locations within the layers and often line scans are performed across the interface from the Si substrate to the gate dielectric. A nice and invaluable example of this technique has been proposed in [31]. The electronic structure at the atomic scale of ultra-thin gate oxides is studied there. It is shown by comparing the O–K edges between interfacial and bulk $SiO_2$ that, for a $1\,nm$ oxide thickness (five silicon atoms across), the fundamental thickness limit of a usable $SiO_2$ gate dielectric is 0.7 nm. Features in the ELNES of the O–K edge are used for this experiment, particularly the modification of the O–K edge at the $Si/SiO_2$ interface is compared with the one of bulk $SiO_2$. This edge represents, in a first approximation, the unoccupied O 2p electronic density of states (DOS) [32]. A reduction in the edge energy onset and a reduction in the intensity of the first bulk peak are measured and attributed respectively to a reduced band gap due to additional electronic states, i.e., induced gap states, and a reduction of O second-nearest neighbours at the interface. HAADF STEM-EELS has also been used to illustrate La diffusion into a thermal oxide of about 2 nm thickness after annealing at 800 °C in $N_2$ ambient. The demonstration was carried out by documenting the decreasing intensity of the La $M_{45}$ edge as the STEM probe moves closer to the interface [9].

Recording a spectrum map that contains a series of EELS spectra acquired simultaneously as a function of a spatial coordinate, for example, the direction perpendicular to the substrate/film interface, is another approach of TEM-EELS. This method has several names: spatially resolved EELS [33, 34], laterally resolved EELS [35] or electron energy-loss spectroscopic profiling [36]. Contrary to STEM, TEM-EELS is intrinsically free from energy shifts that can occur when spectra are acquired sequentially with a small focused probe. Moreover, beam damage can be reduced since a wide lateral area is analyzed. Despite these advantages, it is less used than STEM-based techniques. This approach can be performed with a TEM equipped with an imaging filter (2D CCD detection) operated in the spectroscopy mode, for which a line focus perpendicular to the energy dispersion direction is employed (line-focus energy filtered TEM). An important and sensitive point is that the film/substrate interface must be precisely aligned with the energy-dispersive direction of the filter. In order to overcome this difficulty, a rotation holder can be used [30]. The nature of the IL formed in $Gd_2O_3$ films on clean Si(001) after rapid thermal annealing at 780 °C in an $O_2$ ambient was determined with success on the basis of HRTEM images and a detailed observation of the EELS fine structures observed at the O–K, Si–$L_{23}$ and Gd–$N_{45}$ edges, together with various analysis techniques of the data [30]. Three layers with different contrast are seen on the HRTEM images: a polycrystalline layer, an intermediate amorphous dark layer and a bright band. From the comparison of the fine structures obtained from these regions with the reference spectra from Si, $SiO_2$ and crystalline $Gd_2O_3$, the bright band was assigned to $SiO_2$ and the dark amorphous layer to a mixture of $SiO_2$ and $Gd_2O_3$.

Instead of probing a focused electron beam in HAADF STEM-EELS or of recording a spectrum map in line-focus EFTEM, bidimensional chemical maps with a high spatial resolution can be recorded with a TEM equipped with an energy filter operated in the imaging filtered mode [33]. This image mode allows us to separate the contributions from elastically and inelastically scattered electrons by inserting an energy-selecting slit in the energy-dispersive plane of a filter. Then, zero-loss filtering, or plasmon filtering, or ionization edge filtering is possible. For the case of ionization edge filtering, the contribution of the background is eliminated by recording several filtered images around the edge (referred as the *two-window* or *three-window* methods). For $Pr_2O_3$ films, EFTEM O–K, Pr–M and Si–L maps have been recorded using the *three-window* method and the corresponding line profiles have been extracted across the $Pr_2O_3$ film and the Si substrate. By comparing the filtered images, a two-layer interface is identified, as already confirmed by HRTEM images. By comparing the profiles, these layers have been attributed to a $SiO_2$ layer near the Si substrate and an oxygen-rich bottom layer containing also Pr between the $SiO_2$ layer and the $Pr_2O_3$ film. One noticeable fact is that some Si is present within the entire film and that the corresponding XPS Si 2p signal within the oxygen-rich bottom layer presents three components: $SiO_2$, Si and Si oxidized as a silicate phase. It is concluded that this layer is a "mixed $SiO_2$–$Pr_2O_3$ silicate layer". This layer appears nanocrystalline in dark-field and HRTEM images [37].

## 2.4 Sample Preparation

One essential point for microscopists is the quality of the TEM sample, i.e., when preparation artefacts are minimized. The TEM-EELS experiment will give the best-quality results if the sample preparation is good enough.

Generally, TEM work on high-$\kappa$/Si stacks has been performed on samples prepared by standard procedures (noted (3.) in the following). Among the few studies where different methods were tested, no differences were found between the observations made on the following stack Si/3.5 nm $HfO_2$/1 nm HfSiO/200 nm poly-Si prepared using: 1. Focused Ion Beam (FIB) or 2. cross-sectional tripod polishing or 3. standard grinding, polishing, dimpling and ion milling [38]. On the contrary, in [15], conventional mechanical polishing (case (3.) previously cited) with minimal ion milling (less than 5 min) and plasma cleaning is considered as the best way to obtain satisfactory results for ultra-thin gate dielectric chemical and structural studies. The FIB thinning followed by lift-out with C supporting grid (case (1.) previously cited) was found to introduce detrimental contamination effects on EELS and HAADF analysis results and also to induce major deterioration of the high-resolution HAADF-STEM images. The study was performed on the Si/bare oxide/3 nm $HfO_2$/200 nm poly-Si gate stack.

It is necessary to be careful with the use of water during the polishing treatment because of the possible solubility of REO in water, as mentioned in the case of thin $Gd_2O_3$ layers [17]. $La_2O_3$ is known to be the most unstable REO against ambient atmosphere. This signifies that any ex situ exposure of these films to air will certainly result in an uncontrolled reaction. Nevertheless, no particular remark is made on works dealing with $La_2O_3$ and remarkable device results were shown for these films.

# 3 The Case of $Lu_2O_3$/Si

To our knowledge, the stack $Lu_2O_3$/Si has not yet been investigated by TEM-EELS. Lu oxide films were deposited at $360\,°C$ in an Atomic Layer Deposition (ALD) reactor on a chemical oxide-covered p-type Si(100) wafer. $\{[\eta^5\text{-}C_5H_4SiMe_3]_2LuCl\}_2$ and $H_2O$, kept respectively at $195\,°C$ and $18\,°C$, were injected alternately in the reaction chamber using $N_2$ [39]. Three films have been studied: an as-grown film and two annealed samples at $550\,°C$ and $950\,°C$ in $N_2$ for $60\,s$. Cross-sectional and plan-view TEM samples were prepared by mechanical standard grinding, polishing and minimal $Ar^+$ ion milling (nearly $5\,min$).

## 3.1 Conventional TEM

In order to obtain morphological information on the films, plan-view and cross-section samples were observed with conventional TEM.

Electron diffraction patterns were acquired from a selected area (nearly $300\,nm$ diameter) over the sample (SAED) for all the plan-view samples (the electron beam travels perpendicularly through the $Lu_2O_3$/Si stack). Rings are observed superposed to the discrete diffraction pattern of the Si substrate [001] oriented (Fig. 2). The as-grown film gives rise to diffuse rings, whereas in the annealed films the rings become ever more discontinuous with increasing annealing temperature. This transformation is the consequence of the growing crystallite size. The crystallographic parameters deduced from the measurement of the ring diameter using the Si parameters as a reference correspond to the planes of indices 222, 400, 440 and 622 of the cubic structure of lutetium oxide, as defined in file 40 471 of the Inorganic Crystal Structure Database. Associated to the diffraction pattern are the dark-field images obtained by selecting a portion of the electrons that are diffracted within the first two rings of the $Lu_2O_3$ structure, as indicated by a white small ring on the as-grown diffraction pattern (Fig. 2). All the films are nanocrystalline with grain sizes varying from a few nm for the as-grown film to a few tens of nm for the annealed ones. Though these results are consistent with the grazing incidence X-ray diffraction (GIXRD) experiments performed on similar samples [38], they give a more precise determination of the nanocrystal size which is not possible by XRD at this level. This ability of TEM is important since it could be helpful to understand the role that can play nanocrystals in band gap defect formation through the states localized at grain boundaries between nanocrystals (see the Chapter by *Lucovsky* and *Phillips* in this volume). The measured sizes can be confirmed by the HRTEM observations.

Cross-section images show that the films are homogeneous laterally over large distances with a mean thickness of $10\,nm$ (Fig. 3). The image contrast is no longer uniform in the annealed samples where an IL of brighter contrast than that of the film can be seen.

as-grown          550 °C annealed          950 °C annealed

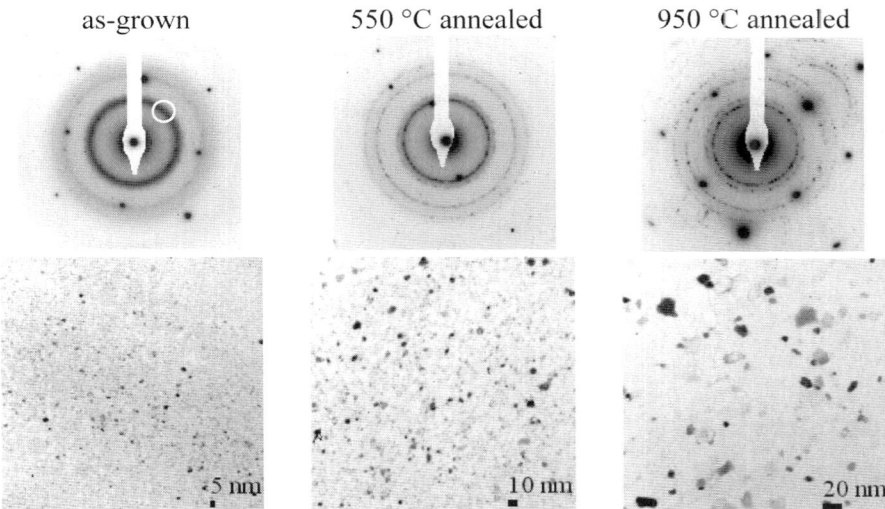

**Fig. 2.** SAED patterns (*top*) and corresponding dark-field images (*bottom*) obtained with selected diffracted electrons as indicated with the *white circle* on the first ring. In the SAED patterns, the rings are associated to the $Lu_2O_3$ films, whereas the discrete squared pattern is due to the [001]-oriented Si substrate. In the dark-field images, nanometre-sized $Lu_2O_3$ crystallites appear with dark contrast

as-grown          550°C annealed          950 °C annealed

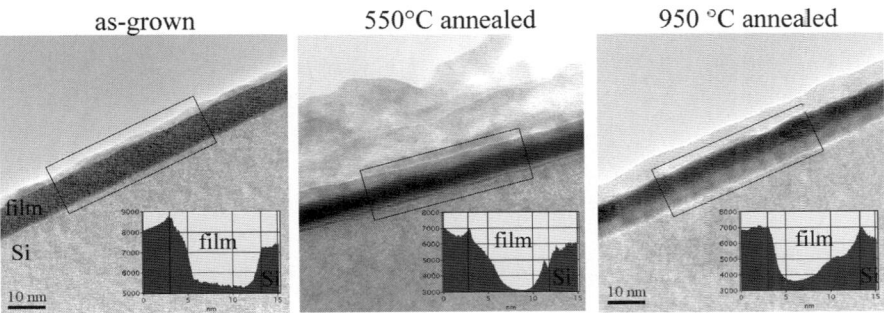

**Fig. 3.** Conventional bright-field images of the cross-sections with the Si substrate [011] oriented. Films are laterally regular and are 10 nm thick

## 3.2 Lattice Imaging

HRTEM experiments have been performed on a TEM-FEG microscope, Tecnai F20 ST, operated at 200 kV and equipped with a corrector of spherical aberration. The attainable point to point resolution is 0.12 nm.

The IA-3 bixbyite structure of $Lu_2O_3$ is confirmed by the quantitative study of the diffractograms (Fast Fourier Transform, FFT, of the image) of crystallized areas in the films imaged by HRTEM. The Si substrate is taken

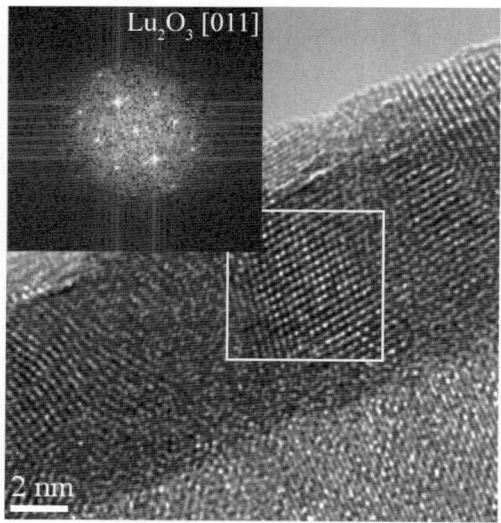

**Fig. 4.** A HRTEM image and the corresponding diffractogram (FFT) of a [011]-projected $Lu_2O_3$ crystallite

as a reference. Figure 4 shows a nanometre-sized crystal [011] oriented in an as-grown film. As in the annealed films, an IL is also observed in the HRTEM images of the as-grown films. The mean thicknesses of the ILs are around 1, 1.5 and 3.5 nm for the as-grown, 550 °C annealed and 950 °C annealed films respectively (Fig. 5). They are in very good agreement with the results of the modelling of the XRR spectra and capacitance–voltage (C–V) characteristics analysis obtained on very similar samples where the as-grown film correlates with a 1.1 nm thick IL and the 950 °C annealed film with a 3.3 nm thick IL [39].

### 3.3 EFTEM-EELS

EELS experiments have been performed on cross-section samples with the Tecnai F20 ST microscope equipped with a last-generation energy filter from Gatan, the TRIDIEM.

The EELS spectrum of a reference $Lu_2O_3$ film has already been discussed in Sect. 2.4 (Fig. 1). The energy positions of the ionization edges of $Lu-O_{23}$, $Lu-N_{45}$, $Lu-N_{23}$ and O–K are 28, 195, 359 and 532 eV respectively. VEELS features and ELNES of Si, $SiO_2$ and $Lu_2O_3$ reference samples have been examined in order to support the study of the $Lu_2O_3$/Si stacks. The cases of Si and $SiO_2$ have been well studied [40–42]. To our knowledge, the case of $Lu_2O_3$ has only been considered 30 years ago for the low-loss part and 20 years ago for the entire EELS spectrum [43, 44].

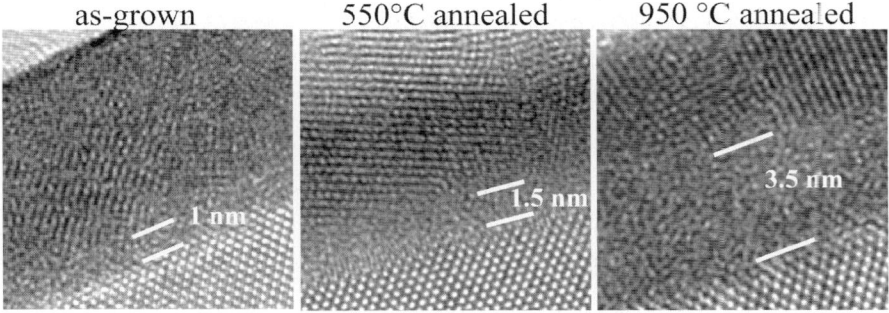

**Fig. 5.** HRTEM images with the [011]-oriented Si substrate. Crystallographic planes of $Lu_2O_3$ can be well seen, together with amorphous ILs of thickness increasing with the temperature of annealing

### 3.3.1 Fine Structure of Reference Samples

*In the Low Energy-Loss Domain: VEELS of Si, $SiO_2$, $Lu_2O_3$*

The spectra presented in Fig. 6 were obtained by correcting the raw data for the point-spread function of the instrumentation and for plural scattering by a Fourier-logarithmic deconvolution [32]. They have been scaled to their maximum. Si is the substrate of the $Lu_2O_3$ films. $SiO_2$ is a chemical oxide film prepared with a standard RCA cleaning ($HCl : H_2O_2 : H_2O = 1 : 1 : 5$ ratio, 10 min at 85 °C) followed by dipping for 30 s in a diluted HF solution ($1 : 50 = HF : H_2O$) at room temperature, and finally repeating the RCA cleaning. A 30 s long rinse in de-ionized water followed each cleaning step for all substrates. $Lu_2O_3$ is a reference thick film prepared with the same parameters as those for the films studied here. The spectrum of Si can be described by a main peak situated at 16.7 eV, the energy of the volume plasmon of Si. Spectra of $SiO_2$ and $Lu_2O_3$ have also plasmons at 22.4 and 15.7 eV respectively, with a larger distribution than the one of Si. Secondary signatures also appear on these spectra, and can be correctly interpreted only through the detailed study of the imaginary part of the dielectric function. The latter can be extracted from the VEELS spectrum. In a rough approximation, this function can be correlated with the Joint Density of States (JDOS) between the valence and conduction bands, and the energies of its maxima can be interpreted as energies of interband transitions. The study of these transitions allows an insight in the band structure diagram near the Fermi level and a comparison to calculations. For $SiO_2$, interband transitions occur in the 10–15 eV energy range. For $Lu_2O_3$, absorption occurs around 7–10 eV and immediately after the plasmon peak, which appears asymmetrical. At energies higher than 25 eV, absorption becomes significant. The $Lu-N_{67}$ transition originating from the Lu $4f$ level might form part of the first domain of absorption since the binding energy of this level is situated at 8.2 eV by XPS [39]

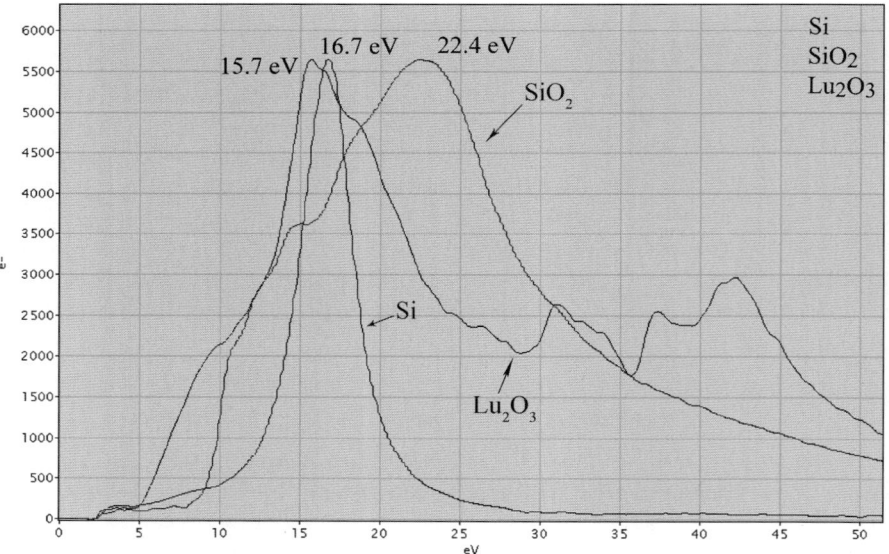

**Fig. 6.** VEELS spectra of reference samples of Si, $SiO_2$ and of $Lu_2O_3$

and the Lu–$O_{23}$ core-loss distribution corresponds to absorption beginning around 28 eV.

The energy of first absorption that corresponds to the energy band gap can be determined by the study of the imaginary part of the dielectric function, based on mathematical treatments of the EELS signal and knowledge of the optical refractive index of the material [45]. Refractive indices for $SiO_2$ and $Lu_2O_3$ found in the literature are 1.45 and 1.9 respectively [46–48]. Only band gaps higher than 3 eV can be determined in the case of our experiment because there is a large contribution of the zero-loss peak tail to the low-loss signal up to this energy that cannot be exactly eliminated by the deconvolution procedure. The band gaps of $SiO_2$ and $Lu_2O_3$ were determined at 9 and 5.4 eV respectively. These values are in good agreement with those proposed in the literature [45, 48–51]. For $Lu_2O_3$, the literature values are between 4.9 and 6 eV. On a single crystal, the most reliable reported value is 5.5 eV [48]. For a 30 nm thick film deposited by beam evaporation, the band gap value determined from the threshold energy for O 1s photoelectron energy loss was found to be 6 eV [50]. On very similar samples to those studied here, the transport gap measured by photoconductivity is 5.8 eV and the optical band gap measured by optical absorption spectroscopy is 4.9 eV [51].

*In the High Energy-Loss Domain: ELNES in Si, $SiO_2$, $Lu_2O_3$*

The Si–L and O–K edges are considered here. These are common edges of the Si, $SiO_2$ and $Lu_2O_3$ phases that are supposed to define the Si/$Lu_2O_3$ stacks.

The study of other references such as silicate phases is in progress. The spectra presented in Fig. 7 were obtained after subtracting the background. They are not deconvoluted for multiple scattering because they were collected on thin enough areas (ratio of the thickness to the mean free path is around 0.3). They have been scaled to their maximum.

The fine structure of Si–L in Si and $SiO_2$ has already been well studied [40]. In the oxide, the edge onset is delayed by 5 eV and the edge shape is completely different compared to that of Si [52, 53] (Fig. 7a). A first small absorption peak is observed at 101 eV for Si, followed by a huge and large absorption band whereas, for $SiO_2$, two sharp peaks at 108 and 115 eV respectively are followed by a larger one centered around 130 eV and a small peak near 156 eV. The Si–L shape in Si is characteristic of a $Si^0$ state that corresponds to a Si atom surrounded by four Si as first neighbours, whereas the Si–L shape in $SiO_2$ is characteristic of a $Si^{4+}$ state that corresponds to a Si atom surrounded by four O first neighbours.

The O–K edge in $SiO_2$ (Fig. 7b) is characterized by a main, sharp absorption peak centered at 538.5 eV and a larger and less intense one centered around 565 eV. Bulk $SiO_2$ is very sensitive to the number of O second neighbours as well as to the Si first neighbours [30]. The O–K edge in $Lu_2O_3$ has a doublet configuration followed by a shoulder around 546 eV and two large peaks centered around 555 and 570 eV respectively. The onset of the edge is 4.5 eV lower than the one in $SiO_2$. The energy splitting of the two first peaks is 5.7 eV. This splitting is systematically observed for rare earth sesquioxides of cubic structure [44]. The feature of the O–K edge in $Lu_2O_3$ is very similar to the one of the isoelectronic compound $Y_2O_3$ [54]. Provided that the excitation mechanism in EELS involves a dipole transition in which an electron is promoted from a ground state to an excited state, the O–K edge can be envisioned to a first-order approximation as an image of the 2p-projected unoccupied density of state, the initial state being the well-defined O 1s state. As in $Y_2O_3$, the final state has in fact a "band-like" character where the O 2p state is hybridized with the Lu unoccupied states, probably the Lu 5d and 6s states [55]. The double peak is associated to the O 2p–Lu 5d hybridization and reflects the splitting of the Lu d-bands by the crystal field due to an eight-fold coordination of Lu.

On the basis of these observations of reference samples, studies of EELS signatures along the $Si/Lu_2O_3$ stacks, as-grown or annealed, are in progress. We aim to define the nature of the interfacial layers formed in these films, as observed on HRTEM images.

### 3.3.2 Si/Lu$_2$O$_3$ Stacks

To start with, particular attention has been paid to the as-deposited and 950 °C annealed films because the results can be compared to those obtained on very similar samples that have already been studied in terms

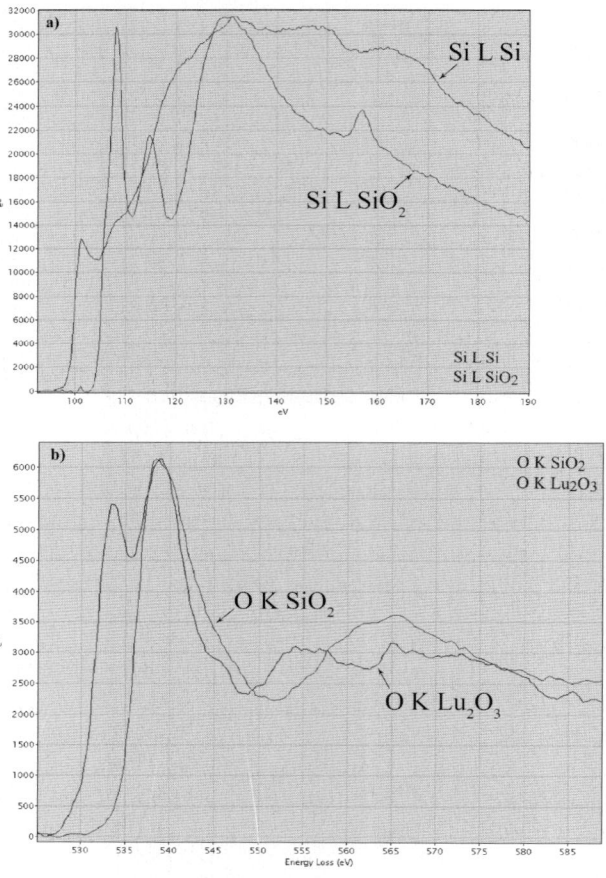

**Fig. 7.** (a) ELNES of the Si L edge in the Si and $SiO_2$ reference samples and (b) of the O–K edge in the $SiO_2$ and $Lu_2O_3$ reference samples

of XRR, atomic force microscopy (AFM), grazing incidence X-ray diffraction (GIXRD), XPS, FT-IR, and C–V characterisctics [39].

Electron diffraction and HRTEM investigations showed that the films are crystalline at the nanometre level with the $Lu_2O_3$ bixbyite structure. We tried to control the elemental composition of the films by comparison with the stoichiometric $Lu_2O_3$ reference sample using the Lu–$N_{45}$ and O–K edges. We encountered two difficulties. First, the films are sensitive to a focused electron beam of sufficient intensity to obtain exploitable spectra. Some deformation accompanied by drilling was observed after focusing a nanometre sized probe for a few seconds on the film (Fig. 8). Second, in the thickest areas, we encountered some difficulties with the extraction of the Lu–$N_{45}$ edge, which is broad and rounded. Thus, it was not possible to perform a precise quantitative analysis. However, one noticeable point is that the Si–L edge

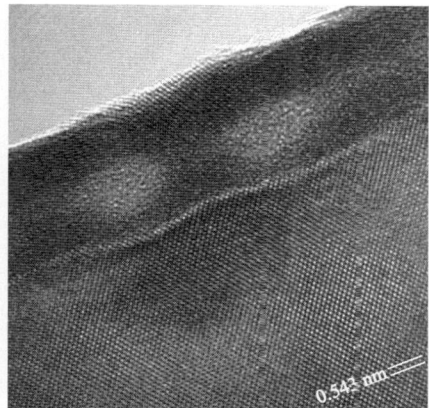

**Fig. 8.** Sensitivity of the sample to a focused electron beam

was always observed within the film at a level of a few at%. This concentration increases at the top of the film up to the $SiO_2$ stoichiometry. This is particularly noticeable for the 950 °C annealed sample. Another point is that the IL near the Si substrate is always a $SiO_2$-rich layer.

To take the analysis a step further and to overcome the sensitivity of the sample to the electron beam, we performed a line-focus EFTEM analysis, where beam damage is reduced and, in the low energy-loss domain, where acquisition times are the shortest. It is important to note that the delocalization of inelastic scattering defines the limit of the spatial resolution attainable in the experiment. In the low energy-loss domain investigated here and considering the experimental conditions used, this value is of the order of 2 nm [32]. In the core-loss domain, the delocalization parameter is better, around 1 nm for the Lu–$O_{23}$ edge and at the sub-nm level for edges of energies higher than 100 eV. The core-loss investigations are in progress. Moreover, near interfaces, bulk plasmons are minimized and surface–interface plasmons appear at lower energies than bulk plasmons (Fig. 9). This is the case for the stacks studied here, where the interfaces are at distances smaller than 5 nm. This distance is the order of magnitude of the distances between the electron probe and an interface for which surface/interface modes can be excited [56]. For this reason and because the energy of the surface/interface plasmon is around 6 eV, the band gap of thin high-$\kappa$ film deposited on $Si/SiO_2$ cannot be studied. There is an overlap of the features associated to the band gap and to the surface/interface modes in the low energy-loss domain (Fig. 9).

The spectrum map of Fig. 10 was acquired on the 950 °C annealed $Lu_2O_3$ sample. The corresponding HREM image is superposed in order to identify the geometrical configuration of the analysis. The distribution of intensity along a horizontal line on the spectrum map corresponds to one EELS spectrum acquired along a line parallel to the Si/film interface. Noticeable modifi-

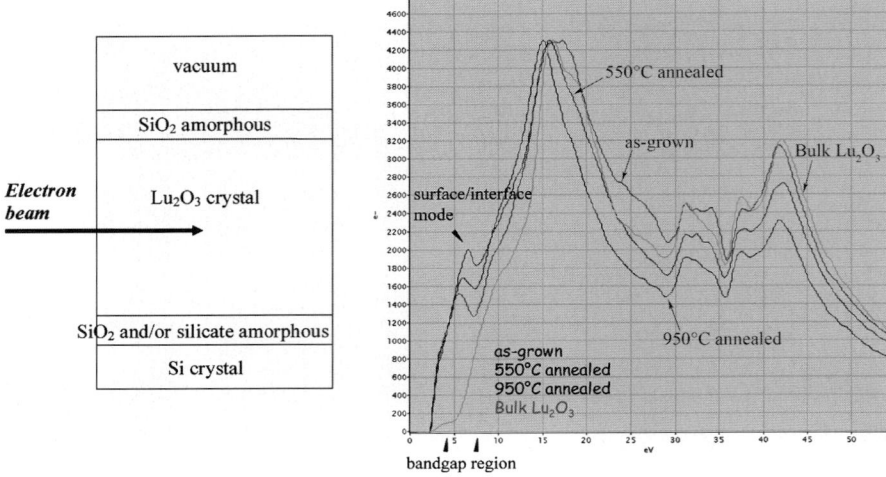

**Fig. 9.** Surface/interface modes overlap the band gap region in thin stacks involving a high-$\kappa$ material deposited on $Si/SiO_2$

cations of the VEELS signatures are seen on passing from the Si substrate, to the IL, to the $Lu_2O_3$ layer, and, finally, to the top of the $SiO_2$ layer. Typical spectra have been extracted from Fig. 10 in order to follow more clearly the evolution of the plasmon and of the secondary low energy-loss features when going from Si to the end of the IL. In Fig. 11, the corresponding spectra have been scaled to their maximum.

The Si plasmon around 17 eV is well identified on the first spectrum of the series (first red), and the mixed signatures of $SiO_2$ and $Lu_2O_3$ are seen in the last spectrum of the series (last blue), corresponding spatially to the $IL/Lu_2O_3$ film interface. In this last spectrum, the $Lu_2O_3$ plasmon is seen superposed to the $SiO_2$ one, together with the Lu $O_{23}$ edge. Qualitatively, the evolution between the first and the last spectra appears continuous with 1. the transition from the Si plasmon to the mixed signature of $Lu_2O_3$ and $SiO_2$ plasmons and 2. the correlated increase of the relative intensity of the Lu–$O_{23}$ edge and of the $Lu_2O_3$ plasmon, compared to that of the $SiO_2$ plasmon. From these data, it is not possible to say if the IL is a mixture of two phases, $SiO_2$ and $Lu_2O_3$, or a silicate, or a mixture of oxides and silicates. A study on the $Lu_2SiO_5$ silicate spectral features is in progress. On the basis of these results, the IL is a silicon oxide-rich phase for the first third part near the substrate and becomes enriched in lutetium oxide as the distance from the substrate increases up to the crystalline $Lu_2O_3$ layer. This result is in qualitative agreement with the result of the modelling of the XRR experiments performed on similar samples where the IL is found as a 2.1 nm thick mixed layer of $SiO_2$ and $Lu_2O_3$ over a 1.2 nm $SiO_2$ layer [39]. The first inves-

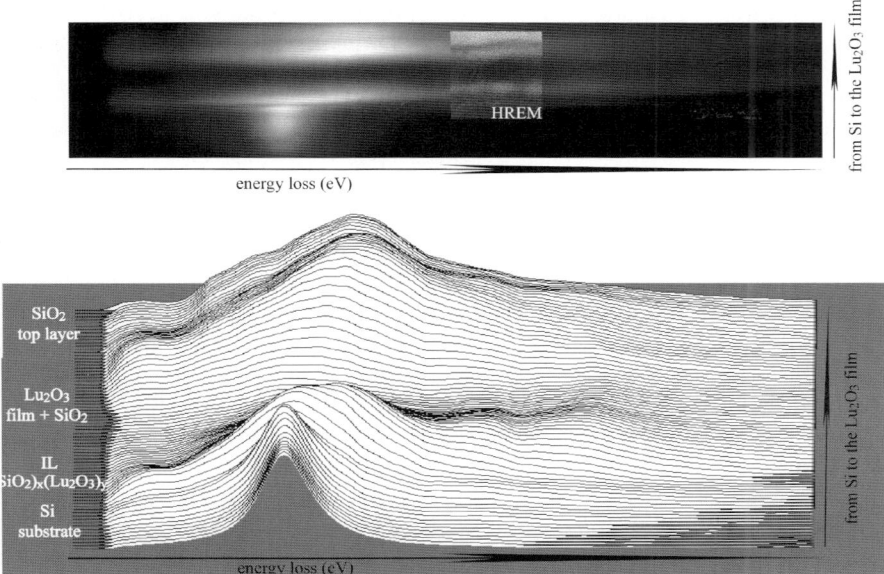

**Fig. 10.** Spectrum map and the corresponding extracted spectra along the direction perpendicular to the Si/film interface

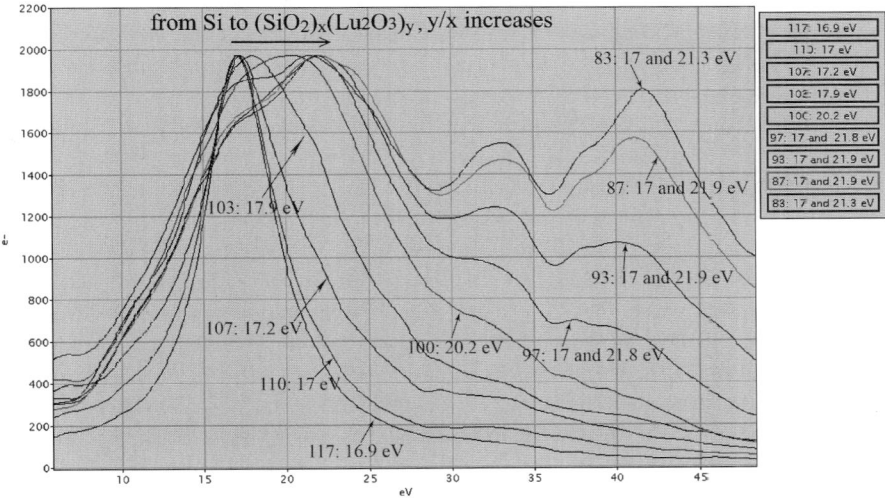

**Fig. 11.** Typical spectra extracted from Fig. 10

tigation on the Si–L signal seems to correlate with the spatial distribution obtained with the low-loss analysis. Complementary analyses are in progress.

# 4 Discussion and Conclusions

Transmission Electron Microscopy and particularly High-Resolution Imaging associated with Electron Energy Loss Spectroscopy are powerful and invaluable tools for the study of ultra-thin films. In the case of REO thin films dedicated for alternative gate dielectrics for Si-based MOSFETs, the thickness of the films, the crystallization state up to the nanometre level, and the chemical nature of the layers that form the stacks can be determined accurately. From earlier work and this study, some systematic results can be underlined: 1. the ultra-thin films are generally amorphous or nanocrystalline up to high temperatures, 2. after post-deposition annealing treatment, depending on temperature and on oxygen partial pressure, a stack with at least two layers is formed. Underneath the REO layer, a $SiO_2$-rich IL is formed near the Si substrate, followed by a mixed $(SiO_2)_x(LnO)_y$ layer. This has been evidenced by TEM-EELS for $La_2O_3$, $Gd_2O_3$, $Pr_2O_3$ and $Lu_2O_3$ films deposited on Si. This chemical behaviour is explained by: a) O diffusion within the REO films and reaction with the substrate to form $SiO_2$-rich ILs [9, 19, 25, 30] and b) Si outdiffusion from the substrate to the film [25, 39], as already discussed for $Y_2O_3$ and $SrTiO_3$ films [57–59]. The Si diffusion is proposed as the origin of an amorphization phenomenon observed in crystalline pulsed laser-deposited $SrTiO_3$ films at growth temperatures above 600 °C. This amorphous state is also observed in all the ILs between the REO and the Si substrate, 3. electronic structure in these systems can be well unravelled through the study of the VEELS and O–K ELNES fine structures. O is present everywhere in the stack with different chemical bondings and, consequently, the O–K edge presents very different features when observed in REO, or $SiO_2$, or at the interface between Si and $SiO_2$.

In the particular case of the $Lu_2O_3$ system, band calculations and projected density of states (DOS) on this system are needed in order to interpret VEELS and ELNES fine structures. Moreover, it is shown here that in the case of the $Lu_2O_3$/IL/Si stack, for small-sized systems where interfaces are at nanometric distances, interface modes present features in the same energy domain as that of the band gap in the VEELS spectrum, and thus band gap determination is not feasible.

# References

[1] T. Wiktorczyk: Rare earth oxide films: their preparation and characterization, Optica Applicata **31**, 5 (2001)
[2] G. D. Wilk, R. M. Wallace, J. M. Anthony: High-k gate dielectrics: current status and materials properties considerations, J. Appl. Phys. **89**, 5243 (2001)

[3] K. J. Hubbard, D. G. Schlom: Thermodynamic stability of binary oxides in contact with silicon, J. Mater. Res. **11**, 2757 (1996)

[4] S. Ohmi, M. Takeda, H. Ishiwara, H. Iwai: Characterization of $Lu_2O_3$ high-k thin films on Si(100) fabricated by e-beam deposition method, in *ISTC* (2002) pp. 251–261

[5] L. Marsella, V. Fiorentini: Structure and stability of rare-earth and transition-metal oxides, Phys. Rev. B **69**, 172103 (2004)

[6] S. Stemmer: Thermodynamic considerations in the stability of binary oxides for alternative gate dielectrics in complementary metal-oxide-semiconductors, J. Vac. Sci. Technol. B **22**, 791 (2004)

[7] M. Fanciulli, S. Spiga, G. Scarel, G. Tallarida, C. Wiemer, G. Seguini: Structural and electrical properties of $HfO_2$ films grown by atomic layer deposition on Si, Ge, GaAs and GaN, Mat. Res. Soc. **786**, 341 (2004)

[8] A. Kingon, J. P. Maria, S. K. Streiffer: Alternative dielectrics to silicon dioxide for memory and logic devices, Nature **406**, 1032 (2000)

[9] S. Stemmer, J. P. Maria, A. I. Kingon: Structure stability of $La_2O_3/SiO_2$ layers on Si(001), Appl. Phys. Lett. **79**, 102 (2001)

[10] D. H. Tryioso, R. I. Hegde, J. Grant, P. Fejes, R. Liu, D. Roan, M. Ramon, D. Werho, R. Rai, L. B. La, J. Baker, C. Garza, T. Guenther, B. E. White, Jr., P. J. Tobin: Film properties of ALD Hf- and La-gate dielectrics grown on Si with various pre-deposition treatments, J. Vac. Sci. Technol. B **22**, 2121 (2004)

[11] J. A. Gupta, D. Landheer, G. I. Sproule, J. P. McCaffrey, M. J. Graham, K.-C. Yang, Z.-H. Lu, W. N. Lennard: Interfacial layer formation in $Gd_2O_3$ films deposited directly on Si (001), Appl. Surf. Sci. **173**, 318 (2001)

[12] B. Mereu, A. Dimoulas, G. Vellianitis, G. Apostolopoulos, R. Scholz, M. Alexe: Interface trap density in amorphous $La_2Hf_2O_7/SiO_2$ high-k gate stacks on Si, Appl. Phys. A **80**, 253 (2004)

[13] D. A. Muller: Gate dielectric metrology using advanced TEM measurements, AIP-Conference-Proceedings **550**, 500 (2001)

[14] A. C. Diebold, B. Foran, C. Kisielowsky, D. A. Muller, S J. Pennycook, E. Principe, S. Stemmer: Thin dielectric film thickness determination by advanced transmission electron microscopy, Microsc. Microanal. **9**, 493 (2003)

[15] A. Y. Du, C. H. Tung, B. H. Freitag, W. Y. Zhang, S. Lim, E H. Ang, D. Ng: Ultra-thin SiON and high-k $HfO_2$ gate dielectric metrology using transmission electron microscopy, in *Proc. 11th Int. Symp. on the Physical and Failure Analysis if Integrated Circuits, IPFA* (2004) p. 135

[16] P. E. Batson: Simultaneous STEM imaging and electron energy-loss spectroscopy with atomic-column sensitivity, Lett. Nature **366**, 727 (1993)

[17] J. Kwo, M. Hong, A. R. Kortan: Properties of high kappa gate dielectrics $Gd_2O_3$ and $Y_2O_3$ for Si, J. Appl. Phys. **89**, 3920 (2002)

[18] D. H. Tryioso, R. I. Hegde, J. Grant, J. K. Schaeffer, D. Roan, B. E. White, Jr., P. J. Tobin: Evaluation of lanthanum based gate dielectrics deposited by atomic layer deposition, J. Vac. Sci. Technol. B **23**, 288 (2005)

[19] R. Lo-Nigro, R. Toro, G. Malandrino, V. Raineri, I. L. Fragalà: Electrical properties of MOCVD praseodymium oxide based MOS structures, in *Proc. of the 33rd European Solid State Device Research (ESSDERC)* (2003) p. 375

[20] H. J. Osten, J. P. Liu, E. Bugiel: Epitaxial growth of praseodymium oxide on silicon, Mat. Sci. Eng. B **87**, 297 (2001)

[21] H. Sim, C. B. Samantaray, T. Lee, H. Yeom, H. Hwang: Electrical characteristics of high-k gate dielectrics with epitaxial $Si_3N_4$ interfacial layer on Si(111), Jap. J. Appl. Phys. **43**, 7926 (2004)

[22] X. Wu, D. Landheer, T. Quance, M. J. Graham, G. A. Botton: Structural comparison of gadolinium and lanthanum silicate films on Si(100) by HRTEM, EELS and SAED, Appl. Surf. Sci. **200**, 15 (2002)

[23] X. B. Lu, Z. G. Liu, Y. P. Wang, Y. Yang, X. P. Wang, H. W. Zhou, B. Y. Nguyen: Structure and dielectric properties of amorphous $LaAlO_3$ and $LaAlO_xN_y$ films as alternative gate dielectric materials, J. Appl. Phys. **94**, 1229 (2003)

[24] A. Li, Q. Shao, H. Ling, J. B. Cheng, D. Wu, Z. G. Liu, N. B. Ming, C. Wang, H. W. Zhou, B. Y. Nguyen: Characteristics of $LaAlO_3$ gate dielectrics on si grown by metalorganic chemical vapor deposition, Appl. Phys. Lett. **83**, 3540 (2003)

[25] H. Ono, T. Katsumata: Interfacial reactions between thin rare-earth-metal oxide films and Si substrates, Appl. Phys. Lett. **78**, 1832 (2001)

[26] L. F. Edge, D. G. Schlom, R. T. Brewer, Y. J. Chabal, J. R. Williams, S. A. Chambers, C. Hinkle, G. Lucovsky, Y. Yang, S. Stemmer, M. Copel, B. Holländer, J. Schubert: Suppression of subcutaneous oxidation during the deposition of amorphous lanthanum aluminate on silicon, Appl. Phys. Lett. **84**, 4629 (2004)

[27] H. Yamada, T. Shimizu, A. Kurokawa, K. Ichii, E. Suzuki: MOCVD of high-dielectric-constant lanthanum oxide thin films, J. Electrochem. Soc. **150**, G429 (2003)

[28] J. P. Maria, D. Wicaksana, A. I. Kingon, et al.: High temperature stability in lanthanum and zirconia-based gate dielectrics, J. Appl. Phys. **90**, 3476 (2001)

[29] S. Guha, E. Cartier, M. A. Gribelyuk, N. A. Gribelyuk, N. A. Bojarczuk, M. C. Copel: Atomic beam deposition of lanthanum- and yttrium-based oxide thin films for gate dielectrics, Appl. Phys. Lett. **77**, 2710 (2000)

[30] G. A. Botton, J. A. Gupta, D. Landheer, J. P. McCaffrey, G. I. Sproule, M. J. Graham: Electron energy loss spectroscopy of interfacial layer formation in $Gd_2O_3$ films deposited directly on Si (001), J. Appl. Phys. **91**, 2921 (2002)

[31] D. A. Muller, T. Sorsch, S. Moccio, F. H. Baumann, K. Evans-Lutterodt, G. Timp: The electronic structure at the atomic scale of ultrathin gate dielectrics, Nature **399**, 758 (1999)

[32] R. F. Egerton: *Electron Energy Loss in the Electron Microscope*, 2nd ed. (Plenum, New York 1996)

[33] L. Reimer (Ed.): *Energy-Filtering Transmission Electron Microscopy* (Springer, Berlin, Heidelberg 1996)

[34] K. Kimoto, K. Kobayashi, T. Aoyama, Y. Mitsui: Analyses of composition and chemical shift of silicon oxynitride film using energy-filtering transmission electron microscope based spatially resolved electron energy loss spectroscopy, Micron **30**, 121 (1999)

[35] U. Golla-Schindler, G. Benner, A. Putnis: Laterally resolved EELS for ELNES mapping of the Fe $L_{23}$- and O K-edge, Ultramicroscopy **96**, 573 (2003)

[36] T. Walther: Electron energy-loss spectroscopic profiling of thin film structures: 0.39nm line resolution and 0.04eV precision measurement of near-edge structure shifts at interfaces, Ultramicroscopy **96**, 401 (2003)

[37] R. Lo-Nigro, R. G. Toro, G. Malandrino, V. Raineri, I. Fragala: A simple route to the synthesis of $Pr_2O_3$ high-k thin films, Adv. Mater. **15**, 1071 (2003)

[38] M. MacKenzie, A. J. Craven, D. W. McComb, D. A. Hamilton, S. McFadzean: Spectrum imaging of high-k dielectric stacks, Inst. Phys. Conf. Ser. **199**, 299 (2003)

[39] G. Scarel, E. Bonera, C. Wiemer, G. Tallarida, S. Spiga, M. Fanciulli, I. L. Fedushkin, H. Schumann, Y. Lebedinskii, A. Zenkevich: Atomic-layer deposition of $Lu_2O_3$, Appl. Phys. Lett. **85**, 630 (2004)

[40] P. Batson, K. L. Kavanagh, C. Y. Wong, J. M. Woodall: Local bonding and electronic structure obtained from electron energy loss scattering, Ultramicroscopy **22**, 89 (1987)

[41] W. M. Skiff, R. W. Carpenter, S. H. Lin: Analysis of valence shell electronic excitations in silicon and its refractory compounds using electron energy loss microspectroscopy, J. Appl. Phys **64**, 6328 (1988)

[42] W. M. Skiff, R. W. Carpenter, S. H. Lin: Si L core edge fine structure in an oxidation series of silicon compounds:a comparison of microelectron energy loss spectra with theory, J. Appl. Phys. **58**, 3463 (1985)

[43] C. Colliex, M. Gasgnier, P. Trebbia: Analysis of the electron excitation spectra in heavy rare earth metals, hydrides and oxides, Le journal de Physique **37**, 397 (1976)

[44] L. M. Brown, C. Colliex, M. Gasgnier: Fine structure analysis in EELS from rare earth sesquioxide thin films, Journal de Physique, Colloque C2, Supp. au n. 2 **45**, 433 (1984)

[45] S. Schamm, G. Zanchi: Study of the dielectric properties near the band gap by VEELS: gap measurements in bulk materials, Ultamicroscop **96**, 559 (2003)

[46] E. Dehan, P. Temple-Boyer, R. Henda, J. J. Pedroviejo, E. Scheid: Optical and structural properties of $SiO_x$ and $SiN_x$ materials, Thin Solid Films **266**, 14 (1995)

[47] O. Medenbach, D. Dettmar, R. D. Shannon, R. X. Fischer, W. M. Yen: Refractive index and optical dispersion of rare earth oxides using a small-prism technique, J. Opt. A: Pure Appl. Opt **3**, 174 (2001)

[48] T. Wiktorczyk: Optical properties of electron beam deposited lutetium oxide thin films, Optica Applicata **XXXI (1)**, 83 (2001)

[49] A. V. Prokofiev, A. I. Shelykh, B. T. Melekh: Periodicity in thre band gap variation of $Ln_2X_3$ (X = O, S, Se) in the lanthanide series, J. Alloys and Compounds **242**, 41 (1996)

[50] H. Nohira, T. Shiraishi, T. Nakamura, K. Takahashi, M. Takeda, S. Ohmi, H. Iwai, T. Hattori: Chemical and electronic structures of $Lu_2O_3$ /Si interfacial transition layer, Appl. Surf. Sci. **76**, 234 (2003)

[51] G. Seguini, E. Bonera, S. Spiga, G. Scarel, M. Fanciulli: Energy-band diagram of metal/$Lu_2O_3$/silicon structures, Appl. Phys. Lett. **85**, 5316 (2004)

[52] K. Kimoto, T. Sekiguchi, T. Ayoama: Chemical shift mapping of Si L and K edges using spatially resolved EELS and energy-filtering TEM, J. Electron. Microsc. **46**, 369 (1997)

[53] S. Schamm, R. Berjoan, P. Barathieu: Study of the chemical and structural organization of SIPOS films at the nanometer scale by TEM-EELS and XPS, Mat. Sci. Eng. B **107**, 58 (2004)

[54] A. Travlos, N. Boukos, G. Apostopoulos, A. Dimoulas, C. Giannakopoulos: EELS study of oxygen supertstructure in epitaxial $Y_2O_3$ layers, Mat. Sci. Eng. B **109**, 52 (2004)

[55] G. Lucovsky, J. G. Hong, C. C. Fulton, Y. Zou, R. J. Nemanich, H. Ade, D. G. Scholm, J. L. Freeouf: Spectroscopic studies of metal high-k dielectrics: transition metal oxides and silicates, and complex rare earth/transition metal oxides, Phys. Stat. Sol. B **241**, 2221 (2004)

[56] D. A. Muller, J. Silcox: Delocalization in inelastic scattering, Ultramicroscopy **59**, 195 (1995)

[57] S. K. Kang, D. H. Ko, E. H. Kim, M. H. Cho, C. N. Whang: Interfacial reactions in the thin film $Y_2O_3$ on chemically oxidized Si(100) substrate systems, Thin Solid Films **353**, 8 (1999)

[58] F. Paumier, R. J. Gaboriaud: Interfacial reactions in $Y_2O_3$ thin films deposited on Si(100), Thin Solid Film **441**, 307 (2003)

[59] P. Ahmet, T. Koida, M. Takakura, K. Nakajima, M.Yoshimoto, H. Koinuma, M. Tanaka, M. Takegushi, T. Chikyow: Diffusion induced amorphization in the crystalline $SrTiO_3$ thin films grown on Si (100) investigated by combinatorial method, Appl. Surf. Sci. **189**, 307 (2002)

# Index

# Strain-Relief at Internal Dielectric Interfaces in High-$k$ Gate Stacks with Transition Metal and Rare Earth Atom Oxide Dielectrics

Gerald Lucovsky[1] and James C. Phillips[2]

[1] Department of Physics, North Carolina State University, Raleigh, North Carolina, 27695-8202, USA
gerry_lucovsky@ncsu.edu

[2] Department of Physics, Rutgers University, Piscataway, New Jersey 08854, USA

**Abstract.** This Chapter addresses the effects of bonding discontinuities at the internal dielectric interfaces in gate stacks that include transition metal and rare earth atom elemental and complex oxides, as well as transition metal silicate alloys. The focus is on the strain-induced defects, and the reduction of defect densities through strain-driven self-organizations that take place during high-temperature post-deposition annealing.

## 1 Introduction

This paper addresses two fundamental issues that impact on the introduction of deposited alternative high-$k$ dielectrics into aggressively scaled Si devices [1, 2]. The first derives from an interfacial transition region (ITR) between the Si substrate and an $SiO_2$ buffer layer that chemically isolates the high-$k$ dielectric from the Si substrate. Of particular importance is the way in which this ITR contributes to device performance and reliability [3–5]. Previous studies demonstrated that replacement of thermally grown $SiO_2$ dielectrics by deposited dielectrics, including $SiO_2$, as well as higher-k alternatives, requires separate and independent steps for formation of an ultra-thin, $\sim 0.5\,\mathrm{nm}$ to $0.6\,\mathrm{nm}$ $SiO_2$ buffer layer, and deposition of an alternative dielectric thin film [6]. The $SiO_2$ buffer layer and its ITR at the Si substrate contribute $\sim 0.35\,\mathrm{nm}$ to the equivalent oxide thickness (EOT), placing a limitation on the ultimate reductions of EOT into the regime below $1.0\,\mathrm{nm}$. EOT is defined as the $SiO_2$ thickness corresponding to the entire gate stack, including the high-$k$ alternative dielectric as well as an interfacial $SiO_2$ layer. The second limitation derives from the conditions for another strain-driven bonding self-organization at the internal dielectric interface between the $SiO_2$ buffer layer and an alternative gate dielectric (see Fig. 1) [7].

Several studies have established that the $SiO_2$ interface with Si produced by thermal oxidation is neither abrupt at an atomic scale, nor strain free [3–5]. In addition, theoretical studies have found that ITRs are an intrinsic bonding property of these interfaces [8, 9]. These are a consequence of interfacial discontinuities in chemical bonding. This includes differences in 1. the average

M. Fanciulli, G. Scarel (Eds.): Rare Earth Oxide Thin Films,
Topics Appl. Physics **106**, 179–202 (2007)
© Springer-Verlag Berlin Heidelberg 2007

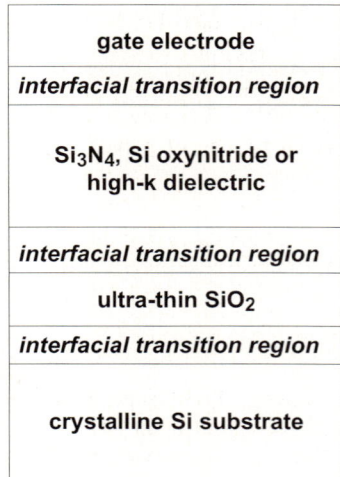

**Fig. 1.** High-$k$ gate stack with $SiO_2$ interfacial buffer layer

number of bonds per atom between Si, 4, and $SiO_2$, 2.67, or equivalently the molar volume mismatch between Si and the $SiO_2$ film created by an oxidation process, $\sim 220\%$, and 2. bond ionicity; the bonding in Si is covalent, whereas the bonding in $SiO_2$ is $\sim 45\%$ ionic. Experimental studies of Si–$SiO_2$ interfaces have applied medium-energy ion scattering (MEIS) [3], spectroscopic ellipsometry (SE) [4], and synchrotron or soft X-ray photoelectron spectroscopy (SXPS) [5], and have identified a very thin ITR with suboxide bonding arrangements. The thickness of this layer was determined in [3] and [4] to be less than 1 nm, and more likely 0.25 nm to 0.5 nm. The thickness of the suboxide ITR has recently been determined by high-resolution transmission electron microscopy (HRTEM) [10], and the SXPS results presented below [11]. These two studies confirm that ITR is $\sim 0.35$ nm thick, and therefore equivalent to about one molecular layer of SiO bonding.

In addition, the ion scattering studies have shown that there is a disordered region in the Si substrate, $Si_D$, with a thickness between 0.5 nm and 1.0 nm, or $\sim$ three atomic layers. This layer was not included in the model of the SE studies of [4], and as direct result of this, the thickness of the SiO ITR was estimated to be approximately 0.75 nm, or about two times thicker than the determinations based on MEIS [3], HRTEM imaging [10] and SXPS [11].

The thin ITR with suboxide bonding provides a strain-reducing buffer layer between the crystalline Si (c-Si) substrate, and the more *elastically compliant* non-crystalline $SiO_2$ dielectric film. As noted above, several factors contribute to interfacial bond-strain and subsequent formation of the ITR. Differences in the Si–Si inter-atomic distances in the c-Si substrate, 0.235 nm, and the $SiO_2$ dielectric, $\sim 0.305$ nm, generate intrinsic compressive stress in the oxide and tensile stress in the substrate [12]. Similar considerations exist at interfaces formed by plasma and chemical oxidation of the Si substrate.

Additionally, differences in linear thermal expansion coefficients between $SiO_2$ ($\sim 0.5 \times 10^{-6}\,C^{-1}$) and the Si substrate ($\sim 2.5 \times 10^{-6}\,C^{-1}$) contribute a component of thermal strain equivalent to several tenths of a percent for oxides grown, or processed at $800\,°C$ to $1000\,°C$, and then returned to room temperature [12]. These levels of interfacial bond-strain can not be relieved elastically; however, this *pins* a relatively high level of stress at the interface, $\sim 5 \times 10^9$ dynes/cm$^2$.

These levels of interfacial stress also result in *plastic deformation*, as indicated by the ion scattering measurements of [3]. Dangling bond formation locally relieves the build-up of in-plane stress, thereby accounting for Si dangling bond formation in the $Si_D$ layer. Typical densities of dangling bonds as determined by electron spin resonance (ESR) are of approximately $2 \times 10^{12}$ cm$^{-2}$ to $3 \times 10^{12}$ cm$^{-2}$, or on average about $0.3\,\%$ of the Si atoms at the surface of a crystal oriented in the $\langle 111 \rangle$ or $\langle 001 \rangle$ directions [13, 14]. The symmetry of the g-tensor establishes that these dangling bonds reside in the Si substrate, and the in-plane bonding of the Si atoms is locally in tensile stress at these defect sites, as evidenced by the relative amplitudes of the $x$, $y$ and $z$ components of the g-tensor.

Defect formation is also consistent with a novel application of bond constraint theory (BCT), originally proposed to account for the ease of glass formation in CRN glasses and thin films such as $As_2S_3$ and $SiO_2$ [15, 16], but also as more recently applied to semiconductor-dielectric, and internal dielectric interfaces [17, 18]. This extension will be addressed in more detail later on in this article.

Single wavelength ellipsometry (SWE) studies were performed on thermally grown $SiO_2$ on Si$\langle 111 \rangle$ faces [12], grown by conventional thermal oxidation in dry $O_2$ at $850\,°C$, and subsequently furnace annealed in Ar at temperatures up to $1100\,°C$. The SWE studies indicated 1. decreases in the optical index of refraction, n, at $632.8\,nm$, with a threshold of about $950\,°C$ to $980\,°C$, and 2. a complementary increase in film thickness also with approximately the same temperature threshold. These changes were attributed to a relaxation of intrinsic in-plane growth stress by a visco-elastic relaxation process.

Section 2 addresses bonding changes at Si–$SiO_2$ interfaces, and proposes a mechanism for a strain-relieving interfacial self-organization that relies on a novel application of BCT. Section 3 addresses bonding changes at internal dielectric interfaces between $SiO_2$ and alternative gate dielectrics including 1. $Si_3N_4$ and Si oxynitride alloys, and 2. high-$k$ dielectrics. Section 4 introduces the concept of gate stack stress/profiling as a pathway to understand defect and defect precursor formation, their cumulative effects on device performance and reliability, and the consequences of interfacial self-organizations at Si–$SiO_2$ and internal dielectric interfaces are addressed in the contexts of 1. the ultimate limits of scaling in bulk complementary metal oxide semiconductor (CMOS) devices, and 2. in a significantly limited number of alternative high-$k$ dielectrics that can meet road-map scaling metrics [1].

## 2 Chemical Bonding Changes at Si–SiO$_2$ Interfaces after High-Temperature Annealing

### 2.1 Spectroscopic Studies

Optical second harmonic generation (SHG) studies identified a transition at $\sim 850\,°C$ to $900\,°C$ in annealed Si–SiO$_2$ interfaces; this has been interpreted in terms of bonding changes [19–22]. SXPS [11] and cathodo-luminescence spectroscopy (CLS) [23] studies have provided direct evidence respectively for interfacial bonding changes, and for defect reduction, each occurring at Si–SiO$_2$ interfaces after a $900\,°C$ anneal.

The SHG response from 1. the bulk of solid, 2. a solid surface, or 3. a buried interface is the generation of coherent light at a photon energy of $2\omega$ when the medium is exposed to intense illumination at a photon energy of $\omega$, as from a high-power laser [19–22]. Applied to Si–SiO$_2$ interfaces, no SHG signal is expected from either 1. the non-crystalline SiO$_2$ layer, due to lack of long-range order, or 2. the bulk c-Si substrate because of the local symmetry at the Si-atom bonding sites. On the other hand, SHG is allowed at Si surfaces and Si–SiO$_2$ interfaces, with different matrix elements and bond resonance energies for different local bonding arrangements and different surface orientations.

An interfacial relaxation has been identified from the temperature dependence of the SHG response of a vicinal Si(111)–SiO$_2$ interface, where the vicinal off-cut is either in the 112 bar, or 1 bar 1 bar 2 direction. This has been obtained by determining the azimuthal anisotropy of the SHG signal, and extracting the temperature dependence of the phase angle, $\theta$, between terrace and step edge Fourier components of that signal; see [19–22] for details.

The plot in Fig. 2 gives $\theta$ as a function of annealing temperature for a Si$\langle 111 \rangle$ interface with an off-cut angle of $5°$ in the 112 bar direction. The largest change in $\theta$ is between $850\,°C$ and $900\,°C$, and this has been interpreted as being due to bonding rearrangements on the terrace and step edges of the vicinal wafers [19–22]. It has also been addressed in the context of a bonding model that includes differences in back-bonding to Si atoms on terraces and step edges [24]. Smaller changes in $\theta$, between $900\,°C$ and $1050\,°C$, are attributed to the visco-elastic relaxation of bulk stress. Interfaces prepared by remote plasma-assisted oxidation at $300\,°C$ give essentially the same as-grown values of $\theta$, to $\pm 2°$, as does the $850\,°C$ thermal oxidation, and essentially the same values of $\theta$, to $\pm 2°$, as does a $900\,°C$ anneal in an inert ambient.

Figure 3 summarizes results of SXPS studies performed on Si$\langle 111 \rangle$ interfaces that have been discussed in detail in [11]. Spectral features identified

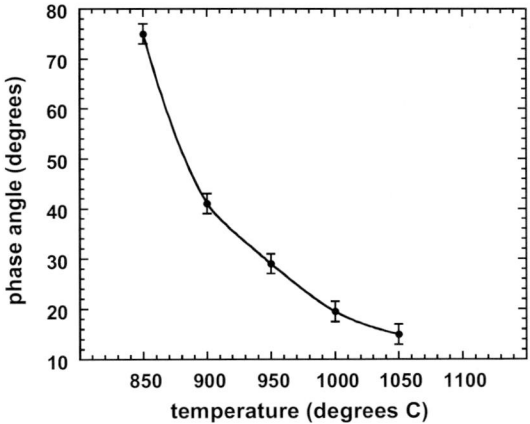

**Fig. 2.** Second harmonic generation phase angle versus annealing temperature for Si–SiO$_2$ interface

as I$_1$, I$_2$, and I$_3$ correspond respectively to bonding arrangements in which Si has one, two and three oxygen neighbors, also designated by formal ionic charges as Si$^{1+}$, Si$^{2+}$, and Si$^{3+}$ [5]. The spectra also include the Si substrate feature, labeled Si$^0$, and a SiO$_2$ feature, labeled Si$^{4+}$. Samples for this study were prepared by remote plasma-assisted oxidation of Si, and remote plasma-enhanced chemical vapor deposition of SiO$_2$, each at 300 °C [6]. These Si–SiO$_2$ structures were then subjected to vacuum annealing at temperatures of 600 °C, 800 °C, 900 °C and 1000 °C.

The most significant change in the spectra after the 900 °C anneal is a marked reduction in the Si$^{2+}$ feature. A quantitative analysis of spectral changes in Fig. 3 indicates that after the 900 °C anneal there is approximately one molecular layer of suboxide bonding with an average composition of SiO. This is in excess of the one mono-layer of Si$^{1+}$ bonding required to form an abrupt Si$\langle 111 \rangle$–SiO$_2$ interface. The suboxide transition region bonding corresponds to a physical thickness of $\sim$ 0.3 nm to 0.4 nm [25], in good agreement with ion scattering measurements of [3], and the HRTEM studies of [10]. Similar bonding rearrangements between the as-grown and 900 °C annealed spectra have been identified for Si$\langle 100 \rangle$–SiO$_2$ interfaces as well.

In the as-grown Si$\langle 111 \rangle$ interfaces, the bonding in the ITR with an SiO composition is approximately *random* with respect to the distribution of the Si$^{1+}$, Si$^{2+}$, and Si$^{3+}$ features with a ratio close to 1 : 1.5 : 1. A random distribution must also include Si$^{4+}$ and Si$^0$ bonding arrangements; however, these cannot be distinguished from the respective SiO$_2$ and Si substrate features. Random SiO bonding has a ratio of 1 : 4 : 6 : 24 : 1 for the arrangements: Si$^0$ : Si$^{1+}$ : Si$^{2+}$ : Si$^{3+}$ : Si$^{4+}$, so that the 1 : 1.5 : 1 ratio is consistent with this description of the interfacial bonding at as-grown interfaces. After the 900 °C anneal, the bonding in the ITR is no longer random, but instead

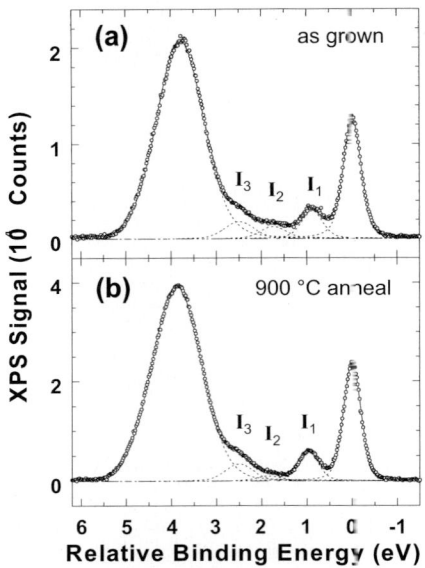

**Fig. 3.** Soft-$x$, synchrotron photoelectron spectra for S(111) interface: (**a**) as-grown at 300 °C, and (**b**) after a 900 °C anneal in an inert ambient

consists primarily of $Si^{1+}$ and $Si^{3+}$ groups in a ratio of approximately $1:1$ and with a significantly reduced $Si^{2+}$ component.

CLS results in Fig. 4 are for an Si–SiO$_2$ gate stack with a SiO$_2$ physical thickness of 5 nm [23]. Spectra were obtained as a function of the electron beam energy between 0.5 keV and 4.5 keV, with those presented in Fig. 4 taken at a 2 keV beam energy. Based on the depth dependence for the generation of electron–hole pairs in the SiO$_2$ film, these spectra are optimized to reveal interface and near-interface features. Spectra are shown for the as-deposited sample in which the interface was formed by 300 °C remote plasma-assisted oxidation ($\sim 0.6$ nm) [6], and the SiO$_2$ layer by 300 °C remote plasma-enhanced chemical vapor deposition. Additional spectra are shown after 1. a 400 °C hydrogenation anneal in forming gas, 10 % H$_2$/90 % N$_2$, 2. a 900 °C rapid thermal anneal (RTA) for approximately 1 min in Ar, and 3. a 900 °C RTA followed by a 400 °C anneal in forming gas. Forming gas anneals terminate Si and O dangling bonds, with the creation, respectively, of H-terminated Si–H and SiO–H groups.

As shown in Fig. 4, the combination of an RTA at 900 °C and a 400 °C anneal in a H-containing ambient significantly reduces interfacial CLS defect features. Defect bands are marked by arrows: DB, centered at about 0.9 eV, is an interfacial Si dangling bond, D$_1$, with a spectral peak at $\sim 1.8$ eV, is a near-interfacial defect, and D$_2$, with a spectral peak at $\sim 3.4$ eV, is a Si substrate band. The spectral peak and asymmetric line shape of the D$_1$ feature are

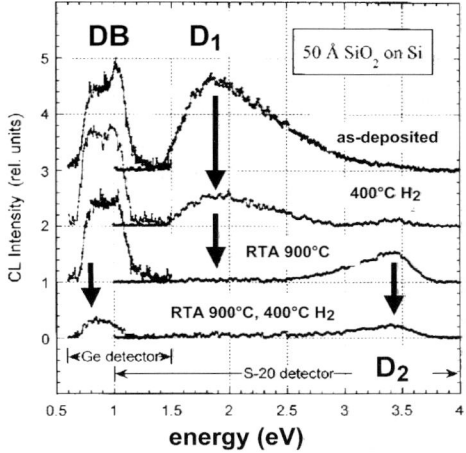

**Fig. 4.** Cathodoluminescence spectra for a 5 nm $SiO_2$ gate stack: as-deposited at 300 °C, after a 400 °C anneal in $H_2$, after a 900 °C rapid thermal anneal in an inert ambient, and after a 900 °C rapid thermal anneal in an inert ambient followed by a 400 °C anneal in $H_2$

essentially the same as for a photoluminescence (PL) feature reported in [26], for a Si suboxide with $x \sim 1.1$. The most significant changes after the 900 °C RTA and 400 °C $H_2$ anneal are the reduction of $D_1$ below the detection limit. The DB defect shows the similar reductions to substrate Si-atom dangling bonds after the 900 °C anneal, followed by the 400 °C $H_2$ anneal [13, 14]. These changes in DB luminescence establish that CLS is sensitive to defect changes at the $10^{11}$ defects $cm^{-2}$ to $10^{12}$ defects $cm^{-2}$ regime.

## 2.2 Kinetics of Interfacial Changes at the Interface Bonding and Defect Levels

A study of the stability of homogeneous bulk Si suboxide films deposited at 300 °C with compositions close to SiO, i.e., $SiO_x$, $x \sim 1$ [27], provides a basis for understanding bonding changing identified by the SXPS studies of [11]. These films contained bonded hydrogen, primarily on Si–H arrangements. The studies of these hydrogenated $SiO_x$ films demonstrated a chemical phase separation into nano-crystalline Si (c-Si) and non-crystalline $SiO_2$ (nc-$SiO_2$) after isochronal annealing in the temperature range from 500 °C to 1000 °C. This is represented in (1) by

$$SiO_x \rightarrow \text{c-Si} + \text{nc-}SiO_2, \quad x \sim 1. \tag{1}$$

The reaction in (1) goes to completion in the same temperature range, 850 °C to 900 °C, as that for the interfacial bonding changes revealed by SXPS.

HRTEM images and Fourier transform infra-red spectroscopy (FTIR) established that the as-deposited homogeneous $SiO_x$, $x \sim 1$ bulk films were phase separated according to (1) after annealing to $\sim 900\,^\circ C$. The kinetics of this bulk transition have been studied by FTIR to determine the *extent of reaction* as a function of annealing temperature [27], and demonstrated that the reaction in (1) was $> 90\%$ complete after a $900\,^\circ C$ anneal in an inert ambient. The kinetics for this reaction proceeded more rapidly for annealing temperatures beween $800\,^\circ C$ and $900\,^\circ C$ than for annealing below $750\,^\circ C$. The ITR chemical bonding changes from random to Si- and O-rich arrangements are qualitatively similar to those in (1), and the dependence of the changes, as extracted from Fig. 2, is similar to that for the thin film chemical phase separation.

The temperature-dependent studies in [27] have been interpreted in terms of a model based on a kinetically limited phase separation reaction in which the separated species have a lower total energy than the homogeneous alloy. The kinetics are determined by a reaction barrier and mass transport to separate the homogeneous suboxides into the chemically and physically separated c-Si and nc-$SiO_2$ reaction products of (1).

This kinetic interpretation is consistent with an equilibrium phase diagram along the tie-line between Si and $SiO_2$, and applies to SiO thin films prepared at low temperatures [27]. Below a temperature of $1000\,^\circ C$, there is no compound phase along the Si–$SiO_2$ tie-line to trap or hinder the phase separation reaction of (1). This means the separated products are more thermodynamically stable than the homogeneous suboxide, and the chemical separation and crystallization of Si nano-crystalline grains is limited by reaction and transport kinetics.

Photoluminescence (PL) measurements on homogeneous $SiO_x$ films reinforce the interpretation of the CLS features in Fig. 3, and in particular the assignment of the 1.8 eV CLS feature to defect bonding in the suboxide ITR. PL has been detected in homogeneous suboxide samples prepared by low-temperature, $200\,^\circ C$ to $300\,^\circ C$, plasma deposition processes [26, 27]. These PL features are composition dependent and have been attributed to recombination through intrinsic suboxide bonding defects involving Si and O atoms [26]. The spectral peak of CLS luminescence-labeled $D_1$ in Fig. 4 is at approximately the same photon energy as that of the spectral peak of the $SiO_x$ feature in [26] for $x \sim 1.1$. Additionally, the studies of [27] demonstrated that the same suboxide PL feature decreased significantly after annealing in inert ambients at temperatures between $650\,^\circ C$ and $750\,^\circ C$, and was not detected after a $900\,^\circ C$ anneal.

## 2.3 Bond Constraint Theory and the $SiO_x$ Bonding Changes

BCT provides a framework for describing the distinction between strain-free bonding in continuous random network (CRN) glasses and thin films [15, 16], and has been extended to describe defect generation in non-ideal or over-constrained networks as well [17,18]. In networks in which the constituent atoms include three- or four-fold non-planar bonding arrangements, BCT yields a linear relationship between the average number of bonds per atom, $N_{av}$, and the total number of valence bond-stretching and bond-bending constraints per atom, $C_{av}$, given by [15, 16]

$$C_{av} = 2.5N_{av} - 3. \tag{2}$$

$C_{av}$ is also proportional to $N_{av}$ if the bonding is planar at either the three- or fold-coordinated bonding sites, as for example in $Si_3N_4$, where the N-atoms are in a planar arrangement with three Si nearest neighbors. The condition for an ideal or strain-free CRN solid is that $C_{av}$ be equal to the network dimensionality of three. This corresponds to a value of 2.4 for $N_{av}$, when the bonding constraints are defined by (2).

BCT provides a remarkably accurate description of network stress in non-ideal CRN amorphous solids in which $C_{av} > 3$, including its consequences with respect to defect formation [17, 18]. This application of BCT to SiO is based on the simple idea that the valence forces in a network amorphous solid can be arranged in a hierarchy from stronger bond-stretching to weaker bond-bending forces. The constraining effects of these forces are a linear function of $N_{av}$, as in (2). For over-constrained networks such as $SiO_x$ with $x \sim 1$ and $N_{av} \sim 3$, the Si and O atom bond-stretching constraints are stronger than the respective Si and O centered bond-bending constraints, so that strain energy accumulates along bending constraints at the atom with the lower coordination, O. This means that Si–O–Si bond angles ($\theta$) are distorted from their average value $\sim 150°$ [28,29] by an amount $\delta\theta$, which is proportional to the difference between $N_{av}$ in this non-ideal strained network and $N_{av}^*$ in a strain-free network, e.g., $SiO_2$ in which a bonding constraint at the O-atoms is broken so that $N_{av}^* = 2.4$;

$$\delta\theta \propto [N_{av} - N_{av}^*]. \tag{3}$$

It is also reasonable to assume that the density of bonding defects is associated with broken bonds that relieve the build-up of local strain. The density of these defects is then proportional to the strain energy [17,18], which in turn is proportional to $[\delta\theta]^2$. The density of defects, $D$, then obeys the following scaling relationship,

$$D \propto [N_{av} - N_{av}^*]^2. \tag{4}$$

If these intrinsic defects are assumed to be the nucleating centers for chemical phase separation reaction in $SiO_x$ bulk films, then the kinetics and dimensional scale of the $SiO_x$ decomposition are expected to correlate with the defect density in (4).

The dominant contribution to strain in the chemically separated $SiO_2$/c-Si system is the strain energy at the c-Si/$SiO_2$ interface. This derives from

1. the molar volume mismatch at that interface, and 2. the difference in thermal expansion coefficients between c-Si and nc-SiO$_2$. This is contrasted with the higher average bond strain in homogeneous SiO films in which $C_{av}$ is increased from 3 in SiO$_2$ to approximately 4.2 in the SiO composition alloy.

This argument for a decreased strain-energy after annealing and chemical phase separation is supported by [27], which demonstrated that the chemical phase-separated Si grains showed no detectable defect luminescence, indicative of a low density of defects in the ITR between the nano-crystalline Si grains and the SiO$_2$ matrix in which they are inbedded. This interpretation for the absence of defect luminescence in the chemical phase-separated films is consistent with the CLS Si–SiO$_2$ interface results of Fig. 3.

The interpretation of photoluminescence results for SiO thin films, combined with the CLS results, suggests that the changes in bonding at Si–SiO$_2$ interfaces after the 900 °C anneal in an inert ambient are also driven by a strain-relief mechanism. In this model, the interfacial defects in the defective region of the Si substrate, Si$_D$, act as nucleating centers for the chemical separation into Si-rich and O-rich domains. In the defect nucleation model, the scale for the nano-scale separation is then correlated with the density of Si dangling bonds in the substrate, $\sim 3 \times 10^{12}\,\mathrm{cm}^{-2}$, or approximately 0.3 % to 0.4 % of the Si atoms on a $\langle 001 \rangle$ surface [13, 14]. Based on this model, the characteristic length scale $r$, for interfacial chemical self-organization is estimated by setting

$$r \sim d_{\text{Si–Si}}]_{\text{Si}(001)} \times [f_{\text{db}}]^{-0.5}, \tag{5}$$

where $d_{\text{Si–Si}}]_{\text{Si}(001)}$ is the spacing between terminal Si atoms on a Si$\langle 001 \rangle$ surface, and $f_{\text{db}}$ is the fraction of Si-atom dangling bonds on that surface. This yields a characteristic length of $\sim 5\,\mathrm{nm}$. This will be shown to be a probable source of the interfacial roughness scattering in the channel transport of electrons and holes in FET devices as extracted from analysis of carrier mobilities [30]. It will also be demonstrated later on in this article that the density of dangling bonds at the Si–SO$_2$ interface is consistent with an extension of BCT to interfaces between Si and SiO$_2$, and between different dielectrics as well.

## 2.4 Self-Organization Transition at Si–SiO$_2$ Interfaces

BCT has provided important insights to the physical mechanisms underlying 1. the formation of ITRs, and 2. defect formation and defect relaxation at these interfaces [7]. The extension of BCT discussed below builds on studies of *Boolchand* and coworkers on the nature of the glass transition, and the compositional dependence of the *floppy to rigidity* transitions that occur in glass-forming binary alloy systems such as a-Se$_{1-x}$Ge$_x$ [31, 32].

These studies have established that 1. there are transitions at two different alloy compositions that are associated with the change from *floppy* or under-constrained bonding in a-Se, to *stressed-rigid* or over-constrained bonding in an alloy with 33 % Ge, $GeSe_2$, and 2. that these compositional transitions span a *self-organized* region that is strain and defect free [31, 32]. The first transition occurs at the onset of average *local bonding rigidity* at a composition of $GeSe_4$ corresponding to an $N_{av}$ =~ 2.4 with $C_{av}$ = 3. Alloy compositions with increasing Ge content are on average over-constrained. However, the onset of *global bonding rigidity*, i.e., the *percolation of rigidity* throughout the entire volume of glass or thin film, is delayed by self-organization into *non-statistical or non-random bonding arrangements* that minimize the *percolation or connection* of local bond-strain. These reorganizations occur up to a composition at which local bond-strain finally *percolates* or connects throughout the entire volume of the alloy. This is the composition at which global bonding rigidity sets in, and the alloy becomes *stressed-rigid* or equivalently, over-constrained.

*Lucovsky, Phillips* and coworkers have pointed out that Si–$SiO_2$ interfaces are heterostructures in which the crystalline substrate Si is effectively *rigid, or over-constrained* with the $N_{av}$ equaling exactly four, and the $SiO_2$ dielectric is ideal or *elastically compliant* with an effective value of $N_{av}$ = 2.4 [18, 33]. Since there is an ITR with a bonding chemistry intermediate between Si and $SiO_2$ that separates these two regions of the gate stack, comparisons with a-$Se_{1-x}Ge_x$ chalcogenide alloys provide additional insights into the ITR properties.

The ITR provides a continuous and smooth *transition* between tensile stress in the Si substrate and compressive stress in the $SiO_2$ dielectric. This suggests that the ITR may play essentially the same role as the *strain-free compositional regime* in the a-$Ge_xSe_{1-x}$ alloys. If this is indeed the case, the bonding changes after the 900 °C anneal can then be construed as a strain-driven self-organization that prevents percolation of in-plane rigidity, and thereby provides a low defect and defect precursor layer that bridges the Si substrate to the $SiO_2$ dielectric.

The bonding on average in the interfacial SiO layer is over-constrained with $N_{av}$ = 3, and $C_{av}$ = 4.5. The SXPS studies presented in Sect. 2 have indicated that in as-grown Si–$SiO_2$ interfaces the bonding in the ITR is random; however, after a 900 °C anneal, the bonding is *non-random* with Si-rich and O-rich regions. These changes *mimic* those occurring in a-$Ge_xSe_{1-x}$ alloys for compositions between the onset of *local rigidity* and the percolation to a state of *global rigidity*. The local bonding in the SiO ITR after the 900 °C anneal is equivalent to the average bonding in SiO, as represented schematically in (6),

$$Si^{1+}(Si_3\mathbf{Si}O) + Si^{3+}(O_3\mathbf{Si}Si) = 2Si^{2+}(O_2\mathbf{Si}Si_2). \qquad (6)$$

The terms in brackets identify the local tetrahedral bonding arrangements with respect to the central Si-atom that is listed second and highlighted. $N_{av}$ = 3.6 for the $Si^{1+}$ groups and 2.8 for the $Si^{3+}$ groups corresponding to $C_{av}$ values of 6 and 4, respectively. However, discounting one bond-bending constraint per O-atom [15, 16] reduces the respective values of $C_{av}$ to 5.8 and 3.5. There are additional reductions of $C_{av}$ associated with bending modes at the Si atom sites. These constraints are broken for bending vibrations whenever two end-member atoms are different, as in Si–Si–O groups in contrast with Si–Si–Si, and O–Si–O groups [34]. These *broken constraints* reduce $C_{av}$ to 5.4 for the $Si^{1+}$ cluster, and to 3 for the $Si^{3+}$ cluster. Non-random bonding after the 900 °C RTA *encapsulates* the locally strained and stressed-rigid Si-rich groups, thereby preventing percolation of in-plane bond-strain. Strain-driven interfacial bonding changes at Si–SiO$_2$ interfaces are then effectively equivalent to the self-organization in the Ge–Se alloys, so that the properties of these ITRs with respect to defects and defect formation are then expected to be similar to those associated with the intermediate state.

Unlike the floppy and over-constrained Ge–Se alloy glasses and thin films that have high defects and defect precursors $> 10^{12}\,\mathrm{cm}^{-2}$, those in the intermediate composition range have low defect concentrations $< 10^{11}\,\mathrm{cm}^{-2}$, and do not age, showing reversible heat flow, and no changes in their glass transition temperatures with time [30, 31]. Similar defect properties have been found by CLS for the interfacial SiO transition regions, and these impact significantly on device performance and reliability. This aspect of the ITR bonding changes at 900 °C is addressed in more detail in Sect. 4.

# 3 Internal Dielectric Interfaces

## 3.1 Fixed Charge at Internal Dielectric Interfaces

Fixed charge has been found at internal interfaces between SiO$_2$ and 1. Si$_3$N$_4$, 2. a Si oxynitride alloy with a composition $(Si_3N_4)_{0.5}(SiO_2)_{0.5}$ [17], 3. Al$_2$O$_3$ [34], 4. Zr silicate alloys, $(ZrO_2)_x(SiO_2)_{1-x}$ [35], 5. Hf silicate alloys $(HfO_2)_x$ $(SiO_2)_{1-x}$ [36], and 6. Y silicate alloys $(Y_2O_{43})_x(SiO_2)_{1-x}$ [37]. The fixed charge has been identified through capacitance–voltage (C–V) studies using standard C–V analysis techniques on MOS capacitors with stacked dielectrics, as shown in Fig. 1 [34]. The fixed charge is positive, except for Al$_2$O$_3$ where it is negative. Fixed charge has been found in as-deposited Hf and Zr silicates, and varies with Hf and Zr content from about $1 \times 10^{12}\,\mathrm{cm}^{-2}$ for alloys with x $\sim 0.25 \times 10^{12}\,\mathrm{cm}^{-2}$ up to $\sim 4 \times 10^{12}\,\mathrm{cm}^{-2}$ for the end-member Hf and Zr oxides. Upon annealing to temperatures of $\sim 800$ °C, fixed charge is reduced by more than one order of magnitude [38]. Additionally, high densities of negative charge, $\sim 10^{13}\,\mathrm{cm}^{-2}$, can be trapped at internal SiO$_2$ interfaces with ZrO$_2$ and HfO$_2$; these are due to processing in O-rich ambients, and are eliminated by annealing at $\sim 500$ °C [39].

## 3.2 BCT and Internal Dielectric Interfaces

The density of defects, $D$, in a constrained network obeys the scaling relationship in (5), and a similar relationship applies at internal dielectric interfaces

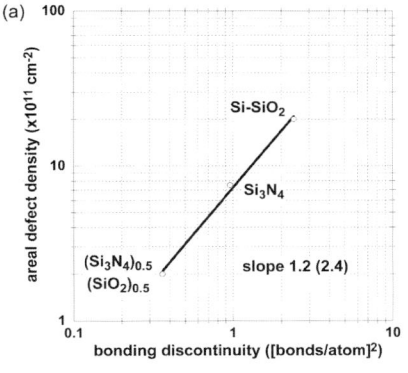 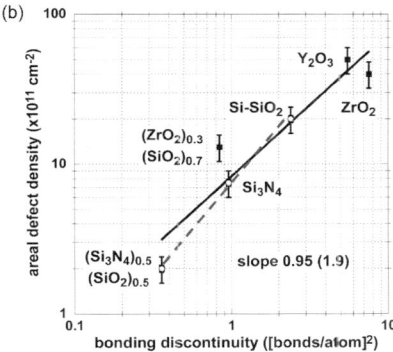

**Fig. 5.** Defect density scaling as a function of $[\Delta N_{av}]^2$ for SiO$_2$ interfaces with (a) Si-based covalent dielectrics and crystalline Si, and (b) Si-based covalent dielectrics, crystalline Si and representative high-$k$ dielectrics

including those between alternative high-$k$ dielectrics and SiO$_2$, as shown in Fig. 1 [18]. The density of interfacial defects, $D_{int}$, scales as the square of the *difference* in the N$_{av}$ of the dielectrics, A and B, which define the interface,

$$D_{int} \propto [N_{av}(A) - N_{av}(B)]^2 = [\Delta(N_{av})]^2 . \tag{7}$$

This approach to scaling is in some ways consistent with the concept of heterovalency applied by *Harrison* et al. to semiconductor $\langle 111 \rangle$ interfaces between Ge and GaAs, and GaAs and ZnSe [40]. At these lattice-matched interfaces, there is a significant mismatch between the number of electrons available for two-electron pair bond formation and the nuclear charge of the interfacial atoms. There is also an average mismatch in bond ionicity of approximately 30 %. For the SiO$_2$/high-$k$ interfaces, the electronic structure mismatches are greater due to the increased bonding coordination of the O- and TM- and RE-atoms associated with increased bond ionicity. The plot in Fig. 5a tests the scaling of (7) for interfaces between SiO$_2$, a Si oxynitride alloy, Si$_3$N$_4$ and the Si–SiO$_2$ interface. These interfaces include CRN solids, and an extension of the concept to crystalline Si. The power law factor for this plot is $2.4 \pm 0.2$.

The plot in Fig. 5b also includes two representative high-$k$ dielectrics, ZrO$_2$ and Y$_2$O$_3$, as well as Zr silicate alloy. The scaling relationship holds for these non-CRN thin film dielectric interfaces as well, and the power law factor is $1.9 \pm 0.2$.

## 3.3 Interfacial Relaxation and Phase Diagrams

It has been proposed, in Sect. 2, that a strain-induced self-organization occurs in the ITR at Si–SiO$_2$ interfaces during 900 °C annealing in inert ambients. Several conditions are necessary for the self-organization to take place: 1. the precursor bonding environments must be consistent with a bonding reorganization that reduces the total bond-strain rearrangements, and 2. it must take place at temperatures consistent with the melting temperature of Si, the decomposition of $SiO_2$ at the Si–$SiO_2$ interface, and/or chemical and structural phase separation within the *bulk* dielectric film. It is equally important that there be no equilibrium phases with congruent melting points between Si and $SiO_2$, or between the end-member oxides at an interface between $SiO_2$ and a TM or RE oxide, silicate or aluminate high-$k$ dielectric. The existence of such a phase would change the end products in the chemical phase separation, and generally not result in a significant reduction in bond-strain energy, and/or impede the kinetics for the chemical phase separation and drive the effective temperature beyond the range of annealing temperatures that meet other process integration restrictions. These conditions are met for Si–$SiO_2$ interfaces subjected to a 900 °C anneal [25]. The temperature of this relaxation is approximately 100 °C lower than the temperature for the onset of visco-elastic relaxation of growth-induced bulk film strain that involves the breaking of Si–O bonds in the bulk $SiO_2$ layer [12].

In contrast, interfaces between 1. $SiO_2$ and 2. $Si_3N_4$ and Si oxynitride alloys display no evidence for strain-driven self-organization for annealing, and/or processing temperatures up to 1000 °C. This is consistent with a compound phase, $Si_2ON_2$, between $SiO_2$ and $Si_3N_4$ with a congruent melting point in excess of 2250 °C [41]. However, the fixed charge at these interfaces is sufficiently small that it does not degrade device performance and reliability in MOS devices [18].

Similar criteria also apply for interfacial self-organizations, including interfacial chemical phase separation of TM and lanthanide RE silicate alloy thin films in contact with $SiO_2$. Those systems that do not have a compound phase with a congruent melting point have the potential to display a separation into the TM or RE oxide, and $SiO_2$, whereas those that do have a compound phase with a congruent melting point do not.

The equilibrium phase diagrams for $SiO_2$ and $ZrO_2$, and $SiO_2$ and $HfO_2$ indicate stable silicate phases but *without congruent melting points* at the respective silicate compound (see Fig. 6) [42]. In addition, the liquidus curves display either stable, or incipient liquid immiscibility characteristics, and therefore are consistent with a spinoidal decomposition for silicate alloys formed by non-equilibrium thin film depositions. Thin film Zr and Hf silicate alloys display a chemical phase separation into $SiO_2$ and either $ZrO_2$ or $HfO_2$ at temperatures of at most 900 °C to 1000 °C [42, 43]. Based on the visco-elastic relaxation temperature of 1000 °C for bond-breaking in $SiO_2$, and the ITR self-organization at 900 °C, interfacial relaxations at Hf and Zr silicate or oxide interfaces with $SiO_2$ are expected to occur at temperatures between 800 °C to 900 °C. This expectation has been realized in devices including $Hf(Zr)O_2$ and $Hf(Zr)$ silicate alloys, where fixed charge has been reduced by more than an order of magnitude for annealing temperatures of about 800 °C [38, 44]. In contrast, devices with Zr and Hf silicate alloys an-

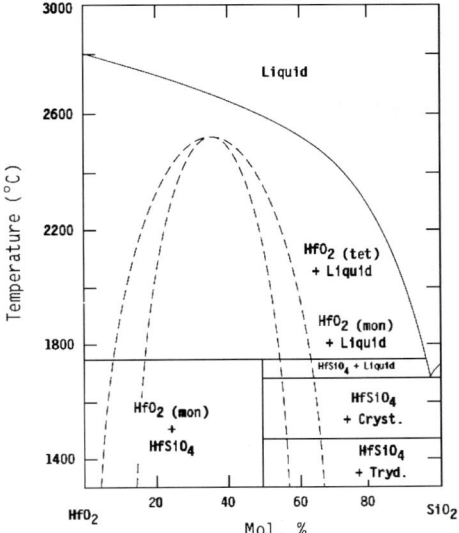

**Fig. 6.** Equlibrium phase diagram for $SiO_2$–$HfO_2$

nealed at 500 °C do not show fixed charge reductions [35, 36]. The reductions of fixed charge of [38, 44] are attributed to an interfacial chemical phase separation into nano-scale crystalline $ZrO_2$ or $HfO_2$ encapsulated by $SiO_2$. This self-organization, like the separation of SiO into nano-scale c-Si encapsulated by $SiO_2$, also results in a decrease in interfacial bond-strain.

In contrast, and consistent with differences in the equilibrium phase diagram between $Al_2O_3$–$SiO_2$ [45] and $ZrO_2$–$SiO_2$ and $HfO_2$–$SiO_2$ [42], there is no reduction of fixed charge at $SiO_2$–$Al_2O_3$ interfaces up to temperatures of at least 900 °C where the $Al_2O_3$ films undergo a bulk crystallization [35]. This is consistent with the existence of a compound composition with congruent melting point in excess of 1800 °C between $SiO_2$ and $Al_2O_3$ [45].

Compound phases with congruent melting points, such as $(Y_2O_3)_1(SiO_2)_2$, and $(Y_2O_3)_2(SiO_2)_1$, exist in the phase diagrams for the group IIIB TM metal atom silicates of Y and La, as well as the lanthanide RE elements in trivalent bonding states (see Fig. 7) [42]. These effectively block self-organizations driven by bond-strain reduction, and account for levels of interfacial fixed charge, typically $> 10^{12} \, cm^{-2}$ even after annealing to temperatures of 900 °C to 1000 °C.

The phase diagrams for $ZrO_2$–$Al_2O_3$ and $HfO_2$–$Al_2O_3$ do not indicate compound compositions between $Al_2O_3$ and the respective transition metal oxides [46, 47], nor do they reveal liquidus features indicative of stable or incipient liquid immiscibility and a driving force for a spinodal decomposition [42]. Eutectic compositions in these systems in the mid-alloy range are at a temperature at least 100 °C higher than in the respective phase diagrams

194     Gerald Lucovsky and James C. Phillips

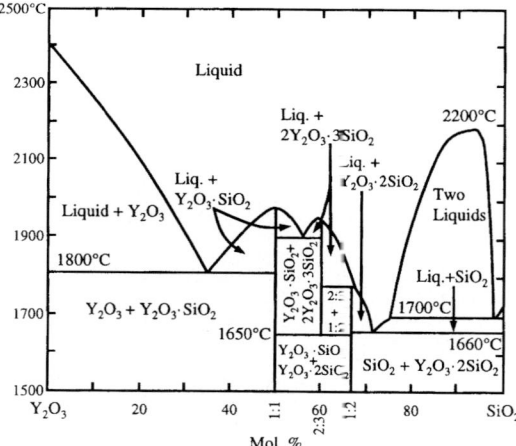

**Fig. 7.** Equilibrium phase diagram for $SiO_2$–$Y_2O_3$

with $SiO_2$. Studies of Hf aluminate alloys with alloy compositions in the range of 35 % to 50 % $HfO_2$ indicated chemical phase separation into $Al_2O_3$ and $HfO_2$ at temperatures in excess of 900 °C to 1000 °C [48]. Electrical studies on capacitors prepared from these alloys and subjected to post-deposition annealing at 800 °C to 900 °C prior to Al metallization displayed significant levels of fixed negative charge similar to those reported for $Al_2O_3$ devices. These were found in both NMOS and PMOS capacitors, confirming that the charge was fixed and not dominated by electron injection and trapping. This suggests that the interfaces of these devices were $Al_2O_3$-like, and that the kinetics for a strain-driven self-organization that would reduce fixed charge were not possible for the range of annealing temperatures explored, or the phase separation products.

# 4 Discussion

## 4.1 Defects in High-$k$ Gate Stacks

Figure 8 includes in a schematic representation the strain profile, defects, and defect precursors that are present in a Si–$SiO_2$ hetero-structure that includes a strain-free, self-organized ITR in which the chemically separated regions are in the nanometer-size regime. Assuming that the nucleation of self-organized regions in the ITR is correlated with the density of dangling bonds in the tensile stressed Si substrate, this scale is estimated to be ∼ 5 nm to 6 nm. This assumption is supported by the interface roughness parameter associated with the universality of the electron and hole channel mobilities in FETs [30]. The product of the correlation distance (L) and roughness

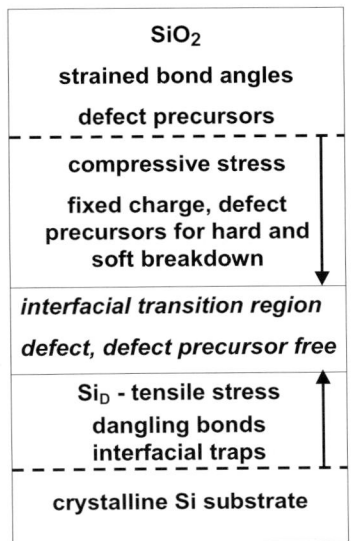

**Fig. 8.** Defect profile for an Si–SiO$_2$ gate stack

"height" ($H$) is $\sim 0.2\,\mathrm{nm}^2$ to $0.3\,\mathrm{nm}^2$. The difference in effective size of Si-rich and O-rich bonding groups is $\sim 0.06\,\mathrm{nm}$ so that $L \sim 5\,\mathrm{nm}$ to $6\,\mathrm{nm}$, in good agreement with the nucleation model of Sect. 2.3.

Based on ESR measurements of [13, 14], the density of Si atom dangling bonds in the strained Si region Si$_\mathrm{D}$ is $\sim 2 \times 10^{12}\,\mathrm{cm}^{-2}$ to $3 \times 10^{12}\,\mathrm{cm}^{-2}$ for Si$\langle 100 \rangle$ (P$_{\mathrm{b}0}$ centers) and increases to $\sim 5 \times 10^{12}\,\mathrm{cm}^{-2}$ for Si$\langle 111 \rangle$ (P$_\mathrm{b}$ centers). There are additional P$_{\mathrm{b}1}$ centers in the ITR for Si$\langle 001 \rangle$ devices, and it is not clear whether these are present after a 900 °C. These dangling bonds are H atom terminated during a post-metallization anneal in an H-containing ambient such as forming gas. In MOS devices, H atom release from these bonds has been proposed as a mechanism for defect generation during device operation [49].

The density of interface traps, $D_{\mathrm{it}}$, has been studied as a function of annealing temperature at Si$\langle 111 \rangle$–SiO$_2$ interfaces prepared by thermal oxidation at 850 °C, furnace annealed at temperatures up to 1100 °C for 30 min in Ar, and then subjected to a PMA for 30 min at 400 °C in forming gas after the initial growth and following each annealing step [18]. The most significant decreases in $D_{\mathrm{it}}$ occur at a temperature of $\sim 980$ °C, very close to the onset of the release of growth-induced stress as in [19]. In marked contrast, there is only a small decrease in $D_{\mathrm{it}}$ after the 900 °C anneal, indicating that these defects are not reduced significantly by the atomic rearrangements that occur during the IRT self-organization [19]. This is consistent with these defects being resident in the Si substrate in the immediate vicinity of the ITR. Increased levels of defects and defect precursors have been correlated with non-optimum interface formation. Increasing the thickness of the interfacial $SiO_2$ layer in devices with Si oxynitride dielectrics results in increases in stress-induced leakage currents (SILC), and decreases in time-dependent dielectric breakdown (TDDB) [33]. This is attributed to forming an ITR with a self-organization that removes defect precursors from the ITR or the $SiO_2$ layer in contact with it.

These arguments, as well as defect modelling of hard and soft breakdown [50], lead us to conclude that 1. Si-atom dangling bonds and interface traps, $D_{it}$, are located in the strained Si substrate region, $Si_D$, and 2. the precursor states for soft and hard dielectric breakdown are in the strained $SiO_2$ film.

Finally, the inclusion of a high-$k$ dielectric generates a second ITR between the $SiO_2$ buffer layer and the high-$k$ dielectric. Fixed charge can occur at this interface, but more importantly, the precursor centers for soft and hard breakdown reside primarily in the high-$k$ part of the stack. In addition, electron trapping in the high-$k$ is generally greater than in $SiO_2$, and this becomes a significant issue for both performance and reliability.

## 4.2 EOT Scaling in Advanced Gate Stacks

As discussed above, ultra-thin $SiO_2$ layers with self-organized ITRs play a key role in optimizing performance and reliability in CMOS devices that include alternative dielectrics. Meeting scaling metrics for channel transport, i.e., electron and hole channel mobilites equivalent to those of Si–$SiO_2$ devices, requires a $SiO_2$ buffer layer ($t_{phys} \sim 0.6\,nm$), and this contributes to the EOT:

$$EOT = EOT_{interface} + EOT_{high\text{-}k} \sim 0.35\,nm + t_{high\text{-}k}(k_{SiO_2}/k), \qquad (8)$$

where $t_{high\text{-}k}$ is the physical thickness of the high-$k$, and $(k_{SiO_2}/k)$ is the ratio of the dielectric constant of $SiO_2$ to that of the high-$k$ dielectric. The addition of $\sim 0.35\,nm$ to EOT from the $SiO_2$ buffer layer and the self-organized ITR places a significant constraint on ultimate EOT scaling. This is estimated by us to be approximately $0.7\,nm$ to $0.8\,nm$, rather than the $0.5\,nm$ target of the ITRS [1].

Deposition of other dielectrics directly on to Si has never yielded interfacial properties that are functionally equivalent to what is formed by either thermal, plasma-assisted or chemical oxidation of Si. Since scaling depends on maintaining high levels of current drive, any interface modification that reduces effective channel transport properties and increases trapping will impact adversely on scaling.

## 4.3 Narrowing the Field of High-$k$ Dielectrics

Reduction of defects at internal dielectric interfaces is crucial. The self-organization in ITRs at Si–$SiO_2$ interfaces is *a gift from nature*, and more importantly is consistent with the Si–SiO$_2$ binary phase diagram. The mismatch in $N_{av}$ and bond-ionicity requires a similar self-organization at internal dielectric interfaces between SiO$_2$ and a high-$k$ oxide or silicate. As demonstrated above, the conditions permitting self-organizations that reduce interfacial strain, and at the same time reduce defects and defect precursors associated with steps in interfacial bond ionicity have been correlated with equilibrium phase diagrams between SiO$_2$ and high-$k$ TM and RE oxides. Based on this criterion, only *two families* of high-$k$ dielectrics have the potential to work. These are the oxides and pseudo-binary silicate alloys of Zr and Hf. Similar conclusions have been obtained by an analysis of direct tunneling including the effects of d-states at conduction band edges [51].

Recent studies of high-$k$ dielectric devices that include internal dielectric interfaces between SiO$_2$ and HfO$_2$ or Hf silicates have confirmed the reduction in fixed charge for these internal interfaces [52]; however, they have identified another source of interfacial traps associated with the interfacial self-organization that reduces fixed charge. These traps are associated with O-atom vacancies that are generated in the HfO$_2$ nano-crystalline-scale grains produced during the chemical phase separation and HfO$_2$ crystallization that occurs at the *Hf silicate interface* between HfO$_2$ and SiO$_2$. This chemical phase separation is in effect a self-organization that reduces interfacial bond-strain, and has two effects on defects: it reduces interfacial fixed charge, but creates O-atom vacancies that can be populated by substrate hole injection [52]. These issues are addressed in more detail in the next chapter of this book.

There are some experimental results that discuss Zr and Hf Si oxynitride alloys [52]. Alloys with high Si$_3$N$_4$ content, $\sim 35\%$ to $40\%$, and approximately equal concentration of SiO$_2$ and either HfO$_2$ or ZrO$_2$, do not display chemical phase separation and crystallization of a HfO$_2$ or ZrO$_2$ phase. This identifies an approach to interfacial engineering at the atomic layer scale that can be carried over to other dielectrics as well, e.g., HfO$_2$ or ZrO$_2$.

Finally, combining the results of this Chapter with the results in the next Chapter (cf. above) identifies two different approaches to integration of high-$k$ dielectrics in advanced Si devices, as well as into devices with other semiconductor substrates such as Ge, Si–Ge alloys, Si–C and the like. The limiting values of EOT scaling projected for these two atomically engineering gate stacks are essentially the same, $\sim 0.7$ nm to $0.8$ nm. Figures 9a and 9b indicate schematic representations of these gate stacks. Each of the stacks includes an ultra-thin SiO$_2$ buffer region, and a nitrided Si–SiO$_2$ self-organized interface; this contributes $\sim 0.35$ nm to EOT. Each stack also includes an engineered interface comprised of a molecular layer of Si$_3$N$_4$, and two molecular layers of the high Si$_3$N$_4$ content, Zr or Hf Si oxynitride alloy discussed above. The remainder of the stack in Fig. 9a is either HfO$_2$ or ZrO$_2$, whilst the remainder of the stack in Fig. 9b is a high Si$_3$N$_4$ content, Zr or Hf Si oxynitride alloy. These EOT are limited by tunneling leakage, and the equivalence between the high-$k$ contributions to EOT is related to higher-$k$ value and physically

a)
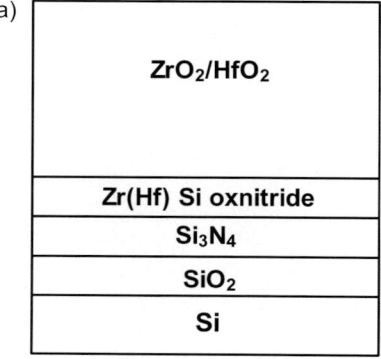

b)
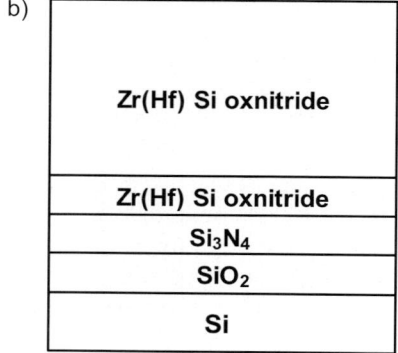

**Fig. 9.** Atomically engineered high-$k$ gate profiles for integration of (**a**) ZrO$_2$ or HfO$_2$, or (**b**) a Zr or Hf Si oxynitride alloy into a high-$k$ gate stack for advanced Si devices

thicker film in Fig. 9a, and to the physically thinner film in Fig. 9b, determined by a lower value of $k$, but with a larger $E_B - m_o^*$ product that gives essentially the same tunneling figure of merit. $\Phi = k(E_B - m_o^*)^{0.5}$, where $E_B$ is the effective conduction band offset energy of the high-$k$ portion of the gate stack relative the conduction band of Si, $\sim 1.5$ eV for ZrO$_2$/HfO$_2$ and SiO$_2$, but $\sim 2.5$ eV for the Zr/Hf Si oxnitirde alloy, and $m_o^*$ is the electron tunneling mass, $\sim 0.15$ mo for ZrO$_2$/HfO$_2$, but 0.35 mo for the Zr/Hf Si oxynitride alloy. Combined with $k$-values of $\sim 20$ for ZrO$_2$/HfO$_2$ and $\sim 10$ for the Zr/Hf Si oxynitride, these give values of $\Phi = 9.5$ and $\Phi = 9.4$, respectively for ZrO$_2$/HfO$_2$ and Zr/Hf Si oxynitride high-$k$ portions of the gate stack.

## Acknowledgements

Supported in part by ONR and SRC. The authors have benefited significantly from on-going discussions and interactions with Punit Boolchand of the University of Cincinnati, and Mike Thorpe of the Arizona State University.

# References

[1] International Technology Roadmap for Semiconductors (2003 ed.)
    URL http://public.itrs.net

[2] G. Wilk, R. W. Wallace, J. M. Anthony: High-$\kappa$ gate dielectrics: Current status
    and materials properties considerations, J. Appl. Phys. **89**, 5243 (2001) and
    references therein

[3] L. C. Feldman, L. Stensgard, P. J. Silverman, T. E. Jackman: in S. T. Pan-
    telides (Ed.): *Proceedings of the International Conference on the Physics of
    SiO₂ and its interfaces* (Pergamon, New York 1978) p. 344

[4] D. E. Aspnes, J. B. Theeten: Optical properties of the interface between Si and
    its thermally grown oxide, Phys. Rev. Lett. **43**, 1046 (1979)

[5] F. T. Himpsel, F. R. McFeely, J. A. Yarmoff, G. Hollinger: Microscopic struc-
    ture of the $SiO_2$/Si interface, Phys. Rev. B **38**, 6084 (1988)

[6] T. Yasuda, Y. Ma, S. Habermehl, G. Lucovsky: Low-temperature prepara-
    tion $SiO_2$/Si(100) interface using a two-step remote plasma-assisted oxidation-
    deposition process, Appl. Phys. Lett. **60**, 434 (1992)

[7] G. Lucovsky, J. C. Phillips: Interfacial strain-induced self-organization in semi-
    conductor dielectric gate stacks. I. Strain relief at the Si–$SiO_2$ interface, J. Vac.
    Sci. Technol. B **22**, 2087 (2004)

[8] Y. Tu, J. Tersoff: Structure and energetics of the Si-$SiO_2$ interface, Phys. Rev.
    Lett. **84**, 2449 (2000)

[9] A. Bongiorno, A. Pasquarello: Atomistic structure of the Si(100)–$SiO_2$ inter-
    face: A synthesis of experimental data, Appl. Phys. Lett. **83**, 1417 (2003)

[10] D. A. Muller, T. Sorsch, S. Moccio, F. H. Baumann, K. Evans-Lutterodt,
     G. Timp: The electronic structure at the atomic scale of ultrathin gate oxides,
     Nature **399**, 758 (1999)

[11] J. W. Keister, J. E. Rowe, J. J. Kolodziej, H. Niimi, H. S. Tao, T. E. Madey,
     G. Lucovsky: Band offsets for ultrathin $SiO_2$ and $Si_3N_4$ films on Si(111) and
     Si(100) from photoemission spectroscopy, J. Vac. Sci. Technol. A **17**, 1250
     (1999)

[12] J. T. Fitch, C. H. Bjorkman, G. Lucovsky, F. H. Pollak, X. Yim: J. Vac. Sci.
     Technol. B **7**, 775 (1988)

[13] E. H. Poindexter, P. Caplan, B. Deal, R. Razouk: Interface states and electron
     spin resonance centers in thermally oxidized (111) and (100) silicon wafers, J.
     Appl. Phys. **52**, 879 (1981)

[14] E. H. Poindexter: MOS interface states: Overview and physicochemical per-
     spective, Semicond. Sci. Technol. **4**, 961 (1989)

[15] J. C. Phillips: Topology of covalent non-crystalline solids. I. Short-range order
     in chalcogenide alloys, J. Non-Cryst. Solids **34**, 153 (1979)

[16] J. C. Phillips: Topology of covalent non-crystalline solids. II. Medium-range
     order in chalcogenide alloys and A-Si(Ge), J. Non-Cryst. Solids **43**, 37 (1981)

[17] G. Lucovsky, H. Yang, H. Niimi, J. W. Keister, J. E. Rowe, M. F. Thorpe,
     J. C. Phillips: Intrinsic limitation on device performance and reliability from
     bond-constraint induced transition region at interfaces of stacked dielectrics,
     J. Vac. Sci. Technol. B **18**, 1742 (2000)

[18] G. Lucovsky, Y. Wu, H. Niimi, V. Misra, J. C. Phillips: Bonding constraints
     and defect formation at interfaces between crystalline silicon and adanced sin-
     gle layer and composite gate dielectrics, Appl. Phys. Lett. **74**, 2005 (1999)

[19]  C. H. Bjorkman, T. Yasuda, C. E. Shearon, Jr., U. Emmerichs, C. Meyer, K. Leo, H. Kurz: Influence of surface roughness on the electrical properties of Si–SiO$_2$ interfaces and on second-harmonic generation at these interfaces, Vac. Sci. Technol. B **11**, 1521 (1993)

[20]  G. Luepke: Surf. Sci. Rep. **35**, 75 (1999)

[21]  C. H. Bjorkman, C. E. Shearon, Jr., Y. Ma, T. Yasuda, G. Lucovsky, U. Emmerichs, C. Meyer, K. Leo, H. Kurz: Second-harmonic generation in Si–SiO$_2$ heterostructure formed by chemical, thermal, and plasma-assisted oxidation and deposition processes, J. Vac. Sci. Technol. A **11**, 964 (1993)

[22]  U. Emmerichs, C. Meyer, H. J. Bakker, F. Wolter, H. Kurz, G. Lucovsky, C. H. Bjorkman, T. Yasuda, Y. Ma, Z. Jing, J. L. Whitten: Optical second harmomic generation: A probe of atomic structure and bonding at Si–SiO$_2$ interfaces, and other chemically modified Si surfaces, J. Vac. Sci. Technol. B **12**, 2484 (1994)

[23]  J. Schafer, A. P. Young, L. J. Brillson, H. Niimi, G. Lucovsky: Depth-dependent spectroscopy defect characterization of the interface between plasma-deposited SiO$_2$ and silicon, Appl. Phys. Lett. **73**, 791 (1998)

[24]  J.-F. T. Wang, G. D. Powell, R. S. Johnson, G. Lucovsky, D. E. Aspnes: J. simplified bond-hyperpolarizability model of second harmonic generation: Application to Si-dielectric interfaces, Vac. Sci. Technol. B **20**, 1699 (2002)

[25]  H. Yang, H. Niimi, J. W. Keister, G. Lucovsky: IEEE Electron. Dev. Lett. **21**, 76 (2000)

[26]  R. Carius, R. Fischer, F. Holzenkampfer, J. Stuke: Photoluminescence in the amorphous SiO$_x$, J. Appl. Phys. **52**, 4241 (1981)

[27]  B. J. Hinds, F. Wang, D. M. Wolfe, C. L. Hinkle, G. Lucovsky: Investigation of postoxidation thermal treatments of Si/SiO$_2$ interface in relationship to the kinetics of amorphous Si suboxide decomposition, J. Vac. Sci. Technol. B **16**, 2171 (1998)

[28]  J. Neuefeind, K. D. Liss: Bond angle distribution in amorphous germania and silica, Ber. Bunsen Phys. Chem. **100**, 1341 (1996)

[29]  J. L. Whitten, Y. Zhang, M. Menon, G. Lucovsky: Electronic structure of SiO$_2$: Charge redistribution contribution to the dynamic dipoles/effective charges of the infrared active normal modes, J. Vac. Sci. Techol. B **20**, 1710 (2002)

[30]  J. R. Hauser: Extraction of experimental mobility data for MOS devices, IEEE Trans. Electron. Dev. **43**, 1981 (1996)

[31]  P. Boolchand: in P. Boolchand (Ed.): *Insulating and Semiconducting Glasses* (World Scientific, Singapore 2000) p. 191

[32]  P. Boolchand, D. G. Georgiev, M. Micoulaut: J. Optoelectron. and Adv. Mater. **4**, 823 (2002)

[33]  G. Lucovsky, J. C. Phillips: Microscopic bonding and macroscopy strain relaxations at Si–SiO$_2$ interfaces, Appl. Phys. A **78**, 453 (2004)

[34]  R. S. Johnson, G. Lucovsky I. Baumvol: Physical and electrical properties of noncrystalline Al$_2$O$_3$ prepared by remove plasma enhanced chemical vapor deposition, J. Vac. Sci. Technol. A **19**, 1353 (2001)

[35]  G. Lucovsky: Electronic structure of trnsition metal/rare earth althernative high-k gate dielectrics: Interfacial band alignments and intrinsic defects, Microeletron. Reliab. **43**, 1417 (2003)

[36]  J. G. Hong: Ph.D. thesis, North Carolina State University, Raleigh, USA (2003)

[37]  J. J. Chambers, G. N. Parsons: Physical and electrical characterization of ultrathin yttrium silicate insulators on silicon, J. Appl. Phys. **90**, 918 (2001)

[38]  R. Chau, S. Datta, M. Doczy, J. Kavalieros, M. Metz: Gate dielectric scaling for high-performance CMOS: From $SiO_2$ to high-k, in *International workshop on gate insulator(s)* (Tokyo, Japan 2003)

[39]  C. C. Fulton, G. Lucovsky, R. J. Nemanich: Process-depended band structure changes of transition-metal (Ti, Zr, Hf) oxides on Si, (100), Appl. Phys. Lett. **84**, 580 (2004)

[40]  W. A. Harrison, E. A. Kraut, J. R. Walthrop, R. W. Grant: Polar heterojunction interfaces, Phys. Rev. B **18**, 4402 (1978)

[41]  H. J. Richter, M. Herrmann, W. Hermel: Calculation of heterogenous phase equilibrial in the system Si–Mg–N–O, J. Eur. Ceram. Soc. **7**, 3 (1991)

[42]  J. P. Maria, D. Wichakana, J. Parrete, A. I. Kingon: Crystallization in $SiO_2$-metal oxides alloys, J. Mater. Res. **17**, 1571 (2002)

[43]  G. B. Rayner, D. Kang, G. Lucovsky: Spectroscopy study of chemical phase separation in zirconium silicate alloys, J. Vac. Sci. Technol. B **21**, 1783 (2003)

[44]  C. L. Hinkle, G. Lucovsky: Remote plasma-assisted nitridation (RPN): Applications to Zr and Hf silicate alloys and $Al_2O_3$, Appl. Surf. Sci. **216**, 124 (2003)

[45]  G. J. Ball, M. A. Mignanelli, J. I. Barry, J. A. Gisby: The calculation of phase equilibria of oxide core-concrete systems, J. Nucl. Mater. **20**, 238 (1993)

[46]  G. Cevales: Ber. Deutsch. Keram. Ges. **45**, 216 (1968)

[47]  V. A. Lysenko: Neorg. Mater. **30**, 930 (1994)

[48]  R. S. Johnson, J. G. Hong, C. L. Hinkle, G. Lucovsky: Electron trapping in noncrystalline remote plasma deposited Hf-aluminate alloys for gate dielectric applications, J. Vac. Sci. Technol. B **20**, 1126 (2002)

[49]  D. Arnold, E. Cartier, D. J. Maria: Theory of high-field electron transport and impact ioninization in silicon dioxide, Phys. Rev. B **49**, 10278 (1994)

[50]  S. Lombardo, J. H. Stathis, B. P. Linder: Breakdown transients in ultrathin gate oxides: Transition in the degradation rate, Phys. Rev. Lett. **90**, 167601 (2003)

[51]  G. Lucovsky, J. G. Hong, C. C. Fulton, Y. Zou, R. J. Nemanich, H. Ade: X-ray absorption spectra for transition metal high-kappa dielectrics: Final state differences for intra- and inter-atomic transitions, J. Vac. Sci. Technol. **22**, 2132 (2004)

[52]  G. Lucovsky, et al.: *Radiation Physics and Chemistry* (2005) in press

# Index

# Electrical Characterization of Rare Earth Oxides Grown by Atomic Layer Deposition

Sabina Spiga, Claudia Wiemer, Giovanna Scarel, Omar Costa, and Marco Fanciulli

CNR-INFM MDM National Laboratory, Via C. Olivetti 2,
20041 Agrate Brianza (MI), Italy
sabina.spiga@mdm.infm.it

**Abstract.** In this contribution, we investigate the electrical properties of thin (3–30 nm) $Lu_2O_3$ and $Yb_2O_3$ oxides grown on silicon by atomic layer depostion. Precursors with various ligands (e.g., cyclopentadienyls, or $\beta$-diketonates) are used as metal source, while water or ozone are used as oxygen source. $Lu_2O_3$ and $Yb_2O_3$ films exhibit a dielectric constant ($\kappa$) of $11\pm1$ and $10\pm1$, respectively. For both rare earth oxides, a low $\kappa$ interlayer (IL) is formed at the film/silicon interface. Gate stacks with capacitance equivalent oxide thickness (CET) down to 3 nm exhibit low leakage current and well-shaped capacitance–voltage curves without frequency dispersion of the accumulation capacitance. A CET of $2.7 \pm 0.1$ nm and leakage of $4.5 \times 10^{-4}$ A cm$^{-2}$ were measured for the thinnest (3.5 nm) $Lu_2O_3$/IL/Si gate stack, and a CET of $3.3 \pm 0.1$ nm and leakage of $1 \times 10^{-4}$ A cm$^{-2}$ for the (4 5 nm) $Yb_2O_3$/IL/Si one. The lowest interface trap density ($D_{it}$), measured for film grown using water, is in the $10^{11}$ eV$^{-1}$ cm$^{-2}$ range for both rare earth oxides. Films grown using ozone exhibit a $D_{it}$ in the $10^{12}$ eV$^{-1}$ cm$^{-2}$ range. The microscopic structure of the electrically active defects is investigated using magnetic resonance spectroscopy.

## 1 Introduction

Binary and ternary rare earth oxides (REOs) are currently being investigated as high dielectric-constant materials to substitute $SiO_2$ in future ultra-scaled devices [1–3]. The interesting properties of several oxides, among the REOs, are high permittivity [4–7] (see also the Chapter by *Scarel* et al. in this volume), large gap and conduction-band offset [8] (see also the Chapter by *Seguini* et al. in this volume), and a predicted thermal stability on Si [9]. Moreover, REOs have a small lattice mismatch with silicon, and they can be grown epitaxially, as already reported for $CeO_2$, $Gd_2O_3$, $Y_2O_3$ and $Pr_2O_3$ [10–13]. Finally, $Gd_2O_3$ is being investigated as promising passivation layer for GaAs [14], and other REOs might be used as gate oxide for high-mobility substrates such as Ge and GaAs (see the Chapter by *Dimoulas* in this volume) and as passivation layers on GaN or SiC.

REOs have been mostly grown using different physical vapor deposition (PVD) techniques, such as e-beam evaporation and molecular beam epitaxy, radio frequency magnetron sputtering, evaporation or sputtering of metal elements followed by oxidation, and pulsed laser deposition. The growth of REOs

M. Fanciulli, G. Scarel (Eds.): Rare Earth Oxide Thin Films,
Topics Appl. Physics **106**, 203–224 (2007)
© Springer-Verlag Berlin Heidelberg 2007

by chemical methods, such as atomic layer deposition (ALD) and metal organic chemical vapor deposition (MOCVD), has received increasing attention in recent years. In particular, ALD allows the deposition of thin films with a precise thickness control and high conformality over large-diameter wafers, as requested by current technological processes. However, there are fewer reports on CVD- than on PVD-grown REOs, due to the limited availability of suitable precursors (for ALD and MOCVD processes) with appropriate volatility, stability and decomposition characteristics. REOs have been mostly grown by CVD using the $\beta$-diketonate complexes and ozone, while recently an increasing effort has been devoted to the development of new precursors, such as those based on cyclopentadienyl (Cp) groups, to improve the film structural and electrical properties [15, 16] (see also the Chapter by *Päiväsaari* et al. in this volume).

The electrical properties of binary REOs grown by PVD have been largely investigated: $Y_2O_3$ [17], $La_2O_3$ [18–22], $CeO_2$ [23,24], $Gd_2O_3$ [25,26], $Pr_2O_3$ [3, 27], $Nd_2O_3$ [3], $Sm_2O_3$ [3, 28], $Dy_2O_3$ [3, 26, 29], $Er_2O_3$ [30], $Lu_2O_3$ [29, 31], $Yb_2O_3$ [22, 32]. The influence of the hygroscopic nature of REOs on the electrical properties has also been discussed [33]. The reported dielectric constant ($\kappa$) for binary REOs grown by PVD ranges mostly between 10 and 20 [3–5]. Higher $\kappa$ values are reported for some oxides by a few authors, e.g., for single-crystal $CeO_2$ ($\kappa = 50$) and polycrystalline $CeO_2$ ($\kappa = 26$) by *Matsushita* et al. [10], $La_2O_3$ ($\kappa = 27$) by *Wu* et al. [18], and $Pr_2O_3$ ($\kappa = 30$) by *Fissel* et al. [12]. Among binary REOs, $La_2O_3$ is considered one of the most attractive materials due to its promising properties such as high $\kappa$ (20–27) [18, 22], large band gap (6 eV) and conduction-band offset (2.3 eV) [34], high breakdown electric field ($> 13\,MV/cm$), and low leakage current [18, 34]. Also $CeO_2$ and $Pr_2O_3$ exhibit a high $\kappa$ value, but the smaller band gap and conduction-band offset of these oxides can cause a larger leakage current, compared to $La_2O_3$. The growth of ternary oxides is also interesting and opens the possibility to tailor the structural and electrical properties of REOs. Ternary REOs deposited by PVD have also been recently investigated: $LaAlO_3$ [35–37], $La_2Hf_2O_7$ [37], scandate compounds ($GdScO_3$, $DyScO_3$ and $LaScO_3$) [38], and $PrTiO$ [3].

Most of the published papers on REOs grown by ALD or MOCVD deal with film growth and structural characterization. Electrical characterization has been reported for several REO thick films (50–200 nm) [1, 6, 15, 39], while the electrical properties of thin films have been addressed only for $La_2O_3$, $LaAlO_3$ and $Lu_2O_3$ oxides [7, 40–42]. The dielectric constant of $La_2O_3$ has been determined to be around 21–25 (film thickness of 100 nm) by *Kang* et al. [43], and 16–19 by *Yamada* et al. [40]. The latter authors studied thin films (3–30 nm range) and they measured a leakage current of $2 \times 10^{-2}\,A\,cm^{-2}$ for $La_2O_3/Si$ stacks with 2.1 nm equivalent oxide thickness (EOT) and of $3 \times 10^{-6}\,A\,cm^{-2}$ for $La_2O_3/SiON$ stacks with EOT = 2.4 nm. This is one of the best results reported for $La_2O_3$ thin films (3–5 nm) grown by MOCVD.

*Jun* et al. [44] report that the ternary compound LaAlO grown by MOCVD shows better thermal stability and leakage current density than do $La_2O_3$ films. *Li* et al. [42] report the growth of LaAlO films by MOCVD with $\kappa = 25$, 6.47 eV band gap and EOT down to 1.2 nm. *Lim* et al. [45] grew $LaAlO_3$ by ALD using the $La(^iPrAMD)_3$ and $H_2O$ precursor combination on HF-last silicon. The authors report an EOT of 2.9 nm for a 9.8 nm film, with $\kappa = 13$, leakage $< 5 \times 10^{-8}$ A cm$^{-2}$ and small hysteresis (20 mV).

The electrical properties of 55–88 nm thick $Y_2O_3$ films grown by ALD using a cyclopentadienyl type of compounds and water as precursors are reported by *Niinisto* et al. [1]. $Al-Y_2O_3-nSi$ capacitors exhibit a dielectric constant of 10 at 500 kHz, and a significative amount of positive fixed charge (flat-band voltage ($V_{FB}$) around 4–5 V.)

A recent review on the electrical and structural properties of 50 nm thick $Ln_2O_3$ (Ln: Nd, Sm, Eu, Gd, Dy, Ho, Er and Tm) grown by ALD is that of *Päiväsaari* et al. [6]. Films were grown at 300 °C (310 °C for $Nd_2O_3$) using $Ln(thd)_3$ and ozone as precursors. The authors present the capacitance–voltage (C–V) characteristics at 500 kHz of Al-50 nm thick $LnO_3-nSi$ capacitors. The presence of positive fixed charge (negative shift of the flat band) is found for $Nd_2O_3$ and $Eu_2O_3$, while $Gd_2O_3$, $Dy_2O_3$, $Ho_2O_3$, $Er_2O_3$, $Tm_2O_3$ exhibit a positive shift of the flat band, indicating the presence of negative fixed charge. Hysteresis width and direction depend on the rare earth (Ln) and on the applied field: counterclockwise hysteresis is measured for $Nd_2O_3$ and $Dy_2O_3$, while the opposite is observed for $Ho_2O_3$, $Er_2O_3$ and $Tm_2O_3$. The measured dielectric constant for the above lanthanide oxides is in the 12–14 range. The differences in the electrical properties of the investigated REOs might be due to differences in impurity concentration (due to the different reactivity and stability of the $Ln(thd)_3$ precursors) and in structural properties.

The electrical properties of thick (around 70 nm) $GdO_x$ and $PrO_x$ grown by ALD at 300 °C are also reported by *Jones* et al. [15]. The authors measured a dielectric constant of 15.9 and 21, respectively, for as-grown and annealed (800 °C, $N_2$, 1 min) $PrO_x$ films. The dielectric constant of $GdO_x$ films was found to be 10.4 and 15.4 for as-grown and annealed (in $O_2$ at 700 °C) films, respectively. Significant hysteresis and a large negative flat band voltage shift ($V_{FB}$: $-2.5$ V) were detected. Post-deposition annealing reduces the amount of hysteresis, but does not modify the $V_{FB}$ position.

*Singh* et al. [39] studied 100 to 700 nm thick $Er_2O_3$ films grown by low-pressure MOCVD (at 450–600 °C). The dielectric constant extracted from capacitance voltage measurements at 1 MHz ranges between 8 and 20. A large positive flat band shift was detected in as-grown films, indicating the presence of negative fixed charges which diminish with increasing growth temperature.

$La_2O_3$ and $LaAlO_3$ films are definitely the most promising CVD-grown REOs in terms of electrical properties (dielectric constant, leakage and EOT). These are close to those reported for films grown by PVD techniques. More work is, however, needed to optimize the growth of other thin ($< 10$ nm) binary or ternary REOs, and to characterize in more detail their electrical properties.

In this Chapter, we investigate thin (3–30 nm) $Lu_2O_3$ and $Yb_2O_3$ films grown by ALD using various precursor combinations. No data are available in the literature related to the electrical and structural properties of $Yb_2O_3$ grown by ALD, and only few reports on $Lu_2O_3$ have been published [7, 8]. These two REOs are also interesting for several reasons. For a start, they are less hygroscopic than $La_2O_3$. Moreover, a large gap and conduction-band offset are expected due to the filling level of the $f$ shell. For $Lu_2O_3$, a band gap of 5.8 eV and a CBO of 2.1 eV were measured [8]. In the following, the structural properties of $Lu_2O_3$ and $Yb_2O_3$ oxides are discussed and correlated with electrical properties, with emphasis on the semiconductor/dielectric interface. Moreover, the microscopic structure and the nature of the electrically active defects are investigated using magnetic resonance spectroscopy. Conventional electron spin resonance spectroscopy, one of the most powerful spectroscopy techniques for the investigation of point defects in semiconductors and insulators, suffers from sensitivity problems when dealing with state-of-the-art test devices due to their small total area as well as low density of defects, both in the bulk and at the interface. Several spin-dependent transport phenomena can be exploited to achieve the high sensitivity and selectivity necessary for the magnetic resonance investigation of interfaces [46]. In this work, we have used the microwave contact-less photoconductive resonance (MWCL-PCR) spectroscopy [47]. This technique is based on the photo-excited free carrier absorption-induced losses of the microwave electric field, which produce a detectable change in the cavity Q-factor [47].

## 2 Experimental Methods

$Lu_2O_3$ and $Yb_2O_3$ films were deposited by ALD in a F-120 ASM-Microchemistry reactor on silicon The Si(100) substrates were n- or p-type (1–5 $\Omega$ cm), or highly resistive (5000–15 000 $\Omega$ cm). RCA cleaning (HCl : $H_2O_2$ : $H_2O$ = 1 : 1 : 5 ratio, 10 min at 85 °C) followed by a 30 s long dip in a diluted HF solution (1 : 50 = HF : $H_2O$) at room temperature was used to prepare the Si surface, which was found to be H-terminated. To eventually form a chemical oxide on the Si substrate, an additional RCA step was applied. A 30 s long rinse in de-ionized water followed each cleaning step.

The $Yb_2O_3$ films were deposited using two Yb precursors: the Yb cyclopentadienyl $Yb(C_5H_5)_3$ and the Yb tetramethyl-heptanedionato $Yb(C_{11}H_{19}O_2)_3$ complexes; in the following text, they will be called, respectively, $Yb(Cp)_3$ and $Yb(thd)_3$. Films from the former compound, which sublimes at 100 °C, were deposited at 360 °C using either $H_2O$ or $O_3$ as oxygen sources [48]. The latter complex, which sublimes at 150 °C, generates an Yb oxide film only when combined with $O_3$ as oxygen source. The films were deposited at 360 °C.

The $Lu_2O_3$ films were deposited from the combination of the newly synthesized bis-cyclopentadienyl [49] complex $[(\eta^5\text{-}C_5H_4SiMe_3)_2LuCl]_2$ with $H_2O$ at 360 °C [7]. The Lu precursor sublimed at 195 °C. A Si-rich $Lu_2O_3$ was instead grown using $[(\eta^5\text{-}C_5H_4SiMe_3)_2LuCl]_2$ as Lu source and $O_3$ as oxygen source [50].

The film structure was analyzed using a tool suited for both X-ray reflectivity (XRR) and grazing incidence X-ray diffraction (GIXRD). The system is equipped with a multilayer monochromator capable of selecting a parallel beam of $Cu\text{-}K_\alpha$ radiation. The XRR spectra are collected on a scintillator detector, and the data are simulated using a model based on a matrix formalism corrected by the Croce–Nevot factor [51]. The GIXRD spectra are collected on a position-sensitive detector (Inel CPS120). Various incidence angles can be chosen, and the diffractograms are simulated using Rietveld refinement [52].

For the electrical measurements, metal-insulator-semiconductor (MIS) capacitors were defined on the films evaporating circular Al dots ($7.8 \times 10^{-4}$ cm$^2$ of area) through a shadow mask. Back contacts are fabricated using an In−Ga alloy in the eutectic composition. Capacitance–voltage (C–V) and current–voltage curves are acquired using, respectively, an HP 4284 A multi-frequency LCR meter and a 4140B pico-amperometer. For all the investigated films, the capacitance equivalent oxide thickness (CET) was extracted from the accumulation capacitance of the gate stacks [53]. The CET is the thickness of a $SiO_2$ layer with the same capacitance density as that of the high-$\kappa$ dielectric, when quantum mechanical corrections are not taken into account. To extract the effective equivalent oxide thickness (EOT), quantum corrections should be taken into account. It is worth noting that it is sometimes difficult to make significant comparisons between the results reported in the literature, since the notation EOT is used by some authors even if quantum corrections are not included.

MWCL-PCR was carried out in an X band (9.2–9.5 GHz) spectrometer equipped with a rectangular $TE_{102}$ cavity. A frequency counter was used to monitor the microwave frequency. The g-factors were determined using the reference signal of $\alpha, \alpha'$-diphenyl-$\beta$-picrylhydrazyl (DPPH). After the growth, samples $4 \times 9$ mm$^2$ in size were cut with the long edge along the [$0\bar{1}1$] direction, and glued onto a quartz rod. During the measurements, the sample under investigation was illuminated with blue light (470 nm) provided by light emitting diodes. The short wavelength was chosen to generate carriers close to the interface, with consequent enhancement of the sensitivity to surface defects. The MWCL-PCR measurements were performed at room temperature and a microwave power of 180 mW, which corresponds to a microwave $H_1$ field of 0.4G [46] and does not lead to a noticeable line broadening. The samples were placed in the resonant cavity with the static magnetic field $H_0$ perpendicular to the surface normal.

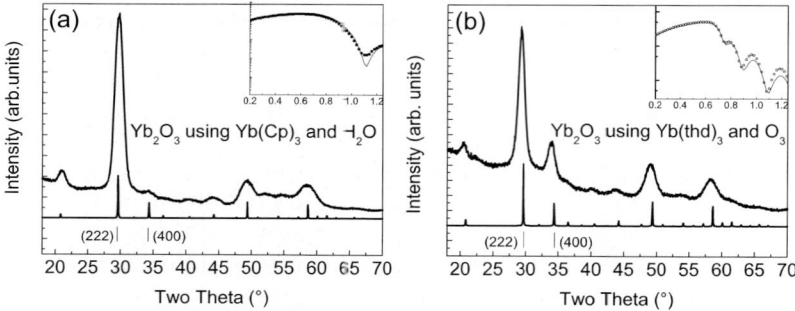

**Fig. 1.** (**a**): GIXRD pattern of 11.3 nm thick $Yb_2O_3$ film grown using $Yb(Cp)_3$ and $H_2O$. (**b**): GIXRD pattern of 30.0 nm thick $Yb_2O_3$ film grown using $Yb(thd)_3$ and $O_3$. The pattern of the bulk bixbyite structure of $Yb_2O_3$ [54] is added for comparison in both (**a**) and (**b**). The (222) and (400) reflections are indicated. The *insets* show the corresponding XRR spectra (*dots*) and fittings (*continuous line*) in the region of the critical angle

# 3 Yb$_2$O$_3$ Grown by ALD

## 3.1 Structural Characterization

$Yb_2O_3$ films grown using $Yb(Cp)_3$ and either $H_2O$ or $O_3$ as oxygen source are investigated in the 3–11 nm thickness range. As-grown films deposited using $H_2O$ or $O_3$ as oxygen precursor crystallize in the cubic bixbyite structure. Figure 1a shows the GIXRD pattern of a 11.3 nm thick film deposited using $Yb(Cp)_3$ and $H_2O$. The diffraction pattern of a random $Yb_2O_3$ powder [54] is also reported for comparison. The intensity of the (400) reflection of the measured diffraction pattern is lower than that of the powder, suggesting that a preferential orientation of the crystallites develops in the deposited films. Actually, Bragg Brentano analysis (not shown) demonstrates that these films have a (111) preferential orientation. The grain size increases from 2 to 13 nm when film thickness increases from 3 to 11 nm. The lattice parameter obtained from Rietveld refinement is 1.034 nm, lower than the value reported for the bulk $(a_{Yb_2O_3} = 1.0436$ nm [54]).

Films grown using $Yb(thd)_3$ and $O_3$ are explored in the 4–30 nm thickness range. Their structural properties differ from those of films grown with $Yb(Cp)_3$ and either $H_2O$ or $O_3$. The diffraction pattern of a 30 nm $Yb_2O_3$ thick film grown using $Yb(thd)_3$ and $O_3$ is reported in Fig. 1b. The lattice parameter obtained by Rietveld refinement is 1.047 nm, very close to the bulk value for $Yb_2O_3$. Assuming the crystallized grains are composed of stoichiometric $Yb_2O_3$, the crystalline fraction of films grown with $Yb(thd)_3$ and $O_3$ is three–four times lower than that of films grown with $Yb(Cp)_3$ and $H_2O$, with grain size always 45–65 % smaller than film thickness. Moreover, no signs of preferential orientation are found in the diffraction pattern.

XRR analysis depicts the structure of the films grown with all these three precursor combinations as a stack formed by a $Yb_2O_3$ layer and an interfacial layer between $Yb_2O_3$ and the silicon substrate. The interfacial layer can be modeled in two ways: either by $1\,nm$ thick $SiO_2$ (electronic density $= 0.67e^-\,\text{Å}^{-3}$) at the interface with silicon plus a thin ($0.5$–$1\,nm$) interdiffused layer between $SiO_2$ and $Yb_2O_3$, or by a single layer, $0.7$–$1.9\,nm$ thick. For films of similar thickness grown with $Yb(Cp)_3$, the interfacial layer is $0.9\,nm$ when the film is grown with $H_2O$, and $1.9\,nm$ when $O_3$ is used as oxygen source. The electronic density of the interdiffusion layer and of the single interfacial layer are in the $0.82$–$1.03$ e$^-$ Å$^{-3}$ range. Finally, a low density cap layer on top of $Yb_2O_3$ is also necessary to correctly reproduce the XRR experimental data.

Different electronic density values of the $Yb_2O_3$ films are found for the layers deposited using either $H_2O$ or $O_3$ (insets of Fig. 1). For films grown using $Yb(Cp)_3$ and $H_2O$, the electronic density is $2.21$ e$^-$ Å$^{-3}$, and does not vary substantially with film thickness down to $3\,nm$. When $O_3$ is used in combination with $Yb(Cp)_3$, the electronic density of the $Yb_2O_3$ layer is $7\%$ higher [48]. These values, taking into account the experimental error, are higher than that of the bulk phase ($\rho_{Yb_2O_3} = 2.11$ e$^-$ Å$^{-3}$ [54]) by $5\%$, and can be partially explained by the isotropic shrinkage of the lattice found by GIXRD. In films grown with $Yb(thd)_3$ and $O_3$, the electronic density, within the explored thickness range, is $1.81$ e$^-$/Å$^3$, lower than the one of bulk $Yb_2O_3$.

Films grown with $Yb(thd)_3$ and $O_3$, annealed at $600\,^\circ C$ in $N_2$, exhibit minor structural differences compared to the as-grown ones. The grain size increases (from 5 to $8\,nm$ for a $7.9\,nm$ thick film), as well as the electronic density (from $1.81$ to $1.92$ e$^-$ Å$^{-3}$). On the other hand, the interfacial layer does not change.

## 3.2 Electrical Characteristics and Interface Defects

Figure 2 shows the C–V curves and leakage currents (inset) of the as-deposited $Yb_2O_3$ films grown using $Yb(Cp)_3$ as metal source and either $H_2O$ (a) or $O_3$ (b) as oxygen source. The two films have similar CET. C–V curves were acquired on Al–$Yb_2O_3$–pSi MIS capacitors in dark ambient, sweeping the voltage from inversion to accumulation to check the amount of hysteresis.

The C–V curves of films grown using $H_2O$ (Fig. 2a) show a very small hysteresis ($< 10\,mV$) and a small dispersion of the accumulation capacitance with increasing frequency. On the other hand, the C–V curves of films grown using $O_3$ (Fig. 2b) exhibit a larger clockwise hysteresis (around $50$–$80\,mV$) than films grown using $H_2O$. Moreover, for films grown using $O_3$, the frequency dispersion in the depletion region and the stretching out of the C–V curves reveal a high interface trap density ($D_{it}$), estimated around $2\times10^{12}\,\text{eV}^{-1}\,\text{cm}^{-2}$ using the Hill–Coleman method [55]. For film grown using $H_2O$, the $D_{it}$ is almost one order of magnitude lower, $3\times10^{11}\,\text{eV}^{-1}\text{cm}^{-2}$. It

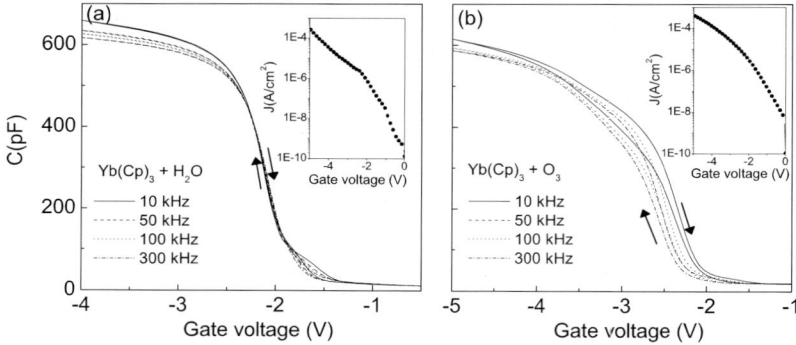

**Fig. 2.** (**a**): C–V curves of $Yb_2O_3$ films (CET = 4.1 nm) grown using $Yb(Cp)_3$ as Yb source and $H_2O$ as oxygen source. In the *inset*, the leakage current. (**b**): C–V curves of $Yb_2O_3$ film (CET 4.3 nm) grown using $Yb(Cp)_3$ and $O_3$. In the *inset*, the J–V curve

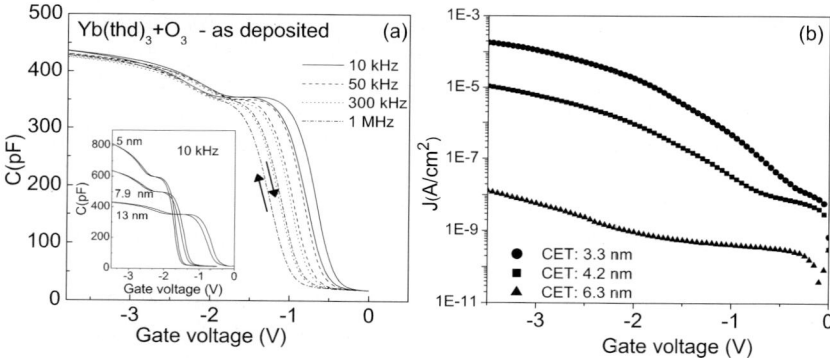

**Fig. 3.** C–V curves acquired at different frequencies (10 kHz to 1 MHz) of 13 nm thick $Yb_2O_3$ film grown using $Yb(thd)_3$ and $O_3$ (**a**). The C–V curves (at 10 kHz) for $Yb_2O_3$ films with different thickness are reported in the *inset* of (**a**), while the corresponding leakage currents are shown in (**b**)

is also worth noting that for both films, the flat band voltage is shifted to negative values with respect to the theoretical one ( $V_{FB}$: $-0.7$ V for Al gate and the Si substrate used in this work), indicating a large amount of positive fixed charges.

Figure 3a shows the C–V curves acquired at different frequencies for $Yb_2O_3$ films grown using the $Yb(thd)_3$ and $O_3$ precursor combination. The shoulder and the large dispersion in the depletion region of the C–V curves evidence a large amount of interface traps, measured to be in the $10^{12}$ $eV^{-1}cm^{-2}$ range. Clockwise hysteresis is also detected, resulting in a positive shift of the flat band voltage. The C–V curves acquired (at 10 kHz) for films with

**Table 1.** Summary of electrical properties for $Yb_2O_3$ films grown using various precursor combinations. *This film was annealed in $N_2$ at $600\,°C$

| Precursor | CET (10 kHz) nm | $V_{FB}$ V | $D_{it}$ $eV^{-1}\,cm^{-2}$ | $J$ $A\,cm^{-2}$ |
|---|---|---|---|---|
| $Yb(Cp)_3 + H_2O$ | $4.1 \pm 0.1$ | $-1.8$ | $3 \times 10^{11}$ | $4.2 \times 10^{-6}$ |
| $Yb(Cp)_3 + O_3$ | $4.3 \pm 0.1$ | $-2.4$ | $2.3 \times 10^{12}$ | $5.9 \times 10^{-5}$ |
| $Yb(thd)_3 + O_3$ | $3.3 \pm 0.1$ | $-1.9$ | $4.5 \times 10^{12}$ | $1 \times 10^{-4}$ |
| | $4.2 \pm 0.1$ | $-1.7$ | $4.3 \times 10^{12}$ | $3.9 \times 10^{-6}$ |
| | $6.3 \pm 0.1$ | $-1.1$ | $3.3 \times 10^{12}$ | $(8 \div 9) \times 10^{-10}$ |
| | $6.3 \pm 0.1^*$ | $-0.7$ | $1 \times 10^{12}$ | $(7 \div 8) \times 10^{-10}$ |

different thickness are reported in the inset of the same figure, while the corresponding leakage currents are shown in Fig. 3b. It is worth noting that the flat-band position and hysteresis width depend on the thickness (see inset of Fig. 3a), and on the voltage range applied to acquire the C–V characteristics. For films 4 to 13 nm thick, the hysteresis is 80–120 mV (voltage sweep between 0 to $-3.5$ V and back), while the flat band voltage varies in the $-1$ to $-2$ range (measured for C–V curves acquired at 300 kHz sweeping the voltage from inversion to accumulation). Clockwise or counterclockwise hysteresis has been reported for various REOs [6, 22, 45]. The hysteresis might be related to charge trapping in the bulk of the dielectric layer or in the interfacial region [17, 56], as well as to charged mobile ions [57, 58]. Electron injection from the gate into the oxide and subsequent trapping might explain the clockwise hysteresis observed for $Yb_2O_3$ films. Another explanation could refer to positive mobile ions [53], but further analyses are necessary to better clarify the origin of hysteresis. The leakage current measured at $|V_{FB}| + 1$ V towards accumulation varies from $8 \times 10^{-10}$ $Acm^{-2}$ for film with 6.3 nm CET to $1 \times 10^{-4}$ $Acm^{-2}$ for the thinnest film with CET = 3.3 nm.

Table 1 summarizes the CET values, the flat band values, the $D_{it}$ and the leakage current ($J$) measured for $Yb_2O_3$ films grown using the three precursor combinations. The $V_{FB}$ values are measured from C–V curves acquired at 300 kHz sweeping the voltage from inversion to accumulation.

$Yb_2O_3$ films grown using $O_3$ exhibit a higher $D_{it}$ than those grown using $H_2O$, independently of the Yb precursor used. For films grown using $Yb(thd)_3$ and $O_3$, a post-deposition annealing at $600\,°C$ in $N_2$ (see Table 1 for film with CET of 6.3 nm) reduces by one third the $D_{it}$, without increasing the leakage current and CET. The decrease of interface trap density is also evidenced by the C–V curves and the conductance signal of the annealed film (Fig. 4 and corresponding inset). Compared to the as-deposited film (Fig. 3a), the frequency dispersion and the shoulder in the depletion region of the C–V curves, as well as the conductance signal, are strongly reduced. Clockwise hysteresis (50–100 mV) is, however, still present. Annealing at $400\,°C$ (data not shown) in $N_2$ or $O_2$ changes neither the C–V characteristics nor the $D_{it}$.

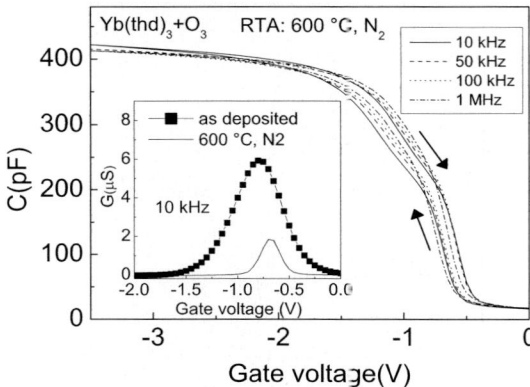

**Fig. 4.** C–V curves of 13 nm thick $Yb_2O_3$ film (CET: 6.3 nm) grown using $Yb(thd)_3$ and $O_3$ after rapid thermal annealing (RTA) at 600 °C in $N_2$. In the *inset*, the variation of the conductance signal after annealing for the same film

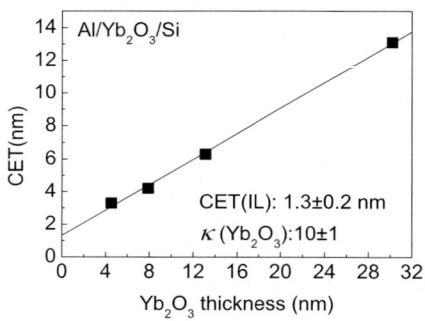

**Fig. 5.** CET of $Yb_2O_3$ films grown using $Yb(thd)_3$ and $O_3$ versus physical thickness. CET values were extracted from the accumulation capacitance at 10 kHz. IL: interfacial layer

The dielectric constant of $Yb_2O_3$ films grown using $Yb(thd)_3$ and $O_3$ was extracted from the plot of the CET values as a function of the $Yb_2O_3$ physical thickness (Fig. 5). From the slope of the linear fit to the data, a $\kappa$ value of $10 \pm 1$ for the $Yb_2O_3$ films is extracted, while the intercept with the $y$-axis gives a CET of $1.3 \pm 0.2$ nm for the interfacial layer. The extracted value for the dielectric constant is lower than the one reported by *Ohmi* et al. [22] for 5 nm thick $Yb_2O_3$ films grown by ultrahigh vacuum e-beam deposition ($\kappa = 14$).

Figure 6 shows the MWCL-PCR room temperature spectra observed with the static field perpendicular to the surface normal ($H_0 \parallel [011]$) for the $Yb_2O_3/Si$ interface. $Yb_2O_3$ films are grown using $Yb(Cp)_3$ as Yb source and either $H_2O$ (a) or $O_3$ (b) as oxygen source. Films grown using $Yb(thd)_3$

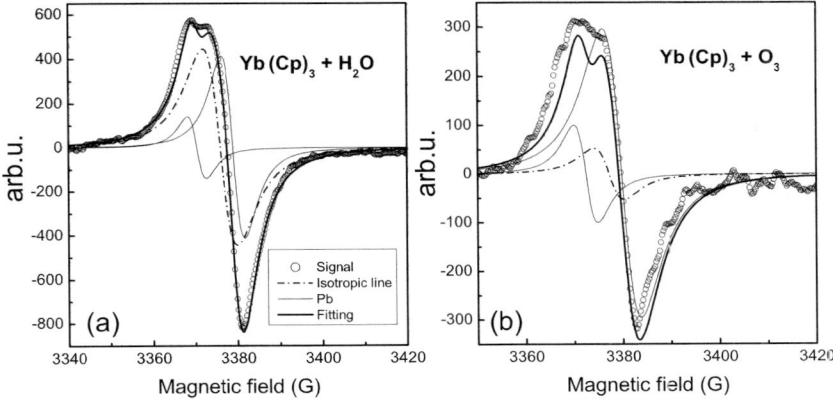

**Fig. 6.** Microwave contact-less photoconductive resonance spectra at room temperature for $Yb_2O_3$ films grown using $Yb(Cp)_3$ as Yb source and either $H_2O$ (**a**) or $O_3$ (**b**) as oxygen source

**Table 2.** $Yb_2O_3$/Si interfaces. Film grown using $Yb(Cp)_3$ as Yb precursor and $H_2O$ or $O_3$ as oxygen precursors

| Precursor | $g_1$ | $\Delta H_1$ | $g_2$ | $\Delta H_2$ | $g_3$ | $\Delta H_3$ |
|---|---|---|---|---|---|---|
| $H_2O$ | 2.0037(1) | 5.5(1) | 2.0088(1) | 4.4(3) | 2.0055(2) | 8.2(2) |
| $O_3$ | 2.0040(2) | 7.9(4) | 2.0083(1) | 4.9(2) | 2.0055(1) | 6.2(1) |

and $O_3$ (not shown) exhibit similar MWCL-PCR spectra to those of films grown using $Yb(Cp)_3$ and $O_3$.

The fitting of the spectrum observed for samples grown with $H_2O$ (see Table 2) reveals resonance features very similar to those reported for the $P_{b0}$ center, the dominant paramagnetic defect at the $Si(001)/SiO_2$ interface [59] showing $g_1 = 2.0039$ and $g_2 = 2.0081$ [60] at this magnetic field orientation, and an isotropic line with $g = 2.0055$ characteristic of Si dangling bonds and attributed to disorder [61]. The interface between the dielectric layer and silicon, for films deposited using $O_3$, reveals also similar resonance features. The observed deviations of the $g$ values from those observed for the standard $P_{b0}$ center may be attributed to a different amount of stress at the interface and to small changes in the local defect microstructure. Further investigations are in progress to clarify this issue. In addition, one can observe that the signal intensity is lower when $O_3$ is used. This is apparently in contrast with the electrical measurements which show a higher $D_{it}$ for the sample grown with $O_3$. However, one should take into account that the signal intensity in MWCL-PCR spectra depends on several parameters in addition to the defects concentration and that not all the electrically active defects are paramagnetic.

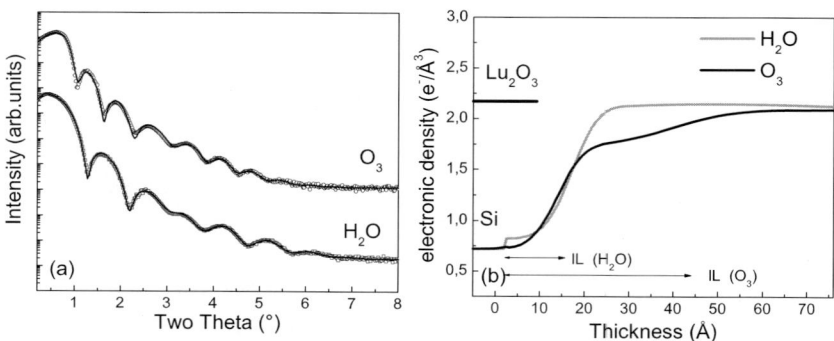

**Fig. 7. (a)**: XRR experimental data and corresponding fitting (*continuous line*) of $Lu_2O_3$ films grown with $H_2O$ (*squares*) and $O_3$ (*circles*). **(b)**: Electronic density profiles corresponding to the simulations presented in **(a)**: *gray*: $H_2O$, *black*: $O_3$. The *bottom arrows* represent the thickness of the interfacial layers (ILs)

On the other hand, the observed reduction of $D_{it}$ upon annealing correlates with a decrease in the MWCL-PCR intensity signal (data not shown).

## 4 $Lu_2O_3$ Grown by ALD

### 4.1 Structural Characterization

$Lu_2O_3$ films deposited using $[(\eta^5\text{-}C_5H_4SiMe_3)_2LuCl]_2$ and either $H_2O$ or $O_3$ as oxygen precursors were investigated. For both precursor combinations, the XRR data (Fig. 7) can be satisfactorily fitted with a top $Lu_2O_3$ layer, and an interfacial layer between $Lu_2O_3$ and Si. The electronic density profile in the region close to the interface with Si is shown in Fig. 7b. For films grown with $H_2O$, this interfacial layer is composed of a 1 nm thick $SiO_2$ layer having a rough interface with $Lu_2O_3$ (r.m.s. roughness of 0.5 nm). On the other hand, for films grown with $O_3$, the interfacial layer is 3–4 nm thick, and its electronic density, with values in the 1.1–1.7 e$^-$ Å$^{-3}$ range, is higher than that of $SiO_2$ (0.67 e$^-$ Å$^{-3}$). The electronic density value of the $Lu_2O_3$ layer is slightly lower for films grown using $O_3$ than for films grown using $H_2O$. The electronic density values in both cases are, however, close to the one of bulk $Lu_2O_3$ (2.18 e$^-$ Å$^{-3}$) [62].

For $Lu_2O_3$ films deposited using $[(\eta^5\text{-}C_5H_4SiMe_3)_2LuCl]_2$ and $H_2O$, crystallization depends on film thickness. Thinner films ($< 5$ nm) are mostly amorphous and thicker films are partially crystallized in the cubic bixbyite structure of $Lu_2O_3$.

Figure 8 shows the XRD patterns obtained for films of different thickness. The grain size is determined by Rietveld refinement. In order to correctly simulate the X-ray scattering in the 30–35 °C $2\theta$ region, a "bump" due to an

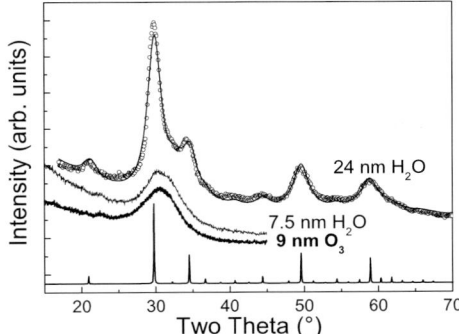

**Fig. 8.** GIXRD pattern of films grown with $[(\eta^5\text{-}C_5H_4SiMe_3)_2LuCl]_2$ and either $H_2O$ (*black line*) or $O_3$ (*bulk line*) as oxygen sources. The simulation obtained by Rietveld refinement (*continuous line*) is added for the 24 nm thick film (*open circles*). The pattern corresponding to the bixbyite structure of $Lu_2O_3$ [62] is also added for comparison

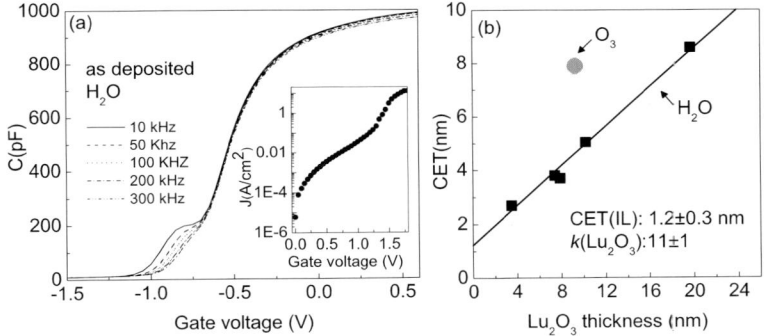

**Fig. 9.** (**a**): C–V of 3.5 nm thick $Lu_2O_3$ films on n-type Si. In the *inset*, the leakage current. The MIS gate area is $7.8 \times 10^{-4}$ cm$^2$. (**b**): CET versus physical thickness measured by XRR of $Lu_2O_3$ films grown using $H_2O$. The *circle* represents the CET value of a 9.3 nm thick $Lu_2O_3$ film grown using $O_3$

amorphous component is added in the simulation. The grain size is found to be always smaller than film thickness. For 20 and 24 nm thick films, the grain size is only 4 nm. For 7.5–9 nm thick films grown using either $H_2O$ or $O_3$, the grain size is the same ($< 3$ nm) (Fig. 8).

## 4.2 Electrical Characteristics and Interface Defects

Figure 9a shows the C–V curves of a 3.5 nm thick $Lu_2O_3$ film grown using $[(\eta^5\text{-}C_5H_4SiMe_3)_2LuCl]_2$ as Lu precursor and $H_2O$ as oxygen precursor. The C–V curves are well shaped, with no frequency dispersion in the accumula-

**Table 3.** Summary of electrical properties for $Lu_2O_3$ films grown using different precursor combinations. The leakage current values $J$ are measured at $|V_{FB}| + 1\,V$ towards accumulation. The flat band voltages are measured from C–V acquired at 300 kHz

| Precursor | Substrate | CET nm | $V_{FB}$ V | $D_{it}$ eV$^{-1}$ cm$^{-2}$ | $J$ A cm$^{-2}$ |
|---|---|---|---|---|---|
| | p-Si | $8.7 \pm 0.1$ | $-1.2$ | $2 \times 10^{11}$ | $1.1 \times 10^{-10}$ |
| $H_2O$ | p-Si | $3.7 \pm 0.1$ | $-1.85$ | $8 \times 10^{11}$ | $3 \times 10^{-7}$ |
| | n-Si | $2.7 \pm 0.1$ | $-0.81$ | $1 \times 10^{12}$ | $4.5 \times 10^{-4}$ |
| $O_3$ | p-Si | $7.9 \pm 0.1$ | $-0.95$ | $3 - 4 \times 10^{12}$ | $1.2 \times 10^{-10}$ |

tion region and no hysteresis. The stretching out of the C–V curves and the shoulder in the depletion region point to a high density of interface traps, estimated around $3 - 4 \times 10^{12}\,eV^{-1}cm^{-2}$ using the Hill–Coleman method [55]. The C–V curves of 8 nm thick films on p-type and n-type Si substrates (not shown) have a shape similar to that of the thin film. Nevertheless, counterclockwise hysteresis is detected, and its amount increases applying a higher negative voltage (hysteresis around 100 mV for a voltage sweep between 0 and $-4\,V$). The flat band voltage (see Table 3) is shifted towards negative values with respect to the theoretical one ($V_{FB} = -0.7\,V$ for Al electrode and the p-type Si substrate), indicating a high density of positive fixed charges. Figure 9b shows the CET values versus film thickness for $Lu_2O_3$ grown using $H_2O$ (square symbols). From the linear fit to the data, a dielectric constant of $11 \pm 1$ for $Lu_2O_3$ and a CET of $1.2 \pm 0.3$ nm for the interfacial layer are extracted. Taking into account the XRR data, the CET value for the interfacial layer is consistent, within the experimental error, with the presence of a $SiO_2$ ($\kappa : 3.9$) dielectric.

The CET value of a 9.3 nm thick $Lu_2O_3$ film grown using $O_3$ is also shown in Fig. 9b. The increased value of CET, with respect to that of the film grown using $H_2O$ and having similar physical thickness, is due to the different stack composition as revealed by XRR analyses. In particular, for film grown using $O_3$, a thick (3–4 nm) low-$\kappa$ interfacial layer is present at the $Lu_2O_3/Si$ interface. The C–V curves of films grown using $O_3$ as oxygen source are reported in Fig. 10. As for $Yb_2O_3$, the use of $O_3$ as oxygen source promotes a higher $D_{it}$ than the use of $H_2O$. Table 3 summarizes the electrical properties of $Lu_2O_3$ films grown using $H_2O$ and $O_3$.

Figure 11 shows the MWCL-PCR room temperature spectra observed at the $Lu_2O_3/Si$ interface for as-grown films obtained using either $H_2O$ (a) or $O_3$ (b) as oxygen source. In both cases (Table 4), we observe two lines with $g$ values very close to those expected for the $P_{b0}$ center and the isotropic line at $g = 2.0055$. However, an inspection of Fig. 11 reveals obvious differences in the overall resonance line shape. This is due to a different intensity of the two lines originated by the $P_{b0}$. This fact has been observed also for other

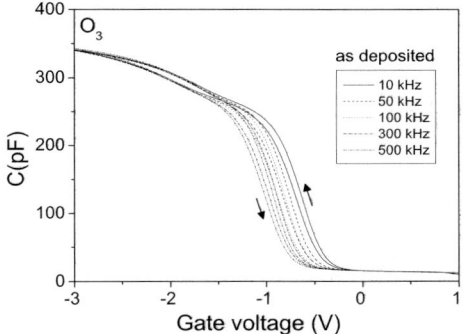

**Fig. 10.** C–V curves acquired at different frequencies of as-deposited $Lu_2O_3$ films grown using $O_3$. The as-grown film was 9.3 nm thick

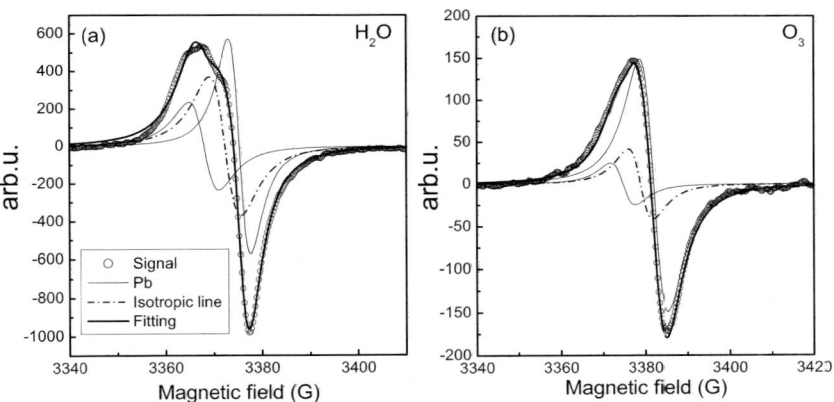

**Fig. 11.** Microwave contact-less photoconductive resonance spectra at room temperature for $Lu_2O_3$ films grown using either $H_2O$ (**a**) or $O_3$ (**b**) as oxygen source

**Table 4.** $Lu_2O_3$/Si interfaces. Film grown using $H_2O$ or $O_3$ as oxygen precursors

| Precursor | $g_1$ | $\Delta H_1$ | $g_2$ | $\Delta H_2$ | $g_3$ | $\Delta H_3$ |
|---|---|---|---|---|---|---|
| $H_2O$ | 2.003 7(1) | 4.6(1) | 2.008 1(1) | 6.2(2) | 2.005 5(1) | 6.6(3) |
| $O_3$ | 2.003 7(1) | 6.7(1) | 2.008 0(1) | 5.8(8) | 2.005 5(1) | 5.9(2) |

oxides, and additional work is in progress to address this issue [63]. As in the $Yb_2O_3$/Si case, the MWCL-PCR signal intensity is lower for films deposited using $O_3$, in contrast with the electrical measurements.

# 5 Conclusions

Thin (4–30 nm) $Yb_2O_3$ and $Lu_2O_3$ films were grown by ALD on Si substrates. The use of various precursor combinations affects both structural and electrical properties. As-grown $Yb_2O_3$ oxides are polycrystalline in the cubic bixbyite phase, also for thin film (4–5 nm), and independently of the precursor combination used. As-grown $Lu_2O_3$ films using Cp and either $H_2O$ or $O_3$ are more amorphous than $Yb_2O_3$ for a similar thickness range. Also 3–5 nm thick $Lu_2O_3$ films grown using $H_2O$ are amorphous while thicker films are partially crystalline in the cubic bixbyite structure. For both REOs, a few nm thick interfacial layer is formed at the film/silicon interface. Its thickness is higher for films grown using ozone, especially for $Lu_2O_3$. The electrical properties were characterized by capacitance–voltage and current–voltage measurements, while the defects at the film/silicon interface were characterized by conductance measurements as well as MWCL-PCR at room temperature. The measured dielectric constant is $11 \pm 1$ for $Lu_2O_3$ oxide, in agreement with theoretical calculations [64] and experimental results obtained for film grown using ultrahigh vacuum e-beam deposition [31]. The dielectric constant measured for $Yb_2O_3$, $10 \pm 1$, is slightly lower than reported theoretical and experimental values [4, 22]. Both REO oxides exhibit well-shaped C–Vs without a significant dispersion of the accumulation capacitance for either n- or p-type silicon substrates. Nevertheless, the large negative shift of the flat band voltage indicates a high density of positive fixed charges. Counterclockwise and clockwise hysteresis loops are detected, respectively, for $Lu_2O_3$ and $Yb_2O_3$ films. The thinnest gate stacks exhibit a leakage current of $4.5 \times 10^{-4}$ Acm$^{-2}$ (CET of $2.7 \pm 0.1$ nm) for $Lu_2O_3$ and $1 \times 10^{-4}$ Acm$^{-2}$ (CET of $3.3. \pm 0.1$ nm) for $Yb_2O_3$ films. In order to reduce further the CET, the formation of a low-$\kappa$ interfacial layer should be avoided. The use of $O_3$ as oxygen source promotes a higher density of interfacial traps, independently of the metal precursor. MWCL-PCR measurements reveal $Pb_0$-like centers at the interface between the deposited dielectric and silicon for all investigated films. While the MWCL-PCR signal intensities do not correlate with the observed $D_{it}$, changes of the resonance intensity signal upon annealing correspond to a decrease of the $D_{it}$. A significant comparison with the electrical properties of other thin rare eath oxides grown by ALD is very difficult due to the few data reported in the literature.

## Acknowledgements

The authors would like to thank Mario Alia of CNR-INFM MDM National Laboratory for thermal treatments and MOS fabrication. The Italian Ministry of Foregin Affairs through a Joint Project between Italy and Russia, and the National Institute for the Physics of Matter (INFM) through the PAIS-REOHK Project, are acknowledged for funding.

# References

[1] L. Niinistö, J. Päiväsaari, J. Niinistö, M. Putkonen, M. Nieminen: Advanced electronic and optoelectronic materials by atomic layer deposition: an overview with special emphasis on recent progress in processing of high-k dielectrics and other oxide materials, Phys. Stat. Sol. (a) **201**, 1443 (2004)

[2] M. Leskelä, M. Ritala: Rare-earth oxide thin films as gate oxide in MOSFET transistors, J. Solid State Chem. **171**, 170 (2003)

[3] S. Jeon, K. Im, H. Yang, H. Lee, H. Sim, S. Choi, T. Jang, H. Hwang: Excellent electrical characteristics of Lanthanide (Pr, Nd, Sm, Gd, and Dy) oxide and Lanthanide-doped oxide for MOS gate dielectric applications, Technical Digest, International Electron Devices Meeting p. 20.6.1 (2001)

[4] D. Xue, K. Betzler, H. Hesse: Dielectric constants of binary rare-earth compounds, J. Phys.: Condens. Matter **12**, 3113 (2000)

[5] V. M. Koleshko, N. V. Babushkina: Properties of rare earth oxide films, Thin Solid Films **62**, 1 (1979)

[6] J. Päiväsaari, M. Putkonen, L. Niinistö: A comparative study on lanthanide oxide thin films grown by atomic layer deposition, Thin Solid Films **472**, 275 (2005)

[7] G. Scarel, E. Bonera, C. Wiemer, G. Tallarida, S. Spiga, M. Fanciulli, I. L. Fedushkin, H. Schumann, Y. Lebedinskii, A. Zenkevich: Atomic layer deposition of $Lu_2O_3$, Appl. Phys. Lett. **85**, 630 (2004)

[8] G. Seguini, E. Bonera, S. Spiga, G. Scarel, M. Fanciulli: Energy-band diagram of metal/$Lu_2O_3$/silicon structures, Appl. Phys. Lett. **85**, 5316 (2004)

[9] G. G. Schlom, J. H. Haeni: A thermodynamic approach to selecting alternative gate dielectrics, MRS Bull. **27**, 128 (2002)

[10] D. Matsushita, Y. Nishikawa, N. Satou, M. Yoshiki, T. Schmizu, T. Yamaguchi, H. Satake, N. Fukushima: Enhancement of dielectric constant due to expansion of lattice spacing in $CeO_2$ directly grown on Si(111), Jpn. J. Appl. Phys. **43**, 1795 (2004)

[11] J. Kwo, M. Hong, A. R. Kortan, K. L. Queeney, Y. J. Chabal, R. L. Opila, D. A. Muller, S. N. G. Chu, B. J. Sapjeta, T. S. Lay, J. P. Mannaerts, T. Boone, H. W. Krautter, J. J. Krajewski, A. M. Sergnt, J. M. Rosamilia: Properties of high k gate dielectrics $Gd_2O_3$ and $Y_2O_3$ for Si, J. Appl. Phys. **89**, 3920 (2001)

[12] A. Fissel, H. J. Osten, E. Bugiel: Towards understandings epitaxial growth of alternative high-k dielectrics on Si: application to praseodymium oxide, J. Vac. Sci. Technol. B **21**, 1765 (2003)

[13] H. Fukumoto, T. Imura, Y. Osaka: Heteroepitaxial growth of $Y_2O_3$ films on silicon, Appl. Phys. Lett. **55**, 360 (1989)

[14] M. Hong, J. Kwo, A. R. Kortan, J. P. Mannaerts, A. M. Sergent: Epitaxial cubic gadolinium oxide as a dielectric for gallium arsenide passivation, Science **283**, 1897 (1999)

[15] A. C. Jones, H. C. Aspinall, P. R. Chalker, R. J. Potter, K. Kukli, A. Rathu, M. Ritala, M. Leskelä: Some recent developments in the MOCVD and ALD of high-k dielectric oxides, J. Mater. Chem. **14**, 3101 (2004)

[16] A. C. Jones, H. C. Aspinall, P. R. Chalker, R. J. Potter, K. Kukli, A. Rathu, M. Ritala, M. Leskelä: Recent developments in the MOCVD and ALD of rare earth oxides and silicates, Mat. Sci. Eng. B **118**, 97 (2005)

[17] V. Ioannou-Sougleridis, G. Vellianitis, A. Dimoulas: Electrical properties of $Y_2O_3$ high-k gate dielectric on Si(001): the influence of postmetallization annealing, J. Appl. Phys. **93**, 3982 (2003)

[18] Y. H. Wu, M. Y. Yang, A. Chin, W. J. Chen, C. M. Kwei: Electrical characteristics of high quality $La_2O_3$ gate dielectric with equivalent oxide thickness of 0.5 nm, IEEE Electron. Dev. Lett. **21**, 341 (2000)

[19] V. Capodieci, F. Wiest, T. Sulima, J. Schulze, I. Eisele: Examination and evaluation of $La_2O_3$ as gate dielectric for sub-100 nm CMOS and DRAM technology, Microelectron. Rel. **45**, 937 (2005)

[20] Y. Kim, K. Miyauchi, S. Ohmi, K. Tsutsui, H. Iwai: Electrical properties of vacuum annealed $La_2O_3$ thin films grown by e-beam evaporation, Microelectron. J. **36**, 41 (2005)

[21] F.-C. Chiu, H.-W. Chou, J. Y.-m. Lee: Electrical conduction mechanisms of metal/$La_2O_3$/Si structure, J. Appl. Phys. **97**, 103503 (2005)

[22] S. Ohmi, C. Kobayashi, I. Kashiwagi, C. Ohshima, H. Ishiwara, H. Iwai: Characterization of $La_2O_3$ and $Yb_2O_3$ thin films for high-k gate insulator application, J. Electrochem. Soc. **150**, F134 (2003)

[23] L. Tye, N. A. El-Masry, T. Chikyow, P. McLarty, S. M. Bedair: Electrical characteristics of epitaxial $CeO_2$ on Si(111), Appl. Phys. Lett. **65**, 3081 (1994)

[24] L. Kim, J. Kim, D. Jung, Y. Roh: Controllable capacitance-voltage hysteresis width in the aluminum-cerium-dioxide-silicon-metal-insulator-semiconductor structure: Application to non-volatile memory devices without ferroelectrics, Appl. Phys. Lett. **76**, 1881 (2000)

[25] D. Landheer, J. A. Gupta, G. I. Sproule, J. P. McCaffrey, M. J. Graham, K.-C. Yang, Z.-H. Lu, W. N. Lennard: Characterization of $Gd_2O_3$ films deposited on Si(100) by electron beam evaporation, J. Electrochem. Soc. **148**, G29 (2001)

[26] C. Ohshima, J. Taguchi, I. Kashiwagi, H. Yamamoto, S. Ohmi, H. Iwai: Effect of surface treatment of Si substrates and annealing condition on high-k rare earth oxide gate dielectrics, Appl. Surf. Sci. **216**, 302 (2003)

[27] R. Lo Nigro, V. Raineri, C. Bongiorno, R. Toro, G. Malandrino, I. L. Fragala: Dielectric properties of $Pr_2O_3$ high-k films grown by metalorganic chemical vapor deposition on silicon, Appl. Phys. Lett. **83**, 129 (2003)

[28] A. A. Dakhel: Dielectric and optical properties of samarium oxide thin films, J. Alloys Compounds **365**, 233 (2004)

[29] N. V. Latukhina, V. A. Rozhkov, N. N. Romanenko: Silicon MOS structures with dysprosium and lutetium oxides and the diffusion of rare-earth elements in silicon, Microelectron. **23**, 28 (1994)

[30] V. Mikhelasvili, G. Eisenstein, F. Edelman: Structural and electrical properties of electron beam gun evaporated $Er_2O_3$ insulator thin films, J. Appl. Phys. **95**, 613 (2004)

[31] S. Ohmi, M. Takeda, H. Ishiwara, H. Iwai: Electrical characteristics for $Lu_2O_3$ thin films fabricated by e-beam deposition method, J. Elecrochem. Soc. **151**, G279 (2004)

[32] T. Wiktorczyk, C. Wesolowska: Some electrical properties of thin $Yb_2O_3$ films produced by different technological methods, Thin Solid Films **91**, 9 (1982)

[33] S. Jeon, H. Hwang: Effect of hygroscopic nature on the electrical characteristics of lanthanide oxide ($Pr_2O_3$, $Sm_2O_3$, $Gd_2O_3$, and $Dy_2O_3$), J. Appl. Phys. **93**, 6393 (2003)

[34] J. Robertson: High dielectric constant oxides, Eur. Phys. J. Appl. Phys. **28**, 265 (2004)

[35] B. Mereu, G. Sarau, A. Dimoulas, G. Apostolopoulos, I. Pintilie, T. Botila, L. Pintilie, M. Alexe: Electrical properties of metal-oxide-silicon structures with $LaAlO_3$ as gate oxide, Mat. Sci. Eng. B **109**, 94 (2004)

[36] X. B. Lu, H. B. Lu, J. Y. Dai, Z. H. Chen, M. He, G. Z. Yang, H. L. W. Chan, C. L. Choy: Oxygen pressure dependence of physical and electrical properties of $LaAlO_3$ gate dielectric, Microelectron. Eng. **77**, 399 (2005)

[37] G. Vellianitis, G. Apostolopoulos, G. Mavrou, K. Argyropoulos, A. Dimoulas, J. C. Hooker, T. Conard, M. Butcher: MBE lanthanum-based high-k dielectrcis as candidates for $SiO_2$ gate oxide replacement, Mat. Sci. Eng. B **109**, 85 (2004)

[38] C. Zhao, T. Witters, B. Brijs, H. Bender, O. Richard, M. Caymax, T. Heeg, J. Schubert, V. V. Afanas'ev, A. Stesmans, D. G. Schlom: Ternary rare-earth metal oxide high-k layers on silicon oxide, Appl. Phys. Lett. **86**, 132903 (2005)

[39] M. P. Singh, C. S. Thakur, K. Shalini, N. Bhat, S. A. Shivashankar: Structural and electrical characterization of erbium oxide films grown on Si(100) by low-pressure metalorganic chemical vapor deposition, Appl. Phys. Lett. **83**, 2889 (2003)

[40] H. Yamada, T. Shimizu, A. Kurokawa, K. Ishii, E. Suzuki: MOCVD of high-dielectric constant lanthanum oxide thin films, J. Electrochem. Soc. **150**, G429 (2003)

[41] D. H. Triyoso, R. I. Hegde, J. Grant, P. Fejes, R. Liu, D. Roan, M. Ramon, D. Werho, R. Rai, L. B. La, J. Baker, C. Garza, T. Guenther, B. E. White, P. J. Tobin: Film properties of ALD $HfO_2$ and $La_2O_3$ gate dielectrics grown on Si with various pre-deposition treatments, J. Vac. Sci. Technol. B **22**, 2121 (2004)

[42] A.-D. Li, Q.-Y. Shao, H.-Q. Ling, J.-B. Cheng, D. Wu, Z.-G. Li, N.-B. Ming, C. Wang, H.-W. Zhou, B.-Y. Nguyen: Characteristics of $LaAlO_3$ gate dielectrics on Si grown by metalorganic chemical vapor deposition, Appl. Phys. Lett. **83**, 3540 (2003)

[43] S.-W. Kang, S.-W. Rhee: Deposition of $La_2O_3$ films by direct liquid injection metallorganic chemical vapor deposition, J. Electrochem. Soc. **149**, C345 (2005)

[44] J. H. Jun, J. Jun, D. J. Choi: Properties of lanthanum aluminate thin film deposited by MOCVD, Electrochem. Solid-State Lett. **6**, F37 (2003)

[45] B. S. Lim, A. Rathu, P. Rouffignac, R. G. Gordon: Atomic layer deposition of lanthanum aluminum oxide nano-laminates for electrical applications, Appl. Phys. Lett. **84**, 3957 (2004)

[46] M. Fanciulli, O. Costa, S. Baldovino, S. Cocco, G. Seguini, E. Prati, G. Scarel: Defects at the high-k/semiconductor interfaces investigated by spin dependent spectroscopies, in E. Gusev (Ed.): *Defects in High-κ Gate Dielectric Stacks* (Springer, Berlin, Heidelberg 2006)

[47] L. S. Vlasenko, Y. V. Martynov, T. Gregorkievwicz, C. A. J. Ammerlaan: Electric paramagnetic resonance versus spin-dependent recombination: Excited triplet states of structural defects in irradiated silicon, Phys. Rev. B **52**, 1144 (1995)

[48] M. Malvestuto, G. Scarel, C. Wiemer, M. Fanciulli, F. D'Acapito, F. Boscherini: X-ray absorption spectroscopy study of $Yb_2O_3$ and $Lu_2O_3$ thin films deposited on Si(100) by atomic layer deposition, Nucl. Instr. Meth. B **246**, 90 (2006)

[49] H. Schumann, I. L. Fedushkin, M. Hummert, G. Scarel, E. Bonera, M. Fanciulli: Crystal and molecular structure of $[(\eta^5\text{-}C_5H_4SiMe_3)_2LuCl]_2$ – suitable precursor for $Lu_2O_3$ films, Z. Naturforsch. **59b**, 1035 (2004)

[50] S. D. Elliott, G. Scarel, C. Wiemer, M. Fanciulli, Y. Lebedinskii, A. Zenkevich, I. L. Fedushkin: Precursor combinations for ALD of rare earth oxides and silicates – a quantum chemical and X-ray study, Electrochem. Soc. Proc. **2005-09**, 605 (2005)

[51] A. Gibaud: X-ray and neutron reflectivity: Principles and applications, in S. Daillant, A. Gibaud (Eds.): *Lecture Notes in Physics. New Series m: Monographs; 58* (Springer, Berlin, Heidelberg 1999) p. 118

[52] L. Lutterotti, S. Matthies, H. R. Wenk: Proceeding of the Twelfth International Conference on Textures of Materials (ICOTOM-12) **1**, 1599 (1999)

[53] K. Kukli, M. Ritala, J. Lu, A. Harsta, M. Leskela: Properties of $HfO_2$ thin films grown by ald from hafnium tetrakis(ethylmethylamide) and water, J. Electrochem. Soc. **151**, F189 (2004)

[54] Inorganic Crystal Structure Database: file 62871, Fachinformationzentrum Karlsruhe (2004)

[55] W. A. Hill, C. C. Coleman: A single-frequency approximation for interface-state density determination, Solid State Electron. **23**, 987 (1980)

[56] R. Degraeve, E. Cartier, T. Kauerauf, R. Carter, L. Pantisano, A. Kerber, G. Groeseneken: On the electrical characterization of high-$\kappa$ dielectrics, MRS Bull. **27**, 222 (2002)

[57] A. Dimoulas, G. Vellianitis, A. Travlos, V. Ioannou-Sougleridis, A. G. Nassiopoulou: Structural and electrical quality of the high-$\kappa$ dielectric $Y_2O_3$ on Si(001): Dependence on growth parameters, J. Appl. Phys. **92**, 426 (2002)

[58] M.-H. Cho, Y. S. Rho, H.-J. Choi, S. W. Nam, D.-H. Ko, J. H. Ku, H. C. Kang, Y. Noh, C. N. Whang, K. Jeong: Annealing effects of aluminum silicate films grown on Si(100), J. Vac. Sci. Technol. A **20**, 865 (2002)

[59] E. H. Pointdexter, P. J. Caplan, B. E. Deal, R. R. Razouk: Interface states and electron spin resonance centers in thermally oxidized (111) and (100) silicon wafers, J. Appl. Phys. **52**, 879 (1981)

[60] A. Stesmans, V. Afanas'ev: Electron spin resonance features of interface defects in thermal $(100)Si/SiO_2$, J. Appl. Phys. **83**, 2449 (1998)

[61] J. L. Cantin, H. J. von Bardeleben: An electron paramagnetic resonance study of the $Si(100)/Al_2O_3$ interface defects, J. Non-Cryst. Solids **303**, 175 (2002)

[62] Inorganic Crystal Structure Database: file 40471, Fachinformationzentrum Karlsruhe (2004)

[63] S. Baldovino, S. Spiga, G. Scarel, M. Fanciulli: unpublished

[64] E. Bonera, G. Scarel, M. Fanciulli, P. Delugas, V. Fiorentini: Dielectric properties of high-k oxides: Theory and experiment for $Lu_2O_3$, Phys. Rev. Lett. **94**, 027602 (2005)

# Index

ALD, 204–206, 208, 214, 218
atomic layer deposition, 204

band gap, 204–206
band offset, 203, 204, 206
$\beta$-diketonate, 203, 204

capacitance–voltage characteristic, 205
CBO, 206
$CeO_2$, 203, 204
charge trap, 211
cyclopentadienyl, 203–207

dangling bond, 213
defect, 203, 206, 207, 209, 213, 215, 218
dielectric constant, 203–205, 212, 216,
    218
$Dy_2O_3$, 204, 205
$DyScO_3$, 204

electron spin resonance, 206
EOT, 204, 205, 207
equivalent oxide thickness, 203, 204,
    207
$Er_2O_3$, 204, 205
$Eu_2O_3$, 205

fixed charge, 205, 210, 216, 218

$Gd_2O_3$, 203–205
$GdScO_3$, 204
GIXRD, 207–209, 215

$Ho_2O_3$, 205

hygroscopic, 204, 206

interface, 203, 206, 207, 209–218
interfacial layer, 209, 212, 214, 216, 218

$La_2O_3$, 204–206
$LaAlO_3$, 204, 205
lanthanide, 205
$LaScO_3$, 204
laser, 203
$Lu_2O_3$, 203, 204, 206, 207, 214–218

microwave contact-less photoconductive
    resonance (MWCL-PCR), 206
MIS, 207, 209, 215
mobility, 203
MOCVD, 204, 205
molecular beam epitaxy, 203

$Nd_2O_3$, 204, 205

ozone, 203–205, 218

$Sm_2O_3$, 204

ternary compound, 205
$Tm_2O_3$, 205

X-ray reflectivity, 207
XRD, 214
XRR, 207–209, 214–216

$Y_2O_3$, 203–205
$Yb_2O_3$, 203, 204, 206, 208–213, 216–218

# Dielectric Properties of Rare-Earth Oxides: General Trends from Theory

Pietro Delugas, Vincenzo Fiorentini, and Alessio Filippetti

CNR-INFM SLACS and Department of Physics, University of Cagliari, Cittadella Universitaria, 09042 Monserrato (CA), Italy
vincenzo.fiorentini@dsf.unica.it

**Abstract.** We present a theoretical perspective on general aspects of the dielectric response and dynamical properties of some rare-earth oxide systems. We deal in particular with sesquioxides and aluminates, the latter both in the amorphous and crystalline phases.

## 1 Introduction and Theoretical Tools

Many oxides are insulators, and as such they have manifold applications. Some of these, such as the use of oxides as insulating layers in transistors or in long-retention-time capacitors and flash memories, tend to require a "high $\kappa$" (as opposed to others which need a "low $\kappa$"), i.e., a large value of the static dielectric constant, and are central to modern microelectronics. A particularly relevant case is the search for novel high-dielectric-constant oxides to replace silica as the gate insulator in the sub-nanometric technology nodes of integrated silicon-based high-performance circuits; the high-$\kappa$ requirement is due to the need to conserve (and indeed increase) the series capacitance of the conducting channel stack upon dimensional scaling-down [1, 2]. This is the same as to reduce the so-called effective oxide thickness $d\kappa_{silica}/\kappa_{oxide}$, without reducing the physical layer thickness so much as to cause important tunneling leakage through the layer. Keeping up with the current and projected scaling-down requires materials with static dielectric constants in the range of about 20.

Oxides with such properties, barring actual or incipient ferroelectrics, are produced only with selected transition-metal and rare-earth cations. While the former have been studied relatively extensively also by ab initio theory [3–6], the latter are rather less commonly studied, especially with respect to their dielectric and dynamical properties. As a consequence, a review of work in this area is somewhat premature. We will nevertheless try to focus on the general aspects of dielectric response in these materials, and on the general trends that can be extracted from first-principles calculations for certain oxide families. In particular, we will address rare-earth sesquioxides, i.e., compounds with the $Ln_2O_3$ formula (Ln: a generic lanthanide species) and aluminates (typical formula $LnAlO_3$), establishing relations between structure and expected dielectric properties.

M. Fanciulli, G. Scarel (Eds.): Rare Earth Oxide Thin Films,
Topics Appl. Physics **106**, 225–246 (2007)
© Springer-Verlag Berlin Heidelberg 2007

Of course, two issues arise in this context. The first is whether ultrathin films behave bulk-like in terms of their dielectric properties. Here we assume that this is the case, and that therefore studying bulk dielectric properties and exporting them to film systems is justified. Besides the obvious considerations that the selection or validation of a candidate dielectric must start from its bulk properties and that films used in microelectronics need not necessarily be extremely thin, this assumption has at least two supporting arguments, related to the observation that in these materials, ionic screening is by far the most relevant. First, modifications and energy shifts of infrared-active phonon modes due to "confinement" in a specific layer of a dielectric stack are expected to be very modest [7]. Second, it has been shown [8] that the (dramatic) changes in the microscopic polarization at Si/silica interfaces occur over short distances of order 4–5 Å, so that oxide layers significantly exceeding that thickness are effectively bulk-like in that respect (this is actually not valid for ultrathin silica layers, as discussed in [8]). Since effective charges and phonon frequencies are what makes up the static response, it is therefore quite justified to use results of studies of bulk dielectric properties in the context of thin films.

The second issue is to what extent the amorphous phase of a given material shares the dielectric properties of the crystal, which is important because amorphous films are electrically more proficient than polycrystalline ones (unless, of course, single-crystal layers can be realized). The answer depends critically on the specific material and its general structural properties. Since the study of amorphous structure is very demanding in itself, there are very few examples of this kind of work, and recent studies have only touched upon zirconia [9] and lanthanum aluminate [10]. The two materials typify rather different behaviors, and we will briefly discuss both.

## 1.1 Linear Response Theory and Dielectric Properties

At the frequencies of interest in present Si-based microelectronic devices (a few GHz), the dielectric permittivity $\varepsilon$ (i.e., $\kappa$) is the sum of an electronic and an ionic contribution. The former, the partial derivative of the polarization with respect to field, is generally a factor 2 to 5 smaller than the latter. We will therefore discuss in more detail ionic screening, which is caused by a polarization produced by ionic displacements. The appearance of a polarization is possible either in the absence of symmetry constraints (e.g., in amorphous materials) or when crystal symmetry allows for polar collective displacement. Interacting with a homogeneous electric field, ions in a crystal may relax their position to minimize their potential energy while preserving crystal symmetry. These collective displacements having lattice periodicity are linear combinations of vectors related to the eigenmodes of the dynamical matrix at the Brillouin zone center ($\Gamma$).

Each normal mode $j$ behaves like a dynamically independent harmonic oscillator and interacts with the electric field $\mathcal{E}$ via a potential energy term

$-\boldsymbol{M}_j \cdot \boldsymbol{\mathcal{E}}$, where the vector $\boldsymbol{M}_j$ is the electric polarization developing per unit cell upon a unit ionic displacement. $\boldsymbol{M}$ is defined as the partial derivative of macroscopic polarization with respect to the normal coordinate $Q_j$ (at vanishing electric field) times the unit cell volume $\Omega_0$:

$$M_{j\alpha} = \Omega_0 \left[ \frac{\partial \mathcal{P}_\alpha}{\partial Q_j} \right]_{\mathcal{E}=0} . \tag{1}$$

Analogously, the partial derivatives of the macroscopic polarization component $\alpha$ with respect to component $\tau_{\kappa\beta}$ of the position vector of atom $\kappa$ in the primitive cell, usually named the Born effective charges, are defined as

$$Z^*_{\kappa,\alpha\beta} = \Omega_0 \cdot \frac{\partial \mathcal{P}_\alpha}{\partial \tau_{\kappa\beta}} . \tag{2}$$

Of course, these polarizations per unit displacement are in general tensor quantities (a $3 \times 3$ matrix for each atom). The displacement of the $\kappa$-th ion, with mass $m_\kappa$, in the $j$-th vibrational mode is the mass-weighted normalized phonon eigenvector

$$U_j(\kappa,\alpha) = \frac{1}{\sqrt{m_\kappa}} \cdot e_j(\kappa,\alpha) . \tag{3}$$

In terms of the Born charges, the mode electric polarization may then be defined as

$$M_{j\alpha} = \sum_{\kappa\beta} Z^*_{\kappa,\alpha\beta} U_j(\kappa,\beta) . \tag{4}$$

The dielectric permittivity is by definition

$$\varepsilon_{\alpha\beta} = \delta_{\alpha\beta} + 4\pi \frac{\mathrm{d}\mathcal{P}_\alpha}{\mathrm{d}\mathcal{E}_\beta} . \tag{5}$$

If no spontaneous (i.e., zero-field) polarization is present, under vanishing strain conditions (see also [4] for another formulation and [11] for the vanishing stress case), the polarization $\mathcal{P}_\alpha$ may be expanded to linear order in the macroscopic electric field as

$$\mathcal{P}_\alpha = \sum_\beta \mathcal{E}_\beta \left( \frac{\partial \mathcal{P}_\alpha}{\partial \mathcal{E}_\beta} + \frac{1}{\Omega_0} \sum_j \frac{M_{j\beta} \cdot M_{j\alpha}}{\omega_j^2} \right) , \tag{6}$$

and hence

$$\varepsilon_{\alpha\beta} = \varepsilon^\infty_{\alpha\beta} + \frac{4\pi}{\Omega_0} \sum_j \frac{M_{j\beta} \cdot M_{j\alpha}}{\omega_j^2} , \tag{7}$$

where $\varepsilon^\infty_{\alpha\beta}$ indicates the dielectric permittivity stemming from electronic polarizability alone while ions are kept fixed. This component is the only one measured at frequencies well above vibrational ones, i.e., when the response of ionic motion to electric fields is completely suppressed. The coupling of each polar mode with electric fields is proportional to the squared modulus of the mode dipole moment. This coupling is usually quantified by the so-called oscillator strengths, which are usefully defined as

$$S_{j,\alpha\beta} = \sqrt{M_{j\alpha}M_{j\beta}/\Omega_0}\,. \tag{8}$$

Thus, $4\pi(S_j/\omega)^2$ yields directly the individual contribution of mode $j$ to the ionic dielectric tensor, and is also known as mode dielectric intensity. Of course, the latter quantity, and therefore the ionic component of the dielectric constant, can become rather large when $M$ is large or $\omega$ is small, or both. Therefore, large $\varepsilon$ will be produced by large effective charges and relatively soft vibrations, and in addition by efficient alignment of contributions to the dipole from the single ions. Effective charges exceeding significantly the nominal ionic charge of the relevant atom in a given compound are usually named anomalous. Anomalies can produce enhancements up to a factor of 4 in ferroelectrics. Due to their effects on the dielectric constant, anomalies are important in high-$\kappa$ materials; an anomaly of 10–15 % can be viewed as modest, one of $\sim 30$–40 % as large.

The phonons involved in optical transitions correspond to the eigenmodes of the dynamical matrix in the limit $q \to 0$. Approaching $\Gamma$, the dynamical matrix is given by its value at $\Gamma$ plus a non-analytical term [12] dependent on the $\hat{q}$ direction whence the $q \to 0$ limit is reached. Since this additive term does not act on modes of which the dipoles are orthogonal to $\hat{q}$, the oscillator strengths of zone center modes may be obtained by infrared absorption and reflectivity measurements revealing only transverse optic (TO) modes. Including the non-analicity, longitudinal optic (LO) modes can also be obtained and compared with those measured at grazing incidence or on films grown on metallic substrates.

All the relevant quantities in the equations above are related to second-order derivatives of the total energy with respect to the ionic positions and to the homogeneous electric field. In the DFT Kohn–Sham approach, the second-order mixed derivatives may be calculated using the linear-response technique, which allows one to calculate the first-order derivatives of the wavefunctions, and hence of the density, with respect to external parameters [13, 14]. A slightly different formulation is based on the so-called $2n + 1$ theorem [15], which enables one to evaluate all mixed derivatives up to third order, knowing the first-order perturbed density [12, 16]. The calculations discussed in this paper are all performed using the abinit[1] and Espresso [18] codes for the linear response, VASP [19] for some of the total-energy calcula-

---

[1] The abinit code is a common project of Université Catholique de Louvain, Corning Incorporated, and other contributors. Available from: http://wwwabinit.org; [17].

tions, and a custom-made code (developed from an early version of the total-energy portion of the current Espresso package) for the pseudo-selfinteraction correction calculations. All are variants of pseudopotential plane wave codes for Kohn–Sham DFT total-energy calculations [20].

## 2 Sesquioxides

Sesquioxides have the formula $X_2O_3$, with X a trivalent metal. We will limit ourselves to the La- and Lu-sesquioxides, and will be pointing out the close analogy of the latter with most of the lanthanide (Ln-) sesquioxides. (As to other cations, we note in passing that many not obviously or generally trivalent $d$-metals [Rh, Fe] form sesquioxides as well; also, oxides of small or relatively small cations such as Al, and Ga to a lesser extent, exhibit a strong tendency to polytypism and complex pressure-induced phase transition (see e.g., [21])). The main motif in Ln-sesquioxides is the competition of cubic bixbyite (C-phase) and hexagonal (A-phase) structures [22]. Indeed, this is to some extent a non-competition, as the hex phase is observed only for La, while it is metastable and with modest abundance for Pr (which exhibits strong mixed-valence behavior) and Nd. Ce-sesquioxide is produced only in strongly reducing atmosphere, and we can consider Ce to act mostly as tetravalent in this context.

Calculations on structural preferences are very scarce; recently, the structures of Ln-sesquioxides were studied from first-principles, but no data on the structural stability of either phase reported [23]. Our own earlier calculations [24] show that for $La_2O_3$ the hex phase is favored by $0.2\,\mathrm{eV}$/formula unit compared to the cubic; for Lu, the opposite happens, with a difference of $0.25\,\mathrm{eV}$. This result agrees with experiment and indicates that the cubic-hex difference increases across the Ln-series in favor of the cubic phase, as one expects. Of course, it is unreliable to predict the crossover point based on these two points alone.

Although for a given material one of the two phases is never actually realized, it is interesting to compare them from the point of view of dielectric behavior, as the two structures do exhibit rather remarkable differences of almost purely structural origin. In particular, it is interesting to analyze the differences or analogies in the main ingredients determining $\varepsilon_{\mathrm{ionic}}$, i.e., the effective charges and vibrational frequencies.

The C-phase or bixbyite structure has a space group $Ia\bar{3}$ (n.206). The body-centered cubic primitive cell contains 40 atoms. Besides the lattice constant, there are four internal parameters $u, x, y, z$. The cations are placed at two inequivalent sites: four are in the $8a$ Wyckoff positions generated applying the group operations to the origin, 12 are placed at $24d$ positions generated from $(u, 0, 1/4)$ in cubic coordinates. The 24 oxygen positions ($48e$) are generated from $(x, y, z)$. For $Lu_2O_3$, we calculated a lattice constant of $10.335\,\text{Å}$, in agreement with the measured value of $10.391\,\text{Å}$ reported in [25], and internal structural parameters are $u = -0.032\,4$ and $(x, y, z) = (0.391, 0.152, 0.379)$. These have been found to agree well with recent XAFS experiments. The A-phase has a hexagonal bravais lattice; the space group is $P\bar{3}m1$. The two cations are placed in $2d$ Wyckoff positions, with components $\pm(1/3, 2/3, u)$ along the lattice vectors; two oxygens are also placed in $2d$ positions at $\pm(1/3, 2/3, v)$, and the remaining one is at $(0, 0, 0)$. The calculated values of the structural parameters are reported in Table 1 for $La_2O_3$. The hex and cubic structures are hypothetical for lutetia and lanthana respectively. We only report the hex structure of lutetia to point out that the volume per formula unit is about 15 % smaller than in lanthana, consistently with lantanide contraction.

**Table 1.** Structural parameters for $La_2O_3$ and hexagonal lutetia

|         | $Lu_2O_3$ (th) | $La_2O_3$ (th) | $La_2O_3$ (ex) |
|---------|---------------|---------------|---------------|
| a (Å)   | 3.63          | 4.00          | 3.94          |
| c/a     | 1.56          | 1.56          | 1.56          |
| $(2d)u$ | 0.248         | 0.243         | 0.247         |
| $(2d)v$ | 0.649         | 0.645         | 0.645         |

## 2.1 Lutetia, Lanthana, and the Hex–Bix Differences

Building on the bixbyite vs. hex structural theme, we now analyze the differences in dielectric response of the two phases and deduce a general trend for all Ln-sesquioxides. We start with byxbyite lutetia, whose static dielectric tensor is diagonal and isotropic because of symmetry, and equals 11.98 (which incidentally hardly qualifies as "high" in the present microelectronics setting). Its electronic part is 4 18, about half of the ionic component, which is consistent with the general hallmark of high-$\kappa$s of having, predominantly, large ionic constants.

For hexagonal lutetia, the dielectric tensor is anisotropic due to symmetry: it has two identical in-plane components, and a third – different – axial component. The total dielectric tensor has planar components of 19.8 and axial component 17.2. Calculating the "inverse average" that would be measured via series capacitance of a stack, i.e.,

$$3/\bar{\varepsilon}_s = 1/\varepsilon_s^{11} + 1/\varepsilon_s^{22} + 1/\varepsilon_s^{3\varepsilon} \tag{9}$$

a value of over 18 is obtained, already in the interesting range. The electronic tensor is 4.65 in-plane and 4.56 axially, which, not unexpectedly, is only marginally larger than that of bixbyite. Therefore, the average ionic part is over 13, a factor 3 larger than the electronic part. Interestingly, the ionic component is nearly twice as large as that of bixbyite. We now analyze this peculiar behavior starting from effective charges. For lutetia, the Born effective charge tensors are

$$\begin{pmatrix} 3.81 & -0.52 & -0.47 \\ -0.47 & 3.81 & -0.52 \\ -0.52 & -0.47 & 3.81 \end{pmatrix} \quad \text{and} \quad \begin{pmatrix} 3.40 & 0.00 & 0.00 \\ 0.00 & 3.89 & -0.39 \\ 0.00 & -0.44 & 3.46 \end{pmatrix}$$

for the cation at the origin and at $(u, 0, 1/4)$ respectively, the analogous tensors for symmetry-equivalent cations being obtained applying space-group rotations. The average over the first group of cations produces an isotropic tensor with a value 3.81, while the isotropic average for the other cation group evaluates to 3.58. For the oxygen, the Born charge is

$$\begin{pmatrix} -2.25 & -0.02 & -0.15 \\ 0.08 & -2.35 & -0.15 \\ 0.30 & -0.22 & -2.68 \end{pmatrix},$$

with an average value of $-2.43$.

Since evidently the Born charges in bixbyite lutetia are not particularly anomalous, to explore the structural dependence we look at charges in the hex phase Here, charge tensors are diagonal. For Lu, we have 3.79 and 3.59 in-plane and axially, while O in $2d$ positions have $-2.49$ and $-2.39$. The remaining oxygen at the origin has $-2.60$ in-plane and $-2.28$ axially. Clearly, these charges are as anomalous as in bixbyite; the differences in the ionic constant must therefore stem from the vibrational spectrum. Also, we can ask the question how general is this behavior for sesquioxides at large?

To answer this, we analyze the vibrational modes. The eigenmodes of the dynamical matrix at the $\Gamma$ point are a basis for irreducible representations of the space group. Modes spanning the same irreducible representation must correspond to the same eigenvalues. Group theory shows that the sum (4) may be non-vanishing only for modes belonging to certain representations, while it must vanish for others, so that one can discriminate between polar and non-polar modes simply on the basis of symmetry arguments (at least in crystals). The irreducible-representation decomposition of ionic displacements in the hex structure is

$$\Gamma_{\text{disp}}^{\text{hex}} = 2A_{1g} \oplus 3A_{2u} \oplus 2E_g \oplus 3E_u . \tag{10}$$

Modes with non-vanishing electric polarization belong to $E_u$ and $A_u$ representations. Rigid translations are also $E_u$ (in-plane) and $A_{2u}$ (axial) representations. So, in the hexagonal structure we have a pair of $E_u$ modes which contribute isotropically to the in-plane component and two $A_{2u}$ contributing to the axial component of the ionic dielectric tensor. The main planar modes are at $221\,\text{cm}^{-1}$, contributing $90\,\%$ of the in-plane component, and $494\,\text{cm}^{-1}$. The axial modes are at $261\,\text{cm}^{-1}$, accounting for over $95\,\%$ of the axial component, and $497\,\text{cm}^{-1}$. The representation decomposition for bixbyite is

$$\Gamma_{\text{disp}}^{\text{byx}} = 4A_g \oplus 5A_u \oplus 4E_g \oplus 5E_u \oplus 17T_u \oplus 14T_g . \tag{11}$$

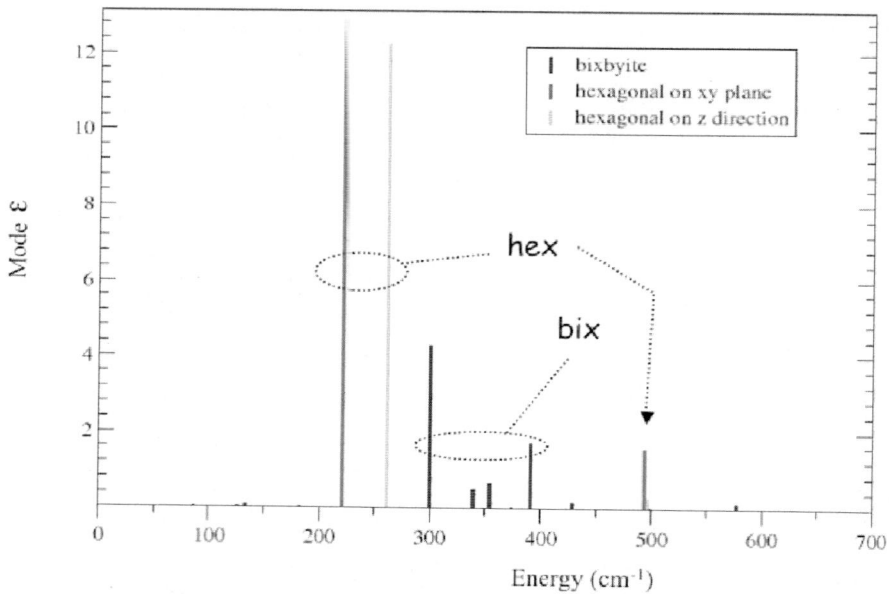

**Fig. 1.** Dielectric mode intensity in bixbyite (*lower*, above $300\,\mathrm{cm}^{-1}$) and hex Lu$_2$O$_3$

$A_g$ and $A_u$ are one-dimensional representations of even ($g$) and odd ($u$) parity with respect to the inversion operation. $T_g$ and $T_u$ are tridimensional representations. $E$ representations may be decomposed into two one-dimensional complex-conjugate irreducible representations. Physical displacements behave like their real and imaginary parts, and as they transform into each other by symmetry operations, they are accordingly considered as bidimensional physically irreducible representations. Only triplet $T_u$ representations may have a non-vanishing dipole. One of these is actually acoustic translations. The members of the 16 remaining vibrational triplets develop electric dipoles along mutually perpendicular directions. The main contribution comes from modes at 300, 339, 354, 391 cm$^{-1}$, which make up 93 % of the total ionic contribution. As bixbyite is the experimental structure, these data can be compared with IR absorption measurements on recently grown films [26]. The comparison, reported in [27], is indeed excellent. The experimental value of $\varepsilon_0 = 12 \pm 1$ obtained both optically and electrically in [27] is also in excellent agreement with the predicted one.

The comparison of the two structures is made more explicit in Fig. 1, which displays the dielectric intensities (the terms in the sum of (7)) as a function of energy for the two structures. It is evident that it is the softer modes supported by the hex structure that produce the near doubling of the ionic component, despite the equally non-anomalous charges in the two structures. Of course, this immediately suggests the question as to whether

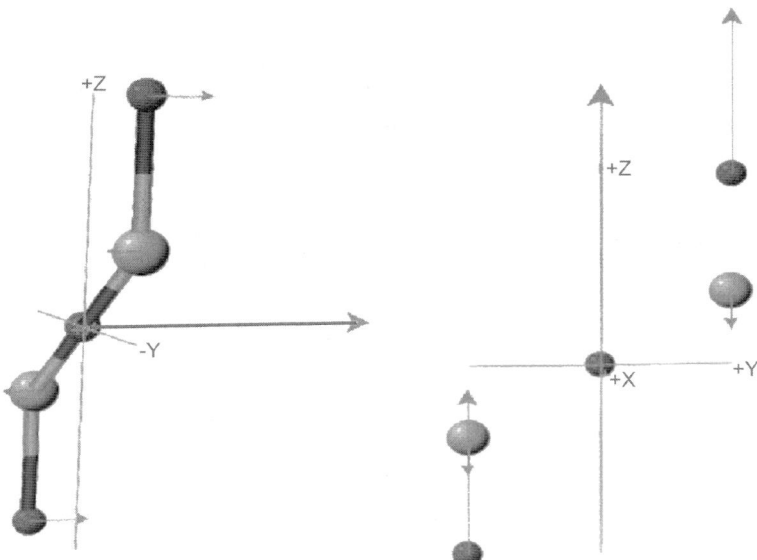

**Fig. 2.** Vibrational motion pattern for hex-$Lu_2O_3$. Small spheres are O, large ones Lu. *Left*: planar mode doublet. *Right*: axial mode singlet

these findings are exportable to other sesquioxides with a different cation? For this to be the case, the vibrational modes should be independent of the cation. In the case of the hex phase, this can be found out by inspection: indeed, the softish modes therein are related to oxygen motions, with the cation just keeping the center of mass still, as shown in Fig. 2.

For bixbyite, the eigenvector pattern cannot be easily visualized. We therefore display in Fig. 3 (from [27]) the mode electric polarization vectors, defined in (4). From top to bottom in the upper panel, we show the modulus of total, O, Lu vectors (the sum in (4) is over atoms, too). In the lower panel, we show the angle between the O and Lu dipoles: if they sum up in phase, i.e., the angle is small, the mode is IR-efficient. As can be seen, O dipoles are dominant, with Lu dipoles about an order of magnitude smaller. Whenever Lu-related modes are of comparable weight, as at low energies, either they are inefficient (large $\alpha$) or have small amplitude. Therefore, one concludes that ionic IR screening in bixbyite is also predominantly due to oxygen motions. Thus, the reduced ionic constant in bixbyite compared to the hex phase is due to a purely structural effect, as the difference in vibrational modes is due to the structure and its energy response pattern (i.e., the vibrational modes). The independence on the cation will, of course, hold as long as the cation is not too light. An indirect countercheck comes from the similarity of the IR spectra of yttria and lutetia in the relevant region. Another indirect evidence of the same effect is that the Raman spectra of

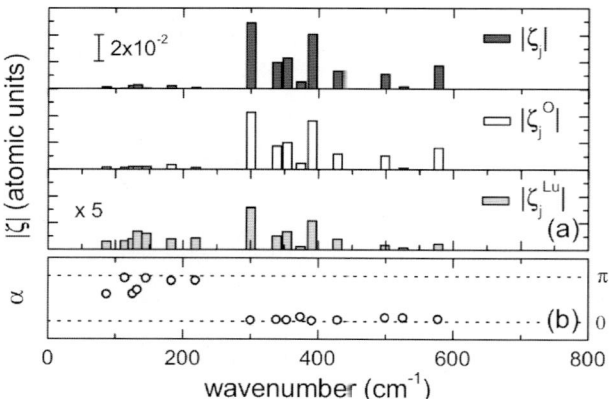

**Fig. 3.** *Upper panel*: moduli of the dipole vectors building up the ionic contribution; *from top to bottom*: total, O, and Lu. *Lower panel*: angle between the O and Lu dipole vectors; *small angles* indicate efficient alignment of the dipoles. The quantities named $\zeta$ are the sum (4) evaluated over cations, anions, or all atoms. Figure reproduced with permission from [27], copyright 2005 American Physical Society

yttria and lutetia are quite similar in the 300–400 cm$^{-1}$ modes, and only differ in some peaks at low frequency: from our calculations, we could identify the former modes with O-related motions, and the latter with cation-related motions. As shown in Fig. 4, our calculated Raman modes compare favorably with the experimental spectrum [28].

Of course, there is a possible direct countercheck with the actual hex-phase material, i.e., $La_2O_3$. To begin with, the effective charges are larger here. The Born charge for La is 4.2, and that of O is −2.8. Also, the main IR modes are, similarly to lutetia, an in-plane doublet at 200 cm$^{-1}$ and an axial singlet at 220 cm$^{-1}$. The high-frequency component is 5.2, a bit larger, but not substantially, than for bixbyite. As a consequence of the softer modes and larger charges, the final value of $\varepsilon_0$ is 22.5, somewhat larger than in hex lutetia – though not as larger as may be suggested by the charge anomaly, due to the larger volume, hence reduced dipole density. Similar results were obtained recently in [29]. Overall, despite the expected quantitative deviations, our earlier inference is confirmed: hex-phase sesquioxides have a larger dielectric constant than their bixbyite counterpart. The differences may vary depending on the cation polarizability, but the main discriminant is structure. Indeed, this conclusion is unfortunate, as most Ln-sesquioxides are bixbyites: Dy and Yb oxides, for instance, have dielectric constants of $\sim 11$, in line with the expectations from lutetia. As we will be discussing next, it is possible that DFT calculations overestimate the effective charge anomalies. This may reduce slightly hex–bix differences related to charge anomalies, but the core of the argument based on the vibrational spectra is not affected.

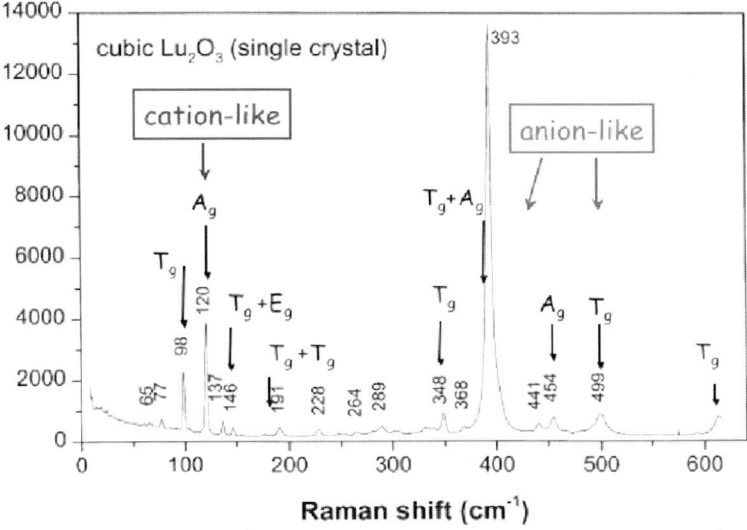

**Fig. 4.** Raman spectrum (data: courtesy of *Laetitia Laversenne* [28]) of $Lu_2O_3$ compared with calculated assignments

# 3 Aluminates

An entirely different class of materials is that of Ln-aluminates. Here too, a number of useful inferences can be drawn from the study of just one of these materials; we concentrate here on lanthanum aluminate, $LaAlO_3$. The key point is that due to the smaller size of Al compared to all Ln, and its preference for six-fold coordination with O, despite the larger bond enthalpies of Ln−O bonds, the primary bonding in $LnAlO_3$ is Al−O, and Al coordination is octahedral. The large Ln ions fit nicely in the interstices of the octahedral network, and a perovskite results: most $LnAlO_3$ are indeed distorted variants of perovskites with Al at the octahedral B-site (indeed, for the same reasons, even $LnScO_3$ seem to obey this rule to a large extent). Distortions have important effects on the details of the dielectric response, but the general picture does not (or is not expected to) change much.

From the analysis presented below of the vibrational spectrum of the crystal, of the structure of the amorphous phase, and the vibrational spectrum thereof, a few general predictions can be made for all Ln-aluminates:

1. In the crystal, the dominant IR modes are at low ($\sim 150\,\mathrm{cm}^{-1}$) frequency, and correspond to vibrations of the Ln cation at the cubic A-site; this is accompanied by a large charge anomaly of the Ln cation, and results in ionic dielectric constants of around 20 and total static values of 25; Al−O motions instead are high-frequency and there is no Al charge anomaly; low-frequency (40 and $130\,\mathrm{cm}^{-1}$) Raman modes are associated to octahe-

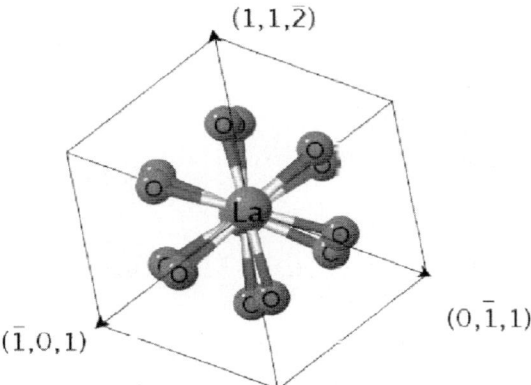

**Fig. 5.** Rhombohedral LaAlO$_3$ in a (111)-axial view

dra rotational distortions in LaAlO$_3$: similar modes may appear in other Ln-aluminates in association with similar low-energy distortions.

2. the simulated amorphous phase conserves Al–O short-range order; Ln charge anomalies are reduced by about 10–15 %; due to disorder, the Ln IR vibrations in the cubic cage and the formerly Raman, but now weakly IR-active modes couple, and a large dielectric intensity at low ($\sim 100\,\mathrm{cm}^{-1}$) frequency appears, related to Ln vibrations. The resulting ionic component of the dielectric tensor is comparable to that of the crystal.

### 3.1 Crystalline LaAlO$_3$

LaAlO$_3$ is observed [30] to be a perovskite with a small rotational rhombohedral distortion. In our calculations [31], we find no hint of ferroelectric distortions, and exchanging the A- and B-cations leads to a metal with 40 % larger volume and higher energy, which excludes the possibility of cation disorder. Simple-cubic perovskite structure has the main IR mode at 167 cm$^{-1}$, carrying 80 % of the intensity. A soft $F_{2u}$ mode with imaginary frequency (130$i$ cm$^{-1}$) at the $R$ point causes a period-doubling instability of the cubic phase towards a rotational volume-conserving distortion producing a rhombohedral phase, whereby octahedra rotate by $\sim 6°$ with respect to their ideal orientation, with a minute energy lowering of 10 meV/formula.

Upon distortion, the soft mode transforms into stable low-energy modes related to the octahedra rotation, namely the Raman modes $E_g$ (33 cm$^{-1}$) and $A_{1g}$ (129 cm$^{-1}$). These Raman modes are in close agreement with experiment [32], and are associated with the rotation of the O octahedra around Al. The IR spectrum is now dominated by the $A_{2u}$ singlet at 168 cm$^{-1}$ and $E_u$ doublet at 179 cm$^{-1}$, which derive from the main IR triplet of the perovskite phase at 167 cm$^{-1}$. These modes provide over 82 % of the total dielectric

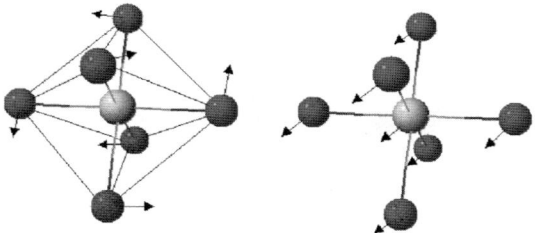

**Fig. 6.** The displacement patterns of the main Raman ($A_{1g}$, *left*) and IR ($A_{2u}$, *right*) modes. Only one Al–O octahedron is shown for clarity

intensity, and hence they are the major responsible of the rather high ionic dielectric constant of $LaAlO_3$, together with the significant charge anomaly induced by the La cation. The distortion-induced Raman doublet repels upwards the IR doublet by approximately $10\,\mathrm{cm}^{-1}$, giving rise to a uniaxial dielectric anisotropy.

In Fig. 6, we display the displacement patterns for the low-energy Raman $A_{1g}$ and IR $A_{2u}$ modes. The Raman mode is a rotation of the O–Al octahedron directly related to the rhombohedral rotational distortion. The dominant IR mode, displayed here in the reference system of La being immobile, is essentially a vibration of La atoms in the cubic cages against the Al–O octahedra backbone.

Neglecting for simplicity their moderate ($\sim$ 1–2 %) axial anisotropy, the GGA-calculated average effective charges are $Z^*_{La} = 4.37$, $Z^*_{Al} = 2.98$, and $Z^*_{O} = -2.45$. The Al effective charges are close to the nominal ionicity, and therefore strictly non-anomalous. On the other hand, compared to the La nominal ionicity of +3, the La charges are appreciably anomalous, and as a consequence so are the O ones. Unlike other perovskites (e.g., $BaTiO_3$), the anomaly does not stem from the primary directional bonding between O and the octahedrally coordinated cation, but rather from the hybridization and charge transfer between the A-cation (here La, but in general a Ln) and O, second-neighbors at about 2.7 Å. This is expected, as the states involved are prevailingly La $d$ and O $p$. The dominance of La anomaly and of octahedra-against-cage modes in the dielectric response are consistent with the high dieletric constant of 25 for $LaGaO_3$ [33], having a closely similar structure. Given the similarity of Al and Ga in this context, this suggests that Ln-gallates should largely share the properties of Ln-aluminates at issue here.

The average electronic dielectric tensor is again small and largely structure independent, $\varepsilon_\infty = 4.77$. As a result of the large charge anomaly and soft modes, the static value is instead large, 25.5 in-plane, and 28.4 axially. The ionic component is again dominant, a factor 4–5 larger than the electronic one, and determines the accuracy of the final static value. The inverse orientational average of the static constant ((9), measured from the series capacitance of polycrystalline samples) equals $\bar{\varepsilon}_s = 26.4$ using the GGA tensor, 11 % larger than experiment [34–36]. It is therefore worth considering the effect of the pseudo-selfinteraction-correction theory [37] on the effective charges, and the resulting changes in the static dielectric constant. It has been previously found [38] that the charge anomaly is reduced by applying the post-DFT pseudo-SIC treatment [37]. As in [38], we find reduced average charges as compared to GGA : $Z_{\mathrm{La}}^{*(\mathrm{SIC})} = 4.06$, $Z_{\mathrm{Al}}^{*(\mathrm{SIC})} = 2.87$, $Z_{\mathrm{O}}^{*(\mathrm{SIC})} = -2.31$. As expected, the non-anomalous Al charge hardly changes at all, while the La charges decrease sizably ($\sim 8\,\%$). This is due to the increased localization, hence reduced polarizability, of hybridized La-$d$/O-$p$ states. The smaller pseudo-SIC charges cause a drop in dielectric constant by roughly 15 % compared with the GGA value, once again entirely due to the ionic part, and in fact to the effective charges appearing therein. The orientational average is now $\bar{\varepsilon}_{\mathrm{s}} = 23.3$, within 2 % of the measured 23.7. This remarkable agreement suggests that DFT calculations supplemented by SIC in the case of strongly anomalous Born charges are predictive of the dielectric constant of high-$\kappa$ oxides.

## 3.2 Conservation of High $\kappa$ in Amorphous Ln-Aluminates?

Recent studies on amorphous $LaAlO_3$ grown by epitaxial techniques have reported dielectric constants in the same range (20–24) as those of the crystalline phase [39–42]. Theoretically, this is puzzling, as the large effective charges co-responsible for such a high $\kappa$ originate from delicate balances of ionicity and covalent hybridization that one expects to be disrupted in amorphous structures. Indeed, a recent calculation [9] for zirconia showed that charge anomaly is reduced by over 10 % upon amorphization; however, it was also reported [9] that this reduction is compensated by an energy downshift of the average IR mode energy (or rather density-of-states centroid), which is lower in the amorphous network than in the crystal: eventually, the dielectric constant of amorphous zirconia is similar to or larger than that of the monoclinic crystalline phase (the one having the smaller dielectric constant). For $LaAlO_3$, it has been suggested that XRD measurements may, due to limited resolution, be observing as amorphous a system which is actually nanocrystalline. In the following, we do not consider this option, and instead analyze the dielectric properties of amorphous aluminate from a theoretical point of view to verify the plausibility of a similarity with the crystal phase.

The study of amorphous systems, and especially of subtle properties such as dielectric and polarization properties, is a formidable task. Here we discuss lanthanum aluminate as a model for all Ln-aluminates, in the spirit of the pioneering work by *Pasquarello* [43] on silica. We create several amorphous models of different sizes by the melt-and-quench technique using a combination of shell model-potentials and ab initio dynamics, and thereafter we compute the phonons, effective charges, and dielectric intensities of those models using linear-response density-functional perturbation theory. The latter calculation is very demanding for the smaller model (80 atoms) and impossible

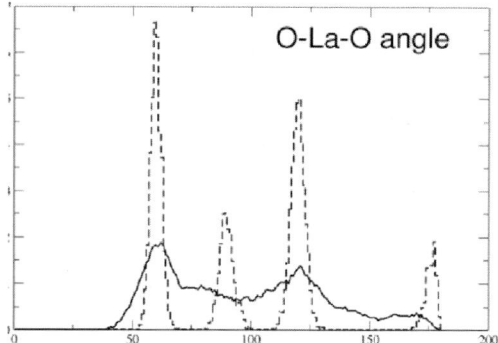

**Fig. 7.** Average O−La−O angle distribution in amorphous (*solid*) and crystalline (*dashed*) LaAlO$_3$ collected in an ab initio dynamics run at 500 K. Similar results are obtained in both small and medium-sized samples

at present for the larger model (320 atoms), which is used as a structural comparison.

The structure of amorphous LaAlO$_3$ can be characterized, e.g., by radial distribution functions and bonding angle distributions. We cannot report all the relevant data, due to space limitations. The main indication is that, although disordering is evident and strong, short-range Al−O order is preserved, i.e., Al−O coordination numbers are around 6, and angles are around 90°, as in the crystal. The crystal–amorphous connection is most blurred in the case of the O−La−O angle. This particular distribution, displayed in Fig. 7, shows that the orientation of Al−O octahedra with respect to La is very disordered. As a result, one expects modifications of the main IR modes (La against Al−O octahedra backbone), mixing thereof with octahedra rotation modes, and finally a reduction of the anomaly in O−La dynamical charge transfer. The charges that were anomalous in the crystal, as shown in Table 2, are indeed reduced sizably and rather uniformly, as indicated by the moderate dispersion over the sample. The non-spherical components of the charge tensors (see [43]) are modest, similarly to silica: this is consistent with the short-range order present in this structure, as it is in silica.

**Table 2.** Average spherical components of effective charge tensors in amorphous (with standard deviation) and crystalline LaAlO$_3$, and tensor decomposition (%) of amorphous charges into spherical, dipolar, and quadrupolar components

| | $Z^*$ | $Z^*_{cry}$ | $s$ | $p$ | $d$ |
|---|---|---|---|---|---|
| Al | $2.96 \pm 0.11$ | 2.97 | 87 | 4 | 8 |
| La | $3.88 \pm 0.04$ | 4.35 | 85 | 3 | 8 |
| O | $-2.38 \pm 0.13$ | $-2.45$ | 83 | 2 | 12 |

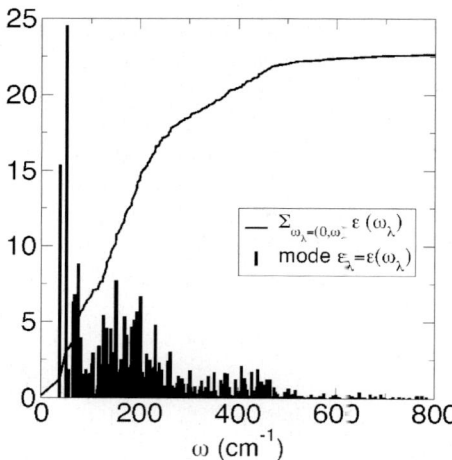

**Fig. 8.** Mode and integrated ionic dielectric constant in amorphous $LaAlO_3$

We now come to dielectric IR intensities and screening. The electronic component is essentially the same as in the crystal, with very small off-diagonal elements and anisotropies. Again, therefore, the dielectric constant is dominated by ionic screening. In Fig. 8, we report the histogram of the dielectric intensities (the mode contributions to $\varepsilon_{ionic}$ and their cumulative sum, i.e., the value of the ionic constant accumulated as a function of frequency). The ionic constant (the asymptotic limit of the upper curve at high frequency) is 22.5; summing the electronic part, we get $\varepsilon_0 = 27$. Allowing for the fact that the effective charges should be reduced by the SIC corrections, applying the same reduction ratio as in the crystal, we end up with an estimate of the ionic part around 20, and hence $\varepsilon_0 = 24$–25, i.e., about the same as in the crystal.

To understand why this is plausible, let us consider Fig. 8. The dielectric intensity in the region of the crystal IR modes has a broad distribution. Also, it is large even at rather low frequencies below the crystal IR energy, with a major peak between around 50 to $100\,\mathrm{cm}^{-1}$. The broadening is disorder-induced, and the softening is related to the imperfect structure around the La ions. The peaks in the $50$–$100\,\mathrm{cm}^{-1}$ range are probably due to the mixing of disordered La translations (crystal frequency $160\,\mathrm{cm}^{-1}$) and Al–O octahedra rotations ($40$ and $130\,\mathrm{cm}^{-1}$); these modes are Raman in the crystal, but symmetry-breaking disorder allows them to mix with polar ones, acquiring a non-zero dipole moment which albeit small, gives a large dielectric intensity because of the inverse quadratic energy dependence.

To support this, we consider the displacement amplitudes in the various modes as a function of energy, displayed in Fig. 9. La amplitudes are large, as in the crystal IR modes at $160\,\mathrm{cm}^{-1}$, all the way from 50 to $200\,\mathrm{cm}^{-1}$. The La amplitudes, in addition, decrease for the very first non-acoustic modes, and O

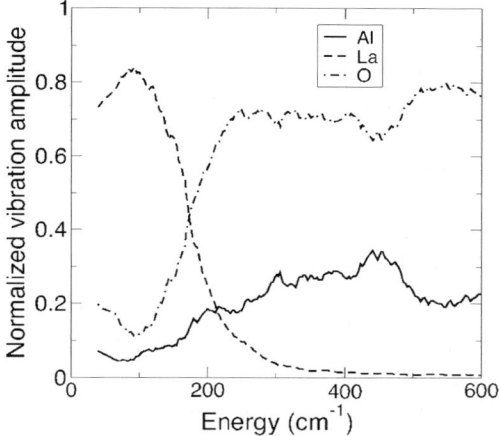

**Fig. 9.** Normalized displacements (smoothed by a running average procedure) of the three species in the IR vibrational modes at a given frequency. The modes below $200\,\mathrm{cm}^{-1}$ are dominated by La displacements, and are therefore related to the dominant IR modes of the crystal. The rise of the O displacement and attendant decrease of La displacement and constancy of Al displacement at low frequency are consistent with the coupling of crystal IR and Raman modes

amplitudes grow somewhat. This is consistent with the possible mixing of low-energy, formerly Raman modes with IR La vibrations. Of course, evaluating accurately the weight of these low-frequency vibrations would require a much larger set of samples in order to accumulate a reliable statistics. Nevertheless, we think our calculation definitely supports the idea that the ionic $\varepsilon$, and therefore the total $\varepsilon$, can well be as large in the amorphous phase as in the crystal phase. This result echoes that of [3], where a broadening and softening of the IR-active density of states was found in amorphous zirconia, and that of [5], where low-frequency backfolded Raman modes became IR-active due to symmetry lowering by epitaxial relation. As the central ingredient is basically the structure of $LaAlO_3$ as an octahedral backbone with interstices in which the large rare-earth ion moves around, it is only natural to infer that any Ln-aluminate with similar structure should behave similarly at the qualitative level. This is supported by the case discussed in the next and final section.

### 3.3 Dielectric Enhancement in Aluminate Alloys

We now mention a recent study by the IBM Zürich group [44], who have used linear response and the virtual crystal approximation to study the behavior of dielectric screening in $La_xY_{1-x}AlO_3$. Their key observation is that yttrium aluminate in the distorted perovskite structure of lanthanum aluminate would have a maximum axial value (along the (111) axis) of the static $\varepsilon$ of 60! This is due to the much lower frequency of the axial $A_{2u}$ singlet IR vibration in distorted-perovskite $YAlO_3$, about $100\,\mathrm{cm}^{-1}$ compared to that of $LaAlO_3$, $170\,\mathrm{cm}^{-1}$. Given that the partner IR doublet stays at about the same position as in $LaAlO_3$, we believe that (although this was not discussed by the authors) this huge downshift may be due to the fact that the repulsive interaction of the IR $A_{2u}$ mode with the Raman $A_{1g}$ mode pushes the former *downward*, rather than upward as in $LaAlO_3$. This is quite plausible, as the Raman downfolded mode is at about the same energy as the IR mode of the undistorted perovskite. Unfortunately, distorted perovskite $YAlO_3$ is unstable, so the authors consider a solid solution with lanthanum aluminate such that the latter dictates the structure and the $YAlO_3$ component may provide an enhanced dielectric constant. They find the behavior of the static $\varepsilon$ as a function of composition shown in Fig. 10. The reported stability range of the solid solution is in the $x \sim 0.4$–$0.5$ composition range; the maximum axial component would then be somewhat higher than for lanthanum aluminate, and hence the enhancement might be exploited for (111)-oriented single crystals. The average $\varepsilon$ which would be observed in polycrystalline material remains at the $LaAlO_3$ level. This work [44] has nevertheless the merit of pointing out another way to exploit the peculiar and, to some extent, unexpected properties of this class of materials.

# 4 Conclusions

We have summarized some recent work on two classes of rare-earth oxides that have attracted attention as dielectrics for microelectronics. The main message of the present work is that rather general conclusions about wide classes of materials (Ln-sesquioxides and Ln-aluminates) can be drawn from a limited, although demanding, set of calculations. In particular, it is predicted that, qualitatively, all Ln-sesquioxides crystallizing in the bixbyite structure will have relatively low dielectric constants (10–12), while their (relatively rare) hexagonal counterparts will show values of around 20. Also, Ln-aluminates are predicted to have large dielectric constants related to their $Al-O$ backbone structure. Due to the nature of their vibrational modes, they are expected to conserve a large dielectric constant in the amorphous phase as well – at least insofar as the simulated amorphous structure can be considered realistic. Some predictions have been confirmed by, or confirm experiments; others await experimental counterchecks, and more theoretical studies for related materials.

**Acknowledgements**

Calculations were done on the SLACS-HPC cluster `ichnusa` at CASPUR Rome, and at CINECA Bologna. AF thanks the Ministry of Research for a "Cervelli per la ricerca" grant.

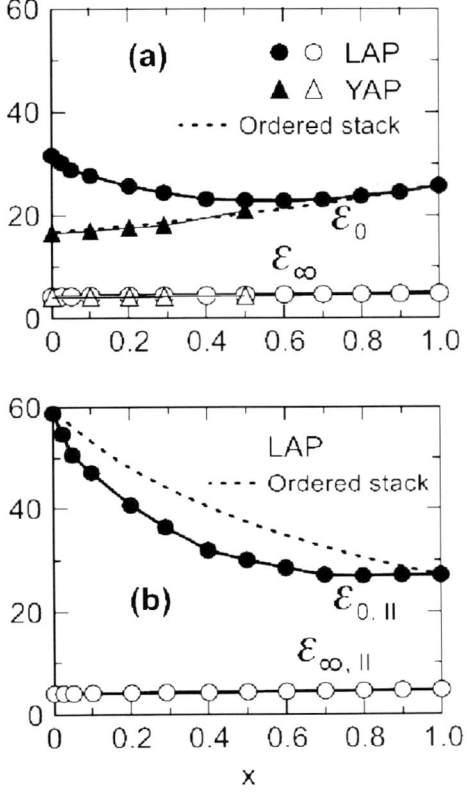

**Fig. 10.** Electronic and static dielectric constant of $La_x Y_{1-x} AlO_3$ in various structures. *Top panel*: average values; *bottom panel*: maximum component. Figure reproduced with permission from [44], copyright 2005 American Physical Society

# References

[1] G. Wilk, R. M. Wallace (Eds.): *Alternative Gate Dielectrics for Microelectronics*, MRS Bulletin **27** (Material Research Society, Warrendale 2002)

[2] G. D. Wilk, R. M. Wallace, J. M. Anthony: High-$\kappa$ gate dielectrics: Current status and materials properties considerations, J. Appl. Phys. **89**, 5243 (2001)

[3] X. Zhao, D. Vanderbilt: Phonons and lattice dielectric properties of zirconia, Phys. Rev. B **65**, 075105 (2002)

[4] X. Zhao, D. Vanderbilt: First-principles study of structural, vibrational, and lattice dielectric properties of hafnium oxide, Phys. Rev. B **65**, 233106 (2002)

[5] V. Fiorentini, G. Gulleri: Theoretical evaluation of zirconia and hafnia as gate oxides for Si microelectronics, Phys. Rev. Lett. **89**, 266101 (2002)

[6] G.-M. Rignanese: Dielectric properties of crystalline and amorphous transition metal oxides and silicates as potential high-$\kappa$ candidates: The contribution of density-functional theory, J. Phys.: Condens. Matter **17**, 357 (2005)

[7] P. Yu, M. Cardona: *Fundamentals of Semiconductors* (Springer, Berlin, Heidelberg 1996)

[8] F. Giustino, P. Umari, A. Pasquarello: Dielectric discontinuity at interfaces in the atomic-scale limit: Permittivity of ultrathin oxide films on silicon, Phys. Rev. Lett. **91**, 267601 (2003)

[9] X. Zhao, D. Ceresoli, D. Vanderbilt: Structural, electronic, and dielectric properties of amorphous $ZrO_2$ from ab initio molecular dynamics, Phys. Rev. B **71**, 0085107 (2005)

[10] P. Delugas, V. Fiorentini: unpublished

[11] F. Bernardini, V. Fiorentini, D. Vanderbilt: Polarization-based calculation of the dielectric tensor of polar crystals, Phys. Rev. Lett. **79**, 3958 (1997)

[12] X. Gonze, C. Lee: Dynamical matrices, born effective charges, dielectric permittivity tensors, and interatomic force constants from density-functional perturbation theory, Phys. Rev. B **55**, 10355 (1997)

[13] S. Baroni, P. Giannozzi, A. Testa: Green's-function approach to linear response in solids, Phys. Rev. Lett. **58**, 1861 (1987)

[14] S. Baroni, S. de Gironcoli, A. Dal Corso, P. Giannozzi: Phonons and related crystal properties from density-functional perturbation theory, Rev. Mod. Phys. **73**, 516 (2001)

[15] X. Gonze: Adiabatic density-functional perturbation theory, Phys. Rev. A **52**, 1096 (1995)

[16] X. Gonze: First-principles responses of solids to atomic displacements and homogeneous electric fields: Implementation of a conjugate-gradient algorithm, Phys. Rev. B **55**, 10337 (1997)

[17] X. Gonze, J.-M. Beuken, R. Caracas, F. Detraux, M. Fuchs, G.-M. Rignanese, L. Sindic, M. Verstaete, G. Zerah, F. Jollet, M. Torrent, A. Roy, M. Mikami, P. Ghosez, J.-Y. Raty, D. C. Allan: First-principles computation of material properties: The ABINIT software project, Comp. Mater. Sci. **25**, 478, (2002)

[18] S. Baroni, A. Dal Corso, S. de Gironcoli, P. Giannozzi, C. Cavazzoni, G. Ballabio, S. Scandolo, G. Chiarotti, P. Focher, A. Pasquarello, K. Laasonen, A. Trave, R. Car, N. Marzari, A. Kokalj:
URL http://www.pwscf.org/

[19] G. Kresse, J. Furthmüller: Efficient iterative schemes for ab initio total-energy calculations using a plane-wave basis set, Phys. Rev. B **54**, 11169 (1996)
URL http://cms.mpi.univie.ac.at/vasp/

[20] W. Pickett: Pseudopotential methods in condensed matter applications, Comput. Phys. Rep. **9**, 115, (1989)

[21] W. Duan, R. M. Wentzcovitch, K. T. Thomson: First-principles study of high-pressure alumina polymorphs, Phys. Rev. B **57**, 10363 (1998)

[22] D. Bloor, J. R. Dean: Spectroscopy of rare earth oxide systems I. Far infrared spectra of the rare earth sesquioxides, cerium dioxide and nonstoichiometric praseodymium and terbium oxides, J. Phys. C: Solid State Phys. **5**, 1237 (1972)

[23] N. Hirosaki, S. Ogata, C. Kocer: Ab initio calculation of the crystal structure of the lanthanide $Ln_2O_3$ sesquioxides, J. Alloys Comp. **35**, 34 (2003)

[24] L. Marsella, V. Fiorentini: Structure and stability of transition-metal and rare-earth oxides, Phys. Rev. B **69**, 172103 (2004)

[25] R. W. G. Wyckoff: *Crystal Structures*, vol. 2 (Interscience, New York 1966)

[26] G. Scarel, E. Bonera, C. Wiemer, G. Tallarida, S. Spiga, M. Fanciulli, I. L. Fedushkin, H. Schumann, Y. Lebedinskii, A. Zenkevich: Atomic-layer deposition of $Lu_2O_3$, Appl. Phys. Lett. **85**, 630 (2004)

[27] E. Bonera, G. Scarel, M. Fanciulli, P. Delugas, V. Fiorentini: Dielectric properties of high-$\kappa$ oxides: Theory and experiment for $Lu_2O_3$, Phys. Rev. Lett. **94**, 027602 (2005)

[28] L. Laversenne, Y. Guyot, C. Goutaudier, M. T. Cohen-Adad, G. Boulon: Optimization of spectroscopic properties of $Yb^{3+}$-doped refractory sesquioxides: Cubic $Y_2O_3$, $Lu_2O_3$, and monoclinic $Gd_2O_3$, Opt. Mater. **16**, 475 (2001)

[29] R. Vali, S. M. Hosseini: First-principles study of structural, dynamical, and dielectric properties of A-$La_2O_3$, Comp. Mat. Sci. **31**, 125 (2004)

[30] D. du Boulay: *Structure, Vibration and Electron Density in Neodymium-Iron-Boride and Some Rare-Earth Perovskite Oxides*, Ph.D. thesis, University of Western Australia, Perth (1999)

[31] P. Delugas, V. Fiorentini, A. Filippetti: Dielectric properties and long-wavelength optical modes of the high-$\kappa$ oxide $LaAlO_3$, Phys. Rev. B **71**, 134302 (2005)

[32] J. F. Scott: Raman study of trigonal-cubic phase transitions in rare-earth aluminates, Phys. Rev. **183**, 823 (1969)

[33] R. L. Sandstrom, E. A. Giess, W. J. Gallagher, A. Segmüller, E. I. Cooper, M. F. Chisholm, A. Gupta, S. Shinde, R. B. Laibowitz: Lanthanum gallate substrates for epitaxial high-temperature superconducting thin films, Appl. Phys. Lett. **53**, 1874 (1988)

[34] J. Krupka, R. G. Geyer, M. Kuhn, J. H. Hinken: Dielectric properties of single crystals of $Al_2O_3$, $LaAlO_3$, $NdGaO_3$, $SrTiO_3$, and MgO at cryogenic temperatures, IEEE Trans. Microwave Theory Tech. **42**, 2418 (1994)

[35] J. Konopka, I. Wolff: Dielectric properties of substrates for deposition of high-Tc films up to 40 GHz, IEEE Trans. Microwave Theory Tech. **40**, 2418 (1992)

[36] Y. G. Makeev, A. P. Motornenko, N. T. Cherpak, I. P. Babiichuk, M. B. Kosmyna: On the anysotropy of dielectric permittivity in single crystal lanthanum aluminate substrates, Tech. Phys. Lett. **28**, 221 (2002)

[37] A. Filippetti, N. A. Spaldin: Self-interaction-corrected pseudopotential scheme for magnetic and strongly-correlated systems, Phys. Rev. B **67**, 125109 (2003)

[38] A. Filippetti, N. A. Spaldin: Strong-correlation effects in born effective charges, Phys. Rev. B **68**, 045111 (2003)

[39] B. E. Park, H. Ishiwara: Formation of $LaAlO_3$ films on Si(100) substrates using molecular beam deposition, Appl. Phys. Lett. **82**, 1197 (2003)

[40] X. B. Lu, X. Zhang, R. Huang, H. B. Li, Z. H. Chen, W. F. Xiang, M. He, B. L. Cheng, H. W. Zhou, X. P. Wang, C. Z. Wang, B. Y. Nguyen: Thermal stability of $LaAlO_3$/Si deposited by laser molecular-beam epitaxy, Appl. Phys. Lett. **84**, 2620 (2004)

[41] W. Xiang, H. Lü, L. Yan, H. Guo, L. Liu, Y. Zhou, G. Yang, J. Jang, H. Cheng, Z. Chen: Characteristics of $LaAlO_3$/Si(100) deposited under various oxygen pressures, J. Appl. Phys. **93**, 533 (2003)

[42] L. F. Edge, D. G. Schlom, S. A. Chambers, E. Cicerrella, J. L. Freeouf, B. Holländer, J. Schubert: Measurement of the band offsets between amorphous $LaAlO_3$ and silicon, Appl. Phys. Lett. **84**, 726 (2004)

[43] A. Pasquarello, R. Car: Dynamical charge tensor and infrared spectrum of amorphous $SiO_2$, Phys. Rev. Lett. **79**, 1766–1769 (1997)

[44] S. Shevlin, A. Curioni, W. Andreoni: Ab initio design of high-$\kappa$ dielectrics: La$_x$Y$_{1-x}$AlO$_3$, Phys. Rev. Lett. **94**, 146401 (2005)

# Index

# Charge Traps in High-$k$ Dielectrics: Ab Initio Study of Defects in Pr-Based Materials

Jarek Dąbrowski[1], Andrzej Fleszar[2], Gunther Lippert[1], Grzegorz Lupina[1], Anil Mane[1], and Christian Wenger[1]

[1] Institute for Semiconductor Physics, IHP, Im Technologiepark 25, D-15236 Frankfurt (Oder), Germany
   dabrowski@ihp-microelectronics.com
[2] University of Würzburg, Am Hubland, 97074 Würzburg, Germany

**Abstract.** In the nearest future, a dielectric with a dielectric constant $k$ several times higher than that of $SiO_2$ will be needed for the fabrication of CMOS (Complementary Metal-Oxide-Semiconductor) devices. Numerous metal oxides and silicates are investigated as candidates and various deposition and annealing techniques are being developed to improve the film quality. These techniques try to utilize the effects attributed to alloying, incorporation of nitrogen, gettering of oxygen, etc. At the same time, the basic knowledge on the microscopic properties of these materials needs improvement, particularly in the case of rare-earth oxides.

We present our fundamental understanding of point defects in Pr-based dielectrics ($PrO_{1.5}$, $PrO_2$, $PrO_{1.75 + \delta}$, and $PrSiO_{3.5}$) in the context of their influence on the electrical properties of the Metal Oxide Semiconductor (MOS) stack. From this point of view, there are three major issues associated with the presence of point defects: bulk charge traps, Trap Assisted Tunneling (TAT) centers, and electrically active interface states. The paper focuses on the first of these issues, as seen from the perspective of ab initio total energy calculations for atomic and electronic structures of point defects. We discuss the dependence of point defect formation on the chemical potential of oxygen and the role of impurities such as moisture, silicon, and boron. In particular, we derive a model of Si-related fixed charge and argue that this model is valid also for typical high-$k$ dielectrics and for thermal $SiO_2/Si$ films.

## 1 Theoretical Approach

The calculations were done with the ab initio pseudopotential plane wave code fhi96md [1, 2]. We applied the Local Density Approximation (LDA) for the exchange and correlation energy [3, 4] and nonlocal pseudopotentials in the Trouller–Martins scheme [5, 6] with 40 Ry cutoff for plane waves. Since numerous defect structures have been investigated in this work and the typical cells have no symmetry but many atoms, a low-symmetry special $k$-point sampling scheme would require a prohibitively high numerical effort. The Brillouin zone was thus sampled at the $\Gamma$ $k$-point corresponding to the cell of dimensions as close as possible to the dimensions of the $PrO_{1.5}$ cell (cube

M. Fanciulli, G. Scarel (Eds.): Rare Earth Oxide Thin Films,
Topics Appl. Physics **106**, 247–268 (2007)
© Springer-Verlag Berlin Heidelberg 2007

with the lattice constant of $\sim 1\,\mathrm{nm}$). Tests with more converged samplings indicate that although more exact calculations are needed to confirm the quantitative results, the qualitative picture presented here is valid.

Because of the open $f$ shell of Pr atoms, a key problem in calculations involving Pr is the construction of a reliable Pr pseudopotential [7, 8]. It turns out that in practice two different Pr pseudopotentials are needed: a pseudopotential with two core $f$ electrons for $PrO_{1.5}$ (trivalent Pr(III), +3 ionic charge), and with only one $f$ electron for $PrO_2$ (tetravalent Pr(IV), +4 ionic charge). Thus, $Pr^{+3}$ and $Pr^{+4}$ are treated by us as distinct species. We calibrate the pseudopotential energy difference such that the experimental difference in the formation enthalpies of $PrO_{1.5}$ and $PrO_2$ is reproduced [9,10]; similarly, the chemical potential of Pr metal is not calculated solely from first principles but obtained from the computed total energy of $PrO_{1.5}$ and the experimental formation enthalpy of this compound, whereby the chemical potential of $O_2$ is adjusted such that the experimental formation enthalpy of $SiO_2$ is reproduced by the calculation (the required correction is of the order of $0.3\,\mathrm{eV}$). The fundamental bulk properties (lattice constant, bulk modulus) of $PrO_{1.5}$ and $PrO_2$ obtained with the Pr pseudopotentials used here are in agreement with experimental data; the discrepancies are well within the range typical for LDA calculations. The drawback of the approach adapted here for Pr pseudopotentials is that the wavefunctions responsible for the $(4+/3+)$ electron transition level of Pr cannot be computed. On the other hand, since all $f$ electrons are now contained in the core, we are not troubled by the fact that, by the electron counting rule, $PrO_{1.5}$ would be an $f$-band metal in LDA (but it is an insulator due to strong electron correlation effects in the $f$ shell).

Defect formation energies in a compound (and, generally, impurity formation energies) depend on the chemical potentials of the components. In particular, for native defects in Pr oxides we have, for $X = (Pr, O)$:

$$\text{X interstitial:} \ G_f(X_I) = G_f^o(X_I) - \mu(X)\,, \tag{1}$$

$$\text{X vacancy:} \ G_f(X_V) = G_f^o(X_I) + \mu(X)\,, \tag{2}$$

$$\text{equilibrium with } PrO_{1.5}: \ \mu(Pr) + 1.5\mu(O) = G_f^o(PrO_{1.5})\,, \tag{3}$$

where $G_f^o$ are the standard (i.e., corresponding to room temperature and atmospheric pressure) free energies of formation. Since we compute total energies at zero Kelvin, we use for calibration and comparison with experiment the corresponding formation enthalpies, rather than free energies.

The important regimes of the chemical potential of oxygen are:

$$PrO_{1.5} \text{ in contact with Pr metal:} \ \mu(O) = G_f^o(PrO_{1.5})/1.5$$

$$PrO_{1.5} \text{ in contact with } SiO_2/Si: \ \mu(O) = G_f^o(SiO_2)/2$$

$$PrO_{1.5} \text{ in contact with } PrO_2: \ \mu(O) = 2(G_f^o(PrO_2) - G_f^o(PrO_{1.5}))$$

$$PrO_2 \text{ in contact with air}: \ \mu(O) = 0$$

Since point defects are usually charged, we must also consider the dependence of the defect formation energy on the electron chemical potential, that is, on the Fermi energy $E_F$:

$$\text{positive charge } n_+ > 0: \quad G_f(n_+, E_{X_F}) = G_f(n_+, 0) + n_+ E_{X_F} \tag{4}$$

$$\text{negative charge } -n_- < 0: \quad G_f(n_-, E_{X_F}) = G_f(n_-, 0) - n_- E_{X_F}. \tag{5}$$

This means that the formation energy of charged defects in a dielectric in electrical contact with the Si substrate is determined by the position of the Fermi level in the substrate and by the valence band offset between Si and the dielectric. Since the latter is affected by the electrical dipole moment at the dielectric/substrate interface, the chemical character and electrical quality of the interface may have a noticeable effect on the defect formation energies and, consequently, on the defect population in the dielectric film.

Here we assume the valence band offset between Si and $PrO_{1.5}$ to be about 1.8 eV. One should keep in mind that this is an estimated value.

## 2 Charge Traps

Defective sites in the atomic structure of a dielectric are often capable of trapping electrical charges. The topology of a perfect lattice of a crystalline or amorphous material is always such that the local charge neutrality condition is fulfilled: if some atoms loose electrons and acquire a positive charge, then other atoms in their immediate neighborhood collect these electrons and acquire a compensating negative charge. The presence of a defect may disturb this balance. For example, if the defect consists of an atom that wants to get rid of one more electron than the neighboring atoms are willing to accept, this additional electron becomes distributed among the available states according to the Fermi statistics. This may mean that the defect site is forced to keep this electron; in this case, the defect remains electrically neutral. Alternatively, the electron may move away from its parent defect to another defect in the host; in this case, the former defect traps a hole and the latter traps an electron. The electron may also be bonded by the Coulomb potential of the trapped hole in the vicinity of the parent defect, or delocalized in the conduction band, or removed altogether from the bulk of the host material.

### 2.1 Trap Assisted Tunneling Centers

As an example, consider an unsaturated, free dangling bond of a substitutional $Si_{Pr}$ atom in $PrO_{1.5}$. The dangling bond is there because, in the surrounding of a PrIII atom in $PrO_{1.5}$, there is enough oxygen just to accept three electrons from the PrIII atom, i.e., there is one oxygen missing to saturate all valences of Si. The dangling bond of $Si_{Pr}$ introduces charge transition states[1], $(+/0)$ and $(0/-)$, at Fermi energies in the range of the Si band gap. This means that $Si_{Pr}$ can change its charge state when external voltage is applied. In other words, it can act as a Trap Assisted Tunneling (TAT) center, enhancing leakage currents across the dielectric. Carriers can tunnel, using such a defect as an intermediate step. The result is that the effective length of the potential barrier is significantly reduced.

## 2.2 Fixed Charges

The charge is considered as fixed when the carrier has been removed from the dielectric (to the substrate or to the gate) and cannot be put back to the defect by applying a voltage across the dielectric. A fixed dipole moment (and a corresponding fixed built-in voltage) is created in this way. Since the magnitude of the built-in voltage does not depend on the applied voltage, the presence of such a fixed charge is revealed by a shift of the CV characteristic.

Fixed charges are problematic because the built-in potential affects the electrical performance of the transistor. Built-in voltages comparable in magnitude to the electrical potential difference needed to switch a classical $SiO_2$-based transistor between its ON and OFF states are unacceptable. In addition, the electrical field produced by randomly distributed charges in the dielectric extends to the MOSFET channel where it enhances the scattering of carriers, reducing the mobility and, consequently, the saturation current and the working speed of the transistor. In practice, the areal density of the charged species should not exceed $10^{11}\,cm^{-2}$ (in industrial $SiO_2$ gate dielectrics, it is of the order of few times $10^{10}\,cm^{-2}$).

In Sect. 3, we identify the conditions when native defects (interstitials, vacancies) are expected to act as an important source of charge traps. The magnitude and sign of the charge due to native defects depend on the presence of a capping layer during post-deposition annealing and on the affinity of this layer to oxygen. In Sect. 4, we consider the influence of silicon impurity on the electrical quality of the film. When the film has a chemical contact to the silicon substrate and at the same time to $O_2$ molecules, silicon contamination may easily occur. Silicon is likely to trap a positive fixed charge. This is because a Si dangling bond has a tendency to pair with an oxygen atom from the network, producing an over-coordinated oxygen atom. Such a defect closely resembles a Valence Alternation (VA) defect $O_3^+$ known in chalcogenides and in $SiO_2$ [11–15]. In silicates it is a native defect, formally corresponding to half an oxygen vacancy, as in $SiO_2$. This dangling bond may become oxidized and then it turns into a negative fixed charge. The sign of the charge introduced by Si contamination thus depends on the balance between the injection of under-oxidized Si from the substrate and the oxidation of this Si by oxygen from the ambient.

---

[1] A charge transition state $(n/m)$ corresponds to the Fermi energy at which charge states $n$ and $m$ have the same formation energy.

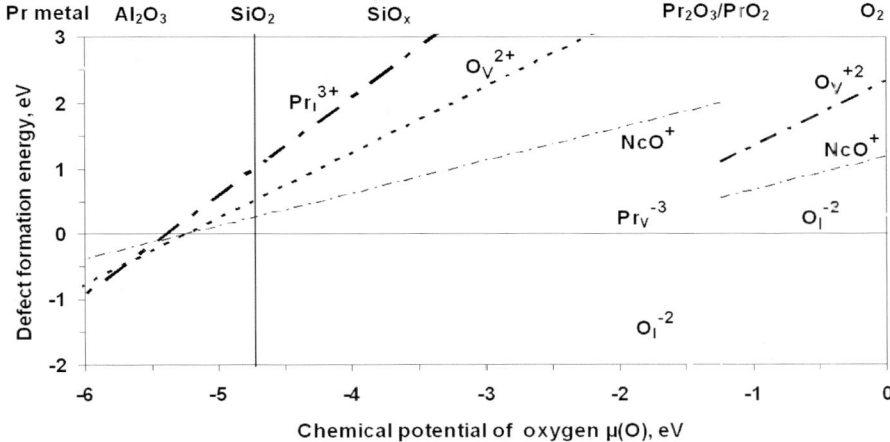

**Fig. 1.** Formation energies of native defects in Pr oxides as a function of the chemical potential of oxygen. The Fermi level corresponds to intrinsic Si. A value of 1.8 eV is assumed for the valence band offset between Si and $PrO_{1.5}$, and 1.0 eV for the valence band offset between Si and $PrO_2$. The energy of the intrinsic $O_3^+$ is estimated as half of that of the oxygen vacancy. Negatively charged defects are indicated by *gray broken* and positively charged defects by *black dash-dot lines*. In $PrO_2$, $O_I^{-2}$ is shown for comparison with $PrO_{1.5}$, but $O_I^0$ has a lower energy

# 3 Native Point Defects

## 3.1 Native Defects in $Pr_2O_3$

The native point defects in Pr oxides are oxygen and praseodymium interstitials and vacancies. In addition, valence alternation defects may be possible in amorphous material or at extended defect sites (grain boundaries, dislocations). The dependence of the defect formation energies on the chemical potential $\mu(O)$ of oxygen is given in Fig. 1. The Fermi level is aligned with its position in intrinsic Si. Slightly doped Si ($10^{16}\,\mathrm{cm}^{-3}$) is nearly intrinsic already for temperatures around 300 °C, that is, Fig. 1 corresponds to thermodynamical equilibrium at the processing temperature.

When $\mu(O)$ is close to the equilibrium between $PrO_{1.5}$, oxygen and Pr metal, oxygen vacancies $O_V^{2+}$ and Pr interstitials $Pr_I^{3+}$ are energetically favorable. This means that Pr is inserted from, e.g., a metallic Pr film deposited on top of $PrO_{1.5}$, and that this film drains oxygen from the oxide, producing oxygen vacancies[2]. The concentration of charged defects increases until the energy loss due to electrostatic repulsion compensates the energy gain due

---

[2] For the Fermi level corresponding to that of intrinsic Si, $O_V$ exists as $O_V^{2+}$. When the Fermi energy increases towards the bottom of the conduction band in Si and beyond, $O_V$ may trap electrons and change to $O_V^+$, $O_V^0$, or maybe even $O_V^-$.

to the formation of $O_V^{2+}$ and $Pr_I^{3+}$. A rough estimate for a 10 nm thick film gives the upper limit for this concentration in the range of $10^{13}$ cm$^{-2}$.

According to (1), the formation energies of these oxygen deficiency defects increase with increasing $\mu(O)$. When $\mu(O)$ approaches the value corresponding to the equilibrium between oxygen, aluminum[3] and $Al_2O_3$, $O_V^{2+}$ formation energy is still negative, meaning that an aluminum film has the potential to create oxygen vacancies in $PrO_{1.5}$ if the temperature is high enough to overcome the energy barriers on the reaction path. When $\mu(O)$ comes into the range close to the equilibrium between oxygen, Si and relaxed $SiO_2$, the formation energies of the positively charged native defects in $PrO_{1.5}$ are positive. This positive formation energy means that it is energetically unfavorable to generate oxygen deficiency defects in $PrO_{1.5}$ and to use the oxygen liberated in this way from $PrO_{1.5}$ to extend the thickness of a relaxed $SiO_2$ film (this process would occur by oxidizing the interface between $SiO_2$ and Si). However, the formation energy of $O_V^{2+}$ is still quite low, of the order of 0.5 eV. If the film is about 10 nm thick and the temperature is around 600 °C, such a formation energy would correspond to about $10^{12}$ cm$^{-2}$ defects, which is a significant amount of positive charge.

The subsequent range of $\mu(O)$ is that of the equilibrium between oxygen, Si and strained $SiO_x$, and extends to about 1 eV above equilibrium with relaxed $SiO_2$. We estimated this range by computing the energies of oxygen atoms inserted into Si–Si bonds at or in the first two Si layers below the interface between Si(001) and $PrO_{1.5}(110)$. At these conditions, $O_V^{2+}$ has a formation energy already too high to be of considerable importance. On the other hand, we may roughly estimate the formation energy of the native VA defect that may appear in non-crystalline regions of the film, an over-coordinated O atom, as half of that of $O_V^{2+}$. Indeed, imagine that a $O_O^0(2-)$ atom[4] is removed from the lattice and two other $O_O^0(2-)$ atoms far away are substituted by two $N_O^-(3-)$ atoms, so that the crystal remains electrically neutral. In other words, we have produced 1. a positively charged $O_V^{2+}$ with under-coordinated Pr atoms and 2. two compensating, negatively charged $N_O^-(3-)$ centers with under-coordinated N atoms. If we now allow the lattice to amorphize, it is plausible that the atoms will re-arrange in such a way that there will be no valence alternation of Pr and N. Simply, the immediate surrounding of each of the two (distant) N atoms will become similar to that in PrN and the oxygen vacancy will disappear. In other words, the $N_O(3-)$ atom feels now like a $N_N^0(3-)$ atom in a network of amorphous PrN. All valances are satisfied. If we now change the $N_N^0(3-)$ atoms back into oxygen, we will obtain a $O_N^0(3-)$ *defect*. This is a typical donor situation, with an over-coordinated oxygen atom occupying a nitrogen site. (For the reasons

---

[3] For example, when Al contact has been deposited on the film.

[4] In order to distinguish between the charge of a defect and a charge (or oxidation state) of an atom, we place the latter charge in round brackets, while the former charge is indicated as usual, as a superscript.

explained in Sect. 3.4, we prefer to call such a defect a Nitrogen-Coordinated Oxygen, NcO in short, rather than an over-coordinated oxygen.) As a result, we have split the original $O_V^{2+}$ into a pair of $NcO^+$, with some energy gain due to the relaxation of the lattice. Thus, by taking the formation energy of NcO as half of that of oxygen vacancy, we are likely to overestimate the energy needed to create this defect, or underestimate the energy gain if the conditions are such that the formation of a certain concentration of oxygen vacancies is energetically favorable.

Returning now to the formation energy diagram in Fig. 1, we see that positive fixed charge may still be created through reduction of $PrO_{1.5}$ by the Si substrate and formation of $NcO^+$ fixed charges even when the potential of oxygen is determined by moderately strained $SiO_x$. However, in the upper range of the strained $SiO_x$ regime, the energy of oxygen interstitial, $O_I^{2-}$, becomes low enough to over-compensate with these negatively charged defects any positive charge due to $NcO^+$.

With further increase of $\mu(O)$, the formation energy of $O_I^{2-}$ is further reduced and becomes strongly negative for any chemical potential achievable under vacuum conditions. This means that any contact to oxygen vapor, even if this is residual oxygen in UHV chamber, results in oxygen incorporation and creation of negative fixed charge in the film. Negatively charged Pr vacancies, $Pr_V^{3-}$, are also stable under such conditions. In other words, when Pr vacancies are created during deposition, they cannot be completely annealed out by Post Deposition Annealing (PDA) of uncapped samples. In $PrO_{1.5}$, these defects ($O_I$, $Pr_V$) do not introduce deep charge transition levels in the region of the Si band gap nor in the region of about $1\,eV$ above and below it.

## 3.2 Selected Native Defects in PrO$_2$

When $\mu(O)$ approaches that of $O_2$ under atmospheric pressure, $PrO_{1.5}$ oxidizes to $PrO_2$ and the defect formation energies change. Here we consider the defects on the oxygen sublattice: oxygen vacancies and oxygen interstitials.

The formation energy of $O_I^{2-}$ is significantly higher in $PrO_2$ than in $PrO_{1.5}$. This is because $O_I$ in $PrO_{1.5}$ occupy these sites from which oxygen is removed when the stoichiometry changes from $PrO_2$ to $PrO_{1.5}$. In $PrO_2$, these low-energy sites are filled with oxygen and the interstitial atoms must go to less convenient locations where they experience a remarkable compressive stress from the crystalline neighborhood. The electrically neutral split-interstitial configuration $O_I^0$ with a peroxy bond to the lattice oxygen becomes more favorable than $O_I^{2-}$ in the whole energy range of Si band gap.

On the other hand, the formation energy of oxygen vacancies $O_V^{+2}$ is significantly lower in $PrO_2$ than in $PrO_{1.5}$. This is apparently associated with the tendency of Pr oxides to form intermediate $PrO_{1.75+\delta}$ phases which are structurally equivalent to $PrO_2$ with an (ordered) array of oxygen vacancies [16]. Annealing in weakly oxidizing ambients might therefore lead to the formation of positive fixed charge, particularly that associated with $NcO^+$.

## 3.3 Selected Native Defects in $PrO_{1.75+\delta}$

Finally, we turn attention to the case when the oxygen potential is close to the boundary between $PrO_{1.5}$ and $PrO_2$. In this range, Pr oxide exists as $PrO_{1.75+\delta}$ in a number of compositions $x = 1.75 + \delta$ corresponding to $n \cdot (PrO_{1.5}) m \cdot (PrO_2)$, that is, there are both PrIII and PrIV ions in the oxide. The atomic structure of the mixed crystals is that of $PrO_2$ with an ordered array of oxygen vacancies [16]. The electrostatic charge of an oxygen vacancy in $PrO_{1.75+\delta}$ is compensated locally by PrIII ions.

A similar self-compensation effect is expected when oxygen-sublattice defects are created in a mixed crystal. Consider the case of O interstitials. (We exclude from this discussion the neutral split-interstitial stable in $PrO_2$, since its atomic and electronic structure is different from that of $O_I^{2-}$ and the self-compensation occurs through formation of a peroxy bond.) Self-compensation of $O_I^{2-}$ by conversion of some $Pr^{+4}$ (PrIII) to $Pr^{+4}$ (PrIV) takes place only when the energy lost due to the electrostatic repulsion between positively charged defects exceeds the energy gain due to the capture of electrons from Pr to oxygen acceptors. More precisely, the electrons captured at the $O_I^{-2}$ acceptor may arrive either locally from PrIII neighbors or from the substrate. In $PrO_{1.5}$ films grown on Si, the latter source is more favorable. As the oxygen content $x$ increases towards 1.75, the $f$ shell of PrIII experiences increasing repulsion from the electrostatic charge of O atoms and becomes an ever more competitive source of electrons. This means that when a $PrO_{1.5}$ film containing a negative fixed charge in form of $O_I^{2-}$ is annealed in a weakly oxidizing atmosphere, the oxygen content in the film increases and a part of $O_I$ is self-compensated: the electrons are taken locally from PrIII ions converting them to PrIV ions and the material becomes $PrO_{1.75+\delta}$.

In principle, one cannot exclude that self-compensation of $O_I$ will produce in $PrO_{1.75+\delta}$ an overall positive fixed charge coming from $NcO^+$ defects. This self-compensation may occur either by PrIII to PrIV conversion discused above, or by the formation of a peroxy bond mentioned in the previous section. Still, if the energy of $NcO^+$ is indeed close to that estimated in Fig. 1, the concentration of these defects in $PrO_{1.75+\delta}$ after PDA should be small. The anticipated formation energy of $NcO^+$ in $PrO_{1.75+\delta}$ is higher than that in $PrO_2$, and already in $PrO_2$ this energy in the PDA regime of $\mu(O)$ barely comes down to the values of practical importance (about $0.5\,eV$).

Thus, the film is expected to remain negatively charged even if it is partially oxidized to $PrO_x$ with $x$ increased towards 1.75, i.e., even when the oxide becomes a mixture of the majority component $PrO_{1.5}$ (with $Pr^{+3}$ ions) and the minority component $PrO_2$ (with $Pr^{+4}$ ions). Such a negative charge, typical for $PrO_{1.5}/Si(001)$ MBE films grown by us, is also consistent with the fact that $PrO_x$ crystals are typically p-type when $x$ is below approximately 1.75 [17]. Only when the chemical potential of oxygen is as low as that determined by the exchange of O atoms with $SiO_2$ (e.g., when the film is annealed under a Si cap) can a positive fixed charge be created in the film (of $PrO_{1.5}$

stoichiometry by now) by classical native point defects. This occurs mostly by oxygen vacancies $O_V^{+2}$. In amorphous material and along grain boundaries and dislocations, oxygen half-vacancies $O_3^+$ (over-coordinated oxygen) may become the major intrinsic source of positive charges. Finally, when the film is brought into contact with Pr metal, positively charged Pr interstitial atoms can be injected into $PrO_{1.5}$. In this metal-rich extreme, there is no significant equilibrium concentration of native defects carrying negative charges.

### 3.4 Selected Native Defects in $Pr_2Si_2O_7$

Figure 2 summarizes the dependence of some native defects in $PrSiO_{3.5}$ on the chemical potential of oxygen. An important difference between native point defects in $PrO_{1.5}$ and in $PrSiO_{3.5}$ is that the oxygen vacancy is not a charged defect in the latter material. This is because O atoms in $PrSiO_{3.5}$ have either only Si neighbors or both Si and Pr neighbors. The O vacancy created by removal of an O atom from between two Si atoms results in a formation of an electrically neutral SiSi bond; this defect is labeled $O_V$(SiSi) in Fig. 2. (One expects that this bridge is a precursor of the family of defects similar to the $E'$ family in $SiO_2$). The removal of an O atom from a site where it had, besides a Si neighbor, also Pr neighbors results in the formation of a Si dangling bond; this defect is labeled $O_V$(SiPr) in Fig. 2. Its formation energy is nearly equal to that of $O_V$(SiSi). Although this dangling bond does capture an electron, the electron comes locally from the metal neighbors: the negative charge localized now at the Si dangling bond is the same as the charge that was collected from the metal atoms by the removed O atom. The electrons in the negatively charged dangling bond are strongly bound in the electric field, occupying an orbital located about 1 eV below the valence band of Si (given the computed valence band offset between Si and $PrSiO_{3.5}$ of 2.7 eV). The dangling bond orbital has thus no states in the energy range of interest.

On the other hand, if an isolated Si dangling bond Si(db) is created far enough from Pr atoms, typical dangling bond transition states are produced. We have built a model of such a dangling bond (Fig. 3, left), starting from the $O_V$(SiSi) Si bridge configuration. We stretch and break the bridge by moving one of the Si neighbors along the SiSi bond until this atom arrives at the $sp^3$ site on the other side of the plane defined by its three O neighbors. We then saturate this atom with H, and we treat the dangling bond on the other Si atom of the broken bridge as a free Si dangling bond.

The computed formation energy of such a defect consists of two major contributions: one is the formation energy of the dangling bond, the other is the formation energy of the $O_3SiH$ defect. In Fig. 2, we adjusted this formation energy to represent the free dangling bond. We did this by adding a corrective term $\Delta E$ bringing it to the value of 1.1 eV at $\mu(0)$ corresponding to the equilibrium with $SiO_2$. Indeed, since the oxygen atom was removed from a configuration closely resembling that in $SiO_2$ and – for this choice

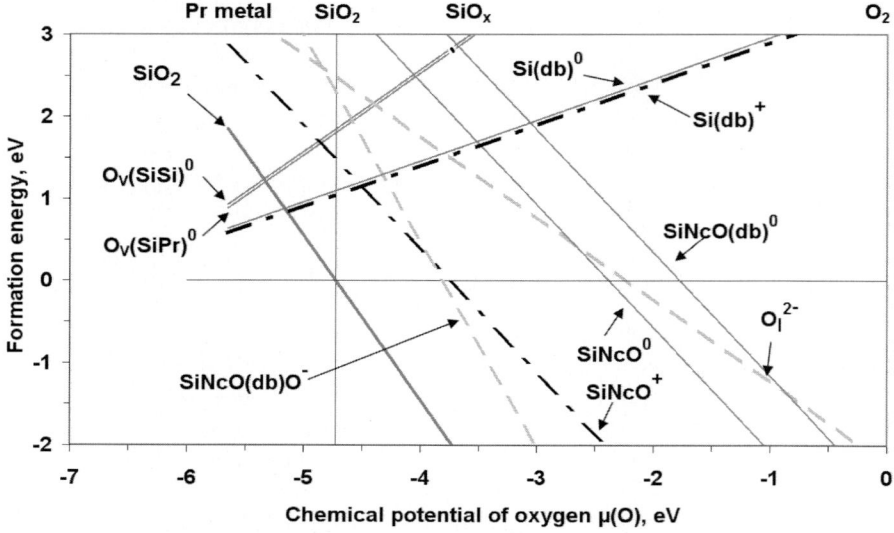

**Fig. 2.** Formation energies of selected native defects in $PrSiO_{3.5}$ as a function of the chemical potential of oxygen. For the chemical potential of Si, equilibrium with bulk Si is assumed. The Fermi level corresponds to that of intrinsic Si. Negatively charged defects are indicated by *gray broken lines*, electrically neutral defects – by *solid lines*, and positively charged defects – by *black dash-dotted lines*. The formation energies of Si(db) and SiNcO have been corrected by $\Delta E$, and $(0/+)$ transition state of Si(db) has been assumed to be located in the energy corresponding to the middle of Si band gap, as explained in the text. The line labeled "$SiO_2$" refers to the no-defect case when a Si atom is taken from bulk Si and oxidized to $SiO_2$

of the chemical potential – was moved to $SiO_2$, the formation energy of the (uncharged) dangling bond should be roughly equal to half of the SiSi bond energy, that is, to 1.1 eV. The same correction $\Delta E$ will be used below to estimate the formation energy of the NcO defect associated with a Si dangling bond. Assuming that our estimate of the Si(db) formation energy is correct, from the energies in Fig. 2 we conclude that, in general, oxygen vacancies in $PrSiO_{3.5}$ do not dissociate into isolated Si(db) dangling bonds. The vacancies may be created when the silicate is in chemical contact with a strongly reducing medium, e.g., with Pr or Ti metal layer.

Another important difference between $PrO_{1.5}$ and $PrSiO_{3.5}$ is that the formation energy of $O_I^{-2}$ is significantly higher in $PrSiO_{3.5}$. The reason is similar as in the case of $PrO_2$: in contrast to $PrO_{1.5}$, there is no interstitial site in $PrSiO_{3.5}$ that is natural for oxygen to fill.

Since silicon is an element native to silicates, Si-related defects (vacancies, interstitials, antisites, VA defects) should be also treated as native defects in $PrSiO_{3.5}$. Above, we considered an isolated Si dangling bond. Now, we turn

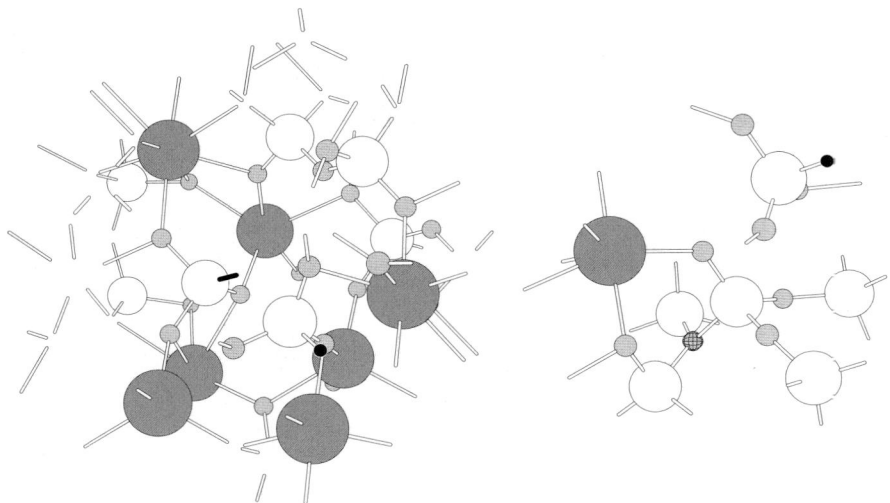

**Fig. 3.** Defects descending from oxygen vacancy in PrSiO$_{3.5}$. (*Left*) Free Si dangling bond, Si(db). (*Right*) Si-related Nitrogen-Coordinated Oxygen (SiNcO$^+$) defect as valence alternation (VA) over-coordinated oxygen O$_3^+$. Si atoms are *white*, O atoms are *light gray*, Pr atoms are *dark gray*, the H atom saturating the spurious dangling bond is *black*, the NcO atom is indicated by *grid lines*, and the dangling bond is rendered in *black*. In order to make the figure more readible, some atoms are removed from the panels (the bonds are retained)

attention to an interstitial SiO associated with valence alternation on the oxygen sublattice.

In PrSiO$_{3.5}$, each Si atom is bonded to four oxygen atoms and each oxygen atom has either one Si neighbor and two Pr neighbors or just two Si neighbors. If the local geometry of an oxygen atom is changed such that the number and location of neighbors ready to give its electrons away to this oxygen (group VI) would satisfy the valences of a nitrogen (group V) atom, a valence alternation defect is created. As noted already in Sect. 3.1, such a defect is expected to behave as a fixed positive charge because it has a surplus electron and the positive electrostatic potential binding this electron is likely to resemble the positive potential binding an electron on a typical "shallow" donor[5]. In SiO$_2$, where each regular O atom in the network has two neighbors, such defects are denoted as O$_3^+$, indicating the over-coordination (3 instead of 2) of the oxygen atom and its donor character (positive charge). In PrSiO$_{3.5}$, one should refer to this defect as O$_3^+$, O$_4^+$ or O$_5^+$, depending on

---

[5] The binding energy of a carrier in Coulomb potential is proportional to the effective mass of the carrier and reversely proportional to the square of the high-frequency dielectric constant. Since in oxides the high-frequency dielectric constant is rather small and the effective masses are rather large, "shallow" states there have much larger binding energies than shallow states in Si or in GaAs.

the oxygen site and on the origin of the valence alternation (a Si atom or two Pr atoms). Instead, we will use the name Nitrogen-Coordinated Oxygen (NcO), as it gives a valid description of such a defect not only in all its configurations in $PrSiO_{3.5}$ but also in other compounds, e.g., in $PrO_{1.5}$. NcO atoms associated with Si dangling bonds will be named SiNcO for short.

A model of SiNcO in the $O_3^+$ geometry (the over-coordinated O atom has only Si neighbors) is easy to construct in $PrSiO_{3.5}$. It suffices to insert a SiO moiety into the interstitial site between SiOSi bridges in the lattice and switch some bonds in the vicinity of the inserted atom. As a result, one obtains $O_3^+$ and a dangling bond on a Si atom located somewhat further away from the defect core (Fig. 3, right). We saturate this dangling bond with hydrogen. This is associated with a distortion similar to that introduced in our model of a free Si dangling bond, Si(db). Therefore, we have added the corrective term $\Delta E$ to the computed formation energy of SiNcO.

Our calculations show that SiNcO is indeed a fixed charge. Although it may in principle capture an electron and transform to the dangling bond configuration SiNcO(db), the energy of the dangling bond in SiNcO(db) geometry is much higher than the energy of a free (relaxed) Si dangling bond in $PrSiO_{3.5}$. This is because the transformation of SiNcO to SiNcO(db) means that a Si atom has to move to the opposite side of its three O neighbors. Since the lattice of $SiO_2$ is relatively dense, such a motion places the dangling bond in an unfavorable position. Electron capture is here more difficult than on a free dangling bond, where no strong distortion is needed. As a consequence, the $(0/+)$ electron transition state of SiNcO is approximately 1 eV above the $(0/+)$ transition state of the free dangling bond; in $PrSiO_{3.5}$ films on Si, the latter lies probably above the middle of Si band gap.

The formation energy of $SiNcO^+$ (Fig. 2) is negative in all oxidizing ambients and is positive but small in most of the $SiO_x$ range. This means that newly oxidized Si tends to dissolve in the film. However, SiNcO may itself be oxidized: $SiNcO^+$ transforms to SiNcO(db) configuration, the oxygen is attached to the dangling bond, and the charge changes to single negative, producing electrically inactive $SiNcO(db)O^-$. On the other hand, the energy gain through oxidation of the substrate to $SiO_2$ is much higher than the energy gain through oxidation of SiNcO, and the concentration of surface Si atoms at the interface to the substrate is much higher than SiNcO concentration in the film. Therefore, it is not clear if efficient oxidation of SiNcO is possible without being accompanied by a substantial oxidation of the substrate.

## 3.5 Selected Native Defects in $SiO_2$

For completeness, we address the issue of the origin of intrinsic positive fixed charge in thermal $SiO_2$ films. One might argue that over-coordinated oxygen, $O_3^+$ (Fig. 4, left), is not a good candidate for a fixed charge in $SiO_2$ because

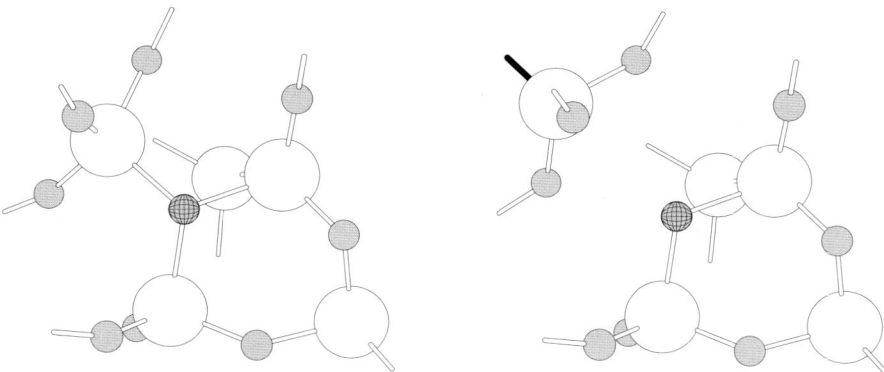

**Fig. 4.** (*Left*) Si-related Nitrogen-Coordinated Oxygen (SiNcO$^+$) defect as valence alternation (VA) over-coordinated oxygen O$_3^+$ in amorphous SiO$_2$ and (*right*) its excited, electrically neutral configuration, SiNcO(db)$^0$, with a Si dangling bond. Si atoms are *white*, O atoms are *gray*, the NcO atom is indicated by *grid lines*, and the dangling bond is rendered in *black*

when an electron is trapped by its electrostatic potential, the defect is expected to transform to a dangling bond [14]. Indeed, our calculations indicate that the energy barrier for such a transformation is small, only about 0.5 eV. Since the (0/+) transition state of an isolated Si dangling bond in SiO$_2$ is within the range of Si band gap, it should be possible to neutralize the charge of O$_3^+$ by electron injection from the substrate, for example, in a tunnel DLTS experiment [18].

Nevertheless, one must note that this reasoning assumes that the dangling bond configuration obtained by neutralization of O$_3^+$ is completely relaxed. As a matter of fact, it is not completely relaxed, and this influences the energy position of the (0/+) transition state. This effect is similar to the one discussed in Sect. 3.4, although it is less pronounced in SiO$_2$ than in PrSiO$_{3.5}$. To quantitify it, we compared the position of the dangling bond (0/+) transition state for two configurations:

1. A fully relaxed dangling bond obtained by removing an oxygen from amorphous SiO$_2$. One of the Si neighbors was then left in its original site and contained the dangling bond, the other neighbor was forced across the plane of its remaining three O neighbors and then saturated with H. This leaves the dangling bond in a large volume, and the neighbors of the Si atom with the dangling bond retain the configuration optimal for a neutral defect.
2. A dangling bond in the ground-state configuration of O$_3^0$ (Fig. 4, right). Here, one of the Si neighbors of the over-coordinated oxygen atom has to move spontaneously across the plane of its three double-coordinated O neighbors. This leaves some strain in the bond angles. The configuration

where all Si neighbors remain on the same side of the oxygen planes has a higher energy (by approximately 0.5 eV) due to Pauli repulsion between the electron in the dangling bond and the closed shell of oxygen.

We find that $(0/+)$ transition state associated with the neutralized $O_3^+$ is approximately 1.0 eV above the $(0/+)$ transition state of the relaxed Si dangling bond in $SiO_2$. Therefore, it is plausible that $O_3^+$ acts as a positive fixed charge when located within the tunneling distance from the interface to Si (20–30 Å [19]). We also note that $O_3^+$ should appear in a natural way as a low-energy state of partially oxidized interstitial Si injected from the oxidizing interface during thermal oxidation.

# 4 Impurities

## 4.1 Moisture

Moisture is definitely the factor that has to be examined in the context of fixed charge formation [20]. As in other rare-earth oxides, $PrO_{1.5}$ readily absorbs water, to the extent that it is easily converted to a hydroxide. When a water molecule is dissolved in $PrO_{1.5}$, it dissociates into $OH_I^-$ interstitial and $H^+$. The latter becomes attached to a lattice oxygen atom, forming a defect which may be termed a substitutional OH group, $OH_O^+$. These defects do not introduce any localized states in the gap of $PrO_{1.5}$ and, since they have the opposite charge, a dissolved $H_2O$ molecule cannot act as a fixed charge.

Nevertheless, it is possible that the charge balance is affected by defect reactions in the film. For example, if the oxygen atom from dissociated $H_2O$ is used to oxidize silicon in the substrate to (fully relaxed) $SiO_2$, the $OH_I^-$ becomes converted to $OH_O^+$. Two positive fixed charges are thus created by each $H_2O$ molecule taking part in such a reaction. We calculated that this oxidation reaction is energetically favorable by about $\simeq 0.3$ eV per H atom in p-type Si. This means that, in principle, one cannot exclude that such processes take place in ultrathin $PrO_{1.5}/Si$ films exposed to moisture. The calculated formation energy of $OH_I^-$ in $PrO_{1.5}$ is also negative within a broad range of O potential in the oxidizing regime[6], meaning that water from air would be a source of negative fixed charge in $PrO_{1.5}$. The formation energy of $OH_O^+$ is sufficiently small only when the chemical potential of oxygen approaches the UHV range. Although the formation energy of $OH_O^+$ is negative in equilibrium with $SiO_2$, this does not seem to have a direct relevance to the fixed charge formation. Positively charged defects might be formed in this

---

[6] We assume here that the chemical potential of H corresponds to that in $H_2O$ remaining in thermodynamic equilibrium with H, that is, the sum of $2\mu(H)$ and $\mu(O)$ yields the formation energy of water.

way if, for example, a water-contaminated oxide is sealed with a Si layer and then annealed.

If single negatively charged defects (e.g., SiNcO(db)O$^-$) are present in the film, the hydrogen atoms from OH$_O^+$ are likely to neutralize them. We have not yet performed the calculations to verify this.

## 4.2 Silicon

High-$k$ dielectrics for MOSFET gate oxides are grown on Si substrates, and for that reason silicon contamination of the dielectric may occur. Since rare-earth oxides readily build silicates, one expects that when oxygen is available, Si can be admixed to the oxide in the form of SiO$_2$ moieties. Indeed, the formation energy of a SiO$_2$ interstitial in PrO$_{1.5}$ is low: the energy of a SiO$_2$ group is only by 1.3 eV higher in PrO$_{1.5}$ than in the network of amorphous SiO$_2$. In other words, it requires 1.3 eV to take a silicon atom from $\mu$(Si) corresponding to Si bulk and two oxygen atoms from $\mu$(O) corresponding to the equilibrium between Si, SiO$_2$ and oxygen, and to place these atoms in the PrO$_{1.5}$ crystal in the form of an interstitial SiO$_2$ molecule. Such a defect is electrically inactive and may be formally viewed as a (SiO$_4$)$^{4-}$ tetrahedron substituting a pair of O$^{2-}$ atoms in the lattice of PrO$_{1.5}$. We have attempted a rough estimate of the diffusion barrier of such a defect and obtained a value of about 2 eV. This value may be an underestimation because at the moment we have too few points on the diffusion path to guarantee that we identified the real barrier configuration.

When a Si atom alone is moved from Si bulk to the interstitial site in PrO$_{1.5}$, it preferably acquires the Si$_I^{2+}$ charge state. The formation energy of Si$_I^{2+}$ at the Fermi energy close to that of intrinsic silicon is approximately 3 eV. A process in which such an intestitial Si atom expulses a Pr interstitial and acquires a substitutional position, Si$_{Pr}^+$, needs then only about 0.5 eV (note that this is the reaction energy, not the energy barrier). When the Pr interstitial is finally brought to the surface of the film and oxidized there, the overall energy balance is determined by the chemical potential of oxygen and varies from energy loss of about 3 eV for $\mu$(O) in equilibrium with SiO$_2$ to energy gain of about 5 eV for $\mu$ in equilibrium with oxygen in air. In other words, the formation of substitutional Si$_{Pr}^+$ is strongly favored in oxidizing ambients and strongly disfavored when the sample is protected by a cap layer opaque to oxygen.

The amount of Si dissolved in PrO$_{1.5}$ may be limited kinetically by the energy barrier on the reaction path, that is, by the energy needed to inject Si interstitials into the film. Assuming that this barrier is close to the energy of Si$_I^{2+}$ and that the attempt frequency is of the order of $10^{13}$ Hz, one estimates that during one second about one atom per $10^4$ surface atoms is injected. A 30 min anneal at 600 °C would thus inject about two monolayers of Si, which translates into Si concentration of about 1 at.% in a 30 nm film. Yet, when the kinetic barrier is by 0.5 eV higher than the reaction energy (3 eV) for interstitial injection, the amount of injected Si drops by three orders of magnitude. This means that relatively small changes in the atomic structure in the interfacial region may result in massive changes in the amount of dissolved Si if the dissolution proceeds through spontaneous injection of interstitials.

Substitutional Si in $PrO_{1.5}$ [20] has a dangling bond which is electrically active: the defect can exist as $Si_{Pr}^+$, $Si_{Pr}^0$ and $Si_{Pr}^-$. Oxygen or OH dissolved in the film may deactivate it. The reaction with $O_I^{2-}$ changes the defect into a negative fixed charge, $SiO_{Pr}^-$, while the reaction with OH changes it into electrically neutral $SiOH_{Pr}^0$.

Another interesting configuration of Si in $PrO_{1.5}$ is analogous to the SiNcO configuration in $PrSiO_{3.5}$ and $O_3^+$ center in $SiO_2$, discussed in Sects. 3.4 and 3.5: three of the valence electrons of Si take part in the bonds with O as in the silicate, while the fourth is removed from the defect and the fourth $sp^3$ hybrid of Si makes a bond with an oxygen atom which already has a complete shell of neighbors. Such a geometry should be possible in amorphous $PrO_{1.5}$ or at extended defects (grain boundaries, dislocation cores). We will now estimate the formation energy of this defect.

Building an atomic structure of such a configuration embedded in amorphous network would be a tedious task. Instead, we have adopted two simplified models representing the limiting case of high and low strain introduced by the defect. We used crystalline $PrO_{1.5}$ and placed an interstitial $SiH_3$ molecule in two configuration: 1. as interstitial in the perfect crystal, and 2. in a $PrO_{1.5}$ void created in an otherwise perfect $PrO_{1.5}$ crystal (Fig. 5). In both cases, the defect behaves as expected: it delivers a positive fixed charge. However, the formation energies $E_{Si-NCO}$ (with respect to Si–Si bond in disilane, $Si_2H_6$) are very different and even their signs differ: when computed for the Fermi level corresponding to that of intrinsic Si, this amounts to 2.4 eV in the first configuration, and to $-1.2$ eV in the second configuration. The difference comes from the high compressive stress in the first configuration. The Si atom is forced to a site close to three Pr atoms (note that in spite of that, a regular bond with the oxygen atom is formed). In contrast to that, the second configuration allows the Si atom to find a place reasonably distant from the Pr neighbors without compromising the Si–O bond length. This leads to a significant lowering of the formation energy in the second configuration. We will treat these two formation energies as the upper- and lower-bound estimates of the bond energy between the dangling bond of Si and NcO atom in $PrO_{1.5}$.

In order to obtain the formation energy of SiNcO, we still need an estimate of the energy of a regular bond between Si atom and an O atom in $PrO_{1.5}$, such as in the case of a substitutional $Si_{Pr}$. It is not straightforward to use the formation energy of $Si_{Pr}$, as it contains a dangling bond. Therefore, we took another approach: we utilized the formation energy of $SiO_2$ interstitial. As noted before, a $SiO_2$ moiety inserted into $PrO_{1.5}$ moiety finds a configuration in which the Si atom is bonded to a tetrahedron of four O atoms. The

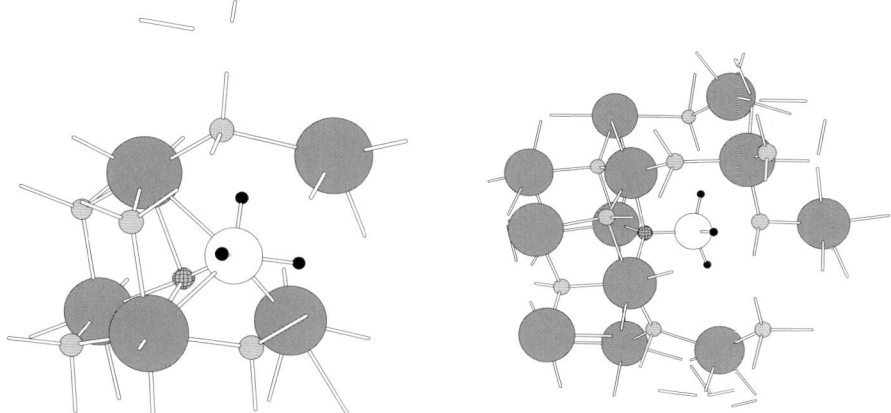

**Fig. 5.** Interstitial SiH$_3$ molecule bonded to lattice oxygen atom and serving as a model for SiNcO defect in PrO$_{1.5}$. (*Left*) SiNcO in a perfect crystal; the Si atom is forced to a close neighborhood with lattice Pr atoms. (*Right*) SiNcO in a void created by removal of a PrO$_{1.5}$ moiety from the crystal; the Si atom has enough free space. Pr atoms are *dark gray*, the Si atom is *white*, O atoms are *light gray* (the NcO atom is marked by *grid lines*), and H atoms in the SiH$_3$ molecule are *black*

formation energy $4E_{\text{Si}-\text{OPr}}$ of this defect with respect to SiO$_2$ is about 1.3 eV and corresponds to the formation of four regular Si$-$O bonds in PrO$_{1.5}$. Since the SiNcO defect has three such bonds and one SiNcO bond, we are now in a position to estimate its formation energy:

$$E_f(\text{SiNcO}) = E_{\text{Si}-\text{NCO}} + 3E_{\text{Si}-\text{OPr}}. \tag{6}$$

The result has been plotted as a function of the chemical potential of oxygen in Fig. 6. The two thin dash-dot lines indicate the upper and lower bounds, while the line SiNcO$^+$ corresponds to their arithmetic average.

We see that SiNcO$^+$ is the energetically most favorable defect in the regime of oxygen chemical potentials already slightly (0.5 eV) above the equilibrium with SiO$_2$, if the average value is taken as the estimate of its formation energy. Even in the pessimistic case (the upper-bound estimate), the formation energy of SiNcO is only slightly higher than the formation energy of O$_i^{2-}$ when the chemical potential of oxygen approaches the equilibrium between PrO$_{1.5}$ and PrO$_2$. Even more significantly, both formation energies (the upper limit for SiNcO$^+$ and the energy of O$_i^{1-}$) are clearly negative in this range of $\mu(\text{O})$, meaning that these defects are formed spontaneously.

Again, oxidation of SiNcO$^+$ to SiNcO(db)O$^-$ should be possible with energy gain similar to that found in PrSiO$_{3.5}$ (Sect. 3.4). As we argued in Sect. 3.4, this oxidation may be difficult without simultaneous oxidation of the silicon substrate. We think that negatively charged SiNcO(db)C$^-$ may be neutralized by hydrogen atoms from OH$_O^-$, but we have not yet performed

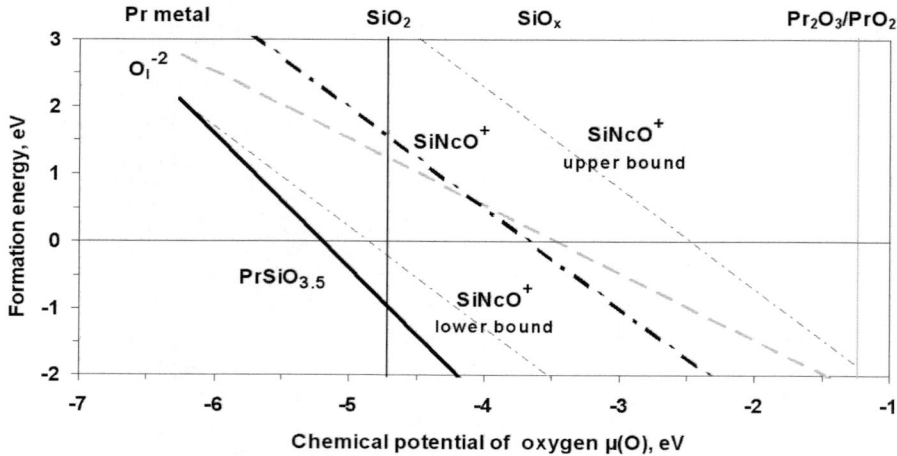

**Fig. 6.** (*Left*) Formation energies of SiNcO defects in $PrO_{1.5}$ as a function of the chemical potential of oxygen. For the chemical potential of Si, equilibrium with bulk Si is assumed. The Fermi level corresponds to that of intrinsic Si. Negatively charged defects are indicated by *gray broken lines*, electrically neutral defects are indicated by *solid lines*, and positively charged defects – by *black dash-dotted lines*. The *solid line* labeled "$PrSiO_{3.5}$" refers to the formation energy of crystalline $PrSiO_{3.5}$

the calculations for this process. The interaction between $SiNcO^+$ and $OH_I^-$ should also be investigated in order to complete the picture.

We expect that SiNcO formation energy is low also in other metal oxides. The reason is that the low energy of SiNcO is caused by the oxidation of the Si atom. Even when the SiNcO formation energy is estimated from quite unfavorable atomic configurations ($SiO_2$ interstitial in $PrO_{1.5}$ and $SiH_3$ interstitial in perfect $PrO_{1.5}$ lattice), the fact that Si−O bonds are formed overweighs the geometrical constraints associated with the particular defect site. Since the presence of metal neighbors to these O atoms affects the strength of this bond only moderately[7], the appearance of SiNcO as a major positive fixed charge source in, e.g., $HfO_2/Si(001)$, seems plausible. More to the point, $HfO_2$ is grown on Si in amorphous form on purpose, and this is the amorphous host that is the natural environment for the SiNcO atomic configuration.

### 4.3 Boron

Boron is the traditional acceptor used in MOSFET channels and in polysilicon gates. It is not a direct source of fixed charge in Pr oxides. The calculations show that boron atoms strongly segregate from Si to $PrO_{1.5}$, but they

---

[7] As proven by the fact that amorphous silicates are stable up to several hundred degrees C can be obtained not only with "silifilic" rare-earth oxides but also with "silifobic" materials such as $HfO_2$ and $TiO_2$.

substitute Pr in the lattice. As $B_{Pr}$, B atoms are isovalent impurities. They are electrically neutral and introduce no localized states, at least not in the hazardous energy region within approximately 1 eV to the band gap of Si.

Nevertheless, boron segregation may be responsible for fixed charge generation by kick-out of Pr interstitials, $Pr_I^{+3}$ by B interstitials, $B_I^+$. The kick-out is energetically favorable by 0.8 eV when the Fermi level is aligned with that of intrinsic Si. This means that each B atom that makes it to the oxide produces one $Pr_I^{+3}$ ion. If the annealing of an uncapped layer takes place in an atmosphere containing enough oxygen to oxidize these ions to $PrO_{1.5}$ (given Fig. 6, this is true also for all "inert" ambients), this effect is irrelevant. However, if the annealing takes place under a capping layer (as is likely to be the case during technological processing), the Pr interstitial atoms may remain unoxidized in the film. Such atoms would produce an electrically fixed charge that is particularly problematic because it may move across the dielectric in the electric field and render the device unreliable by causing a hysteresis in its CV characteristics.

### 4.4 Titanium

Hazardous positively charged interstitial metal atoms may be also injected from such metal films as titanium. The detailed discussion of the interaction between Ti and Pr oxides is beyond the scope of this paper, but we note that this injection becomes energetically unfavorable (by at least 2 eV) when the Ti source is oxidized to $TiO_2$. Nevertheless, nanosize inclusions of $TiO_2$ in $PrO_{1.5}$ and in $PrSiO_{3.5}$ may trap positive charge. Similarly to the charge trapped on $Si_{Pr}$, this positive charge is re-loadable and can be neutralized by electrons. This means that such inclusions act as Trap Assisted Tunneling centers, contributing to the leakage current flowing across the dielectric.

## 5 Summary and Conclusions

Given the results of ab initio total energy calculations, we have analyzed the influence of native point defects and certain impurities (Si, B, Ti, moisture) on the formation of charge traps in dielectric (mostly Pr-related) films on Si. Depending on the environment determining the chemical potential of oxygen, the charge may come from positively charged metal interstitials (fixed), positively charged O vacancies (possibly re-loadable) and half-vacancies, or negatively charged O interstitials (in $PrO_{1.5}$, fixed); moreover, negatively charged metal vacancies created during growth would lead to fixed charge. We argued that the Si substrate is a source of Si atoms which may give rise to positive fixed charge not only in $PrO_{1.5}$ but also in other high-$k$ metal oxides, including $HfO_2$, and in thermal $SiO_2$/Si films. The role of Si is ambivalent: depending on the sample preparation details (availability of oxygen and moisture, growth rate, annealing temperature, thickness of the dielectric film and of the interfacial layer), incorporation of Si may be the cause of positive fixed charge, negative fixed charge, electrically active states in the gap, or have no influence on the electrical properties of the material.

## Acknowledgements

We thank V. Fiorentini, G. Lucovsky, J. Robertson, Th. Schroeder, and R. Sorge for helpful discussions, comments, criticism and suggestions. The calculations have been done on IBM Regatta in von Neumann Institute for Computing, Jülich, Germany (project hfo06).

# References

[1] M. Bockstedte, A. Kley, J. Neugebauer, M. Scheffler: Density-functional theory calculations for poly-atomic systems: Electronic structure, static and elastic properties and ab initio molecular dynamics, Comp. Phys. Comm. **107**, 187 (1997)

[2] A. Fissel, J. Dąbrowski, H. J. Osten: Photoemission and ab initio theoretical study ot interface and film formation during epitaxial growth and annealing of praseodymium oxide on Si(001), J. Appl. Phys. **91**, 8986 (2002)

[3] D. M. Ceperley, B. J. Alder: Ground state of the electron gas by the stochastic method, Phys. Rev. Lett. **45**, 567 (1980)

[4] J. P. Perdew, A. Zunger: Self-interaction correction to density-functional approximation for many-electron systems, Phys. Rev. B **23**, 5048 (1981)

[5] D. R. Haman: Generalized norm-conserving pseudopotentials, Phys. Rev. B **40**, 2980 (1989)

[6] G. B. Bachelet, D. R. Hamann, M. A. Schlüter: Pseudopotentials that work: From H to Pu, Phys. Rev. B **26**, 4199 (1982)

[7] H.-J. Müssig, H. J. Osten, E. Bugiel, J. Dąbrowski, A. Fissel, T. Guminskaya, K. Ignatovich, J. P. Liu, B. P. Zaumsei, V. Zavodinsky: Epitaxial praseodymium oxide: A new high-$k$ dielectric, in *Proc. 2001 IEEE Integrated Reliability Workshop* (South Lake Tahoe, CA (USA) 2001) p. 1

[8] J. Dąbrowski, V. Zavodinsky, A. Fleszar: Pseudopotential study of $PrO_2$ and $HfO_2$ in fluorite phase, Microelectron. Reliab. **41**, 1093 (2001)

[9] H. Bergman: *Gmelin Handbuch der Anorganischen Chemie, Seltenerdelemente, Teil C1* (Springer, Berlin, Heidelberg 1974)

[10] D. R. Lide (Ed.): *Handbook of Chemistry and Physics*, 73rd ed. (CRC Press, Boca Raton, FL (USA) 1993–1994)

[11] N. F. Mott: Charged defects in vitreous silica, J. Non-Cryst. Solids **40**, 1 (1980)

[12] G. N. Greaves: Intrinsic and modified defect states in silica, J. Non-Cryst. Solids **32**, 295 (1979)

[13] G. Lucovsky: Spectroscopic evidence for valence-alternation-pair defect states in vitreous $SiO_2$, Phil. Mag. B **39**, 513 (1979)

[14] E. P. O'Reilly, J. Robertson: Theory of defects in vitreous silicon dioxide, Phys. Rev. B **27**, 3780 (1983)

[15] S. T. Pantelides, R. Buczko, M. Rammamoorthy, S. Rashkeev, G. Duscher, S. J. Pennycook: Local and global bonding at the Si–SiO$_2$ interface, in Y. J. Chabal (Ed.): *Fundamental Aspects of Silicon Oxidation* (Springer, Berlin, Heidelberg 2001) p. 107

[16] H. Inaba, K. Naito: Simultaneous measurements of oxygen pressure, composition, and electrical conductivity of praseodymium oxides: I. Pr$_7$O$_{12}$ and Pr$_9$O$_{16}$ phases, J. Solid State Chem. **50**, 100 (1983)

[17] G. V. Subba Rao, S. Ramdas, P. N. Mehrotra, C. N. R. Rao: Electrical transport in rare-earth oxides, J. Solid State Chem. **2**, 377 (1970)

[18] H. Lakhadari, D. Vuillaume, J. C. Bourgoin: Spatial and energetic distributions of Si–SiO$_2$ near-interface states, Phys. Rev. B **38**, 13124 (1988)

[19] D. Vuillaume, J. C. Bourgoin, M. Lannoo: Oxide traps in Si–SiO$_2$ structures characterized by tunnel emission with deep-level transient spectroscopy, Phys. Rev. B **34**, 1171 (1986)

[20] G. Lippert, J. Dąbrowski, V. Melnik, R. Sorge, C. Wenger, P. Zaumseil, H.-J. Müssig: Si segregation into Pr$_2$O$_3$ and La$_2$O$_3$ high-$k$ gate oxides, Appl. Phys. Lett. **86**, 042902 (2005)

# Index

# Experimental Determination of the Band Offset of Rare Earth Oxides on Various Semiconductors

Gabriele Seguini, Michele Perego, and Marco Fanciulli

CNR-INFM MDM National Laboratory, Via C. Olivetti, 2, 20041 Agrate Brianza (MI), Italy
gabriele.seguini@mdm.infm.it

**Abstract.** The critical role of gate oxide in ultra-scaled devices is being investigated in terms of the properties of rare earth oxides as high dielectric constant (high-$\kappa$) materials to replace $SiO_2$. In particular, the combination of rare earth oxides with high-mobility substrates, like Ge and GaAs, could offer the possibility to improve the interface properties. Among the different properties under investigation, the band alignment at the interface is a key issue because it affects the tunneling behavior of a device. Internal photoemission and X-ray photoelectron spectroscopy are useful techniques to experimentally determine the band offset at the semiconductor/oxide interface. After a detailed description of these two methods, we present a review of the data available in the literature on the interface of different high-$\kappa$ oxides on silicon. Finally, we report our measurements of the $Lu_2O_3$ band alignment on various semiconductor substrates. A conduction band offset value of 2.1 eV has been obtained by internal photoemission for $Lu_2O_3$ films grown on Si, Ge, and GaAs. X-ray photoelectron spectroscopy measurements of the valence band offset were performed on $Ge/Lu_2O_3$ heterojunction. The results are in excellent agreement with those obtained using internal photoemission.

## 1 Introduction

The continuous scaling down of metal-oxide-semiconductor (MOS) devices forces the microelectronic industry to explore the properties of high dielectric constant ($\kappa$) materials to replace $SiO_2$ as gate oxide. There are many requirements that must be fullfilled in order to single out the most suitable canditate. In particular, the material should exhibit a high dielectric constant as well as thermodynamical stability, good interface properties and gate compatibility [1]. Moreover, the effective mass of the tunneling electrons [2] and the band alignment at the interface [3] are very important because they strongly affect the reliability properties of the MOS device. One of the key issues affecting the electron tunneling is the height of the barrier energy at the interface between the oxide and the substrate. The conduction- and valence-band offset, CBO and VBO respectively, must be higher than at least 1 eV. Another problem that can arise due to the substitution of $SiO_2$ with a different gate oxide is the reduction of the mobility in the inversion layer of the substrate due to remote phonon scattering [4]. This effect motivates

M. Fanciulli, G. Scarel (Eds.): Rare Earth Oxide Thin Films,
Topics Appl. Physics **106**, 269–284 (2007)
© Springer-Verlag Berlin Heidelberg 2007

the investigation of high-mobility substrates such as Ge and GaAs. The poor properties of the native oxides of these semiconductors prevent the realization of MOS devices which could compete with those realized on Si. The successful growth of high-$\kappa$ oxides on Ge and III–V justifies the renewed interest for these materials.

During recent years, much effort has been aimed at calculating and/or measuring the CBO and VBO of high-$\kappa$ materials with different semiconductor substrates. Much of theoretical work has been published attempting to predict the band line-up at different dielectric/semiconductor interfaces, e.g., [3]. However, the experimental determination of the barrier energy at the oxide-semiconductor interface of a real MOS device is a key issue. Several experimental methods can be used to measure the bands offsets between the oxide and the substrate (Si, Ge, GaAs). Among the most important are Internal Photoemission (IPE), e.g., [5], X-ray Photoelectron Spectroscopy (XPS), e.g., [6], and Ballistic Electron Emission Spectroscopy (BEEM), e.g., [7].

IPE determines the band alignment by detecting the photocurrent that flows in a biased MOS device, with monochromatic radiation in the range of visible and near-ultraviolet (UV) energies. Furthermore, one can obtain information on the oxide band gap by examining the photoconductivity (PC) part of the spectra. Here, the signal reveals the band to band transitions in the oxide. XPS allows to obtain accurate measurements of the VBO. Studying the shallow core levels and the valence band spectra of the materials in their pure state and using these as a reference, the position of the valence band edges in a heterojunction can be precisely determined. Moreover, information on the band gap can be obtained by studying the loss features of the XPS spectra. BEEM uses the tip of a scanning tunneling microscope (STM) as source of hot electrons to be injected in a MOS device and allows a spatially resolved determination of the barrier energy from the analysis of the collector current.

In this work, we focus on the use of IPE and XPS to determine CBO and VBO of high-$\kappa$ materials with different semiconductor substrates. After a detailed descrition of these two methodologies, we present a summary of the results available in the literature. Finally, we present our own results on the band alignment of $Lu_2O_3$ with different semiconductor substrates.

# 2 Experimental Techniques for Band Offset Determination

## 2.1 Internal Photoemission

The physical process used in IPE spectroscopy is the penetration of the barrier energy at the interface of the optically excited charge carriers in the emitter. Pioneer work was carried out at the end of the 1960s by *Williams* [8], *Goodman* [9–11], *Powell* [12, 13], and *Berglund* et al. [14] on

$SiO_2-Si$ interfaces. A detailed review of the technique was published by *Adamchuk* et al. [15]. The light energy is in the range of the visible and near-UV, so carriers involved in the process come from the continuous bands of the semiconductor. When the barrier fits the image potential model very well, it is possible to determine the barrier energy using the photon energy as a parameter and measuring the current-voltage. This procedure has been used until now only for silicon dioxide-related interfaces [12]. For other interfaces, it is better to use the applied voltage as a parameter, while the characteristic detected is the photocurrent versus the photon energy.

The MOS devices are fabricated with a semitransparent metal layer as gate to allow the light penetration down to the substrate. The test device is polarized with positive gate voltage to detect the CBO. On the other hand to investigate the VBO, the bias is negative [16]. The quantum yield ($Y$) is obtained by normalizing the photocurrent ($I$) on the photon flux incident ($n_{ph}$) on the test device according to formula (1):

$$I(h\nu) = qY(h\nu)n_{ph}(h\nu).$$  (1)

Theoretical work [12] on IPE predicts that in order to measure the barrier energy at a given applied field, one should perform the linear extrapolation to zero yield on the graph of $Y$ elevated to $(1/p)$ versus the photon energy. The coefficient $p$ is 2 if the emitter is a metal and 3 if the emitter is a semiconductor. The difference between the two values is related to the distribution of excited electrons at the emitter surface [12]. Then, the barrier energy at zero field is extracted from a Schottky plot extrapolating to zero the barrier value at different applied fields [15]. The applied voltages are to be chosen high enough to prevent charge carrier emission from the interface opposite to the one being investigated [17]. To identify the emitter, the most useful way is to change the metallization: if the shape of the quantum yield changes with the metal gate, the emitter is located at the metal-oxide interface, while if it does not change, the emission takes place at the interface between the semiconductor and the oxide. An independent proof that the emitter is a semiconductor can be obtained from the energy position of the changes in slope of the quantum yield due to internal absorption across the semiconductor band gap.

The spectral range used in IPE measurements allows also the determination of the oxide band gap by means of the analysis of the photoconductivity region of the spectra. In this case, the photocurrent is due to the band to band transition in the oxide gap. For amorphous oxides, the experimental relationship of the quantum yield on photon energy gives a square-root dependence. The extrapolation to zero yield of this graph gives the band gap value [18].

## 2.2 X-Ray Photoelectron Spectroscopy

The use of photoelectron spectroscopies to study band alignment between different materials is well established. Different excitation sources (X-ray, soft X-ray, ultraviolet) are currently used. In this section, we will focus on the possibility to use a conventional X-ray source (monochromatic Al $K_\alpha$ $h\nu = 1486.6\,\text{eV}$) in order to obtain information on the valence and conduction band offsets of high-$\kappa$ dielectrics on different semiconductor substrates. The XPS energy resolution may not be as good as for other photoelectron spectroscopies using lower kinetic energy photoelectrons, but it has the advantage of a higher photoelectron escape depth. This implies that during XPS measurements, the photoelectron signal is averaged over many atomic layers, minimizing the effects of interface chemical shift and interface-potential variations. The methodology developed by *Kraut* et al. is a hybrid experimental-theoretical method. Starting from the work of *Kraut* et al. [19, 20], a full experimental approach has been proposed by *Chambers* et al. [6]. A different methodology has been developed by *Miyazachi* et al. [21], who also suggested a method to obtain a direct measurements of the band gap of a dielectric and, consequently, to determine (indirectly) the CBO of this oxide with the substrate. A full description of these methodologies is given in the following paragraphs.

Because of the overlap between the valence bands of two materials, it is difficult to determine directly the position of their valence band (VB) maxima in the heterojunction spectrum. Therefore, in order to obtain accurate determination of the VBOs, the main problem is the determination of the VB maxima in a heterojunction with very high accuracy. The methodology proposed by *Kraut* is based on the consideration that at the interface between two different materials, the charge distribution is different from that in the bulk. This implies that, according to Poisson's equation, an electrostatic potential bends all the bands or energy level by the same amount as a function of the distance from the interface. This suggests to use shallow core-level binding energies to determine the VBOs. The energy difference between the VB maxima $E_v$ and a core level $E_{cl}$ which identify the material is measured for each material in its pure state. The core-level binding energies in the heterojunction are then measured and combined with the values of $E_{cl} - E_v$ obtained in the pure materials. The VB discontinuity $\Delta E_v$ at the heterojunction interface between two materials X and Y is therefore given by:

$$\Delta E_v = (E_{cl}^X - E_{cl}^Y)_{X/Y} - [(E_{cl} - E_v)_X - (E_{cl} - E_v)_Y]. \qquad (2)$$

Finally, the problem of the determination of the VBO is reduced to the determination of the quantities $E_{cl} - E_v$. The position of the core level $E_{cl}$ can be determined with high accuracy. *Kraut* proposed to use the center of the peak width at half of the peak height. Alternatively, the position of the core-level energy can be measured by fitting the spectrum with a standard

Voigt function. The main issue is therefore related to the positioning cf the VB maximum. The original method proposed by *Kraut* was based on the fitting of a limited region of the VB spectrum close to the position of $E_v$ with a Gaussian broadened theoretical VB density of states (DOS) $N_v(E)$:

$$N_v(E) = \int_0^\infty n_v(E')g(E - E')\,\mathrm{d}E'\,. \tag{3}$$

where $n_v(E)$ is a nonlocal pseudopotential DOS computed using the method of *Chelikowsky* and *Cohen* [22] and $g(E)$ is a Gaussian function. The VB maximum is identified as the energy at which the DOS goes to zero. The methodology led to excellent results on Si, Ge, and III–V semiconductors [19, 20].

The main limitation of this approach is related to the poor capability of the theory to give a reliable and accurate description of the valence band structure of the materials under analysis. *Chambers* showed that in many cases, due to the inadequacies in the theory, this method can not be used. He suggested a full experimental approach, the so-called linear method [6]. The position of the VB maximum is obtained by the intersection of two straight line segments. One line fits to the background between the VB maximum and the Fermi level, and the other fits the linear portion of the VB leading edge. The results obtained by this methodology on Si, Ge, and GaAs are in excellent agreement (better than 0.1 eV) with those obtained by *Kraut* This new approach can be used when the Kraut method can not be applied.

According to *Miyazaki*'s methods, in order to determine the VB maxima in a heterojunction, the VB spectrum of the heterojunction is deconvoluted into the measured VB spectra of the constituent materials which are used as a reference [23, 24]. The VBO is determined by measuring the distance between the VB maxima of the deconvoluted VB spectra. The maximum of each deconvoluted spectrum is defined as the intercept of the linear extrapolation of the spectrum near the edge of the VB.

The measurement of the band gap is based on the observation that the outgoing photoelectron can suffer inelastic energy losses. The inelastic losses can be due to collective oscillations (plasmons) or single-particle excitations (band transitions). These events introduce typical features in the photoelectron spectra [25]. The collective oscillations generate broad replicas of the photoelectron peaks. These replicas are shifted to a lower binding energy by an amount corresponding to the plasmon energy $\hbar\omega_p$ [26]. The single-particle excitation, corresponding to the excitation of an electron from the conduction to the valence band, can be observed as a step in the spectra at an energy corresponding to the value $E_{gap}$ of the band gap below the core signal. The value of the band gap can therefore be obtained by measuring the distance between the position of the core peak and the linear extrapolation of the segment of maximum negative slope to the background level [21].

# 3 Literature Results

The results obtained to date by IPE and XPS on different high-$\kappa$ oxide/Si heterojunctions are presented in Table 1. Metal oxides, transition metal (TM) oxides, rare earth (RE) oxides, and ternary compounds have been investigated. The band alignment of $Al_2O_3$, $ZrO_2$, $HfO_2$ and their compounds with silicon substrate has been widely investigated using both IPE and XPS. The data reported in the literature are quite scattered, in particular for hafnia and zirconia. This is probably due, on one hand, to the different deposition methods used to grow these materials (the most used are atomic layer (ALD) deposition and molecular beam epitaxy) and, on the other, to the different methodologies that have been used to determine the CBO and VBO. Fewer data is available on the band alignment of these materials with alternative semiconductor (Ge, GaAs) substrates. Although RE oxides and ternary compounds are currently considered as promising candidates for gate oxide applications, a limited set of data are available on the band alignment of these materials with silicon or with other semiconductor substrates.

# 4 $Lu_2O_3$ on Si, Ge and GaAs

In this section, we focus on the study of the band alignment of lutetium oxide with different semiconductor substrates. In particular, we present the IPE results obtained for $Lu_2O_3$ films deposited by ALD on Si, Ge, and GaAs. The IPE data on the $Lu_2O_3$/Ge heterojunctions are compared with those obtained by XPS on similar samples.

## 4.1 Experimental

$Lu_2O_3$ films were deposited by ALD using $[[C_5H_4(SiMe_3)]_2LuCl]_2$ and $H_2O$ as precursors. As substrates, $p$- and $n$-type (100)Si with resistivity $\rho = 1 - 5\,\Omega \cdot cm$, $n$-type (100)Ge, and (100)GaAs were used [43]. MOS devices were prepared by thermally evaporating, through a shadow mask, semitransparent (15 nm thick) aluminum metal contacts. The back connection was fabricated using eutectic InGa. All the samples analyzed in this work did not go through any post-deposition or post-metalization annealing.

The IPE and PC results reported in this section were detected with a 150 W Xe arc lamp as light source. The measurements were performed in air and at room temperature. The spectral range of the photon flux was $h\nu = 2.5$–$6.0\,eV$ with a resolution of $0.02\,eV$. The photon flux was detected with a silicon photodiode while the voltage source and the photocurrent detection was obtained with a sub-femtoamperometer source-meter.

XPS measurements were performed on a PHI 5700 instrument equipped with a monochromatic Al $K_\alpha$ X-ray source and a concentric hemispherical

**Table 1.** Experimental band gap, conduction band offset, and valence band offset for different high-$\kappa$ oxide/Si heterojunctions as reported in the literature. Data have been acquired by IPE, PC, and XPS

| Material | Gap | | CBO | | VBO | | Ref. |
|---|---|---|---|---|---|---|---|
| | PC | XPS | IPE | XPS | IPE | XPS | |
| $Al_2O_3$ | 6.2 | | 2.15 | | | | [5, 18] |
| | | 6.95 | | 2.08 | | 3.75 | [24] |
| $Y_2O_3$ | | 6.0 | | 3.3 | | 1.6 | [27] |
| $La_2O_3$ | | 6.4 | | 2.3 | | 3.0 | [28] |
| $TiO_2$ | 4.4 | | 0.6–0.8 | | | | [29] |
| $ZrO_2$ | 5.4 | | 2.0 | | | | [5, 18] |
| | | 5.65 | | 0.88 | | 3.65 | [30] |
| | | 5.82 | | 1.76 | | 2.95 | [31] |
| | | 5.50 | | 1.23 | | 3.15 | [24] |
| $HfO_2$ | 5.6 | | 2.0 | | 2.5 | | [32] |
| | | 5.7 | | 1.5 | | 3.1 | [33] |
| | | 6.70 | | 1.97 | | 3.61 | [34] |
| | | 5.84 | | 1.47 | | 3.25 | [34] |
| | | 5.10 | | 1.46 | | 3.28 | [34] |
| | | | | | | 3.05 | [35] |
| $Ta_2O_5$ | | 4.65 | | 0.28 | | 3.25 | [24] |
| | 4.2–4.4 | | 0.9 | | | | [36] |
| $Pr_2O_3$ | | | | | | 1.1 | [37] |
| $Sm_2O_3$ | | | 1.6 | | | | [38] |
| $Gd_2O_3$ | 5.8 | | | | | | [39] |
| | | 6.4 | | 3.1 | | 2.2 | [28] |
| $Yb_2O_3$ | | | 2.1 | | | | [38] |
| $Lu_2O_3$ | 5.8 | | 2.1 | | 2.6 | | [16] |
| | | 6.0 | | 1.9 | | 3.0 | [28] |
| $LaAlO_3$ | 5.7 | | 2.0 | | 2.6 | | [40] |
| | | 6.2 | | 1.8 | | 3.2 | [41] |
| $LaScO_3$ | 5.7 | | 2.0 | | 2.5 | | [40] |
| $GdScO_3$ | 5.6 | | 2.0 | | 2.5 | | [40] |
| $DyScO_5$ | 5.7 | | 2.0 | | 2.5 | | [40] |
| $La_2Hf_2O_7$ | 5.6 | | 2.1 | | 2.4 | | [42] |

analyzer. The calibration of the spectrometer was performed using polycrystalline gold, silver, and copper foils. The positions of the Au $4f_{7/2}$, Ag $3d_{5/2}$, and Cu $2p_{3/2}$ peaks were found to be 84.0 eV, 368.3 eV, and 932.7 eV respectively. The data were acquired at a take-off angle of 80° with a band pass energy filter at 11.75 eV.

## 4.2 Results and Discussion

Figure 1 shows the PC spectral dependence, from the square-root quantum yield, for a Lu oxide grown on Si. The band gap is obtained by extrapolating

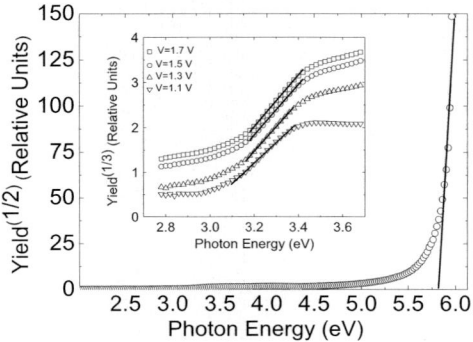

**Fig. 1.** Photoconductivity spectrum of lutetia $Y^{(1/2)}$ versus photon energy for an Al/Lu$_2$O$_3$/n-Si MOS at the negative applied voltage of $-1.5$ V. The *inset* shows the cube root of the quantum yield versus the photon energy at different positive applied voltages ($V$): 1.7 V, 1.5 V, 1.3 V, 1.1 V

this spectrum to zero yield. We found a Lu$_2$O$_3$ band gap of $5.8 \pm 0.1$ eV. The proof that this threshold is an oxide transition and not an interfacial one comes from the observation that the bias polarity and the doping type of the substrate do not affect the value [16]. In the inset of Fig. 1, the bias-dependent behavior of the cube-root quantum yield versus the photon energy is shown for an Al/Lu$_2$O$_3$/n-Si MOS. The applied voltage is positive. The interfacial layer between silicon and lutetia is a chemical silicon oxide of about 1 nm, thin enough to prevent contributions to the barrier height determination [5]. The emitter is the Si substrate, as indicated by the onset at 3.4 eV due to direct optical absorption in Si. The extrapolation at zero field of the voltage-dependent thresholds gives a barrier energy of $3.2 \pm 0.1$ eV. This threshold corresponds to the electronic transitions from the silicon valence band to the lutetia conduction band. Considering the Si band gap (1.1 eV), it is possible to extract a CBO of $2.1 \pm 0.1$ eV. We would like to point out that, in these IPE characteristichs, the threshold is not sharp. There are emissions at lower energies that can be related to tails of localized states in the gap of the amorphous oxide. To obtain the VBO value, we measured (not shown, see [16]) an Al/Lu$_2$O$_3$/p-Si stack with negative applied voltages. The transition investigated in this case is that of holes from the Si conduction band to the oxide valence band. The barrier energy found for this threshold is $3.7 \pm 0.1$, which gives a VBO of $2.6 \pm 0.1$ eV. As already seen for similar transitions [32], this barrier is field independent. The same band gap value obtained by PC is achieved from the sum of the CBO and the VBO measured by IPE and the silicon band gap.

To check the variation of the CBO of the lutetium oxide changing the substrate, we performed IPE measurements on lutetia grown on Ge and on GaAs. In Fig. 2, we show the spectral curves measured with positive applied potential on a lutetia-based MOS structure on Ge. In the inset, the Schottky plot of

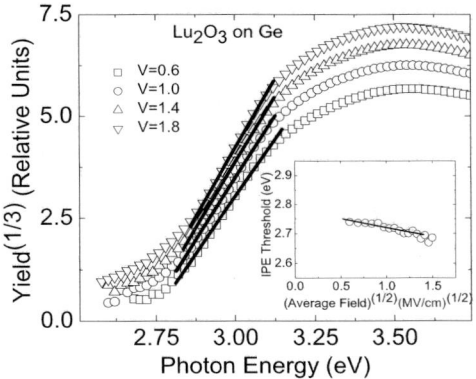

**Fig. 2.** Cube root of IPE yield as function of photon energy for an Al/Lu$_2$O$_3$/n-Ge stack at different positive applied voltages ($V$): 0.6, 1.0, 1.4, 1.8. The *lines* represent the linear fitting. The *inset* shows the Schottky plot of the IPE thresholds

the field-dependent barriers corresponding to the photo emission of electrons from the Ge valence band into the conduction band of the lutetia is shown. The extrapolation at zero electric field gives a barrier of 2.8 ± 0.1 eV. By subtracting the Ge band gap (0.66 eV), we obtained a CBO of 2.1 ± 0.1 eV [44]. The CBO values of the lutetia on Si and on Ge are, within the experimental error, identical. This behavior has been already observed for HfO$_2$ [45,46] and Al$_2$O$_3$ [47] with IPE technique, and for ZrO$_2$ [31] with XPS measurement.

XPS results for a bare Ge(100) (A), a 10 nm thick Lu$_2$O$_3$ film on germanium (B), and a 3 nm thick Lu$_2$O$_3$/Ge heterojunction (C) are shown in Fig. 3. The positions of the VB edges were determined by the intersection between the linear fitting of the leading edge of the valence band and the linear fitting of the flat energy region in the energy gap. The results are shown in the insets of Fig. 3. The positions of the Ge 3$d$ peaks were determined by measuring the center of the peak width at half height. The Lu 4$f$ peaks were fitted using a couple of 3 : 4 doublets after integral background correction. The two doublets were attributed to lutetium in the Lu$_2$O$_3$ form and to lutetium hydroxide [43]. The position of the Lu 4$f_{7/2}$ peak in the doublets associated with the Lu$_2$O$_3$ was used as a core-level reference for the determination of the VBO. More detail about the XPS data analysis can be found in [44]. According to Kraut's method, the VBO was calculated using the following equation:

$$\Delta E_v = (E_{Lu4f} - E_{Ge3d})_{Lu_2O_3/Ge} - (E_{Lu4f} - E_v)_{Lu_2O_3} + (E_{Ge3d} - E_v)_{Ge}. \quad (4)$$

The value of the VBO is 2.9 ± 0.1 eV. Assuming a band gap of 5.8 eV, as derived by PC measurements, we obtain a CBO of 2.2 ± 0.1 eV, in excellent agreement (within the experimental error) with the IPE results.

Figure 4 shows the IPE spectra at the interface between GaAs and lutetia for a MOS structure obtained with an aluminum metallization. The barrier

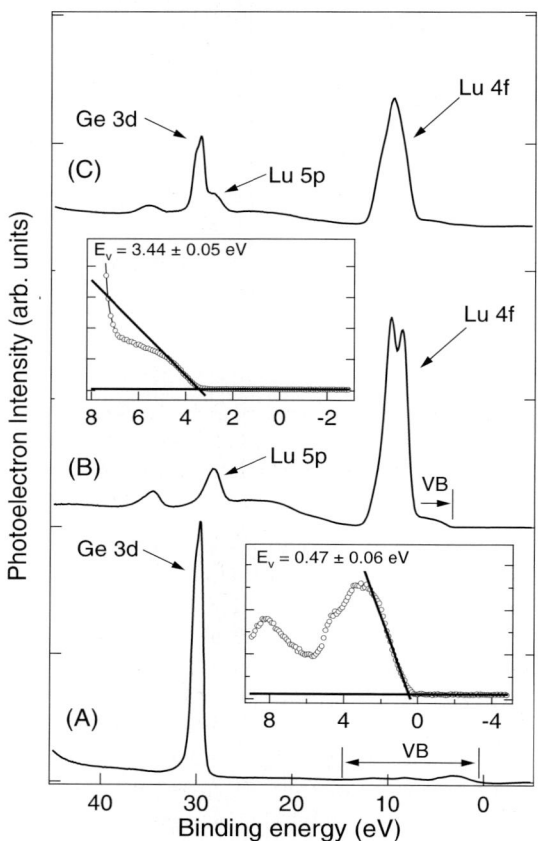

**Fig. 3.** Core-level and valence band spectra for bare Ge(100) (A), a 10 nm thick Lu$_2$O$_3$ on Ge (B), and a 3 nm thick Lu$_2$O$_3$/Ge heterojunction (C). The *insets* show high-resolution scans of the VB region of the Ge(100) and of the 10 nm Lu$_2$O$_3$ sample

extrapolated in a Schottky plot (see the inset in Fig. 4) at zero applied field is $3.5 \pm 0.1$ eV. The transition involved is from the GaAs valence band into the lutetia conduction band; by substracting the semicondutor band gap (1.42 eV), the CBO value was found to be to $2.1 \pm 0.1$ eV. In the Schottky plot, we can observe that the slope of the electric field reduction is bigger than the one observed at the interface with Si or Ge [48]. A similar behavior in IPE mesurements has been already seen for GaAs substrate [39], and can be related to charge at the interface that causes a reduction of the interfacial barrier [49]. It was shown that this interfacial charge affects the barrier reduction, but it does not change the zero-field barrier [49].

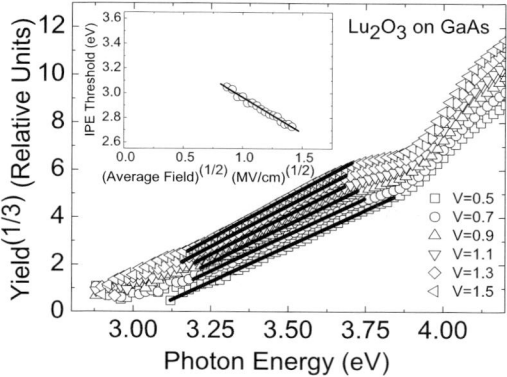

**Fig. 4.** Cube root of IPE yield as function of photon energy for an Al/Lu$_2$O$_3$/n-GaAs stack at different positive applied voltages ($V$): 0.5, 0.7, 0.9, 1.1, 1.3, 1.5, 1.7. The *lines* represent the linear fitting. The *inset* shows the Schottky plot of the IPE thresholds

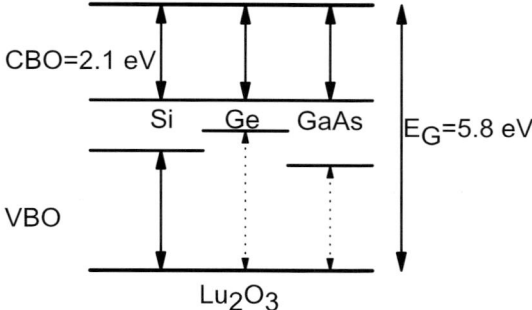

**Fig. 5.** Energy-band diagram for semiconductor-lutetia interface for different semiconductors (Si, Ge, GaAs)

The results of the study of the lutetia band alignment with different semiconductors (Si, Ge, GaAs) are summarized in Fig. 5. We can note that the CBO is basically independent of the semiconductor substrate.

## 5 Conclusions

We have shown that the band alignment of the lutetia with different semiconductors (Si, Ge, GaAs) investigated by IPE gives the same CBO value of 2.1 eV within the experimental error. We can conclude that only the VBO values are affected by the different interfaces. We have also found a good agreement for the VBO values obtained by IPE and XPS on Lu$_2$O$_3$/Ge heterojunctions.

**Acknowledgements**

The authors would like to thank G. Scarel for ALD growth, and S. Spiga and E. Bonera for helpful discussions. M. Alia is acknowledged for help in sample preparation.

# References

[1] G. D. Wilk, R. M. Wallace, J. M. Anthony: High-$\kappa$ gate dielectrics: Current status and materials properties considerations, J. Appl. Phys. **89**, 5243–5275 (2001)

[2] C. L. Hinkle, G. Lucovsky: A novel approach for determining the effective tunneling mass of electrons in $HfO_2$ and other high-$\kappa$ alternative gate dielectrics for advanced CMOS devices, Microelectron. Eng. **72**, 257–262 (2004)

[3] J. Robertson: High dielectric constant oxides, Eur. Phys. J. Appl. Phys. **28**, 265–291 (2004)

[4] M. V. Fischetti, D. A. Neumayer, E. A. Cartier: Effective electron mobility in Si inversion layers in metal/oxide/semiconductor systems with a high-$\kappa$ insulator: The role of remote phonon scattering, J. Appl. Phys. **90**, 4587 (2001)

[5] V. V. Afanas'ev, M. Houssa, A. Stesmans, M. M. Heyns: Electron energy barriers between (100)Si and ultrathin stacks of $SiO_2$, $Al_2O_3$, and $ZrO_2$ insulators, Appl. Phys. Lett. **78**, 3073–3075 (2001)

[6] S. A. Chambers, T. Droubay, T. C. Kaspar, M. Gutowski: Experimental determination of valence band maxima for $SrTiO_3$, $TiO_2$ and SrO and the associated valence band offsets with Si(001), J. Vac. Sci. Technol. B **22**, 2205–2215 (2004)

[7] R. Ludeke, M. T. Cuberes, E. Cartier: Local transport and trapping issues in $Al_2O_3$ gate oxide structures, Appl. Phys. Lett. **76**, 2886–2888 (2000)

[8] R. Williams: Photoemission of electrons from silicon into silicon dioxide, Phys. Rev. **140**, A 569–A 575 (1965)

[9] A. Goodman: Photoemission of electrons from silicon and gold into silicon dioxide, Phys. Rev. **144**, 588–593 (1966)

[10] A. M. Goodman: Photoemission of holes from silicon into silicon dioxide, Phys. Rev. **152**, 780–784 (1966)

[11] A. M. Goodman: Photoemission of electrons from n-type degenerate silicon into silicon dioxide, Phys. Rev. **152**, 785–787 (1966)

[12] R. J. Powell: Interface barrier energy determination from voltage dependence of photoinjected currents, J. Appl. Phys. **41**, 2424–2432 (1970)

[13] R. J. Powell, C. N. Berglund: Photoinjection studies of charge distributions in oxides of MOS structures, J. Appl. Phys. **42**, 4390–4397 (1971)

[14] C. N. Berglund, R. J. Powell: Photoinjection into $SiO_2$: electron scattering in the image force potential well, J. Appl. Phys. **42**, 573–579 (1971)

[15] V. K. Adamchuk, V. V. Afanas'ev: Internal photoemission spectroscopy of semiconductor-insulator interfaces, Prog. Surf. Sci. **41**, 111–211 (1992)

[16] G. Seguini, E. Bonera, S. Spiga, G. Scarel, M. Fanciulli: Energy-band diagram of metal/$Lu_2O_3$/silicon structures, Appl. Phys. Lett. **85**, 5316–5318 (2004)

[17] P. V. Dressendorfer, R. C. Barker: Photoemission measurements of interface barrier energies for tunnel oxides on silicon, Appl. Phys. Lett. **36**, 933–935 (1980)

[18] V. V. Afanas'ev, M. Houssa, A. Stesmans, M. M. Heyns: Band alignments in metal–oxide–silicon structures with atomic-layer deposited $Al_2O_3$, J. Appl. Phys. **91**, 3079–3084 (2002)

[19] E. A. Kraut, R. W. Grant, J. R. Waldrop, S. P. Kowalczyk: Precise determination of the valence-band edge in X-ray photoemission spectra: Application to measurements of semiconductor interface potentials, Phys. Rev. Lett. **44**, 1620–1623 (1980)

[20] E. A. Kraut, R. W. Grant, J. R. Waldrop, S. P. Kowalczyk: Semiconductor core level to valence-band maximum binding-energy differences: Precise determination by X-ray photoelectron spectroscopy, Phys. Rev. B **28**, 1965–1977 (1983)

[21] S. Miyazaki, H. Nishimura, M. Fukuda, L. Ley, J. Ristein: Structure and electronic states of ultrathin $SiO_2$ thermally grown on Si(100) and Si(111) surfaces, Appl. Surf. Sci. **114/114**, 585–589 (1997)

[22] J. R. Chelikowsky, M. L. Cohen: Nonlocal pseudopotential calculations for the electronic structure of eleven diamond and zinc-blende semiconductors, Phys. Rev. B **14**, 556 (1976)

[23] S. Miyazaki, M. Narasaki, M. Ogasawara, M. Hirose: Characterization of ultrathin zirconium oxide films on silicon using photoelectron spectroscopy, Microelectron. Eng. **59**, 373–378 (2001)

[24] S. Miyazaki: Photoemission study of energy-band alignments and gap-state density distributions for high-$\kappa$ gate dielectrics, J. Vac. Sci. Technol. B **19**, 2212–2216 (2001)

[25] F. G. Bell, L. Ley: Photoemission study of $SiO_x$ ($0 \leq x \geq 2$) alloys, Phys. Rev. B **37**, 8383–8393 (1988)

[26] D. Briggs, M. P. Seah: *Pratical Surface Analysis*, vol. 1 (Wiley, New York 1990)

[27] A. Ohta, M. Yamaoka, S. Miyazaki: Photoelectron spectroscopy of ultrathin yttrium oxide films on Si(100), Microelectron. Eng. **72**, 154–159 (2004)

[28] T. Hattori, T. Yoshida, T. Shiraishi, K. Takahashi, H. Nohira, S. Joumori, K. Nakajima, M. Suzuki, K. Kimura, I. Kashiwagi, C. Ohshima, S. Ohmi, H. Iwai: Composition, chemical structure, and electronic band structure of rare earth oxide/Si(100) intefacial transition layer, Microelectron. Eng. **72**, 283–287 (2004)

[29] V. V. Afanas'ev, A. Stesmans, F. Chen, S. A. Campbell: Electrical conduction and band offsets in $Si/Hf_xTi_{(1-x)}O_2$/metal structures, J. Appl. Phys. **95**, 7936–7939 (2004)

[30] R. Puthenkovilakam, J. P. Chang: Valence band structure and band alignment at the $ZrO_2$/Si interface, Appl. Phys. Lett. **84**, 1353–1355 (2004)

[31] S. J. Wang, A. C. H. Huan, Y. L. Foo, J. W. Chai, J. S. Pan, Q. Li, Y. F. Dong, Y. P. Feng, C. K. Ong: Energy-band alignments at $ZrO_2$/Si, SiGe, and Ge intefaces, Appl. Phys. Lett. **85**, 4418–4420 (2004)

[32] V. V. Afanas'ev, A. Stesmans, F. Chen, X. Shi, S. A. Campbell: Internal photoemission of electrons and holes from (100)Si into $HfO_2$, Appl. Phys. Lett. **81**, 1053–1055 (2002)

[33] R. Puthenkovilakam, J. P. Chang: An accurate determination of barrier heights at the $HfO_2$/Si interfaces, J. Appl. Phys. **96**, 2701–2707 (2004)

[34] S. Sayan, T. Emge, E. Garfunkel, X. Zhao, L. Wielunski, A. Bartynski, D. Vanderbilt, J. S. Suehle, S. Suzer, M. Banaszak-Holl: Band alignment issues related to $HfO_2/SiO_2$/p-Si gate stacks, J. Appl. Phys. **96**, 7485–7491 (2004)

[35] Q. Li, S. J. Wang, K. B. Li, A. C. H. Huan, J. W. Chai, J. S. Pan, C. K. Ong: Photoemission study of energy-band alignment for $RuO_x/HfO_2/Si$ system, Appl. Phys. Lett. **85**, 6155–6157 (2004)

[36] V. V. Afanas'ev, A. Stesmans, C. Zhao, M. Caymax, Z. M. Rittersma, J. W. Maes: Band alignment at the interface of (100)Si with $Hf_xTa_{(1-x)}O_y$ high-$\kappa$ dilectric layers, Appl. Phys. Lett. **86**, 072108–1–072108–3 (2005)

[37] H. J. Osten, J. P. Liu, H. J. Müssig: Band gap and band discontinuities at crystalline $Pr_2O_3/Si(001)$ heterojunctions, Appl. Phys. Lett. **80**, 297–299 (2002)

[38] V. A. Rozhkov, A. Y. Trusova, I. G. Berezhnoi: Energy barriers and trapping centers in silicon metal-insulators-semiconductor structures with samarium and ytterbium oxide insulators, Tech. Phys. Lett. **24**, 217–219 (1998)

[39] V. V. Afanas'ev, A. Stesmans, M. Passlack, N. Medendorp: Band offsets at the interfaces of GaAs(100) with $Gd_xGa_{0.4-x}O_{0.6}$ insulators, Appl. Phys. Lett. **85**, 597–599 (2004)

[40] V. V. Afanas'ev, A. Stesmans, C. Zhao, M. Caymax, T. Heeg, J. Schubert, Y. Jia, G. Schlom, G. Lucovsky: Band alignment between (100)Si and complex rare earth/transition metal oxides, Appl. Phys. Lett. **85**, 5917–5919 (2004)

[41] L. F. Edge, D. G. Schlom, S. A. Chambers, E. Cicerrella, J. L. Freeout, B. Holländer, J. Schubert: Measurement of the band offsets between amorphous $LaAlO_3$ and silicon, Appl. Phys. Lett. **84**, 726–728 (2004)

[42] G. Seguini, S. Spiga, E. Bonera, M. Fanciulli, A. Reyes Huamantinco, C. J. Först, C. R. Ashman, P. E. Blöchl, A. Dimoulas, G. Mavrou: Band alignment at the $La_2Hf_2O_7/Si(001)$ interface, Unpublished (2005)

[43] G. Scarel, E. Bonera, C. Wiemer, G. Tallarida, S. Spiga, M. Fanciulli: Atomic-layer deposition of $Lu_2O_3$, Appl. Phys. Lett. **85**, 630–632 (2004)

[44] M. Perego, G. Seguini, G. Scarel, M. Fanciulli: X-ray photoelectron spectroscopy study of energy band alignment of rare earth oxides, Unpublished (2005)

[45] S. Spiga, C. Wiemer, G. Tallarida, G. Scarel, S. Ferrari, G. Seguini, M. Fanciulli: Effects of the oxygen precursor on the electrical and structural properties of $HfO_2$ films grown by atomic layer deposition on Ge, Appl. Phys. Lett. **87**, 112904–1–3 (2005)

[46] V. V. Afanas'ev, A. Stesmans: Energy band alignment at the $(100)Ge/HfO_2$ interface, Appl. Phys. Lett. **84**, 2319–2321 (2004)

[47] G. Seguini: Band alignment of $Al_2O_3$ on Ge, Private Communication (2005)

[48] G. Seguini: Band alignment of $Lu_2O_3$ on GaAs, Private Communication (2005)

[49] V. V. Afanas'ev, A. Stesmans: Trapping of $H^+$ and $Li^+$ ions at the $Si/SiO_2$ inteface, Phys. Rev. B **60**, 5506–5512 (2000)

# Index

# Band Edge Electronic Structure of Transition Metal/Rare Earth Oxide Dielectrics

Gerald Lucovsky

Department of Physics, North Carolina State University, Raleigh, North Carolina 27695-8202, USA
gerry_lucovsky@ncsu.edu

**Abstract.** This Chapter discusses band edge electronic structure of 1. nanocrystalline elemental and complex oxide high-$k$ dielectrics, and 2. non-crystalline Zr and Hf silicates, and Si oxynitride alloys. Experimental approaches include X-ray absorption spectroscopy, photoconductivity, and visible/vacuum ultra-violet and spectroscopic ellipsometry. These measurements are complemented by Fourier transform infra-red absorption, X-ray photoelectron spectroscopy, high-resolution transmission electron microscopy, and X-ray diffraction. Three issues are highlighted: Jahn–Teller term splittings that remove band edge $d$-state degeneracies in nanocrystalline films, intrinsic bonding defects in $ZrO_2$ and $HfO_2$, and chemical phase separation and crystallinity in Zr and Hf silicate and Si oxynitride alloys.

## 1 Introduction

Transition metal (TM) and lanthanide series rare earth (RE) oxides are being considered as alternative gate dielectrics for advanced semiconductor devices. The electronic states at the conduction band edges of these oxide dielectrics play a significant role in the performance and reliability of high-$k$ gate stacks for Si complementary metal oxide semiconductor (CMOS) devices. These band edge states are derived in part from atomic $d$-states of the TM and/or RE atoms, and as such define the lowest conduction band edge states. In particular, these states define the conduction band offset energy which is the barrier, $E_B$, for either thermionic emission from the Si conduction into the high-$k$ dielectric, or for tunneling transport through the high-$k$ dielectric. They also are the electronic states that contribute to the magnitude of the electron tunneling mass, $m^*$. These dependencies are quantified through the introduction of direct tunneling current figure of merit, $\Phi = k(E_B - m^*)^{0.5}$, where $k$ is the dielectric constant. This factor appears a negative multiplier for the physical film thickness in the exponential function for tunneling leakage current.

In general, the minimum optical band gaps, $E_g$, and the values of $E_B$ are smaller for TE/RE elemental and complex oxides than they are for semiconductor or simple metal oxides; e.g., the values of $E_g$ and $E_B$ are respectively $\sim 9\,\text{eV}$ and $3.2\,\text{eV}$ for $SiO_2$, but are reduced to $\sim 5.7\,\text{eV}$ and $\sim 1.5\,\text{eV}$ for $ZrO_2$ and $HfO_2$. These decreases in $E_B$, combined with two- to threefold decreases in $m^*$, lead to significant decreases in the $E_B - m^*$ product, which

M. Fanciulli, G. Scarel (Eds.): Rare Earth Oxide Thin Films,
Topics Appl. Physics **106**, 285–312 (2007)
© Springer-Verlag Berlin Heidelberg 2007

mitigate in part increases in $\Phi$ associated with three- to fivefold increases in $k$, which in turn allow proportional increases in the physical thickness of a dielectric film which yields the same value of capacitance as for a physically thinner $SiO_2$ dielectric. Specific examples of these scaling trade-offs are addressed at the end of the Chapter by *Lucovsky* and *Phillips* in this volume.

The performance and reliability of high-$k$ stacks in advanced Si devices is also determined, and therefore limited by defects. This paper also addresses two types of intrinsic defects – localized states below the conduction band edges of nanocrystalline high-$k$ elemental and compound oxides, and local O-atom vacancy states within nanocrystalline dielectrics, and/or at their interfaces with $SiO_2$. Each of these generic defects is correlated directly with the $d$-state contributions to the electronic states at the top of the valence band and bottom of the conduction band in high-$k$ oxides, whether they are elemental or complex. It is important to note that these states are likely not to be important, or even present in non-crystralline high-$k$ Si oxynitride alloys, although this remains to be established by defect studies of devices employing these non-crystalline Zr/Hf Si oxynitride alloys.

This Chapter presents spectroscopic studies that identify the contribution of TM and RE atom $d$-states to the conduction band edge electronic structure of nanocrystalline elemental and complex oxides, and non-crystalline TM silicate and TM Si oxynitride alloys. These studies are extended to include intrinsic bonding defects as well.

The first two sections of this Chapter address respectively the band edge electronic states in nanocrystalline TM/RE elemental, and TM/RE complex oxides. These have been studied primarily by two complementary spectroscopic techniques: 1. X-ray absorption spectroscopy (XAS), in particular XANES [1], X-ray absorption near edge spectroscopy to distinguish it from EXAFS, extended X-ray absorption fine structure in which X-ray energies extend well beyond the range of electronic transitions that terminate in conduction band states, and 2. vacuum ultra-violet spectroscopic ellipsometry, VUV SE [2]. These measurements have been complemented by other spectroscopic techniques including photoconductivity (PC) [3, 4], and Fourier transform infra-red spectroscopy (FTIR).

The next section of the Chapter deals with intrinsic bonding defects, initially in the elemental oxides $ZrO_2$ and $HfO_2$, and then extended to complex oxides as well. Spectroscopic measurements, XANES and VUV SE, are complemented by PC, and have been used to first identify localized *defect states* approximately $0.6 \pm 0.2\,eV$ below the conduction band edges of $HfO_2$ and $ZrO_2$ [2–4]. This state is assigned to a grain-boundary bonding distortion. A second intrinsic defect state has also been identified by XANES and VUV SE. It is localized within the band gap of $ZrO_2$, and approximately $4.2 \pm 0.15\,eV$ above the valence band edge, or equivalently, $1.4 \pm 0.15\,eV$ below the conduction band edge and is assigned to an oxygen atom vacancy [5]. There is no experimental evidence for a defect state lower in the band gaps of $HfO_2$ and $ZrO_2$, at $\sim 3.3\,eV$ above the valence band edge, which has been predicted by theoretical studies of neutral hydrogen atoms in semiconductors and insulators [6]. This section of the Chapter includes results of electrical measurements that reveal the grain-boundary, and O-atom vacancy defects.

The final portion of this Chapter addresses non-crystalline silicate alloys, where the emphasis is on the chemical stability of these films at the temperatures required for process integration into device structures, i.e., $> 900\,^{\circ}\mathrm{C}$ and generally up to at least $1000\,^{\circ}\mathrm{C}$. This Chapter discusses non-crystalline Zr and Hf silicate pseudo-binary and pseudo-ternary alloys. It identifies ways in which chemical phase separation (CPS), and crystallization of the TM oxide chemically separated phases may be effectively eliminated in pseudo-ternary alloys such as $\mathrm{Zr(Hf)O_2-SiO_2-Si_3N_4}$, provided that alloy compositions are controlled within relatively narrow limits, e.g., high $\mathrm{Si_3N_4}$ content, $\sim 35\text{–}40\,\%$, with approximately equal concentrations of $\mathrm{SiO_2}$ and $\mathrm{Zr(Hf)O_2}$, $\sim 30\text{–}35\,\%$. Proceeding in this way, it is possible to extend EOT scaling to about 0.7 to 0.8 nm, while retaining the ability to use polycrystalline Si, or Si$-$Ge gate electrodes. The use of polycrystalline gate electrodes has well-understood drawbacks, but nevertheless provides a convenient way to identify the intrinsic properties of these pseudo-ternary alloys and their interfaces with $\mathrm{SiO_2}$.

# 2 Experimental Methods

The thin film samples for these studies were prepared by deposition techniques summarized in [1]. These include remote plasma-assisted chemical vapor deposition (RPECVD) for the TM elemental oxides, and reactive evaporation for the TM elemental and TM/RE complex oxides. These TM/RE oxide films are either nanocrystalline as-deposited, or have been annealed at relatively low temperatures, up to $500\,^{\circ}\mathrm{C}$ in inert ambients sufficient to release water, OH or hydrocarbons that impede the formation of nanocrystallites. Nanocrystallinity has been confirmed for these, or similarly prepared samples, by X-ray and/or electron diffraction, and/or infra-red absorption.

XAS measurements were made at the National Synchrotron Light Source (NSLS) at Brookhaven National Laboratory (BNL), and the Stanford Synchrotron Research Laboratory (SSRL) at the Stanford Linear Accelerator Center (SLAC). The SE, XPS, FTIR, HRTEM and XRD measurements were made on conventional and advanced.

# 3 Experimental Results and Discussion: TM Elemental Oxides

In presenting and discussing the XANES results, a distinction is made between 1. *crystal field* (C–F) *d*-state term splittings in *highly symmetric* bonding environments which preserve the two-and threefold degeneracies of the

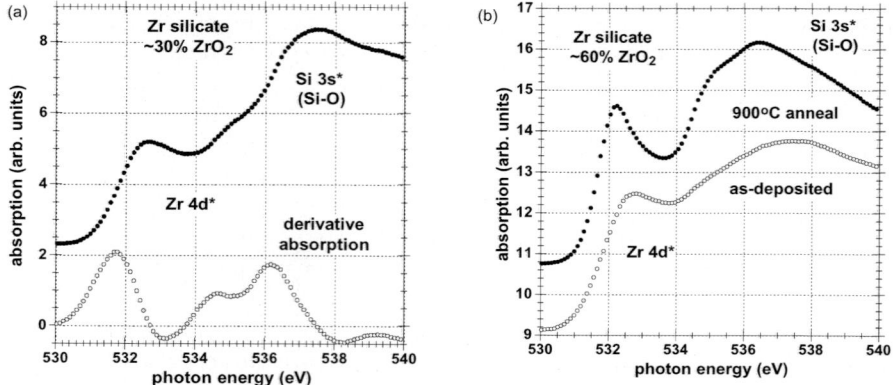

**Fig. 1.** (a) O $K_1$ edge spectrum and derivative for non-crystalline Zr silicate 30 % alloy as-deposited. (b) O $K_1$ edges for 60 % alloy, as-deposited and after a 900 °C anneal which results in chemical phase separation into $SiO_2$ and nanocrystalline $ZrO_2$

respective $E_g$ and $T_{2g}$ d-states, and 2. *Jahn–Teller* (J–T) d-state term-splittings, which are also a C–F splitting, but in *lower symmetry distorted environments* that remove the respective $E_g$ and $T_{2g}$ d-state degeneracies, either partially, or completely [7]. The J–T designation is used to emphasize a contributing mechanism for these bonding distortions.

To clarify the differences been C–F d-state splittings that preserve $E_g$ and $T_{2g}$ d-state degeneracies and J–T term splitting that remove these degeneracies, XANES spectra are presented for Zr silicate alloys as-deposited, and after a 900 °C anneal in Ar. C–F splittings in twofold degenerate $E_g$ and threefold degenerate $T_{2g}$ states are observed in as-deposited non-crystalline Zr and Hf silicate alloy thin films, and representative O $K_1$ edge spectra for Zr silicate alloys with $\sim$ 30 % and 60 % $ZrO_2$ are presented in Fig. 1a and b. There is no evidence in the absorption spectra, or in the derivative of the 30 % alloy spectrum (or 60 % alloy spectrum, not shown), for any additional d-state splittings indicative of a removal of the $E_g$ d-state degeneracy. However, after a 900 °C inert ambient anneal in Ar, the O $K_1$ spectrum in Fig. 1b for the 60 % alloy displays changes in the lower-energy 4d-state in spectral line-width and energy (an $\sim$ 0.5 eV *red shift*) with respect to the as-deposited film that are indicative of d-state degeneracy removal. Similar changes in spectral line-width and energy of the lowest-energy d-state feature also occur for the 30 % alloy (not shown). These changes are attributed to chemical phase separation (CPS) of this alloy into 1. crystalline $ZrO_2$ with a J–T bonding distortion, and 2. non-crystalline $SiO_2$ [1]. Bonding changes in pseudo-binary Zr silicate films, and in pseudo-ternary Zr and Hf Si oxynitrides will be addressed in more detail in Sect. 4.

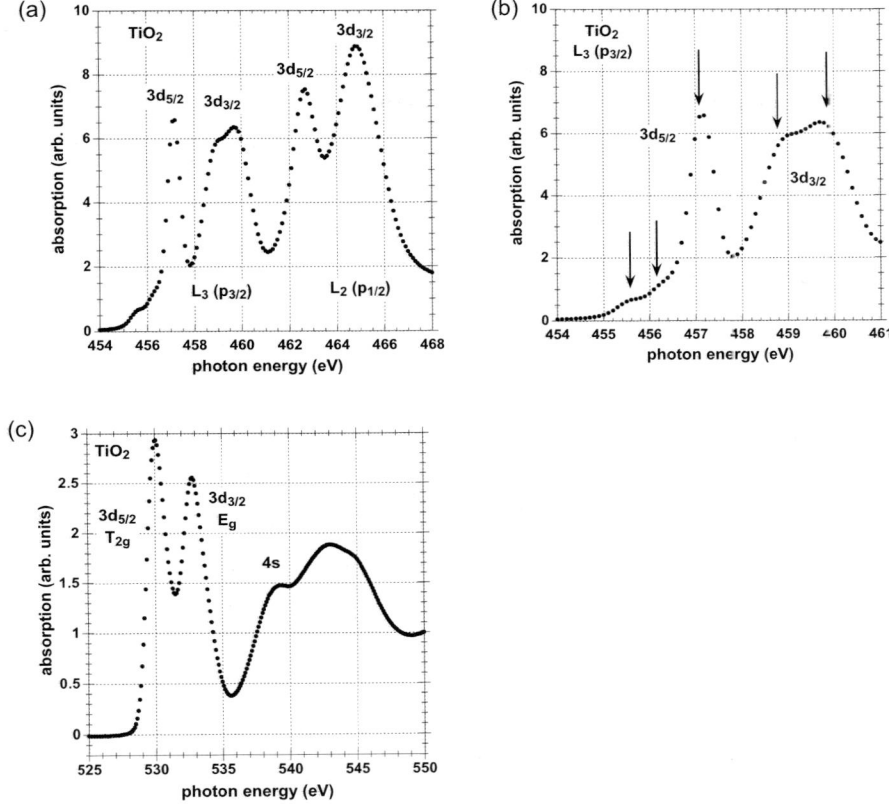

**Fig. 2.** X-ray absorption near-edge spectra for $TiO_2$: (**a**) the Ti $L_{2,3}$ edge spectrum, (**b**) the $L_3$ edge spectrum, (**c**) the O $K_1$ edge spectrum

J–T term splittings are now addressed for nanocrystalline films of the group IVB TM oxides, $TiO_2$, and $ZrO_2$ and $HfO_2$. The bonding coordination of Ti is six-fold in $TiO_2$, while it is typically eight- (and sometimes seven-) fold in $ZrO_2$ and $HfO_2$. This difference is responsible for qualitative differences between the corresponding spectra of 1. $TiO_2$, and 2. $ZrO_2$ and $HfO_2$. Figures 2a, b and c display respectively, in (a) the Ti $L_{2,3}$ XANES spectrum, (b) the $L_3$ edge spectrum of $TiO_2$, and (c) the O $K_1$ edge spectrum.

The $L_{2,3}$ spectrum is *intra-atomic* with transitions originating in occupied and spin-orbit split Ti $2p$ levels, and terminating in empty Ti $3d$ and $4s$ states. These spectral probe a scale of order of $\sim 0.5$ nm that includes the Ti atom, and the symmetry of the local field at this atomic site that results from its six nearest O-atom neighbors. It is important to note that the final state is not a state which mixes Ti-atom and O-atom states, even though mixed O $2p$-Ti $3d$ molecular orbital states define the conduction band edge when probed by other core level transitions, e.g., as in the O $K_1$ edge, and the band edge optical transitions in which the ground state is a spatially extended valence band state. These differences establish increases scales of order for O $K_1$ edge and band spectra. Based on other examples discussed in this Chapter, the scale of order for O $K_1$ transitions is defined from J–T term splittings, and as demonstrated in Zr silicate alloys, this extends to about 2 nm. The scale of order for the band edge transitions is a more complicated issue, and this will be addressed as well in a later section of this Chapter.

The separation between *equivalent features* in the $L_3$ and $L_2$ features is equal to the atomic $2p$ state spin-orbit splitting of Ti. There are three features in the $L_3$ and $L_2$ spectra, the two lower-energy ones being associated with transitions to $3d$ states, and showing additional J–T term splittings, and the third and broader feature being associated with transitions to the Ti $4s$ state. Figure 2b displays five distinct $d$-state features in the $L_3$ spectrum, consistent with a complete removal of $T_{2g}$ (or $d_{5/2}$), and $E_g$ (or $d_{3/2}$) degeneracies. The lower-energy $T_{2g}$ triplet displays two features with equal splittings of $0.75 \pm 0.3$ eV, and the higher-energy $E_g$ doublet has a splitting of $0.8 \pm 0.2$ eV. The $3d$-state splitting between the average $T_{2g}$ and $E_g$ energies is $2.2 \pm 0.2$ eV. Differences between the $L_3$ and $L_2$ spectra are due to symmetry-driven matrix element effects, and these quantitative effects are beyond the scope of this review.

Figure 2c displays the O $K_1$ spectrum for $TiO_2$. Only two $3d$-state features are evident, and their energy separation is $2.7 \pm 0.1$ eV. This $3d$-state separation is larger than *empty* $3d$-state splitting by $\sim 0.5$ eV. There is no evidence for any J–T term splitting of these $d$-state features in the O $K_1$ spectrum, or in a numerical differentiation of that spectrum. As noted above, the final states in the O $K_1$ spectrum are molecular orbital anti-bonding O $2p$ states that are mixed with the Ti $3d$ and $4s$ states. The $d$-state energy differences with respect to the $L_3$ edge reflect this mixing, and are a result of solid state, or equivalent band structure effects. For example, the average $4d$-state energy differences in the $M_{2,3}$ edge of $ZrO_2$, Zr silicate non-crystalline alloys and $ZrO_2-Y_2O_3$ nanocrystalline alloys are about 2.2–2.3 eV and essentially the same as those for Ti in $TiO_2$ and Zr and Hf titanate alloys, whilst the average $4d$-state energy splittings in O $K_1$ edges are larger, $\sim 3.4$ eV. The average Hf $5d$ splittings in O $K_1$ edge spectra continue this trend and are $> 4$ eV.

The XANES spectra are now compared with the conduction band edge spectra obtained from VUV SE measurements [2]. Figures 3a and b display respectively in (a) the imaginary part of the complex constant, $\varepsilon_2$, where $\varepsilon_c = \varepsilon_1 + i\varepsilon_2$, and in (b) the absorption constant, $\alpha$. Differentiation of the $\varepsilon_2$ spectrum indicates five $d$-state features. Their energies are compared to the five $d$-state features in the $L_3$ edge in Fig. 3c. The slope of this plot is 1.1, and indicates a linear relationship between the empty *atomic* $d$-state energies revealed in the $L_3$ edge spectrum, and the $d$-state contributions to the molecular orbital mixed O $2p$-Zr $4d$ states in the lower portion of the

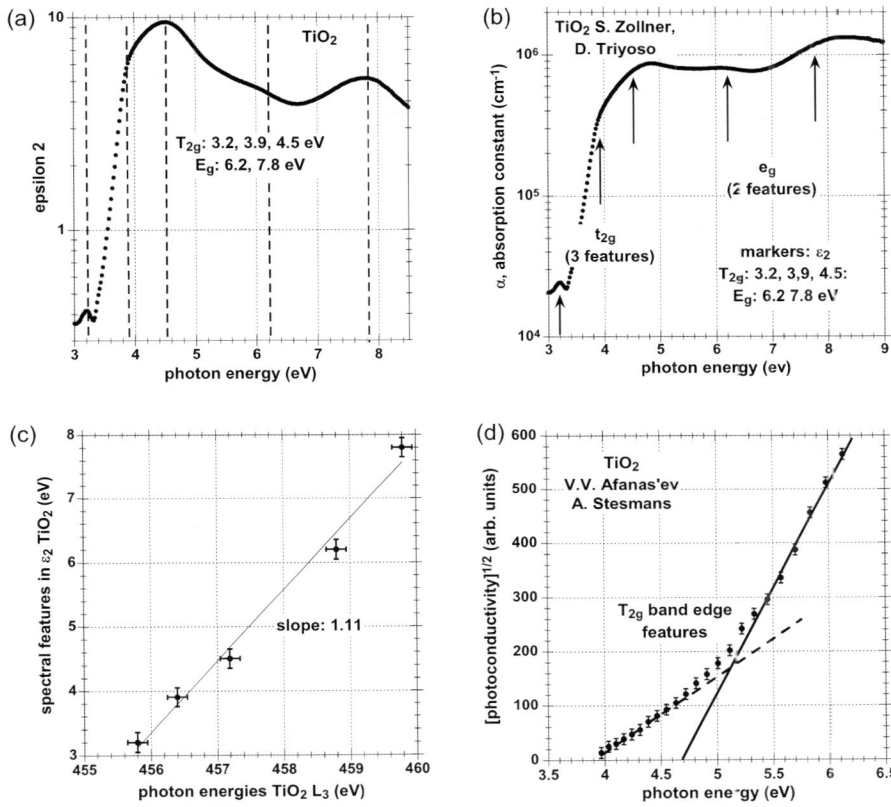

**Fig. 3.** Band edge spectra for TiO$_2$: (**a**) the imaginary part of the complex constant, $\varepsilon_2$, where $\varepsilon_c = \varepsilon_1 + i\varepsilon_2$, (**b**) the absorption constant, $\alpha$, (**c**) comparison between the five $d$-state features in the L$_3$ edge, and in $\varepsilon_2$, and (**d**) the photoconductivity response

conduction band. The final point evident in a comparison between Fig. 3b and Fig. 2b, is the strong matrix element effects for the $d$-state transitions. The lowest energy band edge transition at 3.2 eV is very weak in $\varepsilon_2$ and $\alpha$, and also in the L$_3$ edge. This means the onset of strong absorption occurs at energy about 1 eV greater than the threshold absorption. This is reflected in the photoconductivity edge spectrum in Fig. 3d as well.

There are important qualitative differences between the corresponding XANES, VUV SE and PC spectra of TiO$_2$, and both ZrO$_2$ and HfO$_2$. These are first illustrated for ZrO$_2$, and then summarized more briefly for HfO$_2$. Figures 4a and b display respectively in (a) the Zr M$_{2,3}$ spectra for transitions for Zr spin-orbit split 3$p$ states to empty Zr 4$d$ and 5$s$ states, and in (b) the O K$_1$ edge. In contrast to TiO$_2$, $d$-state term splittings are not detectable in the respective XANES M$_{2,3}$ transitions for ZrO$_2$ due to the short core hole

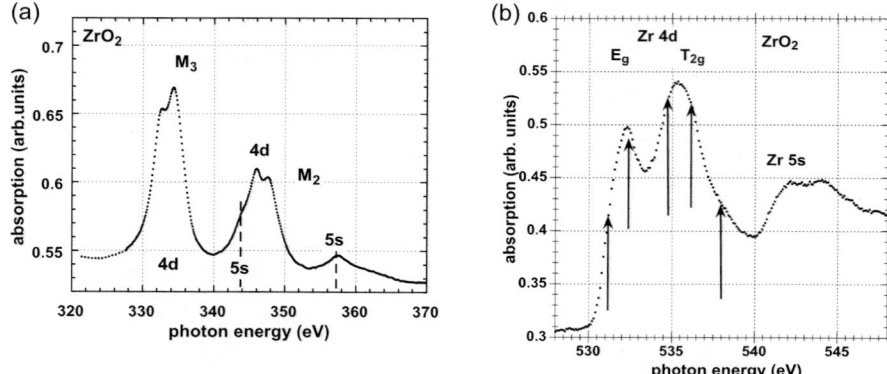

**Fig. 4.** X-ray absorption near-edge spectra for $ZrO_2$: (**a**) the Zr $M_{2,3}$ spectrum, and (**b**) the O $K_1$ edge spectrum

lifetimes associated with the higher $Z$ values of 40 for Zr, and 22 for Ti, but are observable in the $ZrO_2$ O $K_1$ spectra through numerical differentiation.

The average $4d$-state splitting for $ZrO_2$ has been obtained by deconvolution of the spectra in Fig. 4a, and is $2.3 \pm 0.2\,eV$. After differentiation, the average $d$-state splitting in the O $K_1$ edge in Fig. 4b is $3.3 \pm 0.2\,eV$. The differences between these energy separations are attributed to differences for transitions to empty atomic states in the $M_{2,3}$ edge, and molecular orbital states for the O $K_1$ edge.

Figures 5a, b and c are respectively (a) $\varepsilon_2$, (b) $\alpha$ and (c) the PC response for $ZrO_2$. Three $4d$-state features are evident in both $\varepsilon_2$ and $\alpha$, the two $E_g$ states, and the lowest $T_{2g}$ state. Their energy separations are in good quantitative agreement with the corresponding separations in the O $K_1$ edge. In $\varepsilon_2$, the $E_g$ splitting is $0.9 \pm 0.2\,eV$, and the energy separation between the lower $E_g$ state and the lowest $T_{2g}$ state is $2.3 \pm 0.2\,eV$, compared with $1.0 \pm 0.2\,eV$, and $2.6 \pm 0.3\,eV$ in the O $K_1$ edge. The threshold for strong absorption is correlated with the first $d$-state feature in $\alpha$, and in the PC spectrum. This threshold energy defines an effective optical band gap of $\sim 5.6 \pm 0.1\,eV$. Weak features at energies below this effective optical band gap, and extending to about $5.0\,eV$, are addressed in the section that deals with intrinsic defect states.

The core hole lifetime in $HfO_2$ for $4p$ to $5d$ and $6s$ transitions is shorter than it is for $ZrO_2$, since $Z$ has increased to 70. The $N_{2,3}$ spectra show a single $5d$-state feature with no structure that can be extracted by differentiation. Figures 6a, b, c and d display respectively in (a) O $K_1$, (b) $\varepsilon_2$, (c) $\alpha$ and (d) PC spectra for $HfO_2$. Differentiation of the O $K_1$ spectra yields five $d$-state energies, the lower two corresponding to the $E_g$ states, and the upper three to the $T_{2g}$ states. The $E_g$ separation is $1.0 \pm 0.2\,eV$, and the lower $E_g$ state and the lowest $T_{2g}$ state separation is $3.0 \pm 0.3\,eV$. The average $5d$-state splitting

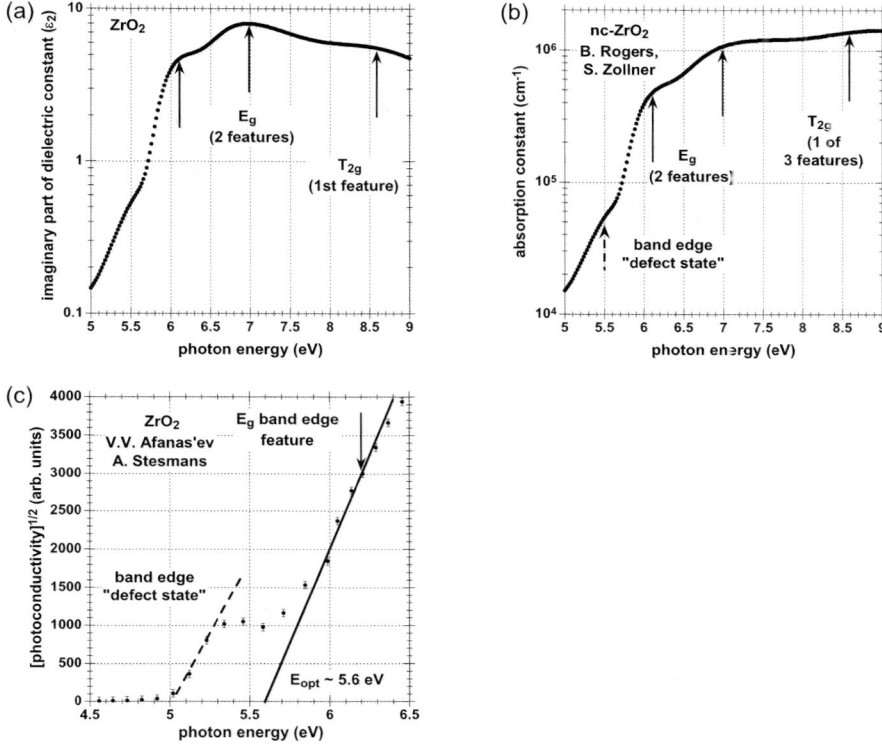

**Fig. 5.** Band edge spectra for $ZrO_2$: (**a**) the imaginary part of the complex constant, $\varepsilon_2$, where $\varepsilon_c = \varepsilon_1 + i\varepsilon_2$, (**b**) the absorption constant, $\alpha$, and (**c**) the photoconductivity response

is increased to $4.6 \pm 0.2\,\mathrm{eV}$, and is larger than the corresponding splitting of $2.7 \pm 0.1\,\mathrm{eV}$ for $TiO_2$ and $3.3 \pm 0.2\,\mathrm{eV}$ for $ZrO_2$, indicating a trend for $3d$-, $4d$-, and $5d$-state mixing with O $2p$ anti-bonding states.

The separation between $E_g$ features in the $\varepsilon_2$ spectrum of $HfO_2$ is larger than in the O $K_1$ edge, $\sim 1.4\,\mathrm{eV}$ as compared to $\sim 1\,\mathrm{eV}$. This may be a matrix element effect, since there is a factor of $> 6$ difference in the values of $\varepsilon_2$ at these $d$-state energies, compared to a difference of $< 2$ for the corresponding features in $ZrO_2$ where the agreement between the $E_g$ splitting between $\varepsilon_2$ is much closer. The effective band gap from $\alpha$ and the PC response is $5.8 \pm 0.2\,\mathrm{eV}$, and this band edge correlates with the lowest-energy $d$-state feature. This optical band gap is slightly larger than for $ZrO_2$, consistent with the $0.3\,\mathrm{eV}$ difference in the atomic Zr $4d$ and Hf $5d$ states [8]. As in the case of $ZrO_2$, there is a sub-band gap feature extending to about $5.3$–$5.4\,\mathrm{eV}$ that is addressed in Sect. 5

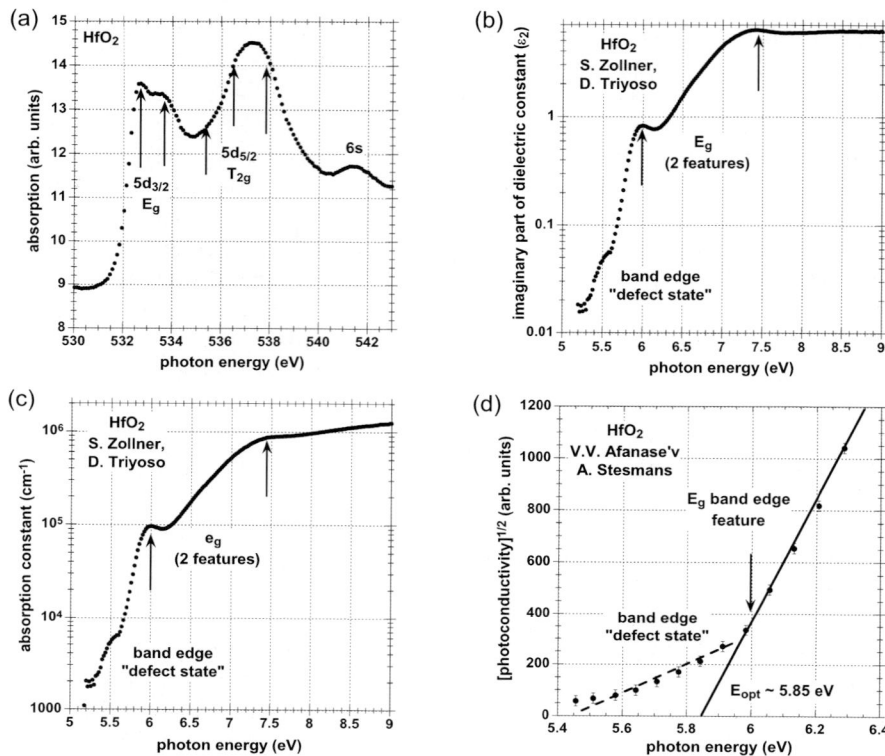

**Fig. 6.** X-ray absorption near-edge spectra for $HfO_2$: (**a**) the O $K_1$ edge spectrum, (**b**) the imaginary part of the complex constant, $\varepsilon_2$, where $\varepsilon_c = \varepsilon_1 + i\varepsilon_2$, (**c**) the absorption constant, $\alpha$, and (**d**) the photoconductivity response

The conclusions reached from the spectroscopic studies presented for TM elemental oxides are the following: 1. as-deposited TM oxides are nanocrystalline as-deposited by remote plasma-assisted chemical vapor deposition or reactive evaporation, and 2. the band edge $d$-states detected by XANES and/or VUV SE indicate a complete removal of $d$-state degeneracies which is attributed to a J–T effect. All of our spectroscopic studies were restricted to film thicknesses in excess of about 5 nm. The $L_{2,3}$, $M_{2,3}$ and $N_{2,3}$ spectra of $TiO_2$, $ZrO_2$ and $HfO_2$, respectively, are intra-atomic spectra that reflect short-range order extending to bonding distances as close as $\sim 0.5$ nm. The $d$-state splittings reflect the bonding symmetry established by the nearest-neighbor O-atoms, but the final states are *effectively atomic $d$-states*, rather than molecular orbital states that have a mixed O $2p$-TM $d$-state character. In contrast, the final states in O $K_1$ edge spectra have a mixed character, and reflect order that can extend to distances as short as 2 nm.

It is possible that films thinner than this may have a non-crystalline or amorphous-like character reflecting intermediate range order, or two-dimensional effects if the scale of intermediate range order becomes less than the film thickness. These issues will not be addressed in this chapter, and require more experimentation that couples different spectra including XANES, EXAFS and VUV SE, combined with X-ray diffraction, and electron diffraction and imaging. In this regard, it is important to recognize that the detection of crystallinity on a scale of less than 5 nm cannot be accomplished by conventional X-ray diffraction, XRD, and other approaches such as XANES, EXAFS, and high-resolution transmission electron microscopy, HRTEM, are required to detect nanocrystallinity on a smaller scale, e.g., the 2 nm scale that is present in many as-deposited films. This final issue is addressed below in the section that treats TM/RE complex oxides, in particular in comparisons between VUV SE spectra of single-crystal and thin film samples with the same composition, but with markedly different scales of crystallinity.

# 4 Experimental Results: TM/RE Complex Oxides

Conduction band edge $d$-states are compared for three different classes of complex or pseudo-binary oxides: 1. mixed tetravalent – trivalent $ZrO_2-Y_2O_3$ alloys, 2. tetravalent $Zr(Hf)O_2-TiO_2$ alloys, and 3. trivalent La scandate and aluminate. Low $Y_2O_3$ content cubic $ZrO_2-Y_2O_3$ alloys display two crystal-field split $4d$-features in O $K_1$ spectra. Alloys with higher $Y_2O_3$ content, as well as $Zr(Hf)O_2-TiO_2$ alloys display increased $d$-state multiplicity. O $K_1$ spectra of perovskite-structured $LaScO_3$ and $LaAlO_3$ indicate Jahn–Teller (J–T) $d$-state term splittings with contributions from both trivalent atomic species, La and Sc, as well as J–T $p$-state splittings for Al.

Figure 7 indicates O $K_1$ edge spectra for (a) $ZrO_2$, (b) and (c) $ZrO_2$ alloyed with 18.5 and 27.5 % $Y_2O_3$, and (d) and (e) cubic, or defective $CaF_2$ with O-atom vacancies of the F sub-lattice, and bixbyite modifications of $Y_2O_3$, respectively. Consistent with eight-fold coordination of both Y and Zr, the lowest-energy $4d$-state features in (a) to (d) are assigned to doubly degenerate $4d_{3/2}$ ($E_g$) states, and the higher-energy features to triply degenerate $4d_{5/2}$ ($T_{2g}$) states. The coordination of Y is six-fold in trace (e) for bixbyite, and the order of the $4d$ states is reversed. Differentiation of spectra indicates symmetric $4d_{3/2}$ and $4d_{5/2}$ features in (b), (c) and (d), i.e., no degeneracy removal, but departures from symmetry indicative of removal of $d$-state degeneracies in (a) and (e). Degeneracy removal in (a) and (e) is attributed to J–T term splittings associated with distorted local bonding of Zr atoms with respect to their O-atom neighbors in (a), and for Y atoms in (e), also with respective to O-atom neighbors. $M_{2,3}$ spectra for Y from 300 and 330 eV, and Zr from 330 and 370 eV are in Fig. 8, in (a) $ZrO_2$, (b) and (c) alloys of $ZrO_2$ with 18.5 and 27.5 % $Y_2O_3$, and (d) bixbyite $Y_2O_3$. Features marked by dashed lines are transitions to Y and Zr $4s$ states; other features

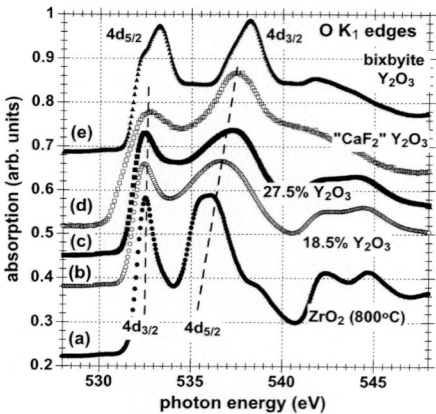

**Fig. 7.** O K$_1$ edge spectra for: (a) ZrO$_2$, (b) and (c) ZrO$_2$ alloyed with 18.5 and 27.5 % Y$_2$O$_3$, and (d) and (e) cubic and bixbyite Y$_2$O$_3$

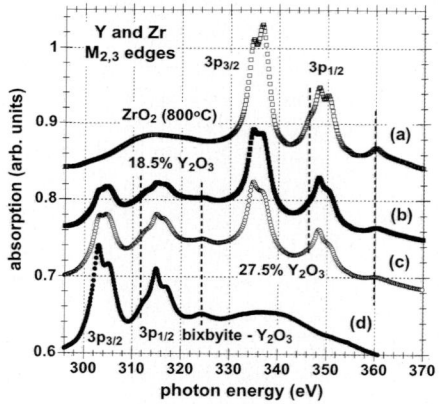

**Fig. 8.** M$_{2,3}$ spectra for (a) ZrO$_2$, (b) and (c) alloys of ZrO$_2$ with 18.5 and 27.5 % Y$_2$O$_3$, and (d) bixbyite Y$_2$O$_3$. Features marked by dashed lines are transitions to Y and Zr 4$s$ states. All other features terminate in 3$d$ states

terminate in 3$d$ states. M$_{2,3}$ transitions are intra-atomic, originating at spin-orbit split Y or Zr 3$p_{3/2}$ and 3$p_{1/2}$ core states, and terminating, respectively in Y or Zr 4$d$ and 5$s$ states. Three features are evident in the Y and Zr M$_3$ and M$_2$ spectra; energy differences between corresponding pairs are $\sim 31$ eV, essentially the same as the respective 3$p$ core state differences. The Y, Zr $d$-state features are too broad to reveal J–T term splittings in the spectra of ZrO$_2$, or bixbyite Y$_2$O$_3$, even though these are evident in the respective O K$_1$ spectra.

Results in Fig. 7a for O $K_1$ edges indicate significant changes in bonding associated to O-atom vacancies, i.e., as an additional band edge feature centered at $\sim 530.5\,\mathrm{eV}$. The end-member traces in (a) for $ZrO_2$, and (d) and (e) for two different crystalline forms of $Y_2O_3$ each indicate features in the O $K_1$ spectra indicative of O-atom vacancies. At this time, it is important to recognize that the spectra in (a) and (e), for $ZrO_2$ and bixbyite $Y_2O_3$, are not dependent on O-atom vacancies even though they are present. In contrast, the cubic form of $Y_2O_3$ is stabilized by these vacancies, as are the O-atom vacancies for the two $Y_2O_3 - ZrO_2$ alloys in traces (b) and (c).

Figure 9 includes O $K_1$ edge spectra of (a) $(HfO_2)_{0.5}(TiO_2)_{0.5}$ and (b) $(ZrO_2)_{0.5}(TiO_2)_{0.5}$. The $(HfO_2)_{0.5}(TiO_2)_{0.5}$ O $K_1$ spectra has three $d$-state features. The lowest-energy feature is assigned to a $TiO_2$ $3d$ state, the middle feature is assigned to mixture of $TiO_2$ $3d$ and $HfO_2$ $5d$ states, and the highest feature to a $HfO_2$ $5d$ state. Differentiation of the spectrum in Fig. 9a indicates symmetric line shapes with no additional $d$-state features. This is consistent with a random alloy of six-fold coordinated Ti, and either six- or eight-fold coordinated Hf. In contrast, other studies have indicated a compound phase for $ZrTiO_4$ with six-fold coordination for both Ti and Zr [9]. The O $K_1$ spectrum for $ZrTiO_4$ is consistent with a compound in which both the Zr and Ti atoms are six-fold coordinated, and are in distorted octahedral bonding arrangements. The Ti $L_3$ spectra for both $(HfO_2)_x(TiO_2)_{1-x}$ and $(ZrO_2)_x(TiO_2)_{1-x}$ display essentially the same five $3d$-state features as those of $TiO_2$. This is consistent with $d$-state degeneracy removal for Ti atoms in a distorted octahedral bonding environment in two titanates discussed above, the $(HfO_2)_{0.5}(TiO_2)_{0.5}$ alloy and the stoichiometric $ZrTiO_4$ compound. The degeneracy removal of the Ti $3d$ states in the $(HfO_2)_{0.5}(TiO_2)_{0.5}$ alloy is indicative of chemical phase separation on a scale of $\sim 2\,\mathrm{nm}$, whereas the degeneracy removal in the stoichiometric $ZrTiO_4$ compound is a manifestation of crystalline order, which may also be on a similar length scale as small as $\sim 2\,\mathrm{nm}$.

Figure 10 displays O $K_1$ spectra: (a) as-deposited $LaAlO_3$, (b) as-deposited $LaScO_3$ and (c) $LaScO_3$ annealed at $1000\,^\circ\mathrm{C}$. The $LaAlO_3$ spectrum displays two band edge features assigned to Al $3p$ states at $\sim 531.5$ and $534\,\mathrm{eV}$, as well as features at $535.5$ and $537.5\,\mathrm{eV}$ assigned to La $5d$ states. The $LaScO_3$ spectra indicate both Sc $3d$ features, and La $5d$ features. The differences between the as-deposited and $1000\,^\circ\mathrm{C}$ annealed samples are the spectral line-widths of these $d$-state features. Figure 10d displays the Sc $L_3$ spectra for as-deposited $LaScO_3$ and Fig. 10e displays the Sc $L_3$ spectra for $LaScO_3$ annealed at $1000\,^\circ\mathrm{C}$. Difference between these films are again in the respective spectral line-widths of the J–T term split features, reflecting changes in crystal grain size increasing significantly between the as-deposited annealed films. These differences are consistent with nanocrystallinity in as-deposited films with a scale of order extending to at least $2\,\mathrm{nm}$ [1], and XRD detection of crystallinity after annealing to $800\,^\circ\mathrm{C}$ and higher, and consistent with a scale of order/crystalline grain size of at least $5\,\mathrm{nm}$.

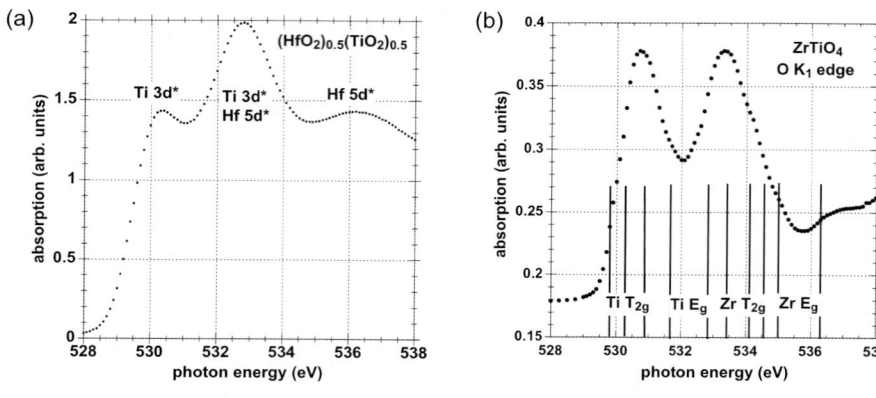

**Fig. 9.** X-ray absorption near-edge spectra for Zr and Hf titanate alloys: O $K_1$ edge spectra of (**a**) a $(HfO_2)_{0.5}(TiO_2)_{0.5}$ alloy and (**b**) the stoichiometric compound $ZrTiO_4$ or equivalently $(ZrO_2)_{0.5}(TiO_2)_{0.5}$

Figures 11a and b display respectively the photoconductive response for $LaAlO_3$ and $LaScO_3$. The extrapolated edge features at $\sim 4.6$ and $5.5\,eV$, each with an experimental uncertainty of at least $\pm 0.15\,eV$, are assigned to the Al $2p$ states. The energy difference between the two features in the O $K_1$ edge of $LaAlO_3$ is $> 2\,eV$, so that a direct comparison between these spectra is not possible. For example, if the PC feature at $\sim 5.5\,eV$ is associated with the O $K_1$ edge feature at $\sim 533.5\,eV$, then the Al feature in the O $K_1$ edge at $\sim 531.5\,eV$ is not associated with either of the features evident in the PC. The issue may be correlated with differences in matrix elements for O $K_1$ edge, and band edge transitions, and needs additional studies, particularly of the term splittings that may be evident in Al $2p$ to $3s$ transitions. These transitions have not as yet been studied by XAS, but are planned by our research group. Additional support for this approach is displayed in Fig. 11c. This shows the energies of the three PC edge features for $LaScO_3$, and the five lowest-energy $d$-state features in the $LaScO_3$ spectrum of annealed sample, as a function of the five atomic $d$-state energies extracted from the Sc $L_3$ spectrum of the annealed sample of $LaScO_3$. The same linear scaling applies to the three states extracted from the PC spectrum, and the five states from the O $K_1$, establishing that all of these features can be assigned to Sc $3d$ states. The decreased slope between the PC and O $K_1$ edges is consistent with the mixed O $2p$–Sc $3d$ character of the PC and O $K_1$ final states, and the atomic character of the $L_3$ final states.

Consider now the lowest-energy Sc $3d$ edge features in the $LaScO_3$ spectra. Based on studies of $Sc_2O_3$ in [10], the lowest-energy band gap in this material is $\sim 4.6\,eV$, and is the same as the lowest-energy feature for a Sc $3d$ state in $LaScO_3$. The onset of strong absorption in $Sc_2O_3$ occurs at $\sim 6\,eV$. This relationship between band edge spectra in elemental oxides, such as $TiO_2$ and

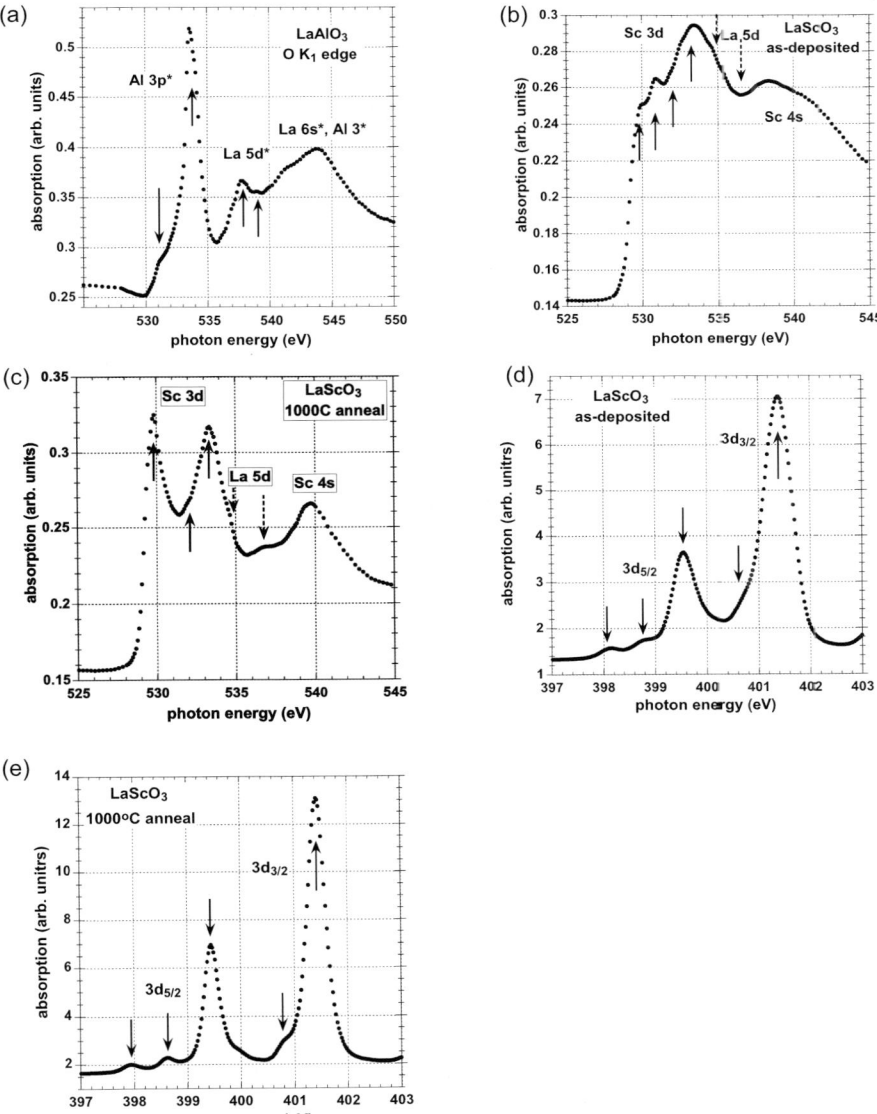

**Fig. 10.** O $K_1$ spectra: (**a**) as-deposited $LaAlO_3$, (**b**) as-deposited $LaScO_3$, and (**c**) $LaScO_3$ annealed at $1000\,^{\circ}C$. 10 displays addition $LaScO_3$ spectra: (**d**) Sc $L_3$ spectra as-deposited, and (**e**) Sc $L_3$ spectra annealed at $1000\,^{\circ}C$

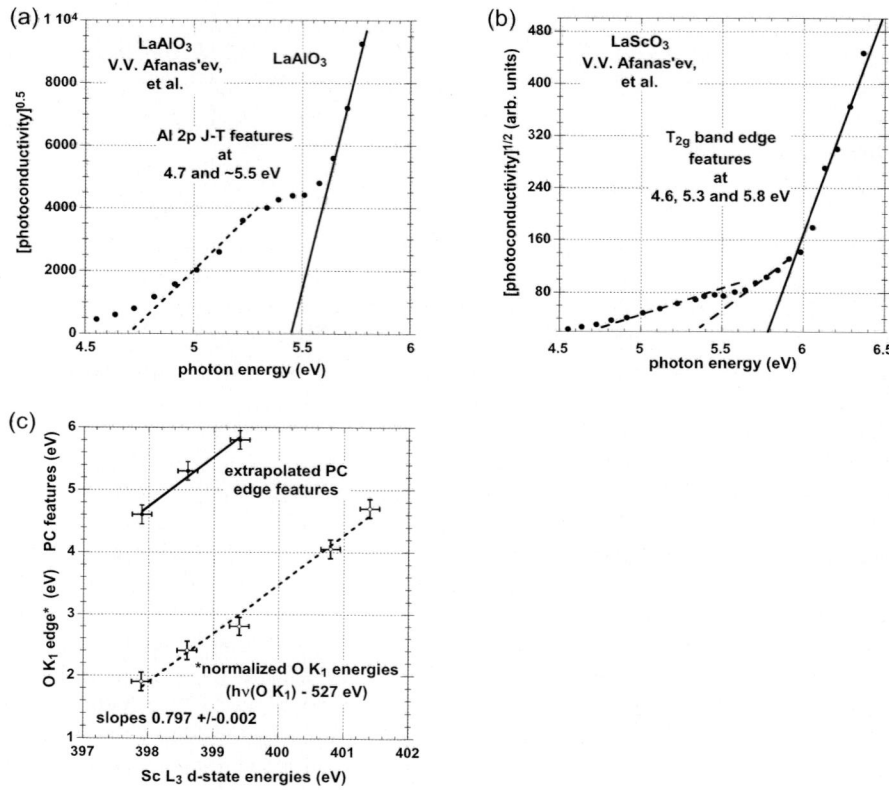

**Fig. 11.** Photoconductive response spectra for: (**a**) $LaAlO_3$, and (**b**) $LaScO_3$. (**c**) photon energies PC and 1000 O $K_1$ Sc $3d$ features as a function the photon energy of the five Sc $3d$ features in the $L_3$ spectrum of $LaScO_3$ annealed at 1000 °C

$Sc_2O_3$, and complex oxides in which they are a constituent elemental oxide is well established and is addressed in [11].

In summary, qualitative and quantitative differences in O $K_1$ *inter-atomic* M–O transitions, and $L_3$, $M_3$ and $N_3$ intra-atomic have been presented for three different groups of complex oxides. These include 1. tetra- and trivalent transition metal (TM) and/or rare earth (RE) atom oxides, 2. two TM tetra-valent oxides, and 3. TM and RE oxides, or Al oxide. These studies indicate that matrix element effects must be considered in the determination of conduction band offset energies between complex oxides, and Si or other semiconductors. These effects can be quantitatively different for all three classes of complex oxides, so considerable care must be taken in combining optical band edge data with valence band edge energy differences obtained from either UV or X-ray photoemission studies, particularly for alloys and compounds that include six-fold coordinated species, as for example RE scandates such as $GdScO_3$ and $DyScO_3$. Care must also be taken to distinguish between transitions to empty conduction band states that have RE/TM $d$-character, and intra $4f$–$4f'$ transitions that can also occur in the visible region of the spectrum. This issue is in part discussed in the introduction to this book.

# 5 Experimental Results: Intrinsic Defect States

This section deals with intrinsic bonding defects. Spectroscopic measurements, including XANES and VUV SE, complemented by photoconductivity, PC, are used to first identify localized *defect states* approximately $0.6 \pm 0.2\,eV$ below the conduction band edges of $HfO_2$ and $ZrO_2$ [2–4]. A second defect state has also been identified; it is localized within the band gap of $ZrO_2$, and approximately $4.2 \pm 0.15\,eV$ above the valence band edge, or equivalently, $1.4 \pm 0.15\,eV$ below the conduction band edge. Based on model electronic structure calculations, the conduction band edge defects are assigned to states localized at grain boundaries between the nanocrystallites (hereafter, GBDs) [1], and the defect states deeper in the gap at about $1.4 \pm 0.15\,eV$ below the conduction band edge are assigned to oxygen atom vacancies (hereafter, OVDs) [5]. It is important to note that there is a one-on-one relationship between GBD defects at grain boundaries that are clearly evident in HRTEM imaging at scales in which lattice planes are evident, and GBDs in between regions with nanocrystallinity for which the scale of order is significantly reduced, $\sim 2\,nm$. These defects arise because of decreases in local site symmetry of the TM atoms with respect to their O-atom neighbors. The differences between the GBDs of these two types of disorder are likely to be quantitative with respect to the energies and/or energy spread of these defects, rather than qualitative with respect to their energy always being below the lowest conduction band states associated with the scale order within the bulk of the crystalline grain.

Figures 5a and c, respectively indicate (a) the absorption constant, $\alpha$, for nanocrystalline $ZrO_2$ thin films as a function of photon energy as derived from spectroscopic ellipsometry measurements, and (c) the square root of the photoconductivity (PC) response versus photon energy. Qualitatively similar results have been obtained for nanocrystalline $HfO_2$ films. The onset of the stronger PC response is defined by a linear extrapolation (solid line) to $5.6\,eV$ in Fig. 5c, and is effectively at a similar extrapolation of the band edge absorption in Fig. 5a. Each of the spectra shows an additional feature at lower photon energies that extrapolates to an energy of $\sim 5\,eV$ in the PC response in Fig. 5c (dashed line), and to a similar energy for $\alpha$ versus photon energy in Fig. 5a. These lower-energy band edge spectral features are assigned to grain-boundary defect states or GBDs that are an *inherent* aspect of the nanocrystallinity.

The O $K_1$ edge spectra for nanocrystalline $ZrO_2$ thin films for two different post-deposition anneals in inert ambients at $500\,°C$ and $800\,°C$ are displayed in Fig. 11a. The strong features with spectral peaks at $\sim 532.5\,eV$ and $535.5\,eV$, respectively, are due to mixing of the O $2p$ anti-bonding states with Zr $E_g$ ($4d_{3/2}$) and $T_{2g}$ ($4d_{5/2}$) states. Differentiation of these features identifies the energies of the respective doublet and triplet term-split states [7]. Figure 11a includes an expanded view of the leading edge of these spectra: the sample annealed at $500\,°C$ does not display any distinct spectral feature in this regime, whereas the one annealed at $800\,°C$ displays a feature at $\sim 530.5\,eV$ that is also present in $(ZrO_2)_{1-x}(Y_2O_3)_x$ cubic zirconia thin film samples for $x$ in the range from 0.15 to 0.25, as indicated in Fig. 11b. This spectral feature is $\sim 2\,eV$ below the first $4d_{3/2}$ ($E_g$) feature of the respective O $K_1$ spectra. This spectral feature is assigned to O-atom vacancies [5]. This is confirmed by the absorption edge determination for a bulk crystal of cubic zirconia.

Figure 11c displays the absorption edge spectrum for a bulk cubic zirconia crystal $[(ZrO_2)_{0.905}(Y_2O_3)_{0.095}]$. The strong sub-band gap absorption with a threshold of $\sim 4.2\,eV$ is assigned to the O-atom vacancy, and is in excellent agreement with the calculations of the Robertson group [5]. It is important to recognize that the energy of the feature in the O $K_1$ edges in Figs. 11a and b to O-vacancies is consistent with the energy of the spectral feature in Fig. 11c. This feature at $4.2\,eV$ in Fig. 11c is $2.0 \pm 0.2\,eV$ below the band edge $4d_{3/2}$ ($E_g$) feature that is obtained by differentiation of the spectrum in Fig. 11c. Figure 11d presents additional evidence for this assignment; it includes the abortion edge spectrum for a thin film sample of cubic zirconia, an alloy of the following composition: $[(ZrO_2)_{0.85}(Y_2O_3)_{0.15}]$. The spectral feature at $\sim 4.2\,eV$ is clearly evident in this spectrum, and occurs at about the same energy as the absorption threshold for the bulk single-crystalline sample in Fig. 11c. There are differences in higher-energy features which reflect the different scales of order in thin film and bulk crystalline samples. These are beyond the scope of the issues addressed in this review.

Figure 12 indicates an energy level diagram that includes a Si substrate, a thin $SiO_2$ interfacial oxide, and then a thicker film of either $ZrO_2$ and $HfO_2$. Defect state energies with the band gap of $ZrO_2$, which are essentially the same for $HfO_2$, are indicated. The energy of the grain-boundary defect has been estimated from model calculations described in [5]; the energy of the O-atom vacancy comes from the Robertson group [5], and the energy of the H-defect from the universal alignment model of Van de Walle in [6].

The electrical activity of these defects has been observed in studies by the IMEC group and their collaborators that are summarized in [12], and by our group at NC State as well. Bulk traps $\sim 0.5$ to $0.8\,eV$ below the high-$k$ conduction band have been reported for injection from $n$-type Si, through thin $SiO_2$ layers ($\sim 1$ to $1.0\,nm$) into thicker $ZrO_2/HfO_2$ films [12]. A hole trap $\sim 0.26\,eV$ above the valence band edge was extracted from C–V data for gate electron injection in the $ZrO_2$ MOSCAPs [12]. Similar results have been

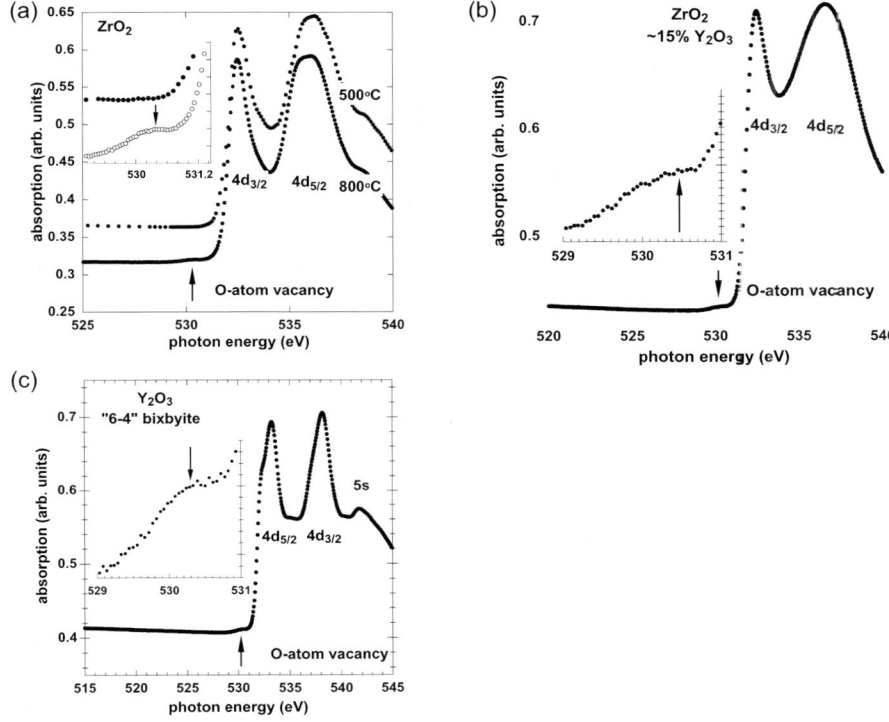

**Fig. 12.** O $K_1$ spectra, with expanded view of the leading edge of these spectra: (a) $ZrO_2$ and (b) an alloy of $ZrO_2$ with 18.5 $Y_2O_3$, and (c) the absorption edge spectrum, $\alpha$, for a bulk cubic zirconia crystal $[(ZrO_2)_{0.905}(Y_2O_3)_{0.095}]$

obtained by us in J–V data for $n$-Si substrate electron injection in MOSCAPS with stacked $SiO_2-ZrO_2$, and $SiO_2-HfO_2$ dielectrics. We have also found an interface trap at $\sim 0.3\,eV$ above the Si valence band edge in C–V traces for gate e-injection in $SiO_2-HfO_2$ MOSCAPs. The energy of the interface trap relative to the $HfO_2$ valence band is $4.2 \pm 0.2\,eV$, agreeing with theory [5] and cathodoluminescence measurements. Our studies, and those in [1], do not indicate an electronically active defect associated with H [6].

To the best of our knowledge, this combined spectroscopic and theory/modeling of intrinsic defects in nanocrystalline $HfO_2$ and $ZrO_2$, coupled with device studies, represents the first unambiguous assignment for electronically active defects in high-$k$ gate stacks. The energies of these defects relative to the forbidden energy gap in Si also explain differences in bias temperature instabilities, BTIs, in NMOS and PMOS FETs, and present a significant obstacle for integration of $HfO_2/ZrO_2$ dielectrics into scaled Si CMOS.

The results presented in this paper, combined with discussions in [5], make it unlikely than any nanocrystalline transition metal or lanthanide series rare earth atom elemental or complex oxide will be integrated into scaled CMOS devices. However, as discussed in the preceding Chapter, it may be possible to perform interface engineering at the atomic scale and thereby utilize nanocrystalline $ZrO_2$ or $HfO_2$. However, because of the necessity of buffer $SiO_2$ layers between a Si substrate and the highest-$k$ component of a stacked dielectric, it is not obvious that these dielectrics will extended EOT scaling to the levels of 0.5 nm below those required by the most aggressive roadmaps. Combinations of defect states associated with fixed charge, band edge grain-boundary defects, and oxygen atom vacancies, as well as defect states associated with the presence or absence of hydrogen [1, 5, 6], will introduce sufficient quantities of defects to prevent devices from displaying targeted performance and/or reliability metrics. In contrast, Zr or Hf Si oxynitride films with approximately equal concentrations of $SiO_2$, $Si_3N_4$ and $ZrO_2$ or $HfO_2$ will remain non-crystalline after annealing to at least 1100 °C, and offer engineering solutions for high-performance devices with EOTs of $\sim 0.7$ to 0.8 nm, and mobile devices with EOTs to $\sim 1.5$ nm. This last statement is predicated on an assumption that defects specifically associated with grain boundaries and/or O-atom vacancies are not generic to non-crystalline Zr and Hf Si oxynitride alloys. Studies are being pursued by my research group to address this issue. In the last Chapter, one aspect of these studies was addressed: prevention of chemical phase separation (CPS) at interfacial Zr of Hf silicate layers that define metallurgical interfaces between $SiO_2$ and 1. $ZrO_2$ or $HfO_2$, and 2. Zr and Hf silicates. It was specifically noted that atomic-scale interface engineering was required to eliminate Zr or Hf silicate bonding at interfaces between $SiO_2$ and Zr or Hf Si oxynitride alloys.

# 6 Experimental Results: Zr Silicate and Si Oxynitride Alloys

Figures 13a and b present FTIR results for Zr silicate alloys with $ZrO_2$ fractions, $x$, equal to 0.23 and 0.5 [13]. The feature at $\sim 950\,cm^{-1}$ in Fig. 13a for the as-deposited films is assigned to a terminal $Si-O^{1-}$ group, an indicator of network disruption by addition of the ionic $ZrO_2$ elemental oxide into the more covalently bonded $SiO_2$ *host* network. The broader $Si-O$ bond-stretching feature in Fig. 13b is assigned to the $SiO_4^{4-}$ molecular ion, and indicates *complete network disruption* at this composition. There are changes in FTIR spectral features in Figs. 13a and b between the as-deposited films and films annealed at 900 °C. These indicate CPS into $SiO_2$ and $ZrO_2$. Separation is confirmed by HRTEM imaging [14]. Spectral features change gradually over the entire range of annealing temperatures, but not with a well-defined activation energy as in an Arrhenius plot. Comparisons between the kinetics for the CPS in the silicate alloys, and the CPS for thin film silicon suboxide,

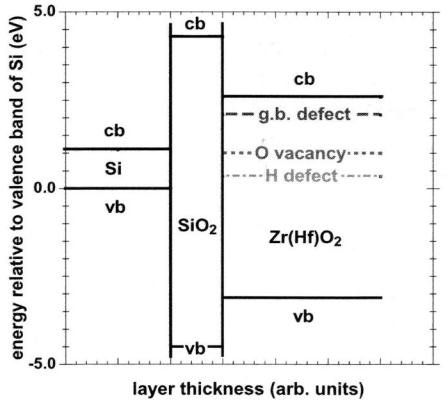

**Fig. 13.** An energy level diagram that includes a Si substrate, a thin $SiO_2$ interfacial oxide, and a thicker film of either $ZrO_2$ or $HfO_2$. Defect state energies within the band gap of $ZrO_2$ are indicated for a grain-boundary defect, an O-atom vacancy, and a H-defect

$SiO_x$, $x \sim 1$, into non-crystalline $SiO_2$ and crystalline Si indicate similar kinetics. This is used to infer a similar mechanism to that in [15] for the CPS of TM silicate thin films.

The corresponding XRD spectra (not shown in this Chapter) indicate no detectable crystallization, except for the $x = 0.5$ sample after a $900\,°C$ anneal. The FTIR spectra for a Zr silicate alloy with $x = 0.61$, in which there is network inversion, are consistent with the $SiO_4^{4-}$ molecular ion as-deposited, and CPS into non-crystalline $SiO_2$ and crystalline $ZrO_2$ after a $900\,°C$ anneal.

Figures 14a and c display FTIR results for two pseudo-ternary $Zr(Hf)O_2$–$SiO_2$–$Si_3N_4$ alloys, and Figs. 14b and d include derivative XPS O $1s$ core level spectra for the same alloys. Spectra in Figs. 14a and c are for an $Si_3N_4$ alloy, $\sim 25\,\%$ $Si_3N_4$, $\sim 45\,\%$ $SiO_2$ and $\sim 30\,\%$ $ZrO_2$. Concentrations are accurate to $\pm 5\,\%$ in these ternary alloys. The FTIR and XPS spectra indicate CPS. In the FTIR spectrum, there are non-crystalline $SiO_2$ features at $\sim 1050\,cm^{-1}$ and $800\,cm^{-1}$, and a nanocrystalline $ZrO_2$ feature at $\sim 450\,cm^{-1}$. In the XPS, there are two spectral features after a $1000\,°C$ anneal. XAS studies of the O $K_1$ and N $K_1$ edges of this alloy composition indicate a loss of bonded nitrogen for annealing temperatures greater than $\sim 750\,°C$.

Spectra in Figs. 14b and d are for an alloy with approximately equal concentrations, $\sim 45\,\%$, of $Si_3N_4$ and $SiO_2$, and a $ZrO_2$ concentration of at least $10\,\%$. After annealing, the FTIR indicates weak Si–O features which are more $SiO_4^{4-}$ ion-like than $SiO_2$ network-like, and a non-crystalline $ZrO_2$ feature. This is supported by derivative XPS that shows a single dominant feature at an energy significantly different from that of either non-crystalline

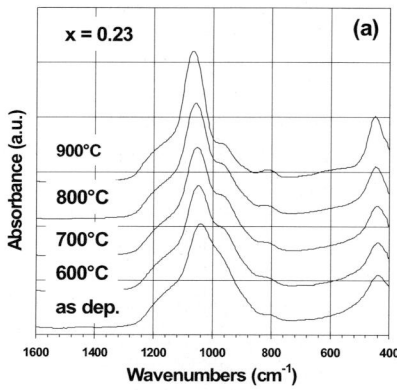 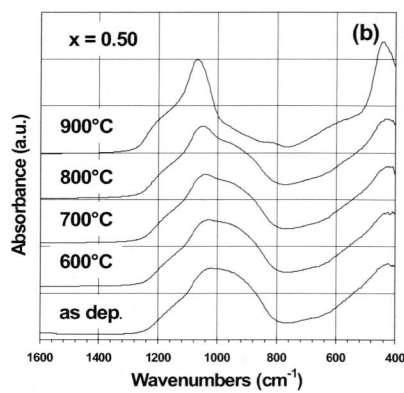

**Fig. 14.** Fourier transform infra-red absorbance spectra for Zr silicate alloys with $ZrO_2$ fractions, $x$, equal to (**a**) 0.23 and (**b**) 0.5

$SiO_2$ or $ZrO_2$, and closer to that of the feature in as-deposited Zr silicate with $50\%$ $ZrO_2$, and therefore with $SiO_4^{4-}$ molecular ions as well. The single O $1s$ XPS feature is also consistent with studies of the O $K_1$ edge which indicate no detectable changes between as-deposited films and those annealed at 500, 750 and $1000\,°C$.

The experimental studies have benefited from the extension of bond constraint theory (BCT) originally developed to explain glass formation [16,17], and extended compositional stability of Na, Ca and Al silicates [18] to transition metal silicates, including $Zr(Hf)O_2$ and $Y(La)_2O_3$. This application of BCT demonstrated that 1. all group IVB and IIIB transition metal (TM) silicates are *over-constrained* upon deposition, with the number of constraints per atom exceeding the network dimensionality of three, and having a *significant break-up* of the $SiO_2$ network, which is a precursor state to chemical phase separation (CPS) into non-crystalline $SiO_2$, and a nanocrystalline TM oxide second phase, and 2. the Zr(Hf) silicon oxynitride alloys are also *unstable* against CPS and crystallization of a TM oxide phase if the $Si_3N_4$ concentration, $y$, is smaller than the $SiO_2$ concentration, $1 - x - y$, independently of the $Zr(Hf)O_2$ concentration, $x$, but 3. if $1 - x - y \approx x \approx 0.3$ to $0.35$, and $y \approx 0.33$ to $0.4$, then CPS and $Zr(Hf)O_2$ crystallizations are suppressed by *chemical self-organizations* resulting in stable bonding arrangements that are effectively defect free [19].

Combinations of FTIR and XPS are used to demonstrate that there are changes in chemical bonding in Zr and Hf silicate alloys that occur with annealing temperatures (in inert ambients) in excess of deposition temperatures of $300\,°C$, and that the rate of bonding changes increases dramatically for annealing temperatures $> 800\,°C$ with CPS and crystallization of the TM oxide readily evident after $900\,°C$ annealing. Similar observations hold for Zr Si oxynitride alloys with low $Si_3N_4$ compositions, $y$, in the range of 0.1 to

0.25. Changes in these alloys were also tracked by XANES, in particular the O $K_1$ and N $K_1$ edge spectra. In addition to CPS and $ZrO_2$ crystallization, the XANES and XPS indicate decomposition with release of N.

FTIR, XPS and XANES studies of Zr silicate alloys in which the $Si_3N_4$ concentration $y \sim 0.4$, and is greater than $x$ and $1 - x - y$, each equal to $\sim 0.3$, demonstrated changes in chemical bonding indicative of a chemical self-organization that 1. prevented CPS, 2. reduced bond strain by encapsulation Zr$-$O$-$Si bonding groups within a $Si_3N_4$ matrix, thereby preventing percolation of bond strain throughout the entire film.

Chemical phase separation has been identified by changes in the C–V characteristics of Zr silicate and Zr Si oxynitride thin film dielectrics as well [20]. In marked contrast, the C–V and I–V characteristics of films with Zr Si oxynitride gate dielectrics, which have about the same composition as those described in Figs. 14b and c, exhibit ideal behavior with dielectric constants in the range of 8 to at least 10. This means that if the relative concentrations of $SiO_2$, $Si_3N_4$ and $Zr(Hf)O_2$ are controlled within narrow limits in *pseudo-ternary* non-crystalline Zr/Hf Si oxynitrides with high $Si_3N_4$ content, e.g., $(Hf(Zr)O_2)_{\sim 0.3}(Si_3N_4)_{\sim 0.4}(SiO_2)_{\sim 0.3}$, then chemical phase separation (CPS) and crystallization are completely suppressed by *chemical self-organizations* which relieve local bond strain [19], resulting in gate dielectrics extend EOT scaling to $\sim 0.7 - 0.8$ nm.

# 7 Conclusions

The conclusions reached from the spectroscopic studies presented for TM elemental oxides are the following: 1. as-deposited TM oxides are nanocrystalline as-deposited by remote plasma-assisted chemical vapor deposition or reactive evaporation, and 2. the band edge $d$-states detected by XANES and/or VUV SE indicate a complete removal of $d$-state degeneracies which is attributed to a J–T effect. All of our spectroscopic studies were restricted to film thicknesses in excess of about 5 nm. It is possible that films thinner than this may have a non-crystalline or amorphous character. This issue has not been addressed. However, it is important to recognize that the detection of crystallinity on a scale of less than 5 nm cannot be determined by conventional X-ray diffraction, XRD, but that other approaches such as XANES, EXAFS, and high-resolution transmission electron microscopy, HRTEM, are required to detect nanocrystallinity on a smaller scale, e.g., the 2 nm scale that is present in as-deposited films.

Qualitative and quantitative differences in O $K_1$ *inter-atomic* M–O transitions, and $L_3$, $M_3$ and $N_3$ intra-atomic have been presented for three different groups of complex oxides: those composed of 1. tetra- and trivalent transition metal (TM) and/or rare earth (RE) atom oxides, 2. two TM tetravalent oxides, and 3. TM and RE oxides, or Al oxide. In addition, matrix element effects must be considered in the determination of conduction band offset energies between complex oxides, and Si or other semiconductors. These effects can be different for all three classes of complex oxides, so considerable care must be taken in combining optical band edge data with valence band edge energy differences obtained from either UV or X-ray photoemission studies.

To the best of our knowledge, this combined spectroscopic and theory/modeling of intrinsic defects in nanocrystalline $HfO_2$ and $ZrO_2$, coupled with device studies, represents the first unambiguous assignment for electronically active defects in high-$k$ gate stacks. The energies of these defects relative to the forbidden energy gap in Si also explain differences in bias temperature instabilities, BTIs, in NMOS and PMOS FETs, and present a significant obstacle for integration of $HfO_2/ZrO_2$ dielectrics into scaled Si CMOS.

Chemical phase separation has been identified by changes in the C–V characteristics of Zr silicate and Zr Si oxynitride thin film dielectrics as well [19]. In marked contrast, the C–V and I–V characteristics of films with Zr Si oxynitride gate dielectrics, that have about the same composition as those described in Figs. 15b and c, exhibit ideal behavior with dielectric constants in the range of 8 to at least 10. This means that if the relative concentrations of $SiO_2$, $Si_3N_4$ and $Zr(Hf)O_2$ are controlled within narrow limits in *pseudo-ternary* non-crystalline Zr/Hf Si oxynitrides with high $Si_3N_4$ content, e.g., $(Hf(Zr)O_2)_{\sim 0.3}(Si_3N_4)_{\sim 0.4}(SiO_2)_{\sim 0.3}$, then chemical phase separation (CPS) and crystallization are completely suppressed by *chemical self-organizations* which relieve local bond strain, resulting in gate dielectrics that have the potential to extend EOT scaling to $\sim 0.7$–$0.8\,nm$.

Finally, the results presented in this paper, combined with discussions in [5], make it highly unlikely than any nano-crystalline transition metal or lanthanide series rare earth atom elemental or complex oxide can be integrated into scaled CMOS devices. Combinations of defect states associated with fixed charge, band edge grain boundary defects, and oxygen atom vacancies, as well as defect states associated with hydrogen [1, 5, 6], will introduce sufficient quantities of defects to prevent devices from displaying targeted performance and/or reliability metrics. In contrast, Zr or Hf Si oxynitride films with approximately equal concentrations of $SiO_2$, $Si_3N_4$ and $ZrO_2$ or $HfO_2$ will remain non-crystalline after annealing to at least $1100\,^\circ C$, and offer engineering solutions for high performance devices with EOTs to $\sim 0.7$ to $0.8\,nm$, and mobile devices with EOTs to $\sim 1.5\,nm$.

## Acknowledgements

The research reported in this article is supported by contracts from the Office of Naval Research (ONR) and the Semiconductor Research Corporation (SRC). In addition, the author wishes to acknowledge research collaborations with graduate students, post-doctoral fellows, and other faculty

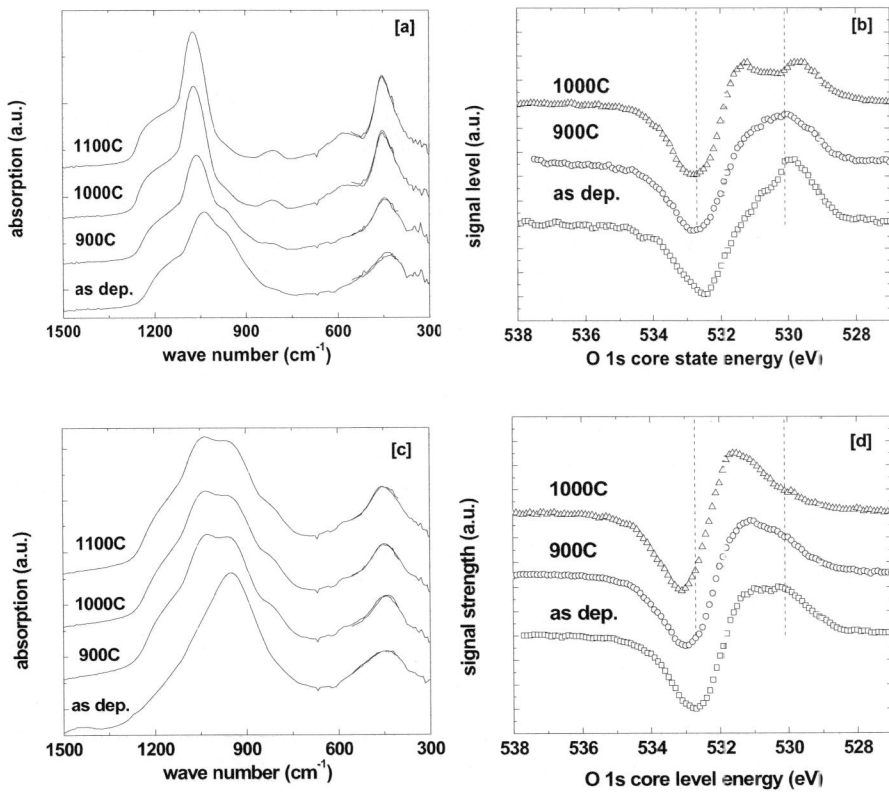

**Fig. 15.** FTIR results for two pseudo-ternary $Zr(Hf)O_2-SiO_2-Si_3N_4$ alloys in (**a**) and (**c**), and include derivative XPS O 1$s$ core level spectra for the same alloys as in (**b**) and (**d**)

and university research staff members, including Charlie Fulton, Chris Hinkle, Nick Stoute, ByongSun Ju, Yu Zhang, Cristiano Krug, Jerry Whitten, Harald Ade, Dave Aspnes and Bob Nemanich from NC State University, Bridget Rogers from Vanderbilt University, Jan Lüning from the Stanford Synchrotron Research Laboratory, and Stefan Zollner and D. Tyrioso from Freescale Semiconductor, Inc.

# References

[1]  G. Lucovsky, C. C. Fulton, Y. Zhang, Y. Zou, J. Luning, L. F. Edge, J. L. Whitten, R. J. Nemanich, H. Ade, D. G. Schlom, V. V. Afanas'ev, A. Stesmans, S. Zollner, D. Triyoso, B. R. Rogers: Conduction band-edge states associated with the removal of $d$-state degeneracies by the Jahn–Teller effect, IEEE Trans. Mat. Dev. Rel. **5**, 65 (2005)

[2]  S. Zollner, D. Tyrioso, B. R. Rogers, S. Zollner: unpublished

[3]  A. Stesmans, V. V. Afanas'ev: Si dangling-bond-type defects at the interface of (100)Si with ultrathin layers of $SiO_x$, $Al_2O_3$, and $ZrO_2$, Appl. Phys. Lett. **80**, 1957 (2002)

[4]  V. V. Afanas'ev, A. Stesmans, C. Zhao, M. Caymax, T. Heeg, J. Schubert, Y. Jia, D. G. Schlom, G. Lucovsky: Band alilgnment between (100)Si and complex rare earth/transition metal oxides, Appl. Phys. Lett. **85**, 5917 (2004)

[5]  J. Robertson, K. Xiong, B. Falabretti: Point defectss in $ZrO_2$ high-$\kappa$ gate oxide, IEEE Trans. on Material and Device Reliability **5**, 84 (2005)

[6]  C. G. Van de Walle, J. Neugebauer: Universal alignment of hydrogen levels in semiconductors, insulators and solutions, Nature **423**, 636 (2003)

[7]  F. A. Cotton, G. Wilkinson: *Advanced Inorganic Chemistry*, 3rd ed. (Wiley Interscience, New York 1972) Chap. 20

[8]  W. A. Harrision: *Elementary Electronic Structure* (World Scientific Publishing, Singapore 1999) , back-cover table

[9]  B. K. Hem, et al.: J. Chem. Thermodynamics **30**, 165 (2001)

[10]  H. H. Tippins: J. Phys. Chem. Solids **27**, 1069 (1966)

[11]  P. A. Cox: *Transition Metal Oxides* (Oxford Science Publications, Oxford 1992) Chap. 2

[12]  J.-L. Autran, D. Munteanu, M. Housa: *High-k Dielectrics* (IOP, Bristol 2004) Chap. 3.4

[13]  G. B. Rayner, D. Kang, G. Lucovsky: Spectroscopic study of chemical phase separation in zirconium silicate alloys, J. Vac. Sci. Technol. B **21**, 1783 (2003)

[14]  G. B. Rayner, D. Kang, G. Lucovsky: Chemical phase in Zr silicate alloys: A spectroscopic study distinguishing between chemical phase separation with different degrees of micro- and nano-crystallinity, J. Non-Cryst. Solids **338**, 151 (2004)

[15]  B. J. Hinds, F. Wang, D. M. Wolfe, C. L. Hinkle, G. Lucovsky: Investigation of postoxidation thermal treatments of $Si/SiO_2$ interface in relationship to the kinetics of amorphous Si suboxide decomposition, J. Vac. Sci. Technol. B **16**, 2171 (1998)

[16]  J. C. Phillips: Topology of covalent non-crystalline solids. I. Short-range order in chalcogenide alloys, J. Non-Cryst. Solids **34**, 153 (1979)

[17]  J. C. Phillips: Topology of covalent non-crystalline solids. II. Medium-range order in chalcogenide alloys and $A-Si(Ge)$, J. Non-Cryst. Solids **43**, 37 (1981)

[18]  R. Kerner, J. C. Phillips: Quantitative principles of silicate glass chemistry, Solid State Commun. **117**, 47 (2001)

[19]  P. Boolchand: *Phase Transitions and Self-Organization in Electronic and Molecular Networks* (Kluwer Academic, New York 2001) p. 65

[20]  J. Byunsun: *Spectroscopic Study of the Interface Chemical and Electronic Properties of High-$\kappa$ Gate Stacks*, Ph.D. thesis, Department of Materials Science and Engineering, North Carolina State University (2005)

# Index

# Electronic Structure and Band Offsets of Lanthanide Oxides

John Robertson and Ka Xiong

Engineering Department, Cambridge University, Trumptington Street,
Cambridge, CB2 1PZ, United Kingdom
jr@eng.cam.ac.uk

**Abstract.** This paper reviews the bulk electronic structures of high dielectric constant oxides, particularly lanthanide oxides. The electronic structures are calculated with methods beyond local density formalism, in order to give correct band gaps and correct energies for $f$ states. The band offsets for the oxide–Si interfaces are an important factor determining the leakage currents, and these are derived from calculations and compared to experimental values. The band offset values for lanthanide oxides are advantageous compared to those of other high-$k$ oxides.

## 1 Introduction

The decrease of dimensions of complementary metal oxide silicon (CMOS) transistors has led to a need to replace the $SiO_2$ gate oxide with an oxide of higher dielectric constant ($k$), in order to maintain a small gate leakage current [1, 2]. The oxide must satisfy various conditions such as be stable in contact with Si [3], and have sufficient band offsets to be a barrier for both electrons and holes [4]. This restricts the choice to the oxides of Hf, Zr, Y, Al and the lanthanides. The most intensively studied high-$k$ oxides are $HfO_2$ and its silicates. Nevertheless, the lanthanide oxides are of interest because early work suggested that they have the lowest leakage of all high-$k$ oxides [5, 6]. This arises because of their larger conduction band offset, so they will ultimately have a lower leakage current than other candidates and may be the leading candidates for the second-generation oxides [7].

Various lanthanide oxides have been studied [8–16]. $La_2O_3$ is also an attractive choice because of its higher $k$ value than $HfO_2$ (Fig. 1). The $k$ decreases, however, along the lanthanide series due to the decrease in metal ionic radius. Also, $La_2O_3$ and the other lanthanide oxides have the disadvantage of being more reactive with water and $CO_2$, implying that their processing is more difficult.

The lanthanides comprise a series of elements with a partially filled $4f$ shell, including the end members $La_2O_3$ and $Lu_2O_3$. We will consider mainly these oxides plus the important perovskite oxide $LaAlO_3$.

In the absence of traps, the leakage current through a thin oxide flows via the Schottky emission mechanism, by injection from the bands of Si into the oxide's bands. Minimising this injection requires that the band offset for

M. Fanciulli, G. Scarel (Eds.): Rare Earth Oxide Thin Films,
Topics Appl. Physics **106**, 313–330 (2007)
© Springer-Verlag Berlin Heidelberg 2007

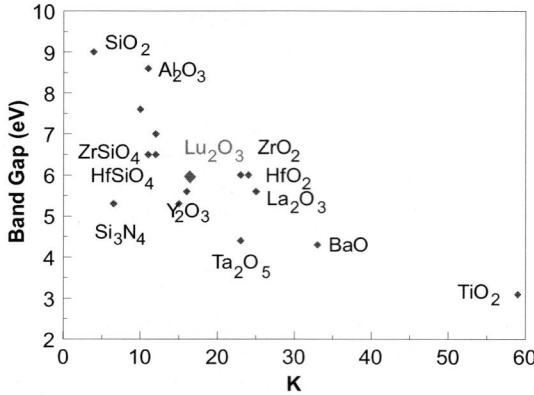

**Fig. 1.** Dielectric constant vs. band gap of possible gate oxides

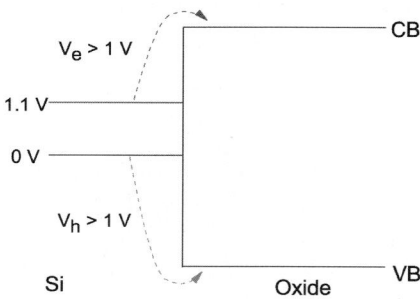

**Fig. 2.** Schematic of valence and conduction band offsets of an oxide on Si

both valence and conduction bands from Si to oxide should be greater than 1 eV [4] (Fig. 1). In $SiO_2$, its very wide band gap of 9 eV means that both of its offsets are over 3 eV. However, most high-$k$ oxides have narrower gaps, so that their offsets tend to be lower. Depending on whether the offsets are disposed symmetrically about the Si gap, this means that one of the offsets could easily be less than 1 eV. Generally, the conduction band (CB) offset turns out to be the lower of the two offsets, and the electron effective mass is lower than the hole mass, so we pay particular attention to the CB offset.

The lanthanides differ from the usual transition metals. They are highly electropositive, their $4f$ shells are chemically inactive even when partially filled, and they generally exert trivalence. They are somewhat analogous chemically to "trivalent" alkaline earth metals.

The band offsets depend on an intrinsic line-up of the Si and oxide bands, and also on the presence of a dipole layer at this interface. To understand these questions in more detail, we must first describe the bulk electronic structure of the oxides, consider a general model of band line-ups, and consider line-ups for particular interface chemistries.

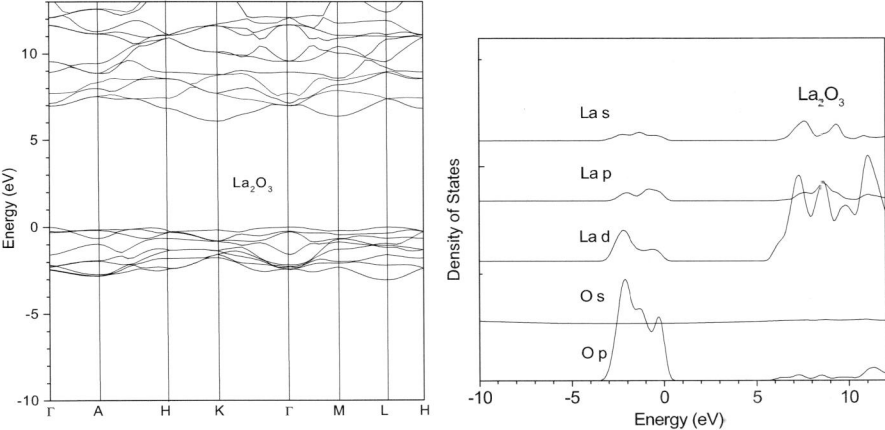

**Fig. 3.** Band structure and density of states of $La_2O_3$ calculated by GGA, with band gap adjusted to the experimental value of about 6.0 eV

## 2 Bulk Electronic Structure

$La_2O_3$ itself crystallises in the hexagonal $La_2O_3$ structure in which La has seven neighbours and oxygen has four neighbours. This structure arises because of the large ionic radius of La, which also gives it its large $k$. The rare earth oxides with smaller metal ionic radius crystallise in the cubic bixbyite structure. This has a large unit cell, in which the metal ions have two types of short-range order, each with six neighbours.

Figure 3 shows the band structure and partial density of states (DOS) of $La_2O_3$ calculated in the local density formalism (LDF), using the generalised gradient approximation (GGA) to the electron exchange-correlation function. The band gap has been adjusted to the experimental value of 5.9 eV. This is because the LDF under-estimates the band gaps. We see that the conduction band consists mainly of metal $d$ states mixed with some metal $s$ states. The valence band consists mainly of oxygen $2p$ states. The high ionicity of the oxide is evident in the narrow valence band width and also the polar nature of the valence and conduction band DOSs (the valence band contains only a small amount of metal character and vice versa).

A key observation is that the top of the valence band is formed from non-bonding O $2p$ states, and the bottom of the conduction band is formed from non-bonding metal $d$ states. These lie at the un-shifted energies of the free atoms, and therefore to a first approximation, the band gap of these oxides will not depend on their crystal structure.

Figure 4 shows the calculated density of states of $Y_2O_3$ [17]. The valence band is slightly narrower than that of $La_2O_3$, which indicates that $Y_2O_3$ is more ionic than $La_2O_3$.

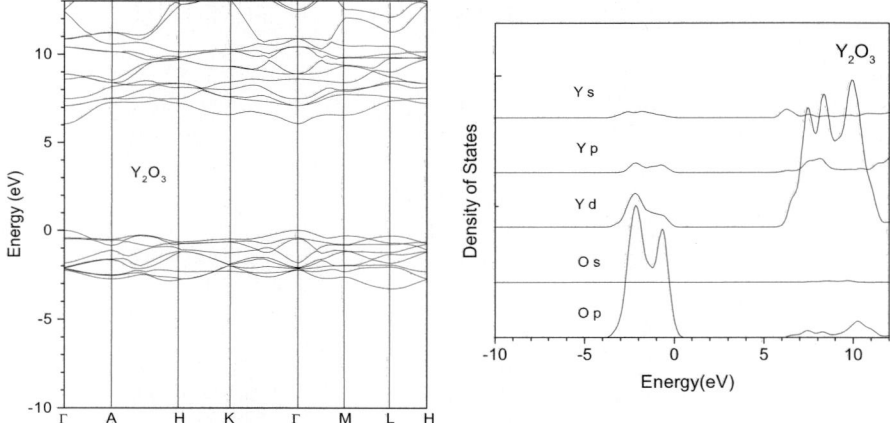

**Fig. 4.** Band structure and density of states of $Y_2O_3$ calculated by GGA, with band gap adjusted to about $6.0\,\mathrm{eV}$

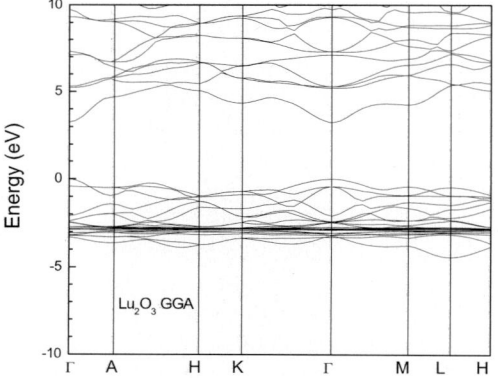

**Fig. 5.** Band structure of $Lu_2O_3$ in the $La_2O_3$ structure, calculated by GGA

$Lu_2O_3$ can exist in both bixbyite and $La_2O_3$ structures [16]. Figure 5 shows the band structure for $Lu_2O_3$ in the $La_2O_3$ structure. This calculation uses the GGA. The main conduction and valence bands are formed of metal $d$ and O $p$ states respectively, as before. However, we see an additional narrow band of Lu $4f$ states, which lie in the O $2p$ valence band. These are fully occupied in Lu.

Figure 6 shows the band structure and DOS of $Lu_2O_3$ in the $La_2O_3$ structure calculated using the weighted density approximation (WDA) [18] for the exchange-correlation energy. Now, the band gap is not under-estimated and the $4f$ states are moved downwards below the O $2p$ valence states. The calculated DOS of the valence band in Fig. 6b is now close to that seen exper-

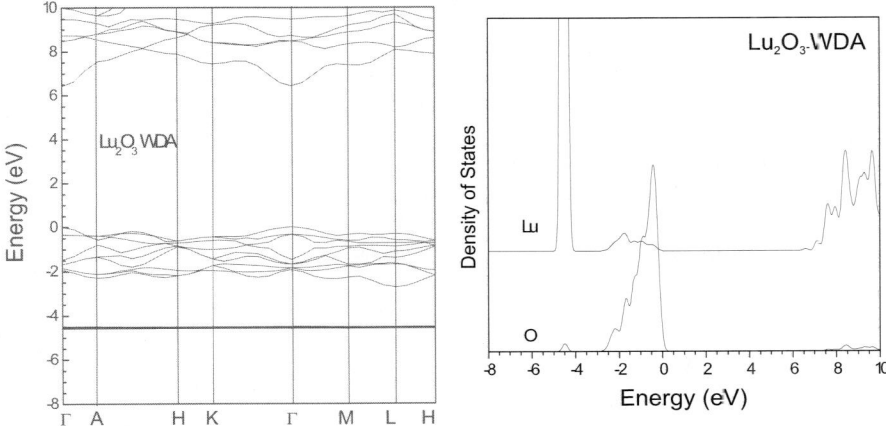

**Fig. 6.** Band structure and density of states of $Lu_2O_3$ in the $La_2O_3$ structure, calculated by WDA

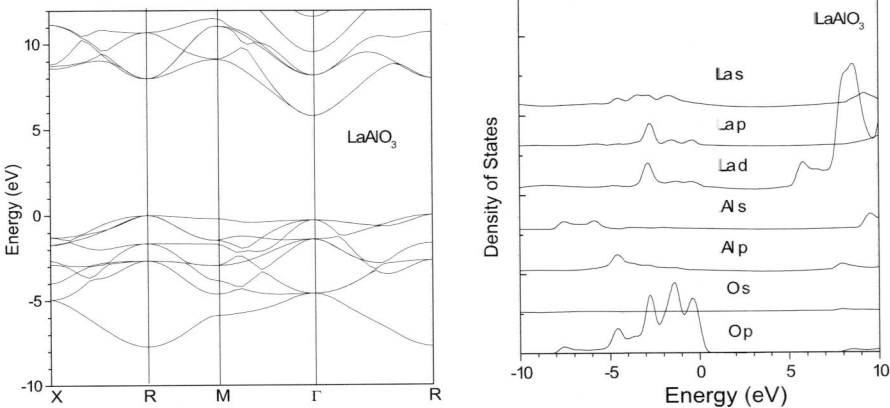

**Fig. 7.** Band structure and density of states of $LaAlO_3$ calculated by GGA with band gap adjusted to the experimental value of $5.8\,\mathrm{eV}$

imentally by photoemission by *Perego* et al. [19]. They found the $4f$ states lying at 5 and 6 eV below the valence band top. The parameter of the WDA has been fitted to Si, and is not available to vary, so the sensitivity of the WDA result cannot be checked.

Figure 7 shows the band structure and partial DOS of $LaAlO_3$ in the cubic structure [17]. There are only minor differences for its rhombohedral phase. The valence band is again formed mainly of O $2p$ states, and the lowest conduction band is formed from La $d$ states. The Al $s, p$ and La $s$ states mix higher up and form highly dispersed bands.

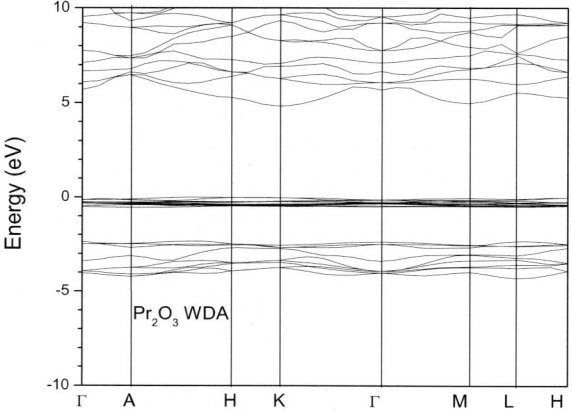

**Fig. 8.** Band structure of $Pr_2O_3$ in $La_2O_3$ structure, calculated by WDA

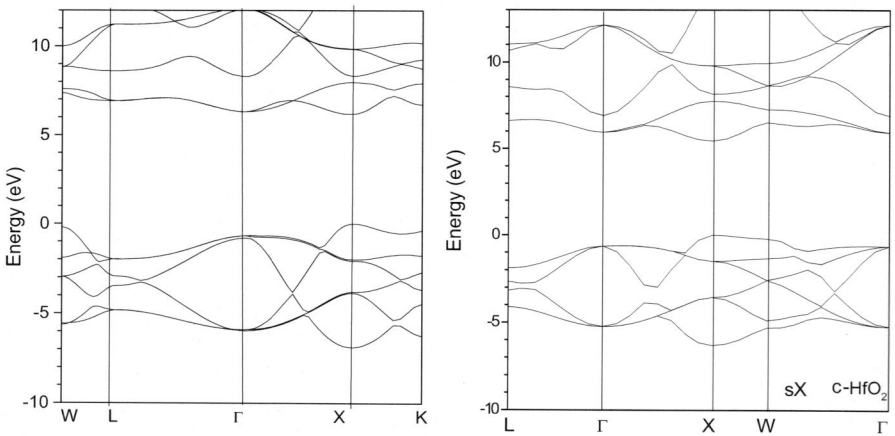

**Fig. 9.** Comparison of band structure of c-$HfO_2$, calculated by GGA and sX

The band structure of the lanthanides with partially filled $f$ states is not well given in the LDA approximation [10], which has problems treating the partial occupancy of the highly localised $4f$ states and their strong intra-atomic correlation energy. This is shown in Fig. 8 for the example of $Pr_2O_3$ in the hexagonal structure. Even the WDA is unable to give $4f$ states in both the valence and conduction bands.

There have been discussions recently that the experimental band gaps of some well-known materials are lower than previously believed. The LDA and GGA methods under-estimate the band gap. The error is often corrected by arbitrarily shifting all conduction band energies upwards to fit the experimental gap, as in Fig. 9 (left) for $HfO_2$. If there is debate about a band gap value, there are theoretical methods to give the correct gap. The most expensive is GW. It has been used for cubic $ZrO_2$ and $HfO_2$ [20, 21], but it is computationally expensive if used on large cells. Lower cost methods are screened exchange (sX) [22] and weighted density approximation (WDA) [17]. Figure 9 (right) shows the screened exchange bands for $HfO_2$. Table 1 compares the experimental optical gaps with those found by calculation.

**Table 1.** Comparison of calculated band gaps by GGA, screened exchange and weighted density approximation

| Gap | GGA (eV) | sX (eV) | WDA (eV) | Experiment |
|---|---|---|---|---|
| Cubic $ZrO_2$ | 3.4 | 5.2 | 6.0 | 5.8 |
| Cubic $HfO_2$ | 3.7 | 5.5 | 6.1 | 6.0 |
| $Al_2O_3$ | 6.5 | | 9.3 | 8.8 |
| a-$LaAlO_3$ | 3.1 | 4.4 | 6.7 | 5.6 |
| $La_2O_3$ | 3.7 | 5.4 | 6.9 | 6 |
| $Y_2O_3$ | 3.3 | | 7.0 | 6 |
| $Lu_2O_3$ | 3.25 | | 6.4 | 5.8 |

## 3 Band Offsets

The band offset between oxide and Si defines the barrier for injection of electrons or holes into the oxide [4, 23]. The electron barrier or conduction band (CB) offset tends to be the smaller of the two. The CB offset is one of the key criteria in the selection of a gate oxide; it must be over $1\,eV$ to give adequately low leakage current.

The band offsets of an oxide on silicon can be found by treating the oxide as a wide band gap semiconductor. It is then the band offset between two semiconductors. The band offset is closely related to the barrier height between the semiconductor and a metal, which is known as the Schottky barrier height. Both these subjects have been intensively studied.

The band offset between two semiconductors is controlled by any charge transfer across the bonds at the interface which would create an interface dipole (Fig. 10). If there is no charge transfer, the band offset is found by placing the energies of each semiconductor on a common energy scale with respect to the vacuum level. This is the Electron Affinity rule, in which the conduction band offset is given by the difference in the electron affinities (EAs). It works for the Si : $SiO_2$ interface.

In practice, there is charge transfer across the interface. Consider a metal-semiconductor interface – the Schottky barrier [24–30] The semiconductor interface now has states within its band gap which decay from the metal into

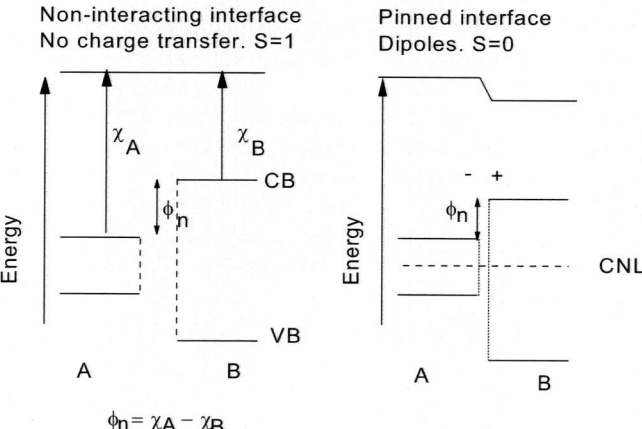

$$\phi_n = \chi_A - \chi_B$$

**Fig. 10.** Effect of interface charge transfer on band line-ups

the semiconductor. There can be charge transfer between the metal states at its Fermi level and the gap states of the semiconductor. This tends to align the metal Fermi level and its equivalent for the interface states, which is called a charge neutrality level (CNL), as shown in Fig. 10 (right). This leads to an equation for the Schottky barrier height of electrons $\phi_n$ between the semiconductor S and the metal M of

$$\phi_n = S(\Phi_M - \Phi_S) + (\Phi_S - \chi_s) \tag{1}$$

or

$$\phi_n = S(\Phi_M - \chi_s) + (1 - S)(\Phi_S - \chi_s) \,.$$

Here, $\Phi_M$ is the metal work function, $\Phi_S$ is the charge neutrality level of the semiconductor and $\chi_s$ is the electron affinity (EA) of the semiconductor. All the energies in (1) are measured from the vacuum level, except $\phi_n$ which is measured from the conduction band edge. $S$ is a dimensionless constant called the pinning factor. It is given by [24]

$$S = \frac{A}{1 + \frac{e^2 N \delta}{\varepsilon \varepsilon_0}} \,, \tag{2}$$

where $e$ is the electronic charge, $\varepsilon_0$ is the permittivity of free space, $N$ is the density of interface states per unit area and $\delta$ is their extent into the semiconductor.

The parameter $S$ is a dimensionless pinning factor, which describes the degree of alignment or "pinning" caused by the interface states. $S = 1$ corresponds to the unpinned Schottky limit, equivalent to the Electron Affinity rule, and $S = 0$ corresponds to the strongly pinned or "Bardeen" limit [25].

There have been a number of models of the states causing the pinning. Intrinsic states are a more general source of pinning than are defects because there are more of the former.

The metal-induced gap state (MIGS) model [26, 29, 30] describes intrinsic pinning. The MIGS can be visualised as the dangling bond states of the broken surface bonds of the semiconductor dispersed across its band gap, or alternatively as the evanescent states of the metal wave-functions continued into the forbidden energy gap of the semiconductor. $\Phi_S$ is the charge neutrality level (CNL) of the interface states. The CNL is like a Fermi level for interface states; it is the energy near mid-gap to which the interface states are filled on a neutral surface. *Monch* [26] found that $S$ empirically obeys

$$S = \frac{1}{1 + 0.1(\varepsilon_\infty - 1)^2}. \tag{3}$$

The band offset at a semiconductor heterojunction is defined in the same way. For two semiconductors $a$ and $b$, the electron barrier $\phi_n$ is the conduction band offset, given by

$$\phi_n = (\chi_a - \Phi_{S,a}) - (\chi_b - \Phi_{S,b}) + S(\Phi_{S,a} - \Phi_{S,b}). \tag{4}$$

$S$ is the pinning factor of the wider-gap semiconductor, that is, the oxide. $S$ can be found from (3) using the experimental value $\varepsilon_0$ from the refractive index.

The CNL is the energy at which the Greens function of the bulk or surface band structure is zero. It can be evaluated by integrating over all bands and over $k$ points in the Brillouin zone,

$$G(E) = \int\limits_{BZ} \int\limits_{-\infty}^{\infty} \frac{N(E')\,\mathrm{d}E'}{E - E'} = 0. \tag{5}$$

This integral can be replaced by a sum over special points of the zone [31]. For tight-binding bands, there is a finite number of bands, whereas for pseudopotential bands we must fix a finite upper limit in integral (5).

Core levels are omitted from the Greens function integration (5) because their diameter is too small for them to contribute to the interface dipole (2). An orbital only contributes to a dipole moment if it is wide enough to extend across the interface, so that its occupancy changes from the isolated-solid case. Thus, localised states such as core states do not contribute. Also, continuous plane waves do not contribute to an interface dipole as they are present uniformly on both sides of the interface.

Table 2 gives the calculated energy of the charge neutrality level for each compound with respect to the valence band maximum. The calculation uses (5) and the GGA band energies, with the conduction band energies shifted upwards by the scissors correction to give the experimental band gaps.

**Table 2.** Tabulation of band gaps [17], experimental electron affinities, experimental $\varepsilon_\infty$, $S$ factors, and calculated values of charge neutrality levels and conduction band offsets. * estimates

| | Gap (eV) exp | EA (eV) exp | $\varepsilon_\infty$ | $S$ | CNL (eV) | CB offset, LDA (eV) |
|---|---|---|---|---|---|---|
| $ZrO_2$ | 5.8 | 2.5* | 4.8 | 0.41 | 3.3 | 1.6 |
| $HfO_2$ | 5.8 | 2.7 | 4 | 0.53 | 4 | 1.3 |
| $Al_2O_3$ | 8.8 | 1* | 3.4 | 0.63 | 6.6 | 2.4 |
| $Y_2O_3$ | 6 | 2* | 4.4 | 0.46 | 2.7 | 2.2 |
| $La_2O_3$ | 6 | 2* | 4 | 0.53 | 2.5 | 2.3 |
| $Lu_2O_3$ | 5.8 | 2 | 4 | 0.53 | 2.5* | 2.3* |
| $Pr_2O_3$ | 3.9 | 2.5 | 4 | 0.53 | 1.9* | 1.2* |
| $Gd_2O_3$ | 5.6 | 2.5 | 4 | 0.53 | 2.5* | 2.2* |
| $LaAlO_3$ | 5.6 | 2.5* | 4* | 0.53 | 3.8 | 1.0 |

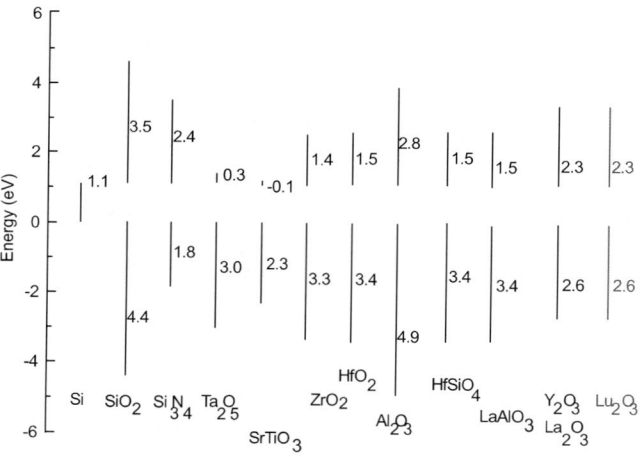

**Fig. 11.** Calculated band offsets of oxides on Si

$S$ is found from (3), $\varepsilon_\infty$ and the experimental value of the refractive index [32]. The CB offset is then found using (4), using the experimental value of the oxide's electron affinity, where known. They are shown in Table 1 and Fig. 11.

Table 3 compares the predicted CB offsets with those determined by experiment. The experimental values come from different types of experiment; photoemission, internal photoemission, or barrier tunnelling. It is seen that the predicted and experimental offsets generally agree quite well. It should be noted that the underlying accuracy of the MIGs model is only about 20 % of the oxide gap, or 1 eV.

The predictions for $HfO_2$ and $ZrO_2$, the most studied, agree well with those from photoemission [33–39], but less well with those from internal pho-

**Table 3.** Comparison of the calculated conduction band offset (by LDA) and experimental values for various gate oxides, by various authors. * estimate

| | Calculated (eV) | Experiment (eV) | Ref. |
|---|---|---|---|
| SrTiO$_3$ | 0.4 | 0 | *Chambers* [45] |
| ZrO$_2$ | 1.6 | 1.4 | *Miyazaki* [33] |
| | | 2.0 | *Afanasev* [40] |
| | | 1.4 | *Rayner* [35] |
| HfO$_2$ | 1.3 | 1.3 | *Sayan* [34] |
| | | 2.0 | *Afansev* [41] |
| Al$_2$O$_3$ | 2.4 | 2.8 | *Ludeke* [46], *DiMaria* [47] |
| | | 2.2 | *Afansev* [41] |
| a-LaAlO$_3$ | 1.0 | 1.8 | *Edge* [7] |
| La$_2$O$_3$ | 2.3 | 2.3 | *Hattori* [43] |
| Y$_2$O$_3$ | 2.3 | 1.6 | *Ohta* [44] |
| Lu$_2$O$_3$ | 2.3 | 2.1 | *Seguini* [15] |
| Pr$_2$O$_3$ | 1.2* | 1.1 | *Osten* [11] |
| Gd$_2$O$_3$ | 2.1* | 2.1 | Osten, private comm. |

toemission by *Afanasev* [40, 41]. The value of $S \sim 0.5$ was also confirmed experimentally [42].

For lanthanides, the predicted value for La$_2$O$_3$ agrees very well with that of 2.3 eV [43] found by photoemission. There is reasonable agreement for LaAlO$_3$ [7]. The CB offset derived for Y$_2$O$_3$ from photoemission is 1.6 eV [44], slightly less than that predicted.

The DOS of Lu$_2$O$_3$ is essentially the same as those of La$_2$O$_3$ and Y$_2$O$_3$, so the calculated CNL energies are the same. Note that the $4d$ levels should be ignored in the Greens function integration (5) because their small size, like for core levels, prevents them from creating an interface dipole.

Photoemission [15] gives a CB offsets of 2.1 eV for Lu$_2$O$_3$, compared to 2.3 eV predicted. A similar experimental value is also found by internal photoemission.

The calculated band offsets show clear chemical trends. The CB offset becomes larger for oxides with a smaller metal valence [4]. This occurs because the CNL in (5) is an average of the valence and conduction band density of states (DOS) weighted as $1/E$, so

$$E_{CNL} \approx \frac{N_1 E_2 + N_2 E_1}{N_1 + N_2} . \tag{6}$$

The CNL is a balance point of the weights of the valence and conduction band DOSs. Thus, a small density of states in the valence band tends to

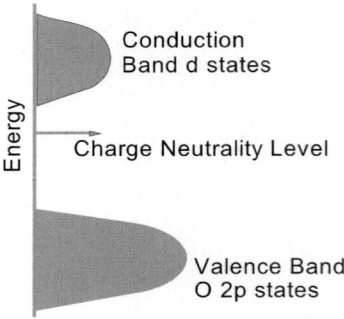

**Fig. 12.** Illustration of how the density of states controls the CNL energy

%marginparAuthor: Fig. 3 is not cited in the text. Should it not be cited here? push the CNL towards the valence band and vice versa. A small metal valence means fewer oxygens per formula unit, fewer O states in the valence band, pushing CNL down in the gap, and thus increasing the CB offset. Thus, we expect all trivalent oxides like $La_2O_3$ and the other lanthanides to have larger CB offsets than $HfO_2$ or $ZrO_2$. This is seen in general terms. It is one of the main reasons for choosing a lanthanide oxide [4].

However, internal photoemission (IP) gives mixed results. It should be said that IP directly measures the CB offset, plus the Si gap, whereas normal photoemission measures the VB offset, from which the CB offset is derived via the band gaps. However, IP gives almost the same CB offset of 2.1 eV for $HfO_2$ and $Lu_2O_3$. A larger CB offset for $Lu_2O_3$ is expected.

## 4 Explicit Calculations

There are in general two components to the charge transfer across an interface, one due to the potential difference between the average electronegativity of the bulk solids, and the other due to the specific atomic configuration at the interface.

The MIGS model gives a reasonable description of the bulk contribution to the charge transfer. The advantage of this method is that the CNLs are properties which are defined by the band structure of the bulk materials, it is applicable to both covalent and ionic bonding, it has been tested over a wide range of band gaps, and it does not require an explicit model of the interface bonding. This is an advantage because the oxides can be amorphous and the interface bonding is often not known.

Band offsets can also be calculated by methods using explicit models of the interface, and the offset is found from the potential step at the interface. This is easiest for systems of which the bonding continues across the interface, for example, at zincblende heterojunctions, as in the work of *Baldereschi* et al. [48]. Alternatively, the potential step can be calculated from the bonding using the model solid method of *van der Walle* [49]. These methods need

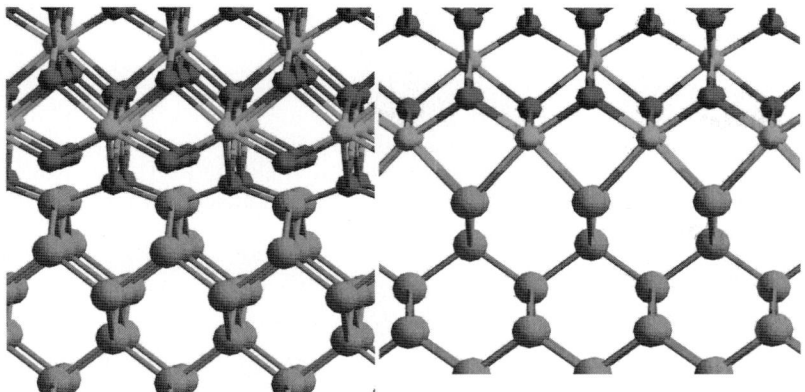

**Fig. 13.** The $O_4$- and $Zr_6$-terminated interfaces for (100)Si : $ZrO_2$

a calculation for each interface structure. These direct calculations allow one to study the variation of interface dipoles with interface structure and oxide termination.

*Fissel* et al. [10] have carried out calculations of the change of band offset with change of interface bonding for $Pr_2O_3$ on Si. A 0.6 eV change was found, a sizeable amount. *Forst* [50] found a 0.5 eV shift of offset for the (100)Si : $SrTiO_3$ interface for a change of termination.

*Peacock* and *Robertson* [51–53] have carried out calculations of the dependence of band offsets on interface bonding only for the case of (100)Si : $HfO_2$ and Si : $ZrO_2$ interfaces, not lanthanides. However, the causes are general. Figure 13 shows two interfaces which satisfy bonding requirements and which are also insulating, the oxygen-terminated $O_4$ interface and the Zr-terminated $Zr_6$ interface. Both are shown for the relaxed structures.

It is seen that there is a change in VB offset for the three cases. Broadly, the O-terminated interface has a dipole $Si^+ - O^-$ which raises the oxide levels and lowers the VB offset (in the oxide). On the other hand, the Zr-terminated interface has a dipole of $Si^- - Zr^+$ which lowers the levels in the oxide and raises the VB offset, as shown in Table 4.

**Table 4.** Valence band offset of various abrupt (100)Si : $ZrO_2$ interfaces

| Interface | VB offset |
|---|---|
| $O_4$ | 2.9 |
| $O_3$ | 2.9 |
| $Zr_6$ | 3.3 |
| Bulk CNL (LDA) | 3.1 |

# 5 Conclusions

In conclusion, the electronic structures of the lanthanide oxides are closely related to that of $La_2O_3$ itself. The band offsets are the key factor which controls the leakage currents. The band offsets of those lanthanide oxides with band gaps near 6 eV will be very similar to that of $Si : La_2O_3$ itself, that is, a CB offset of 2.3 eV. This is because the $4d$ levels are too tightly bound to contribute to any interface dipole.

## Acknowledgements

The authors are very grateful to S. J. Clark of Durham University for use of his computer codes and to M. Perego, G. Scarel and H. J. Osten for sending data before publication.

# References

[1]  G. D. Wilk, R. M. Wallace, J. M. Anthony: High-$\kappa$ gate dielectrics: Current status and materials properties considerations, J. Appl. Phys. **89**, 5243 (2001)

[2]  J. Robertson: Energy levels of point defects in $SrTiO_3$ and related oxides, J. Appl. Phys. **93**, 1054 (2003)

[3]  K. J. Hubbard, D. G. Schlom: Thermodynamic stability of binary oxides in contact with silicon, J. Mater. Res. **11**, 2757 (1996)

[4]  J. Robertson: Band offsets of wide-band-gap oxides and implications for future electronic devices, J. Vac. Sci. Technol. B **18**, 1785 (2000)

[5]  Y. H. Wu, M. Y. Yang, A. Chin, W. J. Chen, C. M. Kwei: Electrical characteristics of high quality $La_2O_3$ gate dielectric with equivalent oxide thickness of 5 A, IEEE E. D. Lett. **21**, 341 (2000)

[6]  H. Iwai, S. Ohmi, S. Akama, C. Ohshima, A. Kikuchi, I. Kashiwagi, J. Taguchi, H. Yamamoto, J. Tonotani, Y. Kim, I. Ueda, A. Kuriyama, Y. Yoshihara: Advanced gate dielectric materials for sub-100 nm CMOS, Tech. Digest IEDM IEEE p. 625 (2002)

[7]  L. F. Edge, D. G. Schlom, S. A. Chambers, E. Cicerrella, J. L. Freeouf, B. Hollander, J. Schubert: Measurement of the band offsets between amorphous $LaAlO_3$ and silicon, Appl. Phys. Lett. **84**, 726 (2004)

[8]  J. Kwo, M. Hong, A. R. Kortan, K. T. Queeney, Y. J. Chabal, J. P. Mannaerts, T. Boone, J. J. Krajewski, A. M. Sergent, J. M. Rosamilia: High epsiv gate dielectrics $Gd_2O_3$ and $Y_2O_3$ for silicon, Appl. Phys. Lett. **77**, 130 (2000)

[9]  J. Kwo, M. Hong, A. R. Kortan, K. L. Queeney, Y. J. Chabal, R. L. Opila, D. A. Muller, S. N. G. Chu, B. J. Sapjeta, T. S. Lay, J. P. Mannaerts, T. Boone, H. W. Krautter, J. J. Krajewski, A. M. Sergnt, J. M. Rosamilia: Properties of high $\kappa$ gate dielectrics $Gd_2O_3$ and $Y_2O_3$, J. Appl. Phys. **89**, 3920 (2001)

[10]  A. Fissel, J. Dabrowski, H. J. Osten: Photoemission and ab initio theoretical study of interface and film formation during epitaxial growth and annealing of praseodymium oxide on Si(001), J. Appl. Phys. **91**, 8986 (2002)

[11] H. J. Osten, J. P. Liu, H. J. Mussig: Band gap and band discontinuities at crystalline $Pr_2O_3$/Si(001) heterojunctions, Appl. Phys. Lett. **80**, 297 (2002)

[12] S. Ohmi, M. Takeda, H. Ishiwara, H. Iwai: Electrical characteristics for $Lu_2O_3$ thin films fabricated by E-beam deposition method, J. Electrochem. Soc. **151**, G279 (2004)

[13] J. X. Wu, Z. M. Wang, S. Li, M. S. Ma: In situ photoemission study of a $Pr_2O_3$ thin film on GaAs(111), J. Vac. Sci. Technol. A **22**, 594 (2004)

[14] D. H. Triyoso, R. I. Hedge, J. Grant, P. Fejes, R. Liu, D. Roan, M. Ramon, D. Werho, R. Rai, L. B. La, J. Baker, C. Garza, T. Guenther, B. E. White Jr., P. J. Tobin: Film properties of ALD $HfO_2$ and $La_2O_3$ gate dielectrics grown on Si with various pre-deposition treatments, J. Vac. Sci. Technol. B **22**, 2121 (2004)

[15] G. Seguini, E. Bonera, S. Spiga, G. Scarel, M. Fanciulli: Energy-band diagram of metal/$Lu_2O_3$/silicon structures, Appl. Phys. Lett. **85**, 5316 (2004)

[16] E. Bonera, G. Scarel, M. Fanciulli, P. Delugas, V. Fiorentini: Dielectric properties of high-$\kappa$ oxides: Theory and experiment for $Lu_2O_3$, Phys. Rev. Lett. **94**, 027602 (2005)

[17] P. W. Peacock, J. Robertson: Band offsets and Schottky barrier heights of high dielectric constant oxides, J. Appl. Phys. **92**, 4712 (2002)

[18] P. P. Rushton, D. J. Tozer, S. J. Clark: Nonlocal density-functional description of exchange and correlation in silicon, Phys. Rev. B **65**, 235203 (2002)

[19] M. Perego, G. Scarel: private communication

[20] B. M. Bylander, L. Kleinman: Good semiconductor band gaps with a modified local-density approximation, Phys. Rev. B **41**, 7868 (1990)

[21] B. Kralik, E. K. Chang, S. G. Louie: Structural properties and quasiparticle band structure of zirconia, Phys. Rev. B **57**, 7027 (1998)

[22] J. Dabrowski, V. Zavodinsky, A. Fleszar: Pseudopotential study of $PrO_2$ and $HfO_2$ in fluorite phase, Microelectron. Reliab. **41**, 1093 (2001)

[23] J. Robertson, C. W. Chen: Schottky barrier heights of tantalum oxide, barium strontium titanate, lead titanate, and strontium bismuth tantalate, Appl. Phys. Lett. **74**, 1168 (1999)

[24] A. M. Cowley, S. M. Sze: Surface states and barrier height of metal-semiconductor systems, J. Appl. Phys. **36**, 3212 (1965)

[25] W. Mönch: Metal-semiconductor contacts: Electronic properties, Surf. Sci. **299–300**, 928 (1994)

[26] W. Mönch: Role of virtual gap states and defects in metal-semiconductor contacts, Phys. Rev. Lett. **58**, 1260 (1987)

[27] M. Schluter: Chemical trends in metal-semiconductor barrier heights, Phys. Rev. B **17**, 5044 (1978)

[28] M. Schluter: Theoretical models of Schottky barriers, Thin Solid Films **93**, 3 (1982)

[29] J. Tersoff: Theory of semiconductor heterojunctions: The role of quantum dipoles, Phys. Rev. B **30**, 4874 (1984)

[30] J. Tersoff: Schottky barriers and semiconductor band structures, Phys. Rev. B **32**, 6968 (1985)

[31] M. Cardona, N. E. Christensen: Acoustic deformation potentials and heterostructure band offsets in semiconductors, Phys. Rev. B **35**, 6182 (1987)

[32] E. D. Palik: *Handbook of Optical Properties of Solids*, vol. 1–3 (Academic Press, San Diego 1985)

[33] S. Miyazaki: Photoemission study of energy-band alignments and gap-state density distributions for high-$\kappa$ gate dielectrics, J. Vac. Sci. Technol. B **19**, 2212 (2001)

[34] S. Sayan, E. Garfunkel, S. Suzer: Soft X-ray photoemission studies of the $HfO_2/SiO_2/Si$ system, Appl. Phys. Lett. **80**, 2135 (2002)

[35] G. B. Rayner Jr., D. Kang, Y. Zhang, G. Lucovsky: Nonlinear composition dependence of X-ray photoelectron spectroscopy and Auger electron spectroscopy features in plasma-deposited zirconium silicate alloy thin films, J. Vac. Sci. Technol. B **20**, 1748 (2002)

[36] W. J. Zhu, T. P. Ma, T. Tamagawa, J. Kim, Y. Di: Current transport in metal/hafnium oxide/silicon structure, IEEE E. D. Lett. **23**, 97 (2002)

[37] S. J. Wang, A. C. H. Huan, Y. L. Foo, J. W. Chai, J. S. Pan, Q. Li, Y. F. Dong, Y. P. Feng, C. K. Ong: Energy-band alignments at $ZrO_2/Si$, SiGe, and Ge interfaces, Appl. Phys. Lett. **85**, 4418 (2004)

[38] O. Renault, N. T. Barrett, D. Samour, S. Quiais-Marthon: Electronics of the $SiO_2/HfO_2$ interface by soft X-ray photoemission spectroscopy, Surf. Sci. **566**, 526 (2004)

[39] M. Oshima, S. Toyoda, T. Okumura, J. Okabayashi, H. Kumigashira, K. Ono, M. Niwa, K. Usuda, N. Hirashita: Chemistry and band offsets of $HfO_2$ thin films for gate insulators, Appl. Phys. Lett. **83**, 2172 (2003)

[40] Y.-C. Yeo, T.-J. King, C. Hu: Metal-dielectric band alignment and its implications for metal gate complementary metal-oxide-semiconductor technology, J. Appl. Phys. **92**, 7266 (2002)

[41] V. V. Afanas'ev, M. Houssa, A. Stesmans, M. M. Heyns: Electron energy barriers between (100)Si and ultrathin stacks of $SiO_2$, $Al_2O_3$, and $ZrO_2$ insulators, Appl. Phys. Lett. **78**, 3073 (2001)

[42] V. V. Afanas'ev, M. Houssa, A. Stesmans, M. M. Heyns: Band alignments in metal-oxide-silicon structures with atomic-layer deposited $Al_2O_3$ and $ZrO_2$, J. Appl. Phys. **91**, 3079 (2002)

[43] T. Hattori, T. Yoshida, T. Shiraishi, K. Takahashi, H. Nohira, S. Joumori, K. Nakajima, M. Suzuki, K. Kimura, I. Kashiwagi, C. Ohshima, S. Ohmi, H. Iwai: Composition, chemical structure, and electronic band structure of rare earth oxide/Si(100) interfacial transition layer, Microelectron. Eng. **72**, 283 (2004)

[44] A. Ohta, M. Yamaoka, S. Miyazaki: Photoelectron spectroscopy of ultrathin yttrium oxide films on Si(100), Microelectron. Eng. **72**, 154 (2004)

[45] S. A. Chambers, Y. Liang, Z. Yu, R. Droopad, J. Ramdani, K. Eisenbeiser: Band discontinuities at epitaxial $SrTiO_3/Si(001)$, Appl. Phys. Lett. **77**, 1662 (2000)

[46] R. Ludeke, M. T. Cuberes, E. Cartier: Local transport and trapping issues in $Al_2O_3$ gate oxide structures, Appl. Phys. Lett. **76**, 2886 (2000)

[47] D. J. DiMaria: Effects on interface barrier energies of metal-aluminum oxide-semiconductor (MAS) structures as a function of metal electrode material, charge trapping, and annealing, J. Appl. Phys. **45**, 5454 (1974)

[48] A. Baldereschi, S. Baroni, R. Resta: Band offsets in lattice-matched heterojunctions: A model and first-principles calculations for GaAs/AlAs, Phys. Rev. Lett. **61**, 734 (1988)

[49] C. G. van de Walle: Band lineups and deformation potentials in the model-solid theory, Phys. Rev. B **39**, 1871 (1989)

[50] C. J. Forst, C. R. Ashman, K. Schwarz, P. E. Blöchl: The interface between silicon and a high-$\kappa$ oxide, Nature **427**, 53 (2004)

[51] P. W. Peacock, J. Robertson: Structure, bonding, and band offsets of (100)SrTiO$_3$-silicon interfaces, Appl. Phys. Lett. **83**, 5497 (2003)

[52] P. W. Peacock, J. Robertson: Bonding, energies, and band offsets of Si$-$ZrO$_2$ and HfO$_2$ gate oxide interfaces, Phys. Rev. Lett. **92**, 057601 (2004)

[53] J. Robertson: Interfaces and defects of high-$\kappa$ oxides on silicon, Solid State Electron. **49**, 283 (2005)

# Index

# Electronic Structure of Rare Earth Oxides

Leon Petit[1], Axel Svane[2], Zdzislawa Szotek[3], and Walter M. Temmerman[3]

[1] Computer Science and Mathematics Division and Center for Computational
Sciences, Oak Ridge National Laboratory, Oak Ridge, Tennessee 37831, USA
[2] Department of Physics and Astronomy, University of Ǻårhus, Ny Munkegade,
DK 8000 Ǻårhus, Denmark
svane@phys.au.dk
[3] Daresbury Laboratory, Daresbury, Warrington WA4 4AD, United Kingdom

**Abstract.** The electronic structures of dioxides, $REO_2$, and sesquioxides, $RE_2O_3$, of the rare earths, RE = Ce, Pr, Nd, Pm, Sm, Eu, Gd, Tb, Dy and Ho, are calculated with the self-interaction-corrected local-spin-density approximation. The valencies of the rare earth ions are determined from total-energy minimization. Ce, Pr, and Tb are found to have tetravalent configurations in their dioxides, while for all the sesquioxides the trivalent ground-state configuration is most favourable. Tetravalent $NdO_2$ is predicted to exist as a metastable phase – unstable towards the formation of hexagonal $Nd_2O_3$. The trends of the band gap structure are discussed.

## 1 Introduction

The rare earth (RE) oxides find important applications in the catalysis, lighting and electronics industries. In particular, the design of advanced devices based upon the integration of rare earth oxides with silicon and other semiconductors calls for detailed understanding of the bonding, electronic and dielectric properties of these materials. The present Chapter reports ab initio electronic structure calculations of the RE dioxides and sesquioxides using the self-interaction-corrected local-spin-density (SIC-LSD) approximation, which is particularly well suited to treat the issue of rare earth valency in the solid state [1]. Specifically, the dioxides $REO_2$ and sesquioxides $RE_2O_3$ of Ce, Pr, Nd, Pm, Sm, Eu, Gd, Tb, Dy and Ho are considered [2].

The rare earth elements oxidize with varying strength [3]. Under suitable conditions, all the rare earth elements form a sesquioxide [4], and there is general agreement that, in the corresponding ground state, the rare earth atoms are in the trivalent, $RE^{3+}$, configuration [5,6]. Ce metal oxidizes completely to $CeO_2$ in the presence of air, i.e., Ce adopts in this case the tetravalent $Ce^{4+}$ configuration. Pr occurs naturally as $Pr_6O_{11}$, exhibiting an oxygen-deficient fluorite structure, while the stoichiometric fluorite structure $PrO_2$ exists under positive oxygen pressure. From Nd onwards, all the rare earth oxides, with the exception of Tb, occur naturally as sesquioxides, $RE_2O_3$. Tb oxide occurs naturally as $Tb_4O_7$, but transforms into $TbO_2$ under positive oxygen pressure.

The description of the rare earth ionic configuration in a basic theory necessitates a special treatment of the $f$-electrons. Electronic structure calcu-

M. Fanciulli, G. Scarel (Eds.): Rare Earth Oxide Thin Films,
Topics Appl. Physics **106**, 331–344 (2007)
© Springer-Verlag Berlin Heidelberg 2007

lations, based on density functional theory (DFT), have in general been very successful in describing the cohesive properties of solid-state systems with itinerant valence electrons. However, the strong correlations experienced by the localized $4f$-electrons in the rare earth compounds are not sufficiently accounted for by the exchange and correlation effects of the homogeneous electron gas, which underpin the local-spin-density (LSD) approximation of DFT. With respect to the electronic structure of rare earth oxides, there has been a number of band structure calculations for $CeO_2$ [7–9], $PrO_2$ [7, 10], $Ce_2O_3$ [8, 9], and RE sesquioxides in general [11]. From these studies, it has become clear that, for example, the electronic structure of $CeO_2$ is best described in terms of itinerant $f$-electrons, since in this compound the $f$-shell is empty, and the $f$-degrees of freedom enter only through hybridization. For $Ce_2O_3$, on the other hand, considering one Ce $f$-electron as part of the core leads to better agreement with the experimental lattice parameter [8]. Obviously, in this case, the $f$-electron ground-state configuration, and consequently the choice of calculational approach, have been determined from a comparison to empirical data. *Fabris* et al. [9] show that the LDA+U approach gives good agreement with experiment for both Ce polymorphs, but the method requires input of a Hubbard U parameter (3 eV for Ce). In $PrO_2$, the LSD description reveals a large $f$-peak at the Fermi level, in disagreement with the fact that $PrO_2$ is an insulator.

The SIC-LSD method [12, 13], applied in the present study, allows for a separation of the $f$-electron manifold into localized and delocalized subsets. The decision whether an $f$-electron is treated as localized, i.e., benefiting from SIC energy, or delocalized, i.e., hybridizing and participating in the bonding, is made on the ground of energetics. The total energies corresponding to different configurations of localized states are compared, and consequently the ground-state configuration of the RE ion may be determined from the global minimum of the total energy. The method has previously been succesfully applied to describe the valencies of the RE elements [1] and compounds (see, for example, [14, 15] and references therein). The method is briefly discussed in the next section, while more elaborate accounts of the method may be found in [12,13]. Results for rare earth dioxides and sesquioxides are presented in Sects. 3 and 4, respectively, while Sect. 5 presents our conclusions.

## 2 SIC-LSD Formalism

The SIC-LSD method [16] is a total-energy approach derived within the general framework of density functional theory [17, 18]. The basic ingredient is a functional defining the total energy of the electronic system, given the positions of the atomic nuclei:

$$E^{SIC}\left[\{\psi_\alpha\}\right] = T[n] + U[n] + E_{xc}^{LSD}[n] + V_{ext}[n] - \sum_\alpha^{occ.} \delta_\alpha^{SIC}. \tag{1}$$

Here, the first four terms are the kinetic, Hartree, exchange-correlation and external potential energies of the electron system, whose spin- and charge-density is denoted by $n(\boldsymbol{r})$. Taken together, these four terms constitute the LSD total-energy functional[1]. The last term is the self-interaction correction [16]

$$\delta_\alpha^{\mathrm{SIC}} = U[n_\alpha] + E_{\mathrm{xc}}^{\mathrm{LSD}}[n_\alpha], \tag{2}$$

where for each occupied orbital, $\psi_\alpha$, the self-Hartree and self-exchange-correlation energies are subtracted from the LSD energy functional. For states of Bloch symmetry in an infinite crystal, the self-interaction vanishes. To benefit from the self-interaction term, a spatially localized state must be formed. The ensuing many-electron wavefunction becomes a mixture of Bloch-type and Heitler–London type one-particle states, i.e., delocalized and localized $f$-electrons coexist. The delocalized $f$-electrons, like the $s$-, $p$-, and $d$-electrons, move in the LSD potential, meaning that their exchange and correlation energies are those derived on the basis of the homogeneous electron gas, and their ability to form bands leads to a gain in hybridization energy. The localized $f$-levels, on the other hand, acquire core-like character due to an additional negative potential term, arising from the self-interaction correction contribution, which effectively suppresses hybridization. The physical rationale behind this distinction is that the localized electrons reside on each atomic site long enough for the surroundings to respond to their presence. The itinerant electrons, on the other hand, move quickly through the crystal, thus experiencing only the LSD mean field potential. In the SIC-LSD approach, the localized and itinerant $f$-states are treated on an equal footing, in the sense that they are expanded in the same set of basis functions. The minimization of $E^{\mathrm{SIC}}$ becomes a question of optimizing the expansion coefficients. In particular, RE configurations with varying numbers of localized $f$-states – in general, different "localization scenarios" – can be explored and compared with respect to their total energies. Consequently, the preferred ground-state configuration, as far as the number of localized $f$-electrons is concerned, can be determined from the global total-energy minimum. The SIC-LSD approach has been successfully applied to describe pressure-dependent valency changes in rare earth compounds [14].

The SIC-LSD approach has been implemented using the tight-binding linear muffin-tin orbital (LMTO) method [20] in the atomic sphere approximation (ASA). The spin-orbit interaction is included in the Hamiltonian. Empty spheres are inserted on high-symmetry interstitial sites. The valence panel includes the $6s$, $5p$, $5d$ and $4f$ orbitals on the rare earth atom, and the $2s$ and $2p$ on the O atom, while the higher-order $\ell$ orbitals of O ($3d$ and $4f$), as well as all degrees of freedom of the empty spheres are treated as downfolded [20]. A separate energy panel is used to describe the semicore

---

[1] We have used the parametrization of $E_{\mathrm{xc}}^{\mathrm{LSD}}$ in (1) as given by [19]. The results presented are not sensitive to this particular choice.

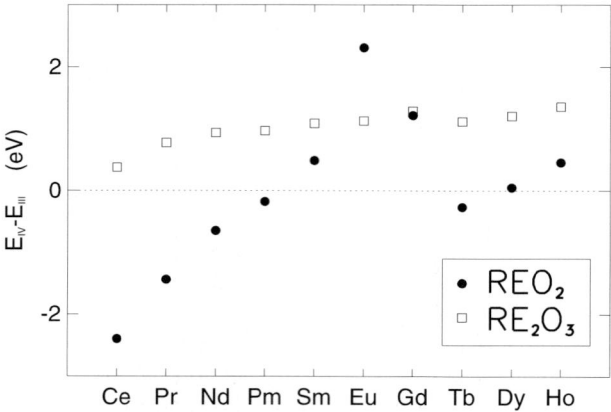

**Fig. 1.** Energy difference $E_{IV} - E_{III}$ (in eV per rare earth ion) between the tetravalent and trivalent rare earth configurations in dioxides (*solid circles*) and A-type sesquioxides (*squares*). A negative value of the energy difference implies that the tetravalent configuration is the more favourable one

$5s$ states of the rare earth. In the valence band, one distinguishes between the $f$-electrons that are SIC-localized, and the delocalized $f$-electrons which, together with the $s$-, $p$- and $d$-electrons, contribute to bonding. Different configurations of these localized and delocalized electrons will give rise to different nominal valencies which, in the SIC-LSD approach, are defined as: $N_{val} = Z - N_{core} - N_{SIC}$, where $Z$ is the atomic number, $N_{core}$ is the number of atomic core and semicore electrons, and $N_{SIC}$ is the number of localized $f$-electrons. According to this definition, a localized $f^1$ configuration of the RE ion will be referred to as trivalent in the case of Ce, tetravalent in the case of Pr, pentavalent in the case of Nd, etc.

## 3 Rare Earth Dioxides

The total energies of all the RE dioxides from $CeO_2$ to $HoO_2$ were calculated with a rare earth valency assumed to be either 3, 4 or 5. In all these cases, the crystal structure was taken to be the cubic fluorite structure (space group $Fm\bar{3}m$, no. 225) [21], and the crystal volume was varied to determine the equilibrium lattice spacing. The results for the energy differences between the tetravalent and trivalent configurations, $E_{IV} - E_{III}$, at their respective theoretical lattice spacings, are shown in Fig. 1. A negative energy difference indicates that the compound prefers the tetravalent ground-state configuration. The energy difference for the pentavalent configuration is not shown, since this configuration is found to be unfavourable in all the dioxides. A decreasing affinity for the tetravalent configuration is seen from $CeO_2$ through to $PmO_2$, and from $SmO_2$ the energy difference changes sign, indicating that

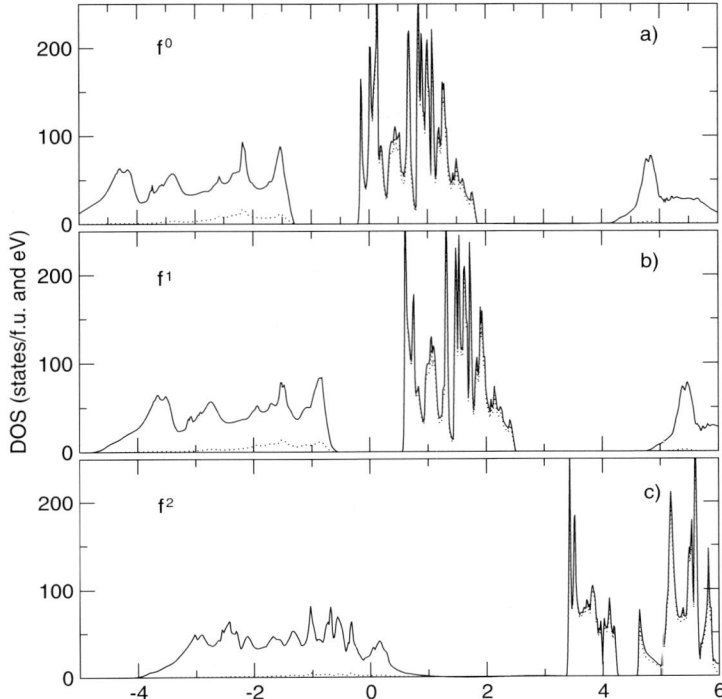

**Fig. 2.** Total DOS (*solid line*) and $f$-projected DOS (*dotted line*) for $PrO_2$, with the Pr ions in (**a**) the pentavalent ($f^0$) configuration (**b**) the tetravalent ($f^1$) configuration and (**c**) the trivalent ($f^2$) configuration. The energy is in units of eV, with zero marking the Fermi level

now the trivalent configuration is energetically more favourable. With increasing nuclear charge, the $f$-electrons become more tightly bound to the RE atom, and the smaller overlap with neigbouring atoms results in a reduced gain in hybridization energy. Eventually, it becomes more favourable to localize an extra electron, and gain the corresponding SIC energy, which is what happens from $SmO_2$ onwards, with a resulting trivalent ground-state configuration. In the late RE dioxides, the tetravalent configuration is preferred only for $TbO_2$, due to the stability of the half-filled $f$-shell of $Tb^{4+}$.

A closer examination of the electronic structure of $PrO_2$ reveals why the trivalent configuration is energetically unfavourable in this compound. In Fig. 2, both the total and $f$-projected density of states (DOS) of $PrO_2$ are shown for the pentavalent (Fig. 2a), tetravalent (Fig. 2b), and trivalent (Fig. 2c) Pr configurations. In Fig. 2a with all the $f$-electrons treated as delocalized, the Fermi level is situated in the $f$-peak, in accordance with the LSD calculations by *Koelling* et al. [7], and in disagreement with the observed insulating nature of $PrO_2$. As witnessed by the DOS of Fig. 2b, localizing one $f$-electron leads to an insulating compound, with the Fermi level situated between the occupied O $p$ states and the unoccupied Pr $f$-band states. In the trivalent scenario of Fig. 2c, a further $f$-electron becomes localized, which results in the Fermi level moving down into the O $p$-band. In this process of localizing an additional electron to form Pr $f^2$ ions, the O $p$-band states are depopulated, i.e., charge transfer is imposed on the system. The associated cost in energy is significantly larger than the gain in $f$-localization energy. In the heavier RE dioxides, this energy cost is reduced, due to the lowering of the $f$-band with respect to the O $p$-bands, and the trivalent configuration becomes energetically more favourable. The calculated band gap for $PrO_2$ is approximately $1.1\,eV$ in the $f^1$ ground state of Fig. 2b, which is, however, considerably larger than the $0.262\,eV$ derived from conductivity measurements by *Gardiner* et al. [22].

For $CeO_2$, we find a clearly preferred tetravalent ground-state configuration (the trend towards delocalization of the $f$-electrons is even stronger than in $PrO_2$), i.e., $CeO_2$ is best described in the LSD approximation, with $f$-orbitals of Bloch symmetry, in line with the results from earlier band structure calculations [7, 8]. The calculated volume in this configuration is in excellent agreement with the experimental value. In $TbO_2$, the tetravalent ground-state configuration is also energetically most favourable, although the energy difference between the trivalent and tetravalent configurations is decreased in comparison with $CeO_2$ and $PrO_2$. Again, the volume calculated for the tetravalent configuration is in good agreement with the experimental value. The remaining RE dioxides do not form in nature, and the determination of their ground-state configuration might seem of academic interest only. The energy differences, $E_{IV} - E_{III}$ in Fig. 1, indicate that $NdO_2$ and $PmO_2$ would also prefer the $RE^{4+}$ ion configuration. However, a closer examination suggests that these compounds are unstable with respect to the reduction to their respective sesquioxide [2].

# 4 Sesquioxides

At temperatures below $2000\,°C$, the rare earth sesquioxides adopt three different structure types [4]. The light REs crystallize in the hexagonal $La_2O_3$ structure (A-type, spacegroup $P\bar{3}m1$, no. 164) [21], and the heavy REs assume the cubic $Mn_2O_3$ structure (C-type, spacegroup $Ia\bar{3}$, no. 206) [21], also known as the bixbyite structure. The middle REs can be found in either the C-type structure, or the B-type, which is a monoclinic distortion of the C-type structure. Conversions between the different structure types are induced under specific temperature and pressure conditions [23]. The electronic structure of both the A-type and C-type sesquioxides has been investigated in the present work. The C-type bixbyite structure was approximated by the fluorite $REO_2$ structure, with $1/4$ of the O atoms removed [24], i.e., from four fluorite formula units in a conventional simple cubic supercell, the oxygen

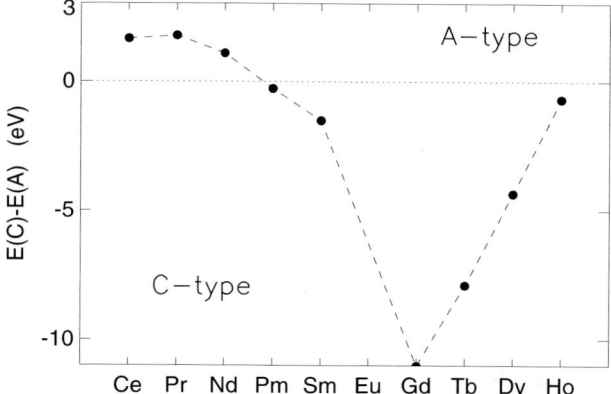

**Fig. 3.** Calculated structural energy difference (in eV) between the A-type and C-type sesquioxide for trivalent rare earth. A negative energy difference implies that the C-type structure is lower in energy

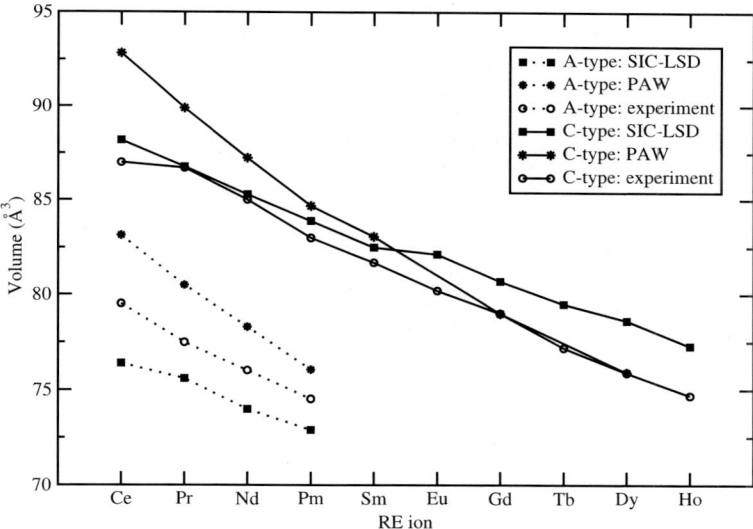

**Fig. 4.** Calculated equilibrium volumes (in units of $\text{Å}^3$) of the rare earth sesquioxides, crystallizing in the hexagonal A-type structure (*dotted line*), and the cubic C-type structure (*continuous line*). The *stars*, *squares*, and *circles* refer to the PAW [11], SIC-LSD (present), and experimental results, respectively [21]

atoms at the origin and at the cube center were replaced with empty spheres. No relaxation of the remaining oxygen positions was attempted. For the A-type structure, either the experimental $c/a$ ratio was taken or, for compounds not observed in A-type structure, an interpolated $c/a$ ratio was inferred. The charge density sampling used 64 and 35 k-points in the irreducible part of Brillouin zone for A-type and C-type, respectively.

The valency energy difference between tetravalent and trivalent RE configurations in the A-type structure is displayed in Fig. 1. In all cases, the trivalent configuration is found to be the ground state. A similar behaviour is seen in the simulated C-type structure. Again, the small anomaly around Gd can be attributed to the half-filling of the $f$-shell. Figure 3 shows the energy difference between the two structures of the sesquioxides in their trivalent ground states [2]. The A-type structure is lower in energy than the C-type structure for the first three rare earths, i.e., Ce, Pr and Nd, while the C-type structure is lowest in energy in the remainder of the series. The oxidation process from the sesquioxide to the dioxide was discussed in [2] by comparison of total energies of these phases, which amounts to the neglect of effects of finite temperature. The chemical potential of free oxygen is an essential parameter for obtaining quantitatively accurate results. It was revealed that the oxidation energy indeed would be lowest for $Ce_2O_3 \rightarrow CeO_2$, $Pr_2O_3 \rightarrow PrO_2$ and $Tb_2O_3 \rightarrow TbO_2$, in accordance with the observation of precisely these three dioxides in nature. For Nd, the dioxide was found stable with respect to oxydation into the C-type structure, but unstable with respect to oxydation into the A-type structure. Hence, if steric hindrances or substrate interactions would not allow transformation of reduced $NdO_2$ into the hexagonal form of the sesquioxide, the Nd dioxide phase might exist as a metastable phase. This is a unique situation for Nd, since for all the later REs the C-type is the more stable phase of the sesquioxide (cf. Fig. 3).

The lanthanide contraction of the sesquioxides is illustrated in Fig. 4. As one moves across the lanthanide series, the attractive force of the nucleus is only partially screened by the added $f$-electrons, leading to a steady increase of the effective nuclear charge with $f$-shell filling, and hence increased bonding. The agreement between the present calculations and the experimental values of the crystal volumes of the rare earth sesquioxides is seen to be excellent, both with respect to the absolute values and the overall trends. Figure 4 also includes the equilibrium volumes calculated by *Hirosaki* et al. [11], using the projector augmented-wave (PAW) method, with the localized partly filled $f$-shell being treated as part of the core. For the A-type structure, the SIC-LSD calculations consistently yield lower volumes than observed, while the PAW values are consistently higher. For the C-type structure, the present SIC-LSD calculations agree somewhat better with observations than the PAW results for the earlier REs, which may be due to the core approximation used for the $f$-electrons in the work of [11], and which may be too restrictive for the earlier RE sesquioxides.

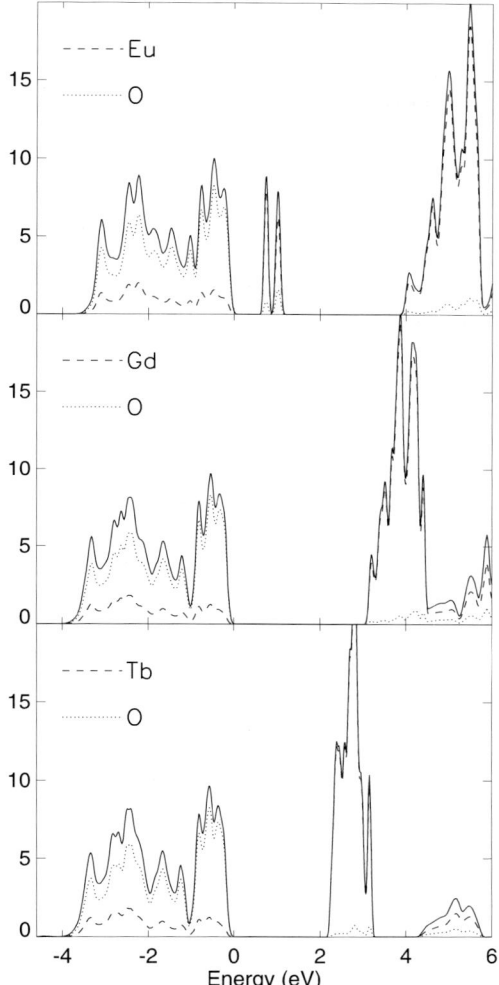

**Fig. 5.** Total density of states (*continuous line*) of Eu$_2$O$_3$, Gd$_2$O$_3$ and Tb$_2$O$_3$ in the trivalent configuration in the A-type hexagonal structure at the calculated equilibrium volume. The RE and O partial densities of states are shown with *dashed* and *dotted lines*, respectively. Energy is in eV, with 0 at the top of the valence bands, while DOS is in states per formula unit and eV

The densities of states and band structures for all the sesquioxides are quite similar and deviate mostly with respect to the unoccupied electron states. The latter are not very accurately determined in density-functional-based approaches, but here we review the trends of the band gap structures as found in the SIC-LSD approach, with the expectation that they might coincide with the experimental trends, even if the absolute values of the gaps are not fully correct. As representative examples, the DOSs of $Eu_2O_3$, $Gd_2O_3$ and $Tb_2O_3$ are shown in Fig. 5, calculated in the trivalent ground state, namely $f^6$, $f^7$ and $f^8$ configurations, respectively. The energy zero has been put at the top of the valence bands. The valence bands originate from the O-$p$-states, and are quite similar in all compounds, of the order of 3.5 eV width. Hybridization and charge transfer result in the O $p$-band being completely filled, with an empty $f$-band situated in the gap between the valence and (non-$f$) conduction bands. Figure 5 illustrates the low position of the $f$-band in $Eu_2O_3$, which is caused by the exchange attraction with the localized $f^6$ shell of Eu. In $Gd_2O_3$ and $Tb_2O_3$, there are only unoccupied $f$-states of minority spin, and their position is significantly higher. All the sesquioxides are found to be insulators, with the exception of C-type $Eu_2O_3$, for which the gap between the O $p$-band and the unoccupied majority $f$-states just closes. The position of the unoccupied $f$-band varies across the series. This is illustrated in Fig. 6, which shows the evolution of the unoccupied $f$-band (hatched area) and the non-$f$ conduction edge with respect to the top of the O $p$-bands (at zero energy) through the series of A-type sesquioxides. One observes that the gap between valence and non-$f$ conduction bands ($v \rightarrow c$) shows a rather slow increase through the RE series, whereas the gap between valence and $f$-band ($v \rightarrow f$) decreases from $Ce_2O_3$ to $Eu_2O_3$, and again from $Gd_2O_3$ to $Dy_2O_3$. For $Ce_2O_3$, $Pr_2O_3$ and $Gd_2O_3$, no separate edge for the non-$f$ conduction bands was found in the calculations. This trend of the unoccupied $f$-band is in agreement with the picture emerging from high-temperature conductivity measurements [25], although the calculated $f$-band positions are lower relative to the conduction edge than the experimentally observed ones. Note that our theory assumes that the excited state of $f$-character may be described by bands, even if the ground state has localized $f$-electrons. Alternatively, one could imagine also excited states of localized $f$-character corresponding to intra-atomic transitions, which would be given by largely free ionic multiplets. The present theory also did not consider the possibility of $f \rightarrow c$ transitions, which may occur at lower excitation energy than for the $v \rightarrow c$ bands depicted in Fig. 6. This situation is likely what happens in the early lanthanide sesquioxides, as evidenced by the relatively small gaps in $Ce_2O_3$ and $Pr_2O_3$ [26].

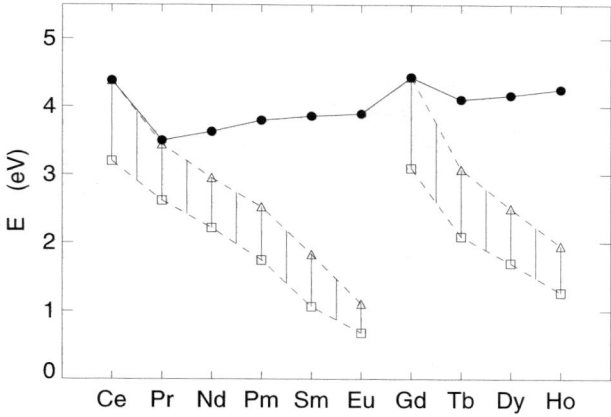

**Fig. 6.** Evolution of the SIC-LSD gap structure through the rare earth sesquioxide series (A-type). The valence band maximum is at zero energy, the unoccupied $f$-band (of majority spin character only from Ce to Eu) is positioned between *squares* and *triangles* (hatched area). The non-$f$ conduction edge is marked with *solid circles*. Energy is in eV

# 5 Conclusions

The electronic structures of rare earth dioxides and sesquioxides have been investigated based on calculations in the self-interaction–corrected local-spin-density approximation. We have provided a simple understanding of these systems in terms of the electronegativity of the oxygen, which takes as many electrons as needed to fill its $p$-levels from the RE atom [2]. We have shown that the formation of the dioxide is possible precisely for those rare earths where the energetics allows the formation of a tetravalent rare earth ion.

**Acknowledgements**

This work has been partially funded by the Training and Mobility Network on "Electronic Structure Calculation of Materials Properties and Processes for Industry and Basic Sciences" (contract: FMRX-CT98-0178) and by the Research Training Network on "Ab-initio Computation of Electronic Properties of f-electron Materials" (contract: HPRN-CT-2002-00295). Work of L.P. was sponsored in part by the Office of Basic Energy Sciences, U.S. Department of Energy. The Oak Ridge National Laboratory is managed by UT-Battelle LLC for the Department of Energy under Contract No. DE-AC05-00OR22725.

# References

[1] P. Strange, A. Svane, W. M. Temmerman, Z. Szotek, H. Winter: Understanding the valency of rare earths from first-principles theory, Nature (London) **399**, 756 (1999)

[2] L. Petit, A. Svane, Z. Szotek, W. M. Temmerman: First-principles study of rare earth oxides, Phys. Rev. B **72**, 205118 (2005)

[3] E. Holland-Moritz: Coexistence of valence fluctuating and stable Pr ions in $Pr_6O_{11}$, Z. Phys. B **89**, 285 (1992)

[4] L. Eyring: The binary rare earth oxides, in K. A. Gschneider Jr., L. Eyring (Eds.): *Handbook on the Physics and Chemistry of Rare Earths*, vol. 3 (North-Holland, Amsterdam 1979) p. 337

[5] S. Tanaka, H. Ogasawara, K. Okada, A. Kotani: Theory of the $4d \rightarrow 2p$ X-ray emission spectroscopy in $Ce_2O_3$, $Pr_2O_3$ and $Dy_2O_3$, J. Phys. Soc. Jpn. **64**, 2225 (1995)

[6] A. Moewes, D. L. Ederer, M. M. Grush, T. A. Callcott: Probing electron correlation, charge transfer, and Coster-Kronig transitions at the $3d$ and $4d$ thresholds of Nd by resonant inelastic scattering, Phys. Rev. B **59**, 5452 (1999)

[7] D. D. Koelling, A. M. Boring, J. H. Wood: The electronic structure of $CeO_2$ and $PrO_2$, Solid State Commun. **47**, 227 (1983)

[8] N. V. Skorodumova, R. Ahuja, S. I. Simak, I. A. Abrikosov, B. Johansson, B. I. Lundqvist: Electronic, bonding, and optical properties of $CeO_2$ and $Ce_2O_3$ from first principles, Phys. Rev. B **64**, 115108 (2001)

[9] S. Fabris, S. de Gironcoli, S. Baroni: Taming multiple valency with density functionals: A case study of defective ceria, Phys. Rev. B **71**, 041102 (2005)

[10] J. Dabrowski, V. Zavodinski, A. Fleszar: Pseudopotential study of $PrO_2$ and $HfO_2$ in fluorite phase, Microelectron. Reliab. **41**, 1093 (2001)

[11] N. Hirosaki, S. Ogata, C. Kocer: Ab-initio calculation of the crystal structure of the lanthanide $Ln_2O_3$ sesquioxides, J. Alloys Comp. **351**, 31 (2003)

[12] W. M. Temmerman, A. Svane, Z. Szotek, H. Winter: Applications of self-interaction corrections to localized states in solids, in J. F. Dobson, G. Vignale, M. P. Das (Eds.): *Electronic Density Functional Theory: Recent Progress and New Directions* (Plenum, New York 1998)

[13] W. M. Temmerman, A. Svane, Z. Szotek, H. Winter, S. Beiden: On the implementation of the self-interaction corrected local spin density approximation for d- and f-electron systems, in H. Dreyssé (Ed.): *Electronic Structure and Physical Properties of Solids: The Uses of the LMTO Method*, Lecture Notes in Physics **535** (Springer, Berlin, Heidelberg 2000)

[14] A. Svane, G. Santi, Z. Szotek, W. M. Temmerman, P. Strange, M. Horne, G. Vaitheeswaran, V. Kanchana, L. Petit, H. Winter: Electronic structure of Sm and Eu chalcogenides, Phys. Stat. Sol. B **241**, 3185 (2004)

[15] G. Vaitheeswaran, L. Petit, A. Svane, V. Kanchana, M. Rajagopalan: Electronic structure of praseodymium monopnictide and monochalcogenides under pressure, J. Phys. Condens. Matter **16**, 4429 (2004)

[16] J. P. Perdew, A. Zunger: Self-interaction correction to density-functional approximations from any-electron systems, Phys. Rev. B **23**, 5048 (1981)

[17] W. Kohn, L. J. Sham: Self-consistent equations including exchange and correlation effects, Phys. Rev. **140**, A1133 (1965)

[18] L. J. Sham, W. Kohn: One-particle properties of an inhomogeneous interacting electron gas, Phys. Rev. **145**, 561 (1966)

[19] S. H. Vosko, L. Wilk, M. Nusair: Accurate spin-dependent electron liquid correlation energies for local spin-density calculations – a critical analysis, Can. J. Phys. **58**, 1200 (1980)

[20] O. K. Andersen, O. Jepsen, D. Glötzel: Canonical description of the band structures of metals, in F. Bassani, F. Fumi, M. P. Tosi (Eds.): *Canonical Description of the Band Structures of Metals* (North-Holland, Amsterdam 1985)

[21] P. Villars, L. D. Calvert: *Pearson's Handbook of Crystallographic Data for Intermetallic Phases*, 2nd ed. (ASM International, Ohio 1991)

[22] C. H. Gardiner, A. T. Boothroyd, P. Pattison, M. J. McKelvy, G. J. McIntyre, S. J. S. Lister: Cooperative Jahn–Teller distortion in $PrO_2$, Phys. Rev. B **70**, 024415 (2004)

[23] H. R. Hoekstra, K. A. Gingerich: High-pressure B-type polymorphs of some rare-earth sesquioxides, Science **146**, 1163 (1964)

[24] R. W. G. Wyckoff: *Crystal Structures*, vol. 2 (Wiley, New York 1967)

[25] H. B. Lal, K. Gaur: Electrical conduction in non-metallic rare-earth solids, J. Mater. Sci. **23**, 919 (1988)

[26] A. V. Prokofiev, A. I. Shelykh, B. T. Melekh: Periodicity in the band gap variation of $Ln_2X_3$ (X = O, S, Se) in the lanthanide series, J. Alloys Comp. **242**, 41 (1996)

# Index

# Rare Earth Oxides in Microelectronics

Kuniyuki Kakushima[1], Kazuo Tsutsui[1], Sun-Ichiro Ohmi[1], Parhat Ahmet[2], V. Ramgopal Rao[3], and Hiroshi Iwai[2]

[1] Interdisciplinary Graduate School of Sciences and Engineering, Tokyo Institute of Technology, 4259, Nagatsuta, Midori-ku, Yokohama, 226-8502, Japan
[2] Frontier Collaborative Research Center, Tokyo Institute of Technology, 4259 Nagatsuta, Midori-ku, Yokohama, 226-8502, Japan
h.iwai@ieee.org
[3] Indian Institute of Technology, Bombay 400 076, India

**Abstract.** A feasibility study of rare earth oxides for replacing $SiO_2$ gate oxide for CMOS integrated circuits has been conducted. Rare earth oxides have relatively higher dielectric constants and are suitable as gate dielectrics. Two-dimensional device simulations reveal that the desirable dielectric constant, without affecting the short-channel performance, is less than 50. The dielectric constant values of rare earth oxides satisfy this condition. One of the issues in rare earth oxides that needs to be addressed is the hygroscopic properties of the film. These physical and electrical changes in the oxides caused by this moisture absorption, which are not suitable in electronics applications, can be suppressed by coating a metal film or by using a passivation layer.

Of all the rare earth oxides, it is found that $La_2O_3$, after a proper heat treatment, has the best electrical properties for gate insulator applications in MOSFETs, because of its higher barrier height for the conduction band electrons and valence band holes as well as its higher dielectric constant. The smallest gate leakage current demonstrated through experimental results was $3 \times 10^{-4}$ /cm$^3$ at 1 V for an EOT of 0.6 nm. The conduction mechanism through $La_2O_3$ has been modeled, and has been shown to be mainly by SCLC.

With post-metallization annealing (PMA) after Al gate electrode formation, $La_2O_3$ gated MISFET has shown high effective electron mobility of 319 cm$^2$/V s. This value is lower compared to $SiO_2$, but still one of the highest among all the high-$\kappa$ MISFETs, for a gate oxide EOT value which is slightly larger than 2.3 nm. The PMA recovers the flat band shift and improves the mobility presumably by the diffusion of Al into $La_2O_3$ which compensates the positive charges in the film. However, the growth of an interfacial $Al_2O_3$ layer, which results in an increase in EOT, is still a problem. The solution could be to replace the Al gate electrode with a less reactive metal with $La_2O_3$, and doping $La_2O_3$ films with some elements which compensate the negative charge.

Interfacial layer suppression between the silicon substrate and the $La_2O_3$ film has been accomplished by using $Y_2O_3$ as a buffer layer, which is necessary to achieve an EOT less than 1 nm.

Reliability and yield of rare earth oxides still need to be investigated, but the results shown here hold promise for rare earth oxides, especially $La_2O_3$, as a suitable candidate for the post-Hf-based gate insulator in advanced CMOS integrated circuits.

M. Fanciulli, G. Scarel (Eds.): Rare Earth Oxide Thin Films,
Topics Appl. Physics **106**, 345–366 (2007)
© Springer-Verlag Berlin Heidelberg 2007

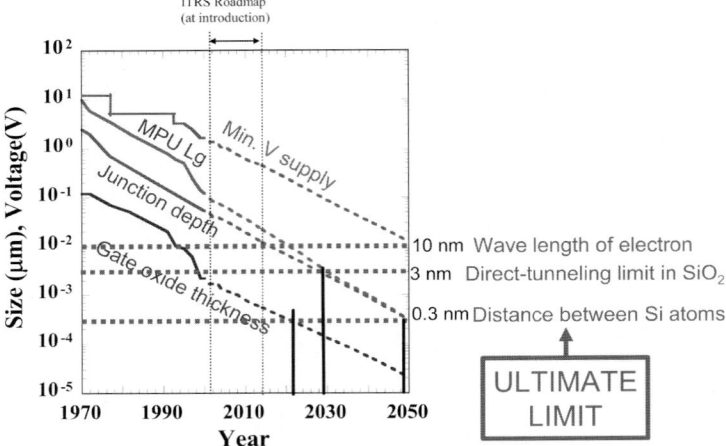

**Fig. 1.** Trend of size reduction in MOSFET. The ultimate limit seems to be the size of atoms

# 1 Introduction

Among many inventions made in the last century, electronics or electronic circuits can be considered as the most important, because of the fact that most human activities are nowadays controlled by electronic circuits. Progress in electronics or electronic circuits has been driven by the downsizing of its components. Downsizing of the devices has not only increased the functions of the circuits, but also has raised the switching speed and reduced the power consumption of the circuits. Electronic devices or components started as a vacuum tube in the early 1900s, and evolved to transistors. It is amazing that the feature sizes of the electronic components – such as filament size of the vacuum tubes and the gate length of MOSFETs – reduced by one million in 100 years. In 2005, the gate length of the most advanced commercial CMOS LSI devices is already sub-40 nm. Now, the downsizing trend is approaching to its ultimate limit, because of the size of atoms.

Figure 1 shows the trend of the featured size reduction in a MOSFET since 1970. The first expected fundamental limit was the gate insulator thickness due to the direct tunneling. Note that gate oxide thickness is about 2 orders of magnitude smaller than other feature sizes such as gate length and junction depth. Gate oxide thickness of 3 nm was thought to be the limit in downsizing [1]. Fortunately, this issue was solved in small gate length MOSFETs in 1994 [2]. Second, the downsizing limit of the gate length was considered to exist at the wave length of electrons, which is about 10 nm. However, 5 nm length-CMOS transistors were reported by *Wakabayashi* et al. in 2003 [3], and fairly good transistor operation for both n- and p-MOSFETs was demonstrated. The next issue that we will need to face is again the thickness of the

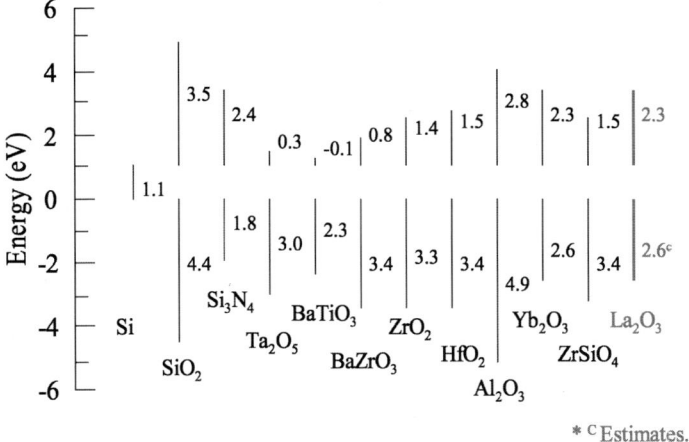

*$^c$Estimates.

**Fig. 2.** Energy barrier height of various insulators [5]

gate insulator, which is close to the distance of silicon atoms (0.3 nm) in the silicon crystal. A MOSFET operation with gate oxide thickness of 0.8 nm was reported [4], where 0.8 nm is already less than the distance between four silicon atoms. Because of this aggressive scaling of gate dielectrics, the total amount of the gate leakage current in an LSI will not be negligible anymore, even in the highest performance microprocessors. The allowable gate leakage current in a chip would be about 100–1000 A/cm$^2$ maximum, considering the total gate area in an LSI chip and the fact that maximum consumable power in a MPU is about 100 W. To overcome this problem, a material with a higher dielectric constant (high-$\kappa$) to substitute conventional SiO$_2$ is necessary, as it can provide an equivalent capacitance per unit area even with physically thicker dielectric layers, required for reducing the leakage current. For practical device operation, integration of high-$\kappa$ gate insulator is essential for MOSFETs with gate lengths below 50 nm.

## 2 Issues in High-$\kappa$ Materials

To date, various high-$\kappa$ materials have been reported for replacing SiO$_2$. The most promising material for 65 or 45 nm commercial technology nodes is Hf-based oxides or their silicates with dielectric constants of 10 to 15 [6, 7], which means that 1 nm of EOT is achievable with a physical thickness of 4 nm, thus decreasing the leakage current. There have been several serious problems for high-$\kappa$ gate insulator technology developments. Those are 1. interfacial layer growth during the deposition or heat treatments after the deposition, 2. micro-crystallization formation and phase separation in silicates during the heat treatment, and the resulting increase in the gate leakage

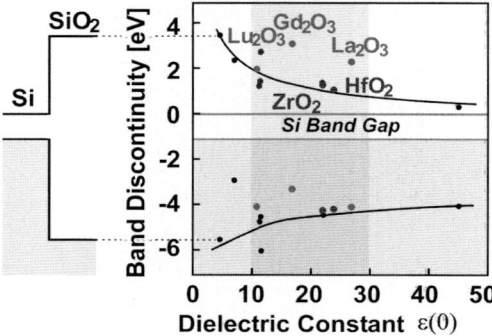

**Fig. 3.** Band discontinuity of rare earth oxides and silicon [8]

current, 3. lateral oxidation or growth of silicates at the edge of the gate stack, 4. fixed charges inside the high-$\kappa$ film, inducing large flat band voltage shift and mobility degradation for MOSFETs, 5. high density of the interface states causing mobility degradation, 6. non-uniformity and defects in the film, 7. contamination from the precursor used for ALD or MOCVD, 8. poor reliability and yield of the film.

Fortunately, these problems are now being solved by using nitrogen-doped Hf-silicate or $HfSi_xO_yN_z$, and these dielectrics therefore are expected to be introduced into commercial products within the next few years [6]. However, because of the relatively lower $\kappa$ value of the $HfSi_xO_yN_z$ and also because of the interfacial layer with thickness of a few to several tenths of nm between Si and high-$\kappa$ film, which are required for solving the above problems, it seems difficult to scale this kind of film practically for EOT values below 1 nm.

To realize an EOT less than 1 nm, we have focused on rare earth oxides, especially on $La_2O_3$, as it has a dielectric constant value higher than for $HfO_2$ and higher energy band offset to the conduction of Si, as shown in Fig. 3 [5]. *Hattori* et al. measured the band discontinuity of Si and high-$\kappa$ materials experimentally by X-ray photon spectroscopy (XPS) [8], and showed the offset of $La_2O_3$ to be 2.3 eV, which is much higher than for $HfO_2$.

Figure 4 summarizes the leakage current densities of several rare earth capacitors with EOTs identical to those of other reported high-$\kappa$ materials [9]. Due to the high offset barrier, $La_2O_3$ exhibited the smallest leakage current at low EOTs.

## 3 Short Channel Effect Enhancement

With a higher dielectric constant, one can further reduce the leakage current, but there is a practical limit for application in electronic devices. Let us look at a MOS transistor with a gate length of 40 nm with two kinds of gate insulators, one with $SiO_2$ ($\kappa = 3.9$) and the other with an extreme value of

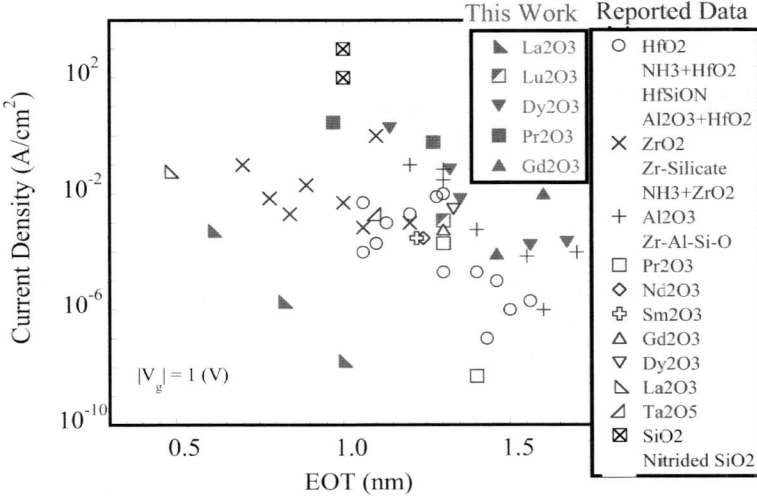

**Fig. 4.** Summary of leakage current densities of rare earth capacitors. La$_2$O$_3$ exhibited the lowest leakage current at low EOT

dielectric constant ($\kappa = 390$). EOT of both transistors is set to 2 nm. The 2D device simulation results of $I_d$–$V_g$ characteristics at a constant drain voltage of 0.4 V are shown in Fig. 5. The threshold voltage of 0.4 V on SiO$_2$ gated transistor was shifted to $-1.35$ V in the case of high-$\kappa$. The $V_{th}$ shift for the high-$\kappa$ gate insulator can be explained as follows. If we consider an electrical distance, defined as physical distance divided by the permittivity, the gate electrodes of the two transistors are exactly at the same distance to their respective channels. On the other hand, the electrical distances between source and drain are completely different due to the presence of the high-$\kappa$ insulator. The electric flux is proportional to the permittivity, so the electrical distance of the source and drain is reduced. The drain voltage turns on the channel, and thus negative gate voltage is required to turn off the transistor. The equipotential drawings show the electric field penetration from the drain region to the channel through the gate insulator, as shown in Fig. 5. Therefore, an extremely high-$\kappa$ material aggravates the short channel effects. To avoid the short channel effects, additional halo doping is needed to control the $V_{th}$. The practical upper limit of the dielectric constant can be estimated to be 50. Special designs to prevent the short channel effects are required to use the higher dielectric constant materials above 50 [10].

# 4 Hygroscopic Properties of Rare Earth Oxide Materials

One of the issues of rare earth oxides for electronics is the stability against water and organic gas absorption. It is reported that lanthanide (Ln) ox-

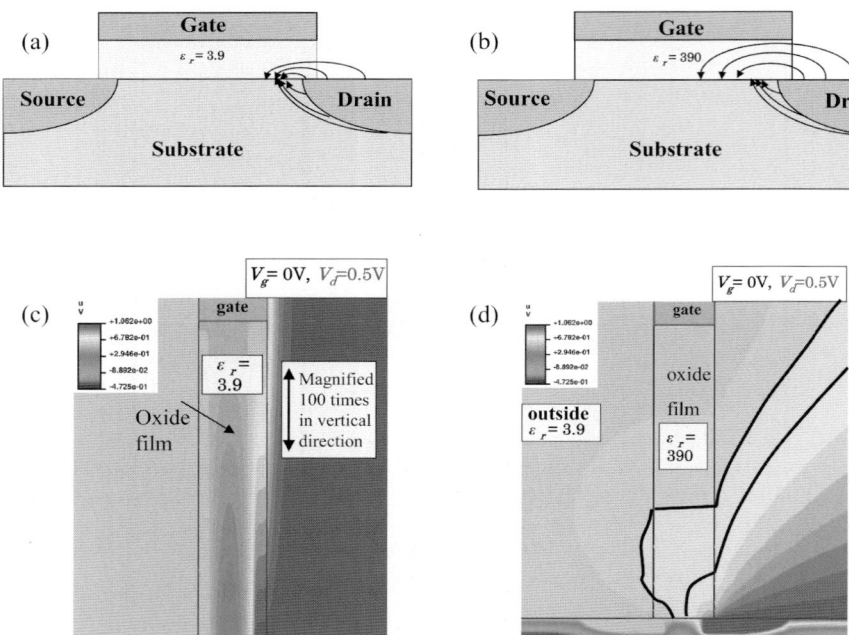

**Fig. 5.** Electric flux lines from drain region with dielectric constant of (**a**) 3.9 and (**b**) 390. Corresponding equi-potential lines are shown in (**c**) and (**d**)

ide forms hydride $Ln_2O_3$–$H_2O$ after absorbing moisture, and eventually the whole material will form hydroxide, $Ln(OH)_3$. On the contrary, lanthanide carbonate is formed at the surface with exposure to $CO_2$ gas. In addition to the physical thickness growth by moisture absorption, usually these hydride, hydroxide and carbonate materials have much lower dielectric constants. In order to apply rare earth oxides for electronics, their properties therefore should be investigated. In this section, the electrical properties of the rare earth oxides, $ZrO_2$ and $SiO_2$, on exposure to moisture are described.

Thin films of rare earth oxide on silicon substrate were kept in glass and acrylic containers with a humidity of 80 % at room temperature. The C–V curves of the films showed significant decrease in the capacitance densities, including two orders of magnitude reduction in the current densities. This indicates a film thickness growth due to water absorption. Capacitance equivalent thickness (CET) of various high-$\kappa$ materials after exposure to humidity are shown in Fig. 6a and b. Except for $SiO_2$, all other oxides including $ZrO_2$ absorbed moisture and therefore need treatments or passivations to avoid these degradations. From the AFM measurements, it was observed that the films which showed degradation in the surface roughness after moisture ab-

(a)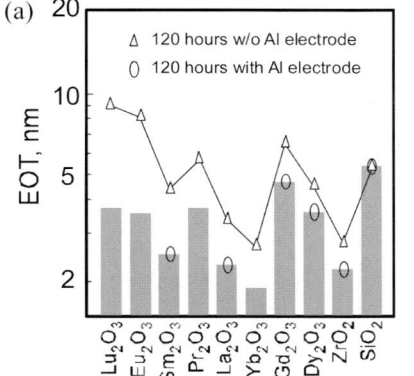

(b)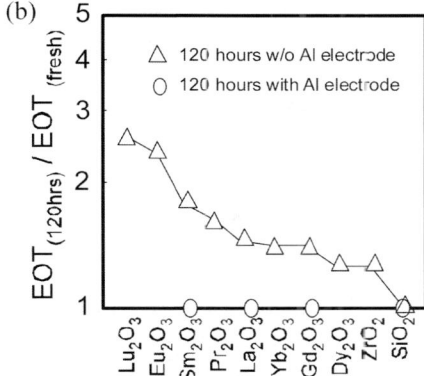

**Fig. 6.** (a) Equivalent oxide thickness changes of rare earth oxides after 120 h exposure to moisture. $ZrO_2$ and $SiO_2$ are also shown. (b) The ratio in EOT after moisture absorption with and without Al electrode. Using Al metal as capping, the degradation of the film can be suppressed

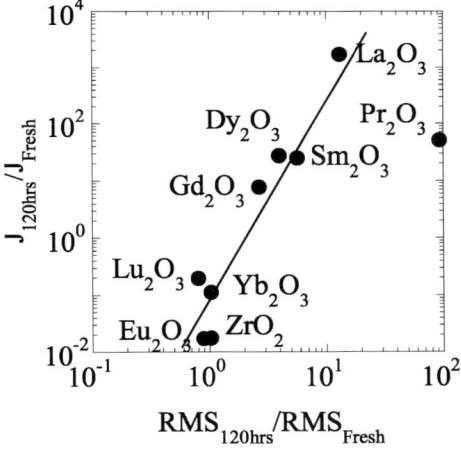

**Fig. 7.** Leakage current density changes of various rare earth oxides on the ratio of the surface roughness after 120 h moisture exposure

sorption also showed a higher leakage current (Fig. 7). One way to avoid this hygroscopic property is to coat the high-$\kappa$ film by a passivation film, in situ after the deposition. All the films, including $La_2O_3$ film with Al electrode coated after the deposition, showed no degradation in the capacitance, as shown in Fig. 6a.

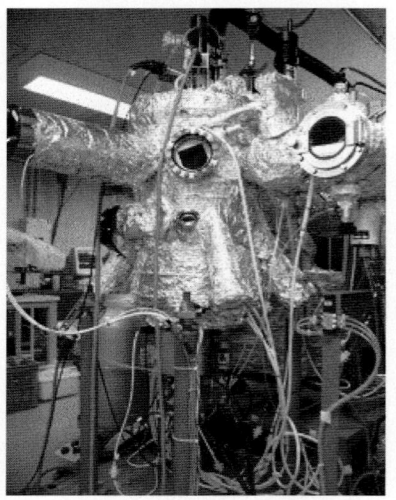

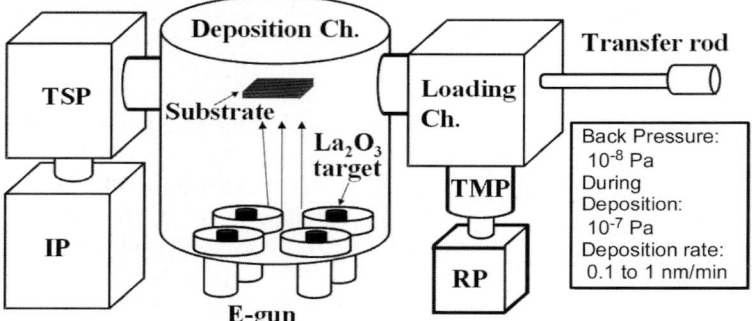

**Fig. 8.** Electron beam deposition setup for rare earth oxide deposition. The chamber is kept under vacuum in the order of $10^{-7}$ Pa through deposition

## 5 La$_2$O$_3$ Deposition

Unlike thermally grown SiO$_2$ for gate insulators, high-$\kappa$ materials are usually deposited either by physical or chemical methods, namely PVD (Physical Vapor Deposition) or CVD (Chemical Vapor Deposition). CVD-deposited films contain high concentrations of carbon, hydrogen, or chlorine, which are difficult to remove. Thus, in this study, we have chosen electron-beam vapor deposition in ultra-high vacuum chamber so that clean and pure films can be obtained. Figure 8 depicts the e-beam evaporation system. The chamber was kept in ultra-high vacuum ($10^{-7}$ Pa) using an ion pump during the deposition of the materials. Four sources can be mounted on hearths for electron beam heating for successive depositions.

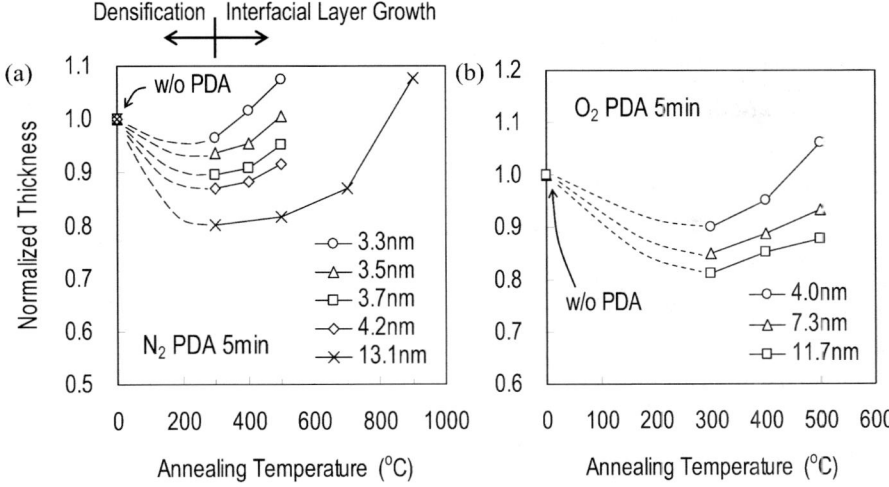

**Fig. 9.** Thickness of $La_2O_3$ layer depending on post-deposition annealing (PDA) temperature. (**a**) $N_2$ ambient and (**b**) $O_2$ ambient

$La_2O_3$ was deposited on a Si wafer after $H_2SO_4/H_2O_2$ and diluted HF dip cleaning. The substrate temperature and the deposition rate were set to 250 °C and 0.5 nm/min, respectively. The samples were then annealed in either oxygen or nitrogen ambient at various temperatures, by rapid thermal annealing using infrared. The thicknesses of the films after different heat treatments were measured by ellipsometry, which is shown in Fig. 9.

In this figure, the thickness was normalized by the as-deposited sample thickness. The film thickness reductions after low-temperature annealing indicate the densification of the film, and these trends continued up to 400 °C for both nitrogen and oxygen ambients. Evaluating the densities of the films, the as-deposited film had a density of $5.3\,g/cm^3$, while the density of the bulk $La_2O_3$ is $6.5\,g/cm^3$. The maximum density of $5.9\,g/cm^3$ was obtained at 300 °C annealing in nitrogen. The transmission electron micrograph (TEM) of the $La_2O_3$ film, which was annealed in nitrogen at 300 °C, is shown in Fig. 10. The thickness of the $La_2O_3$ was 10.2 nm, which is consistent with the thickness of 10.5 nm measured by ellipsometer. The thickness of the interfacial layer was found to be 0.3 nm which can, however, be minimized by optimizing the annealing condition. There was a 1.8 nm thick $Al_2O_3$ layer between $La_2O_3$ and the Al electrode, which was formed by post-metallization annealing (PMA) after Al electrode deposition.

On the other hand, higher annealing temperatures tend to increase the thickness of the films, especially with oxygen. The increases in thickness were caused by the growth of the interfacial layer of La-silicate and $SiO_2$ without phase separation. The origin of the interfacial layer growth is not yet fully

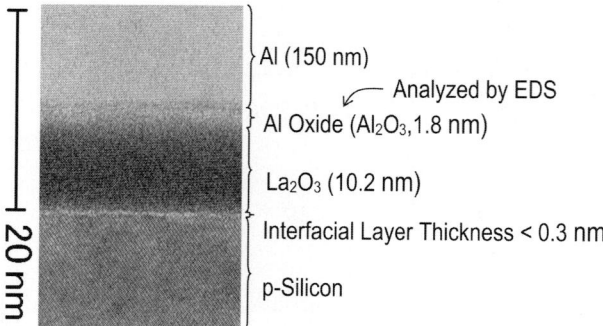

**Fig. 10.** High-resolution TEM micrograph of the $L_2O_3$ film annealed at 300 °C in nitrogen ambient

analyzed, but we believe it may be associated with Si oxidation and La reduction processes near the interface. This is in good agreement with the fact that the interfacial layer growth was also observed in vacuum annealed samples. Therefore, the density of the film decreased due to the interfacial layer growth, which has a lower dielectric constant.

# 6 Electrical Properties of $La_2O_3$ MIS Capacitors

Figure 11 shows the typical C–V characteristics of annealed $La_2O_3$ films in nitrogen and oxygen ambients at various temperatures [11]. An Al layer was thermally evaporated to form the gate electrode on the surface through stencil mask. Solid lines represent the theoretical capacitances using the NCSU-CV program including quantum effects [12]. The measured C–Vs of as-deposited films and the one annealed at 300 °C exactly fit the simulated curves, which indicates that the interface state densities are negligible. Beyond 400 °C and annealing in oxygen, on the other hand, significant mismatches were observed between the measured and the simulated characteristics due to interface states.

Flatband voltage ($V_{FB}$) shift is one of the important reliability issues for high-$\kappa$ devices. The temperature dependence of the $V_{FB}$ is plotted in Fig. 12, where negative shifts were observed both in nitrogen and oxygen ambients annealed at high temperature. $V_{FB}$ shift of the samples after high-temperature nitrogen annealing was found to be around $-0.9\,\mathrm{V}$.

On the other hand, annealing after Al electrode deposition recovers the $V_{FB}$ shift towards a positive direction, shown in Fig. 13. This recovery in the $V_{FB}$ shift can be explained from the diffusion of Al into the $La_2O_3$ film, which compensates the positive charge. This positive shift is in good agreement with the report from *Buchanan* et al. where they have shown that $Al_2O_3$ shifts the $V_{FB}$ to a positive direction [13]. Increase in capacitance value after PDA is due

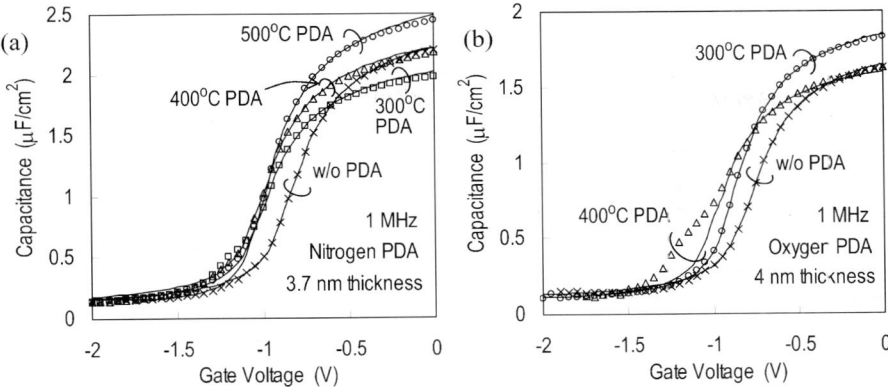

**Fig. 11.** High-frequency PMOS capacitance as a function of gate voltage at 1 MHz for **(a)** nitrogen and **(b)** oxygen ambient PDAs. No frequency dispersion was observed at the frequency range from 10 kHz to 1 MHz. Symbols represent experimental results for films without PDA (*cross*), and with 300 °C (*circle*), and 400 °C (*triangle*) PDA temperatures. *Solid lines* represent theoretical values without consideration of interface state density using the NCSU-CV program

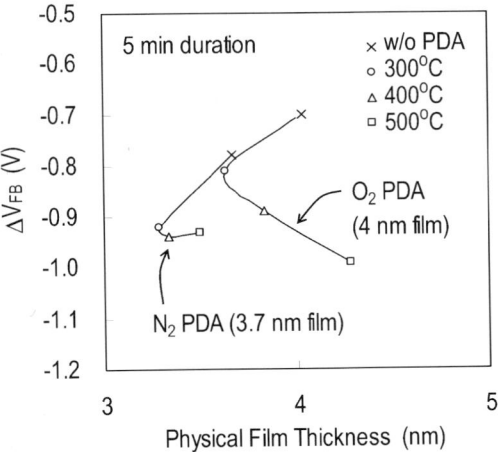

**Fig. 12.** Flat band volatge ($V_{FB}$) shift of $La_2O_3$ PDA samples in nitrogen and oxygen gas

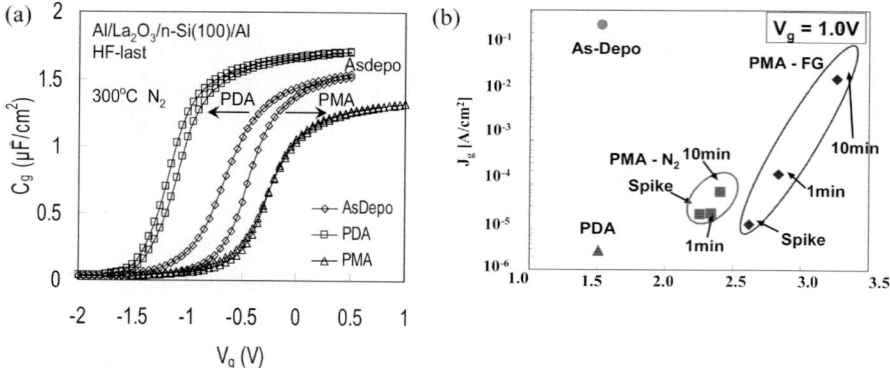

**Fig. 13.** (a) High-frequency CV measurement of $L_2O_3$ film with Al electrode. PDA and PMA in nitrogen ambient at 300 °C. (b) J–V measurement of $L_2O_3$ film with Al electrode at various annealing conditions

to the densification effect of the $La_2O_3$ film, whereas the decrease after PMA is due to the interfacial $Al_2O_3$ layer formation between $La_2O_3$ and the Al electrode. Long PMA increases the $Al_2O_3$ interfacial layer growth, resulting in large EOT, especially for the case of forming gas ($N_2 : H_2 = 97 : 3$), as shown in Fig. 13.

# 7 Conduction Mechanisms of $La_2O_3$ MISCAP

Various conduction models are summarized in Table 1, and are used to examine the leakage current through the $La_2O_3$ film. Figure 14 shows the leakage current–voltage characteristics of $La_2O_3$ samples with Al as gate electrode. At very low voltages, the conduction mechanism was found to be ohmic. At the gate voltage of 0.1 V, space charge-limited current (SCLC) was observed, where the current followed the square gate voltage law. By further increasing the gate voltage, the currents were found to be associated with the n power law, as shown in the Fig. 14. From the report of *Sussman* [14], the extrapolations of the fitted lines corresponding to the SCLC conduction of the exponential trap level distribution, with different temperatures, meet at a single point. This is a good indication of SCLC with the exponential trap distribution. Traps might be introduced during the annealing, as the $V_{FB}$ shift of the annealed film showed a higher negative value compared to the as-deposited film. Furthermore, it is suggested that the traps can be generated by the deposition process itself, as the equilibrium vapor pressure of La is greater than that of $La_2O_3$ at a given temperature. If La atoms are incorporated in the $La_2O_3$ film, the atoms act as positive charge. When the gate voltage is around 1 V, which corresponds to a field of 4.1 MV/cm, F–N conduction was observed to be dominant. The barrier height, defined as the

**Table 1.** Theoretical expressions of various conduction mechanisms

| Bulk-limited | |
| --- | --- |
| Mechanisms | Expressions |

| | |
| --- | --- |
| $\Omega$ | $J = qn\mu\dfrac{V}{s} = qN_C \exp\left[-\dfrac{q}{kT}(F - E_C)\right]\mu\dfrac{V}{s}$ |
| Poole | $J = A\exp\left(-\dfrac{S_P E}{k_B T}\right) = A\exp\left(-\dfrac{qd}{2}\dfrac{E}{k_B T}\right) \quad S_P = \dfrac{1}{2}qd$ |
| Poole–Frenkel | $J = qN_C\mu(T)\exp\left(-\dfrac{\phi_0}{k_B T}\right)\exp\left(\dfrac{\beta_{PF}}{k_B T}E^{\frac{1}{2}}\right)E \quad \beta_{PF} = 2\beta_S$ |
| Hopping | $J = \dfrac{q^2}{k_B T}\dfrac{d^2}{\tau_0}n^*(T)E\exp\left(-\dfrac{4\pi m^*}{h}\phi_m d\right) \quad \beta_{PF} = \left(\dfrac{q}{\pi\epsilon_0\epsilon_i}\right)^{\frac{1}{2}}$ |
| VRH[a] | $J = J_0\sinh\left(\dfrac{qRE}{kT}\right)$ |
| SCLC[b] | |
| shallow | $J_{\text{shallow}} = \dfrac{9}{8}\mu\epsilon_i\epsilon_0\theta(T)\dfrac{V^2}{s^3}$ |
| uniform | $J_{\text{uniform}} = 2QN_\mu\dfrac{V}{s}\exp\left(\dfrac{2\epsilon_0\epsilon_i V}{N_t kTqs^2}\right)$ |
| exponential | $J_{\text{exponential}} = N_c\mu q^{1-l}\left[\dfrac{\epsilon l}{N_t he(l+1)}\right]\left(\dfrac{2l+1}{l+1}\right)^{l+1}\dfrac{V^{(l+1)}}{s^{(2l+1)}} \quad l = T_c/T$ |

[a] Variable Range Hopping, [b] Space Charge-Limited Current: single, uniform and exponential/Gaussian distributions

energy difference of the conduction bands of $La_2O_3$ and Si, was found to be 2.14 eV with 0.25 effective mass. This barrier height is closer to the bulk $La_2O_3$ value of 2.3 eV, rather than that for $SiO_2$ of 3.1 eV [15]. The interfacial layer of $SiO_2$ would be almost transparent because of tunneling. When the gate voltage is over 1 V, the leakage current becomes saturated due to the series resistance.

# 8 Electrical Properties of MISFET with La₂O₃ Gate Insulator

The fabrication process of $La_2O_3$ gated MISFET is shown in Fig. 15. The isolation and source/drain region were formed prior to the gate formation. $La_2O_3$ film and Al electrodes were deposited on the surface and patterned by lithography. Al contacts for source/drain were formed using the lift-off process. The gate length and width were 2.5 μm and 27 μm, respectively. Finally, the backside of the wafer was coated with Al.

Figure 16 shows the $I_d$–$V_g$ characteristics of MISFET with nitrogen-annealed $La_2O_3$ at 300 °C and 500 °C [16]. It is found that the PDA at higher

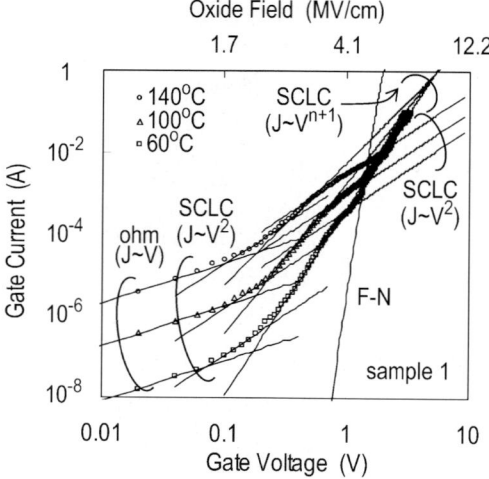

**Fig. 14.** Leakage current–voltage characteristics of $La_2O_3$ film at different temperature. The thickness of the $La_2O_3$ was 4 nm

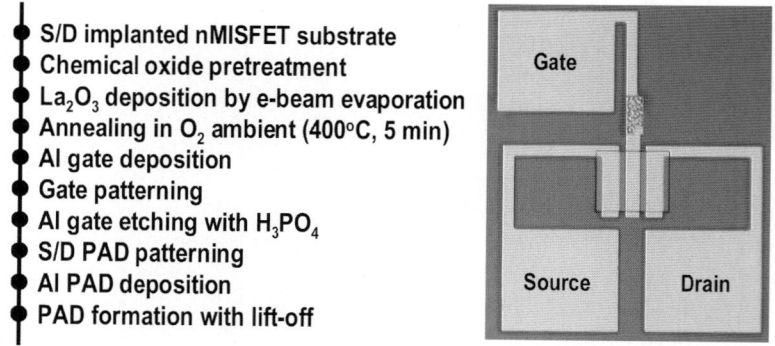

**Fig. 15.** Fabrication process of $La_2O_3$ MISFET

temperatures will lead to higher $I_{dsat}$ degradation. This degradation is mainly due to the channel mobility degradation. Similar effects were observed in $HfO_2$ films [17]. Compared to as-deposited $La_2O_3$ MISFET, improvement in mobility would be related to the forming of the interfacial silicate layer and the removal of oxide trap. The MOSFET with 300 °C PDA yielded a peak mobility of $261\,cm^2/V \cdot s$.

Figure 17 compares the $I_d$–$V_g$ characteristics of MOSFET that underwent 300 °C PDA and 300 °C PMA. PMA samples show a better turn-off characteristic than that of PDA. Peak effective mobility of $319\,cm^2/V \cdot s$ is obtained from the PMA sample. Reduction in fixed charge after PMA, consistent with

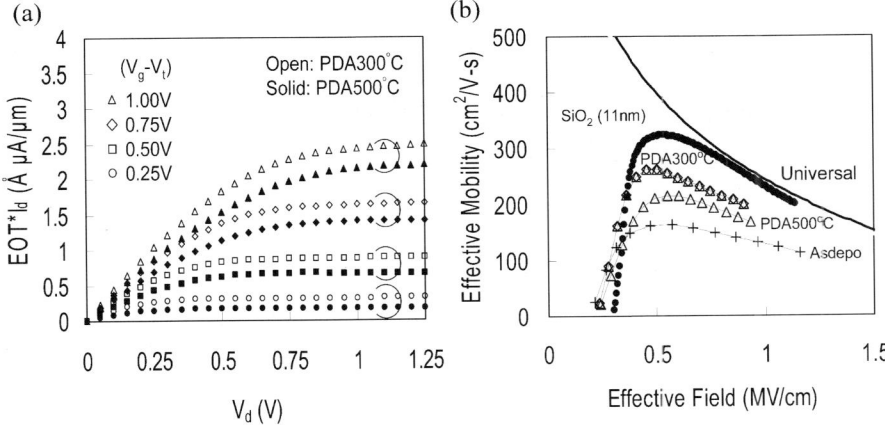

**Fig. 16.** (a) Drain current voltage characteristics of La$_2$O$_3$ gated MIFSET with PDA at 300 °C and 500 °C with (b) effective mobility PDA at 300 °C showing higher effective mobility that at 500 °C

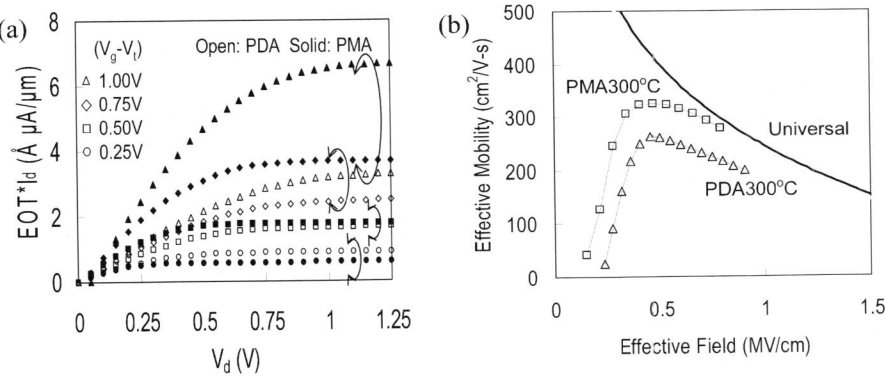

**Fig. 17.** (a) Drain current–voltage characteristics of La$_2$O$_3$ gated MISFET with PMA. (b) Higher mobility of 319 cm$^2$/V · s was achieved, but the EOT increased from 1.29 nm to 2.33 nm

the observed V$_{FB}$ shift, may be responsible for reduction in fixed charge-induced Coulomb scattering and increases the channel mobility [18]. It should be noted that EOT of the gate insulator of this MISFET is relatively thick at 2.3 nm, caused by PMA.

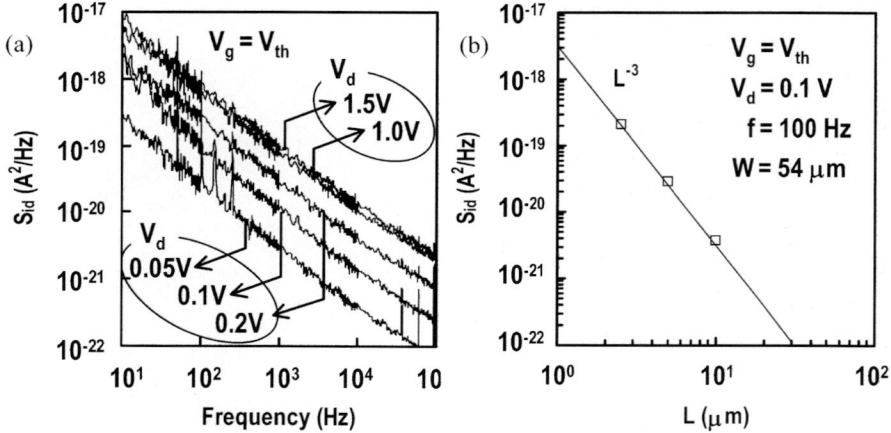

**Fig. 18.** (a) Spectrum power density on different bias conditions. (b) Noise power density with different channel length at 100 Hz

## 9 Low Noise Frequency Measurement

Low frequency noise, which is dominated by 1/f noise, is one of the important characteristics in analog applications, as it limits their performance and functionality [19]. The 1/f noise spectrum is up-converted to phase noise at high frequency in oscillators, mixers and modulators. The noise level requirement as per the ITRS2004 should be as low as $1 \times 10^{-4} \, V^2/Hz$ [20].

The 1/f noise measurement of MISFET with $La_2O_3$ gate insulator is shown in Fig. 18a. The device dimensions are $W \times L = 54 \, \mu m \times 2.5 \, \mu m$. The spectrum power densities of the drain current were dependent on the drain bias voltages. The noise spectrum increased, as the bias voltage increased, and the noise saturation was obtained as the transistor entered the saturation region.

The drain current noise power densities at 100 Hz as a function of channel length are shown in Fig. 18b. Drain bias was set to 0.1 V while the gate voltage was equal to the threshold voltage. The noise power densities scaled as per $L^{-3}$ rule with the channel length scaling.

Figure 19 shows the normalized drain noise level of MISFETs with thermally grown $SiO_2$ and deposited $La_2O_3$ gate insulators as a function of normalized square of transconductance. The $La_2O_3$ gate MISFET showed a noise about one order of magnitude higher than that of $SiO_2$. Linearity of the curve near the threshold voltage indicates that the noise is due to the carrier trapping and detrapping events. To clarify the mechanism in the low frequency range, the normalized drain current noise power density and square transconductance are plotted as a function of drain current. Channel mobility fluctuation followed the $1/I_d$ rule, while number fluctuation followed the normalized square of transconductance. Fitting the measured noise to unified

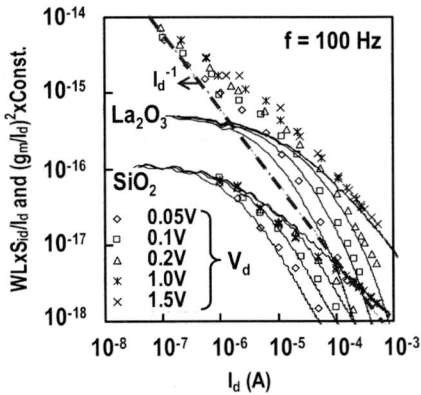

**Fig. 19.** Normalized drain noise level of MISFETs with thermally grown $SiO_2$ and $La_2O_3$ gate insulator as a function of normalized square transconductance

model, the $La_2O_3$ MISFET had a trap density of $N_t = 4 \times 10^{19}\,eV^{-1}/cm^3$, while for thermal $SiO_2$, it was $N_t = 2 \times 10^{18}\,eV^{-1}/cm^3$. These trap density values were in good agreement with the values obtained from $V_{FB}$ shift. The measured $La_2O_3$ gate MISFET exhibited 1 to 2 orders higher 1/f noise, and further process optimizations need to be done in order to reduce the trap densities at the $Si/La_2O_3$ interface.

## 10 Interfacial Layer Suppression

High-temperature annealing of $La_2O_3$ film on Si results in the formation of $SiO_2$ at the interface, which will increase the EOT. To avoid the formation of the $SiO_2$ interfacial layer, thin films of $Y_2O_3$ layer were deposited prior to $La_2O_3$ deposition. Generally, $Y_2O_3$ is known to have strong affinity to oxygen, high thermal stability and a relatively high dielectric constant of 18 with a high energy band gap of 5.5 eV. In the present case, 2 nm of $Y_2O_3$ and 2 nm of $La_2O_3$ in thickness were deposited and annealed at 600 °C in oxygen gas. To compare the effect of $Y_2O_3$, $La_2O_3$ and $Y_2O_3$, MIS capacitors were also fabricated. From the XPS analyses, shown in Fig. 20, the $La_2O_3$ sample showed a peak at 154 eV, which suggests the formation of $SiO_2$ at the interface, whereas the intensity at this energy for $Y_2O_3$ was small. By adding a $Y_2O_3$ layer underneath the $La_2O_3$ layer, the peak intensity remained low after heat treatment. The reaction at the interface is supposed to be Y silicate, showing a dielectric constant still higher than that of $SiO_2$. Therefore, use of $Y_2O_3$ as a buffer layer appears feasible for achieving EOTs less than 1 nm.

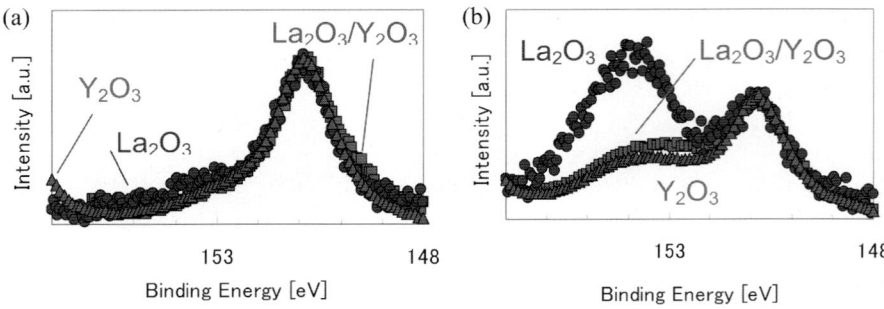

**Fig. 20.** XPS analyses on Si-2p (**a**) as-deposited and (**b**) annealed in oxygen ambient film. The peak at 154 eV was clearly supressed by using $Y_2O_3$ as a buffer layer

## 11 Conclusions

With $SiO_2$-based gate insulators reaching their ultimate limit, feasibility studies of rare earth oxides for replacing the $SiO_2$ have been conducted from many different aspects. From two-dimensional device simulations, it is found that a dielectric constant value of less than 50 is desirable to suppress the short channel effects of MOSFETs, and rare earth oxides seem to meet this requirement. The first technology challenge that needs to be addressed is the highly hygroscopic nature of the rare earth oxides. It is confirmed that moisture absorption can be suppressed by coating the films by passivation layers, such as gate electrode and sidewall.

Among the rare earth oxides, it is found that $La_2O_3$, after a proper heat treatment, has the best electrical properties for gate insulator applications in MOSFETs, due to its higher barrier height for the conduction band electrons and valence band holes as well as its higher dielectric constant. The smallest gate leakage current demonstrated through experimental results was $3 \times 10^{-4}$ cm$^3$ at 1 V for an EOT of 0.6 nm. The conduction mechanism through $La_2O_3$ has been modeled, and has been shown to be mainly by SCLC.

With post-metal annealing after Al gate electrode formation, $La_2O_3$ MISFET has shown high effective electron mobility of 319 cm$^2$/V · s. This value is lower compared to $SiO_2$, but still one of the highest among all high-$\kappa$ MISFETs, for a gate oxide EOT value which is slightly larger than 2.3 nm. Post-metal annealing recovers the flat band voltage shift and improves the mobility presumably by the diffusion of Al into $La_2O_3$, which compensates the positive charges in the film. However, the growth of an interfacial $Al_2O_3$ layer, which results in an increase in EOT, is still a problem. The solution could be to replace the Al gate electrode with a less reactive metal with $La_2O_3$, and doping $La_2O_3$ films with some elements which compensate the negative charge.

Interfacial layer suppression between the silicon substrate and the $La_2O_3$ film has been achieved by using $Y_2O_3$ as a buffer layer, which is necessary to achieve an EOT less than 1 nm.

Reliability and yield of rare earth oxides still need to be investigated, but the results shown above hold promise for rare earth oxides, especially $La_2O_3$, as a suitable candidate for the post-Hf-based gate insulator in advanced CMOS integrated circuits.

## Acknowledgements

This study was partially supported by the Semiconductor Technology Academic Research Center (STARC) and Special Coordination Funds for Promoting Science and Technology by the Ministry of Education, Culture, Sports, Science, and Technology, Japan.

## References

[1] C. Mead, L. Conway: *Introduction to VLSI Systems* (Addison-Wesley 1979)
[2] H. S. Momose, M. Ono, T. Yoshitomi, T. Ohguro, S. Nakamura, M. Saito, H. Iwai: Tunneling gate oxide approach to ultra-high current drive in small-geometry MOSFETs, IEDM Tech. Dig. **593** (1994)
[3] H. Wakabayashi, S. Yamagami, N. Ikezawa, A. Ogura, M. Narihiro, K. Arai, Y. Ochiai, K. Takeuchi, T. Yamamoto, T. Mogami: Sub-10-nm planar-bulk-CMOS devices usgin lateral junction control, IEDM Tech. Dig. **989** (2003)
[4] R. Chau, J. Kavalieros, B. Roberds, B. Schenker, D. Lionberger, D. Barlage, B. Doyle, R. Arghavani, A. Murthy, G. Dewey: 30 nm physical gate length CMOS transistors with 1.0 ps n-MOS and 1.7 ps p-MOS gate delays, IEDM Tech. Dig. **45** (2000)
[5] J. Robertson: Band offsets of high dielectric constant gate oxides on silicon, J. Non-Cryst. Solids **303**, 94 (2002)
[6] T. Aoyama, T. Maeda, K. Torii, K. Yamashita, Y. Kobayashi, S. Kamiyama, T. Miura, H. Kitajima, T. Arikado: Proposal of new HfSiON CMOS fabrication process (HAMDAMA) for low standby power device, IEDM Tech. Dig. **95** (2004)
[7] C. Choi, C. Y. Kang, S. J. Rhee, M. S. Abkar, S. A. Krishn, M. Zhang, H. Kim, T. Lee, F. Zhu, I. Ok, S. Koveshnikov, J. C. Lee: Fabrication of TaN-gated ultra-thin MOSFETs (EOT = 1.0nm) with $HfO_2$ using a novel oxygen scavenging process for sub 65 nm application, Symp. of VLSI Technology **226** (2005)
[8] T. Hattori, T. Yoshida, T. Shiraishi, K. Takahashi, H. Nohira, S. Joumori, K. Nakajima, M. Suzuki, K. Kimura, I. Kashiwagi, C. Ohshima, S. Ohmi, H. Iwai: Composition, chemical structure, and electronic band stucture of rare earth oxide/Si(100) interfacial transition layer, Microelectron. Eng. **72**, 283 (2004)

[9] H. Iwai, S. Ohmi, S. Akama, C. Ohshima, A. Kikuchi, I. Kashiwagi, J. Taguchi, H. Yamamoto, J. Tonotani, Y. Kim, I. Ueda, A. Kuriyama, Y. Yoshihara: Advanced gate dielectric materials for sub-100 nm CMOS, IEDM Tech. Dig. **625** (2002)

[10] N. R. Mohapatra, M. P. Desai, S. G. Narendra, V. R. Rao: The effect of high-k gate dielectrics on deep submicrometer CMOS device and circuit performance, IEEE Trans. Elec. Dev. **49**, 826 (2002)

[11] Y. Kim, S. I. Ohmi, K. Tsutsui, H. Iwai: Analysis of variation in leakage currents of lanthana thin films, Solid-State Electron. **49**, 825 (2005)

[12] N. Yang, K. Henson, J. Hauser, J. Wortman: Modeling of study of ultrahin gate oxides using direct tunneling current and capacitance-voltage measurments in MOS devices, IEEE Trans. Elec. Dev. **46**, 1464 (1999)

[13] D. A. Buchanan, E. P. Gusev, E. Cartier, H. Okorn-Schmidt, K. Rim, M. A. Gribelyuk, A. Mocuta, A. Jamison, J. Brown, R. Arndt: 80 nm poly-silicon gated n-FETs with ultra-thin $Al_2O_3$ gate dieletric for ULSI applications, IEDM Tech. Dig. **223** (2000)

[14] A. Sussman: Space-charge-limited currents in copper phthalocyanine thin films, J. Appl. Phys. **38**, 2738 (1967)

[15] Y. Yeo, T. King, C. Hu: Direct tunneling leakage current and scalability of alternative gate dielectrics, Appl. Phys. Lett. **81**, 2091 (2002)

[16] J. A. Ng, Y. Kuroki, N. Sugii, K. Kakushima, S. I. Ohmi, K. Tsutsui, T. Hattori, H. Iwai, H. Wong: Effects of low temperature annealing on the ultrathin $La_2O_3$ gate dielectric; comparison of post deposition annealing and post metallization annealing, Microelectron. Eng. **80**, 206 (2005)

[17] H. Wong, K. L. Ng, N. Zhan, M. C. Poon, C. W. Kok: Interface bonding structure of hafnium oxide prepared by direct sputtering of hafnium in oxygen, J. Vac. Sci. Technol. B **22**, 1094 (2004)

[18] K. Torii, Y. Shimamoto, S. Saito, K. Obata, T. Yamauchi, D. Hisamoto, T. Onai, M. Hiratani: Effect of interfacial oxide on electron mobility in MIS-FETs with $Al_2O_3$ gate dielectrics, Microelectron. Eng. **65**, 447 (2003)

[19] E. Simoen, C. Claeys: On the flicker noise in submicron silicon MOSFETs, Solid-State Electron. **43**, 865 (1999)

[20] International Technology Roadmap for Semiconductors URL http://public.itrs.net

# Index

# Requirements of Oxides as Gate Dielectrics for CMOS Devices

Gennadi Bersuker and Peter Zeitzoff

SEMATECH, 2706 Montropolis Drive, Austin, Texas 78741, USA
gennadi.bersuker@sematech.org

**Abstract.** Analysis of fundamental material properties of the $d$-electron-based high-$k$ dielectrics, which are under consideration for the application as gate dielectrics, indicates that this class of materials has serious limitations. However, the material intrinsic properties do not necessarily have to satisfy all performance requirements: certain properties can be modified by proper engineering. We formulate a technological approach to the requirements for gate dielectrics, which is specific to a given class of materials and aims to address negative aspects of intrinsic material properties.

## 1 Introduction

Gate dielectrics are one of the most critical components of the transistor: as such, they must lead to maximum density of the channel carriers without compromising their mobility, while, at the same time, maintaining low parasitic gate leakage current to keep power consumption under control. These requirements impose rather strict constraints on the properties of the dielectric materials. While operating under relatively high bias and temperature conditions, the electrical characteristics, in particular, the threshold voltage and the gate leakage current, must remain stable in order to meet reliability criteria. The dielectric material should be manufacturable, i.e., its properties need to be compatible with and remain unaffected by the complex multi-step CMOS fabrication process. All these constraints have been met admirably by silicon dioxide [1], due to characteristics such as wide band gap, low density of as-grown defects, and amorphous structure, which helps to accommodate mismatch with the Si substrate with minimum density of dangling bonds. However, performance goals for the 45 nm technology node and beyond specified by the ITRS [2] require both very high specific capacitance translated to the equivalent oxide thickness (EOT) of 1.3 nm or less, and low gate leakage current (see Table 1). Silicon dioxide is unable to meet the gate leakage requirements for such thin films.

Indeed, the scaling goal is to increase drive current $I_d \propto (k/t)$ without increasing gate leakage $I_g \propto \exp(-t)$, where $t$ is the gate dielectric thickness, and $k$ is its relative dielectric constant. Simulations show that if scaling proceeds according to the recent ITRS specifications, in the near future the

M. Fanciulli, G. Scarel (Eds.): Rare Earth Oxide Thin Films,
Topics Appl. Physics **106**, 367–378 (2007)
© Springer-Verlag Berlin Heidelberg 2007

**Table 1.** Technological requirements for future CMOS devices

| Year of Production Technology Node | 2010 hp45 | 2011 | 2012 | 2013 hp32 | 2014 | 2015 | 2016 hp22 | 2017 | 2018 |
|---|---|---|---|---|---|---|---|---|---|
| Equivalent physical oxide thickness for MPU/ASIC $T_{ox}$ (nm) [A, Al] | 0.7 | **0.7** | 0.7 | 0.6 | **0.6** | 0.6 | 0.5 | **0.5** | 1.5 |
| Gate dielectric leakage at 100 °C (μA/μm) high-performance [B, B1, B2] | 0.33 | **0.33** | 0.33 | 1 | **1** | 1 | 1.67 | **1.67** | 1.67 |
| Equivalent physical oxide thickness for low operating | 0.9 | 0.9 | 0.9 | 0.8 | 0.8 | 0.8 | 0.7 | 0.7 | 0.7 |
| Gate dielectric leakage at 100 °C (nA/μm) LOP [B, B1, B2] | 2.33 | **2.33** | 2.33 | 3.33 | **3.33** | 3.33 | 10 | **10** | 10 |
| Equivalent physical oxide thickness for low standby power $T_{ox}$ (nm) [A, Al] | 1.3 | **1.3** | 1.2 | 1.1 | **1.1** | 1.1 | 1 | **1** | 0.9 |
| Gate dielectric leakage at 100 °C (pA/μm) LSTP [B, B1, B2] | 20 | **20** | 20 | 27 | **27** | 27 | 33 | **33** | 33 |
| Thickness control EOT (% $3\sigma$) [C] | <±4 | < **±4** | <±4 | <±4 | < **±4** | <±4 | <±4 | < **±4** | < ±4 |

gate leakage current will exceed the limits acceptable for both low-power and high-performance logic applications, as shown in Fig. 1. Since the option of thinning $SiO_2$ is exhausted, the current trend is to introduce materials with higher $k$ values. The class of high-$k$ materials currently under investigation is transition metal oxides and, more recently, rare-earth oxides. The specific feature of transition metal and rare-earth ions is the presence of $d$-orbitals, which are characterized by low-symmetry spatial distribution [3]. Certain optically active off-center displacements of the metal ions were shown to increase the overlap between the metal $d$- and oxygen $p$-orbitals, which lowers the total energy of the system. Correspondingly, these types of ion displacements tend to have lower frequency and greater ionic contribution to the polarizability, hence greater $k$ values. However, this class of dielectrics has certain limitations, which may force us to reconsider some well-established traditional requirements for materials intended for the gate dielectric application.

## 2 General Properties of Metal Oxides

Compared to $SiO_2$, $d$-electron materials have smaller band gaps (due to smaller overlap between the $d$ and $p$ states) and, therefore, lower band offsets (lower energy barrier for the leakage current), higher density of as-grown defects and a tendency to long-range ordering that leads to poly-crystalline structure of the dielectric. In addition, there are serious issues associated with the fabrication of gate stacks with the metal oxide dielectrics. Due to high diffusivity of oxygen in metal oxides, a $SiO_2$-like layer usually growth at the interface with the Si substrate during the high-$k$ deposition and subsequent anneal. Formation of this layer, usually 5–11 A of physical thickness (see Fig. 2), may limit the scalability of the high-$k$ gate dielectrics.

The high-$k$ dielectric structural features have significant implications for transistor electrical characteristics. Some of the structural defects may represent electron traps; electron capture by these traps, which may occur via short (microseconds) [4] and long (seconds) [5] processes, leads to lower drainage current (DC) mobility and threshold voltage ($V_t$) instability (see Fig. 3), respectively. While the nature of the electron trapping defects is still unclear and is currently under intense investigation [6, 7], it was found that scaling of physical thickness of high-$k$ films results in reduction of both fast transient trapping and slow trapping [8] (see Fig. 4). In the case of Hf-based dielectrics, this could be associated with lower trap density due to suppression of grain formation in thin films, assuming that the electron trapping defects are associated with the grain boundaries (see Fig. 5). However, electron trapping characteristics of the high-$k$ films does not scale with the density of the grain boundaries since nano-crystalline films (seemingly amorphous) exhibit significantly lower electron trapping rate. This suggests that the electron traps are associated with the defects in the bulk of the grains; in particular, with a specific structural feature present in the crystalline film. As was demonstrated by the O-edge EELS data [9], the stoichiometry of the $HfO_2$ grains improves by high-temperature annealing, which also results in increased electron trapping. It was proposed that the electrically active defects are formed by the negatively charged three-fold coordinated oxygen vacancies in the crystallized $HfO_2$, which has monoclinic structure [6]. Highly delocalized $d$-states associated with this defect are unable to confine the trapped electron within the dielectric film when its thickness is comparable to the dimension of the traps. Crystallization of the Hf-based films, as well as phase separation phe-

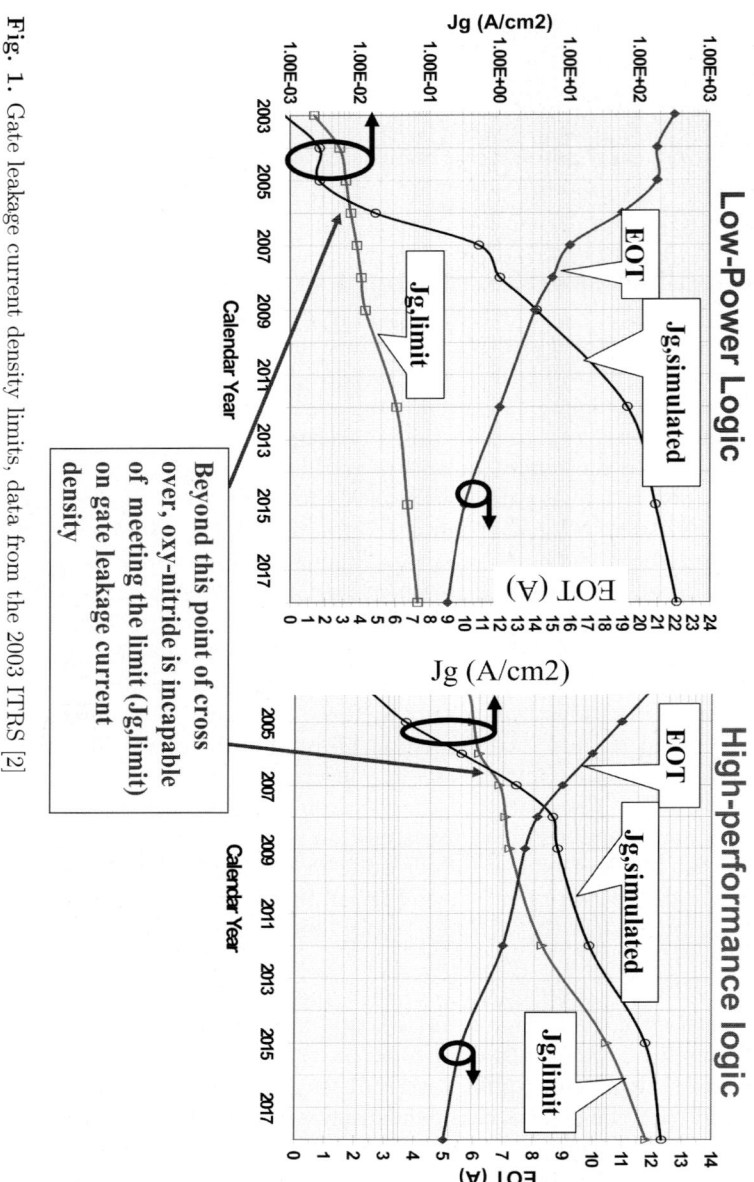

Fig. 1. Gate leakage current density limits, data from the 2003 ITRS [2]

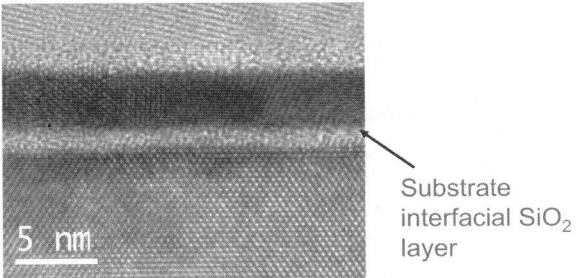

**Fig. 2.** TEM image of the fully processed (including 1000 °C/10 s dopant activation anneal) $HfO_2$ transistor gate stack

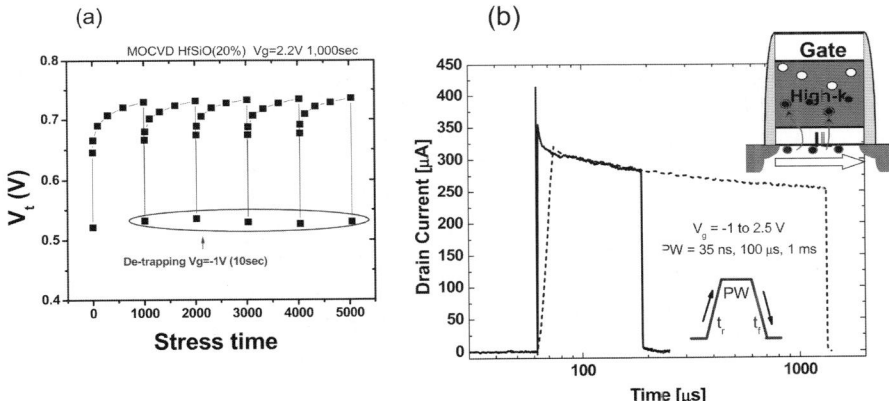

**Fig. 3.** (a) Variation of the HfSiO (20 % $SiO_2$) NMOS transistor $V_t$ during stress cycles, which include 1000 s substrate injection stress followed by 10 s stress of the opposite bias under the specified voltage conditions. Stress cycle conditions are as labeled. (b) Example of the pulsed drain current vs. pulse time ($PW = 35$ ns, 100 ms, 1 ms) characteristics, illustrating the degradation of the drive current over time due to channel electron trapping in the high-$k$ film

nomena observed in Hf silicates, introduces non-uniformity of electrical properties along the transistor channel, which is detrimental to mobility [10] and $V_t$ uniformity in short-channel devices [11]. This could be at least partially alleviated by incorporation of N into the dielectric structure (see Fig. 6).

The interfacial $SiO_2$ layer exhibits an effective relative dielectric constant significantly higher than that of a $SiO_2$ film of comparable thickness ($k$ values of the $SiO_2$ layer of various thicknesses were calculated in [12]), presumably due to the additional oxygen deficiency in the $SiO_2$ induced by the overlaying $HfO_2$ film [13]. For example, an interfacial $SiO_2$ layer of about 10–11 A of physical thickness, which is usually formed during high-$k$ deposition by

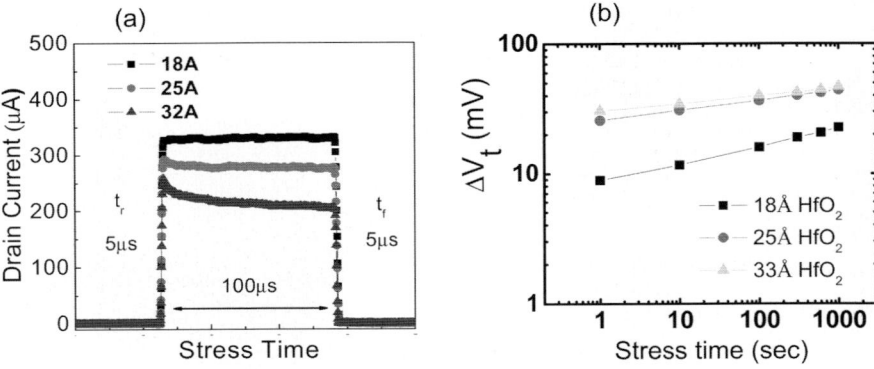

**Fig. 4.** Reduction of high-$k$ thickness improves both (**a**) transient (pulsed measurements) and (**b**) long-term (constant voltage stress) $V_t$ stability

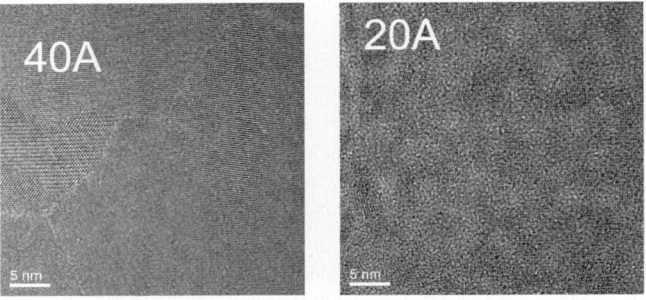

**Fig. 5.** Top-view TEM images of $HfO_2$ films of thickness 4 nm (**left**) and 2 nm (**right**)

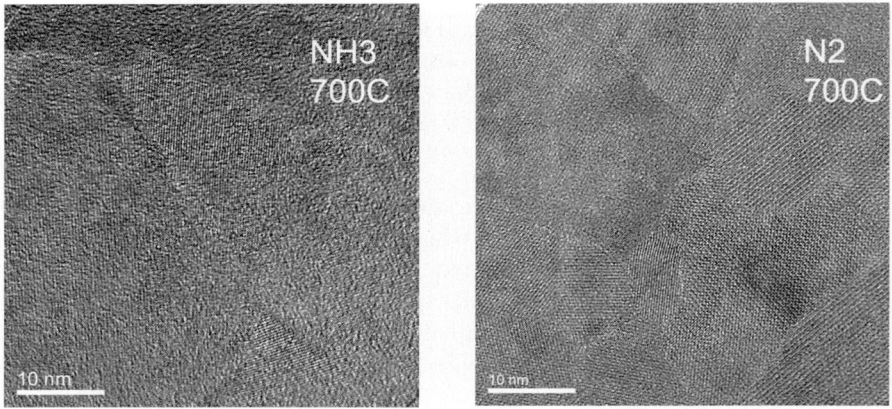

**Fig. 6.** Top-view TEM images of 4 nm $HfO_2$ films subjected to the post-deposition anneal at 700 °C in $NH_3$ (**left**) and $N_2$ (**right**) ambient

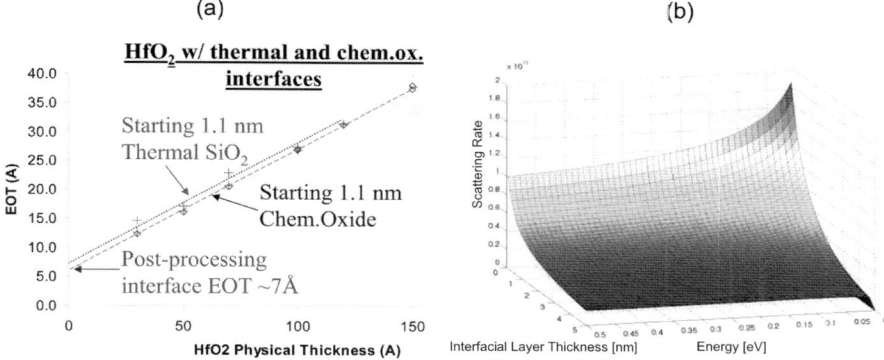

**Fig. 7.** (a) EOT of the gate stack vs. physical thickness of the HfO$_2$ films deposited on two types of starting surfaces – thermal SiO$_2$ and chemical oxide of similar physical thickness. (b) Dependence of the channel electron scattering rate on the phonon energy and thickness of the interfacial layer

almost any technique and substrate preparation, exhibits $k > 6$ (see Fig. 7a), while a calculated effective $k$ value of a thermal SiO$_2$ film of similar thickness is around 5. Effective $k$ values of thin SiO$_2$ films are higher than the 3.9 value of the bulk SiO$_2$ due to oxygen deficiency of the transitional layer formed near the dielectric interface with the Si substrate to accommodate the mismatch between the lattice constants of Si and SiO$_2$. High values of the dielectric constant of the interfacial layer allow for scaling of the gate stack beyond 1 nm EOT. In addition, this interfacial layer effectively suppresses the coupling between the high-$k$ soft phonons and channel carriers (see Fig. 7b), improving mobility to the level of currently manufactured transistors with SiON gate dielectrics.

Another issue specific to high-$k$ dielectric materials is their susceptibility to micro-contamination. Due to the above-mentioned high diffusivity of various ions in the high-$k$ materials, they can be contaminated by the process-related oxygen species even after the high-$k$ dielectric is capped by the gate electrode. Contaminants can be introduced along the edges of the gate stack during/after the gate stack definition and then diffuse into the dielectric film during subsequent high-temperature steps, as illustrated in Fig. 8a, affecting $V_t$ values in the short-channel devices [14] (see Fig. 8b). It was shown that this micro-contamination of the high-$k$ dielectrics can be effectively suppressed by employing a thin SiN "seal" layer immediately after the gate etch step (see Fig. 8c).

## 3 Technological Requirements for Novel Gate Dielectrics

Certain well-established general criteria for the gate dielectric materials developed based on the SiO$_2$ technology may not be directly applicable to the

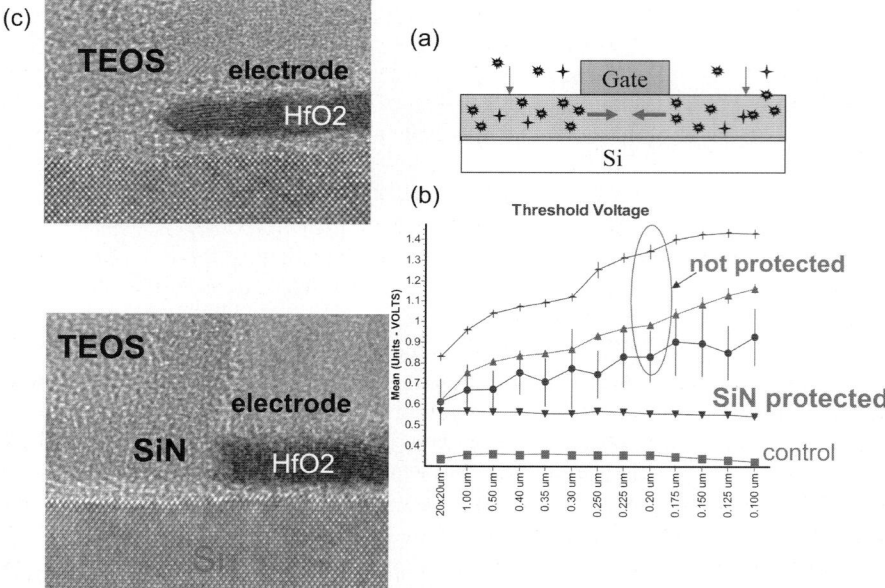

**Fig. 8.** (a) Schematic of process-induced charging in high-$k$ films. (b) Threshold voltage vs. channel length dependence in transistors with 4.5 nm MOCVD HfSi$_x$O$_y$ (20 % SiO$_2$) gate dielectric fabricated with different process schemes with and without the SiN protective layer. (c) TEM images of the gate edge areas of transistors fabricated with 4.5 nm MOCVD HfSiO film and without SiN layer

gate stacks fabricated with novel materials. For instance, the requirement of stability of high-$k$ dielectrics with Si seems to be, to some degree, irrelevant since, on the one hand, the bottom interfacial SiO$_2$ layer separating high-$k$ film from the substrate is apparently needed to meet performance requirements, as discussed above, and, on the other, these dielectrics are intended to be used with metal or fully silicided electrodes. The presence of the interfacial layer in the scaled devices with operational voltage $V_{dd} < 1$ V partially alleviates the requirement for the dielectric band offset to be $> 1$ eV. Dielectric material homogeneity on a scale of the device channel dimension seems to represent a critical device performance factor, which was not explicitly formulated initially.

In view of the above-mentioned fundamental material features of the $d$-electron-based high-$k$ dielectrics, one may conclude that this class of materials has very serious limitations from the device performance standpoint and, thus, may not be suitable for the gate dielectric application. However, the dielectric material intrinsic properties do not necessarily have to satisfy all performance requirements: as was illustrated above, certain properties can be modified by proper engineering, in particular, by changing material composition, deposition and post-deposition processing conditions, physical dimensions, etc. Therefore, technological solutions need to provide manufacturable processes, which mitigate negative aspects of intrinsic material properties. We may formulate technological requirements for gate dielectrics specific to this class of materials: neutralize electrically active defects (electron traps, fixed and mobile charges); suppress crystallization and phase separation; control stoichiometry of the bottom interfacial layer. Here, we avoided discussion on the effects of the gate electrode processing on dielectrics.

While performance requirements present significant technological challenges, reliability requirements may be easier to meet. Indeed, the $d$-electron materials can be expected to be more stable than $SiO_2$ with respect to electrical stress-induced defect generation, since the charge trapped on $d$-orbitals is delocalized over the crystal cell and, therefore, cannot perturb the bonds to the point of their significant weakening and breakdown. On the other hand, instability of transistor parameters (i.e., threshold voltage) caused by high density of pre-existing defects can be addressed via the above-mentioned technological solutions. In this sense, the interfacial $SiO_2$ layer may represent a weak link in the high-$k$ gate stack from a reliability standpoint.

Meeting performance and reliability specifications for new gate dielectrics requires optimization in the multidimensional space of various factors (EOT, $k$ values, mobility, physical thickness, offset energy, etc.), achievable via understanding of the relations between the material structure and its electrical properties that can guide process engineering.

# References

[1] A. M. Stoneham, J. L. Gavartin, A. L. Shluger: The oxide gate dielectrics: Do we know all we should?, J. Phys.: Condens. Matter **17**, S2027 (2005)

[2] *International Technology Roadmap for Semiconductors (ITRS)*, 2003 Edition (Semiconductor Industry Association)
URL http///www.itrs.net/

[3] G. Bersuker, P. Zeitzoff, G. Brown, H. R. Huff: Novel dielectric materials for future transistor generations, Mat. Today **6**, 26 (2004)

[4] J. H. Sim, R. Choi, B. H. Lee, C. Young, P. Zeitzoff, D.-L. Kwong, G. Bersuker: Trapping/de-trapping gate bias dependence of Hf-silicate dielectrics with poly and TiN gate electrode, Jpn. J. Appl. Phys. **44**, 2420 (2005)

[5] C. Young, Y. Zhao, M. Pendley, B. H. Lee, K. Matthews, J. H. Sim, R. Choi, G. A. Brown, R. W. Murto, G. Bersuker: Ultra-short pulse current-voltage characterization of the intrinsic characteristics of high-k devices, Jpn. J. Appl. Phys. **44**, 2437 (2005)

[6] G. Bersuker, B. H. Lee, H. R. Huff, J. Gavartin, A. Shluger: Mechanism of charge trapping reduction in scaled high-k gate stacks, in E. Gusev (Ed.): *Proc. NATO Workshop on Defects in Advanced High-k Dielectric Devices* (Springer, Berlin, Heidelberg 2006) p. 227

[7]  G. Lucovsky, C. C. Fulton, Y. Zhang, Y. Zou, J. Luning, L. F. Edge, J. L. Whitten, R. J. Nemanich, H. Ade, D. C. Schlom, V. V. Afanase'v, A. Stesmans, S. Zollner, D. Triyoso, B. R. Rogers: Conduction band-edge states associated with the removal of $d$-state degeneracies by the Jahn–Teller effect, IEEE TDMR **5**, 67 (2005)

[8]  J. H. Sim, S. C. Song, P. D. Kirsch, C. D. Young, R. Choi, D. L. Kwong, B. H. Lee, G. Bersuker: Effects of ALD $HfO_2$ thickness on charge trapping and mobility, Microelectronics Eng. **80**, 218 (2005)

[9]  G. D. Wilk, D. A. Muller: Correlation of annealing effects on local electronic structure and macroscopic properties for $HfO_2$ deposited by atomic layer deposition, Appl. Phys. Lett. **83**, 3984 (2003)

[10] G. Bersuker, P. Zeitzoff, J. H. Sim, B. H. Lee, R. Choi, G. Brown, C. D. Young: Mobility evaluation in transistors with charge trapping gate dielectrics, Appl. Phys. Lett. **87**, 042905 (2005)

[11] A. R. Brown, J. R. Watling, A. Asenov, G. Bersuker, P. Zeitzoff: Intrinsic parameter fluctuation in MOSFETs due to structural non-uniformity of high-k gate stack materials, SISPAD (2005) Abstract of SISPAD 2005 International Conference on simulation of semiconductor processes and devices, september 1–3 2005, Komaba Eminence, Tokyo, Japan

[12] F. Giustino, P. Umari, A. Pasquarello: Dielectric effect of a thin $SiO_2$ interlayer at the interface between silicon and high-k oxides, Microelectron. Eng. **72**, 299 (2004)

[13] G. Bersuker, J. Peterson, J. Barnett, A. Korkin, J. H. Sim, R. Choi, B. H. Lee, J. Greer, P. Lysaght, H. R. Huff: Properties of the interfacial layer in the high-k gate stack and transistor performance, Electrochem. Soc. Proc. **PV2005-05**, 147 (2005)

[14] G. Bersuker, J. Gutt, N. Chaudhary, N. Moumen, B. H. Lee, J. Barnett, S. Gopalan, G. A. Brown, Y. Kim, C. D. Young, J. Peterson, H.-J. Li, P. M. Zeitzoff, J. H. Sim, P. Lysaght, M. Gardner, R. W. Murto, H. R. Huff: Integration issues of high-k gate stack: Process-induced charging, IEEE-IRPS Proc. **691**, 479 (2004)

# Index

# Rare Earth Oxides Grown by Molecular Beam Epitaxy for Ultimate Scaling

Athanasios Dimoulas

MBE Laboratory, Institute of Materials Science, National Center for Scientific Research – DEMOKRITOS, 153 10 Athens, Greece
dimoulas@ims.demokritos.gr

**Abstract.** Combination of lanthanum and other rare earth oxides with $HfO_2$ yields stable compounds which withstand transistor processing and have low gate leakage and negligible threshold voltage instabilities. This could make these compounds good candidates for future low-operating power transistors with high-$k$ gates, although they may not be suitable for aggressive scaling of Si-based devices due to rather high equivalent oxide thickness. On the other hand, rare earth oxides could share a good role in future nanoelectronics as part of the gate stack in high-mobility Ge and III–V compound semiconductor MOSFETs. Several of the rare earth oxides could easily change their oxidation state so that they could promote catalytic reactions to effectively passivate semiconductor surfaces and improve the electrical characteristics of devices.

## 1 Introduction

The traditional downscaling of transistor devices based on the shrinking of gate length has benefited both density and high-frequency performance of integrated circuits and devices. Today's 90 nm CMOS transistors have gate length of 50 nm and $SiO_2$ gate dielectric which is only 1.2 nm thick. Further scaling along the same route is very challenging. While the advent of immersion lithography will get us down to 45 nm and perhaps 32 nm technology nodes, more expensive and cumbersome lithographic tools may be needed for high volume production in the 22 nm node at the end of the roadmap and beyond. In addition, in order to reach the ultimate limits of scaling [1] with equivalent oxide thickness (EOT) $\sim 0.5$ nm and achieve sub-picosecond gate delay times, it is required that either the $SiO_2$ gate dielectric become thinner than 1 nm or alternative high-$\kappa$ gate dielectrics will have to replace $SiO_2$ [2, 3]. In both cases, the downscaling is difficult either due to excess gate leakage ($SiO_2$) or due to mobility degradation and threshold voltage instabilities (high-$\kappa$ dielectrics) which affect high-frequency performance and reliability.

The use of high-mobility channels and substrates is an alternative route to downscaling. In fact, mobility-enhancing channels such as strained Si have already been introduced in today's 90 nm CMOS and are considered to be necessary components in the next-generation devices at the 65 and 45 nm nodes. The high-mobility channels directly affect the drive current, so that

M. Fanciulli, G. Scarel (Eds.): Rare Earth Oxide Thin Films,
Topics Appl. Physics **106**, 379–390 (2007)
© Springer-Verlag Berlin Heidelberg 2007

they could deliver high speed without the need for an aggressive scaling of the gate length and the oxide thickness. To maximize the benefit of enhanced carrier mobility, the aim is to fabricate transistors with high-$\kappa$ gate dielectrics directly on high-mobility Ge [4] and III–V compound [5] semiconductor substrates.

A significant effort has been devoted to the development of $HfO_2$-based materials [6] and hafnium silicates as alternative high-$k$ gate dielectrics on Si substrates, mainly because of their compatibility with CMOS processing. However, the dielectric permittivity $\kappa$ of $HfO_2$ films is typically between 15 and 20, which is significantly lower than the bulk value of 25. For ultimate oxide scaling down to EOT $\sim$ 0.5 nm, $\kappa$ must be around 30, which means that new oxides with higher $\kappa$ values are needed. Rare earth oxides are attractive because there is a good compromise between the energy gap and the dielectric constant so as to combine good insulating properties with potential for oxide scaling. The band gap for most of the rare earth oxides is about 5.8 eV, slightly higher than that of $ZrO_2$ and $HfO_2$ which is around 5.5 eV. $La_2O_3$ has high dielectric permittivity [2] close to 30, giving EOT values [7] as low as 4.8 Å and exhibiting interesting electrical characteristics [7–9] and in the Chapter by *Kakushima* et al. in this volume. However, this material is known to be hygroscopic, which imposes limits on the use of wet chemistry during transistor processing. Combination of $La_2O_3$ with other oxides such as $Al_2O_3$, $ZrO_2$ or $HfO_2$ makes more stable compounds which could withstand high-temperature processing. $LaAlO_3$ (LAO) has received attention [10–13] as a gate dielectric for both Si and Ge devices. In fact, LAO has produced Ge MOSFETs [14] which are better than any other material, including $HfO_2$. One of the drawbacks is that the measured dielectric permittivity $\kappa$ is often substantially lower than the expected bulk value, ranging between 14 and 17 [10, 11]. Since LAO crystallizes in pseudo-perovskite crystal structure, there has been some effort [13] to achieve epitaxial growth directly on Si(100), similarly to the case of the other perovskite $SrTiO_3$ on Si [15], but without much success [13]. LAO is obtained typically in the amorphous state when deposited on Si substrates. Recently, perovskite rare earth scandates ($LaScO_3$, $GdScO_3$, $DyScO_3$) have been studied either in amorphous [16] or epitaxial form on oxide substrates or on Si using appropriate buffer oxide templates [17]. These scandate oxides with a measured $\kappa$ around 22 [16] give good electrical characteristics and have been shown to withstand high-temperature processing [16], remaining amorphous up to 1000 °C. Several of the rare earth oxides in the form of either binary or ternary compounds are cubic and have commensurate lattices with Si, Ge and GaAs semiconductors, offering the possibility for epitaxial growth directly on these semiconductors with well-controlled interfaces. For example, bixbiyte $Sm_2O_3$, $Eu_2O_3$, $Gd_2O_3$ and $(La_xY_{1-x})_2O_3$ [18] as well as pyrochlores $La_2Zr_2O_7$ (LZO) and $La_2Hf_2O_7$ (LHO) have lattice mismatch with Si less than 1 % at room temperature, while fluorite $CeO_2$ is nearly lattice matched to Ge and GaAs. It has been shown [19] that LZO can been grown epitaxially on Si(111) while LHO can be grown [20] on Si(001) in a local cube-on-cube epitaxial mode with a strong preferential orientation along (001) direction when the substrate temperature is high enough (around 770 °C). Finally, most of the oxides [21–23] with the bixbyite crystal structure $M_2O_3$, including $Pr_2O_3$ [24, 25] and other rare earth oxides, adopt an undesired epitaxial orientation $(110)_{M_2O_3}//(001)_{Si}$ which is attributed to the matching of the oxygen sublattice along one in-plane orientation. This creates extensive twining and other defects, which can be minimized by using misoriented Si(100) substrates [23, 26]. Finally, it should be noted that recent experimental and theoretical work [27] based on rare earth $Lu_2O_3$/Si has deepened our understanding of the physical properties of high-$\kappa$ dielectrics.

Some of the rare earth oxides have an additional property which distinguishes them from other metal oxides: they can be found in two different stable oxidation states +3 and +4 in the case of ceria and praseodymium oxide or +2 and +3 in the case of europium and samarium oxides. This multiple valency of the rare earths is often regarded as a disadvantage, since it may modify the properties of the bulk material in an uncontrollable way or inhibit epitaxial growth. However, this could also turn into an advantage since mixed valency oxides could promote catalytic reactions on semiconductor surfaces, effectively passivating them and reducing electrically active defects. This could be particularly useful for Ge and GaAs (or other III–V compound semiconductors), for which good surface passivation layers and procedures have not been identified yet.

In this work, I will first review the properties of the $La_2Hf_2O_7$ compound on silicon(001) substrates both in the crystalline and the amorphous state. Subsequently, I will review the evidence we have so far that rare earth oxides match with Ge substrates better than the $HfO_2$/GeON combination which is more commonly used. I will present the case of $CeO_2$ and compare side by side the results on Ge and GaAs substrates. This oxide could by used as a model material to understand the interface chemistry of the rare earth oxides with semiconductors, which could lead us to the best passivating materials of the semiconductor surfaces.

# 2 Substrate Cleaning and Thin Film Preparation Methodology

The oxide molecular beam epitaxy/deposition (MBE/MBD) methodology has been described in detail elsewhere [3]. Here, I summarize the main characteristics of this technique. Since no precursors are needed, MBE is a flexible research tool for fast screening of a large number of complex metal oxides. In addition, being an ultrahigh vacuum (UHV) technique, it offers the possibility to desorb the native oxide in situ by heating the Si or Ge semiconductor substrate to temperatures around 900 °C and 360 °C, respectively. The process

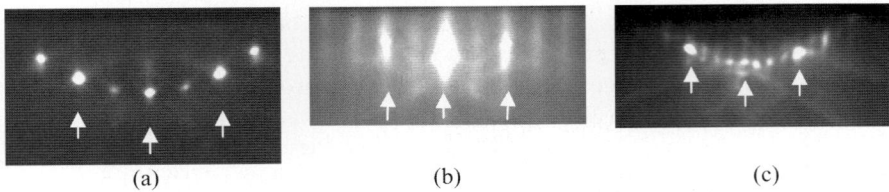

**Fig. 1.** RHEED patterns of clean (001) semiconductor surfaces along the (110) azimuth after the native oxide has been removed. (**a**) Si, (**b**) Ge, (**c**) GaAs. The *arrows* indicate the position of the main diffraction spots (*streaks*). The *weaker spots* in between indicate $(2 \times 1)$ reconstruction for Si and Ge surface and $(4 \times 2)$ reconstruction for GaAs surface

is monitored by high-energy electron diffraction (RHEED) until $(2 \times 1)$ reconstruction spots appear on the screen, indicative of a clean surface. For the case of GaAs, the procedure is slightly different. Native oxides in GaAs can be removed at temperatures higher than $600\,°C$ only under arsenic (As) overpressure, in order to prevent volatile As from leaving the substrate surface. Alternatively, the native oxides on the GaAs surface can be reduced at temperatures lower than $400\,°C$ by exposing the surface to atomic hydrogen for 5–10 min, until a $(4 \times 2)$ reconstruction pattern appears on the RHEED screen. The removal of the native oxide in situ means that the dielectric growth can start from a clean semiconductor surface, which cannot be easily achieved by conventional CVD techniques.

Another advantage of the MBE/MBD technique is that the oxides can be prepared using a wide temperature window which includes room temperature, not readily available by CVD-based techniques. The deposition temperature in many cases affects the interface quality, the microstructure, the density and the electrical quality of the films. Therefore, MBE is an excellent tool to optimize the quality of the dielectric. In addition, according to the evidence we have so far, the MBE/MBD films directly deposited on a clean semiconductor surface present excellent nucleation, producing smooth continuous films. Finally, MBE offers the possibility of epitaxial growth of oxides combined with epitaxial semiconductor overgrowth, which could lead to integration of heterogeneous metal oxide/semiconductor device structures such as fully epitaxial semiconductor-on-insulator or resonant tunnelling structures. In the present work, La is evaporated from an effusion cell; Hf and Ce are evaporated from an e-beam evaporator. The oxygen source is a remote RF plasma generator which produces reactive oxygen atomic beams [3].

## 3 Lanthanum-Based Compounds on Silicon Substrates

As already mentioned in the introduction, $La_2Zr_2O_7$ (LZO) [19] and $La_2Hf_2O_7$ (LHO) [20] can be grown epitaxially on Si(111) [19] and Si(001) [20], respec-

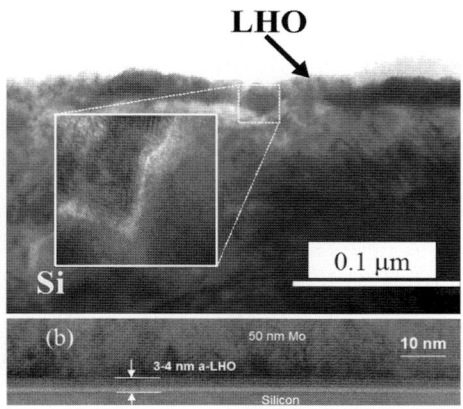

**Fig. 2.** (a) Epitaxial 15 nm thick $La_2Hf_2O_7$ (LHO) grown at a high temperature of 770 °C. The inset shows {111} faceting of the Si substrate which induces roughness to the epilayer. (b) As-deposited amorphous LHO layer showing smooth morphology with a 1 nm thick interfacial layer. The deposition temperature is 450 °C

tively. These oxide compounds have similar group symmetry (Fd3m) with Si and small lattice mismatch ($\sim 0$ for LZO and $-0.74\%$ for LHO at room temperature and $\sim 0$ at 800 °C), which is in favour of epitaxial growth. Local cube-on-cube epitaxy has been demonstrated for LHO only at very high growth temperatures of $\sim 770$ °C. However, the epitaxial layer is rough mainly due to the rough morphology of the Si substrate, as shown in Fig. 2a. As shown in the inset, the starting Si surface is defective, showing {111} faceting. This could be understood by considering that Si(001) clean surface is exposed to $O_2$ at high substrate temperature just before LHO growth. $O_2$ reacts with Si, producing volatile Si sub-oxide species, effectively etching the surface.

Because of these problems, the epitaxial LHO prepared at high temperatures cannot be used for devices unless the morphology is improved. On the other hand, the LHO can be deposited amorphous [20] at temperatures lower than 650 °C with very smooth morphology, as shown in Fig. 2b, albeit with an interfacial layer which is formed either during growth or after forming gas annealing (FGA). MIS capacitors with Mo gate electrode and 4 nm thick amorphous LHO exhibit capacitance and gate leakage currents which depend on the deposition temperature $T_d$, as shown in Fig. 3. The accumulation capacitance increases with temperature and reaches its highest value for $T_d = 600$ °C (Fig. 3a), corresponding to sub-1 nm EOT. The sample grown at highest temperature also exhibits the largest value of the leakage current $J_g$. This behaviour can be understood by considering that under UHV conditions, any $SiO_x$ interfacial layer which is spontaneously formed during deposition becomes less stable at higher temperatures and desorbs, resulting

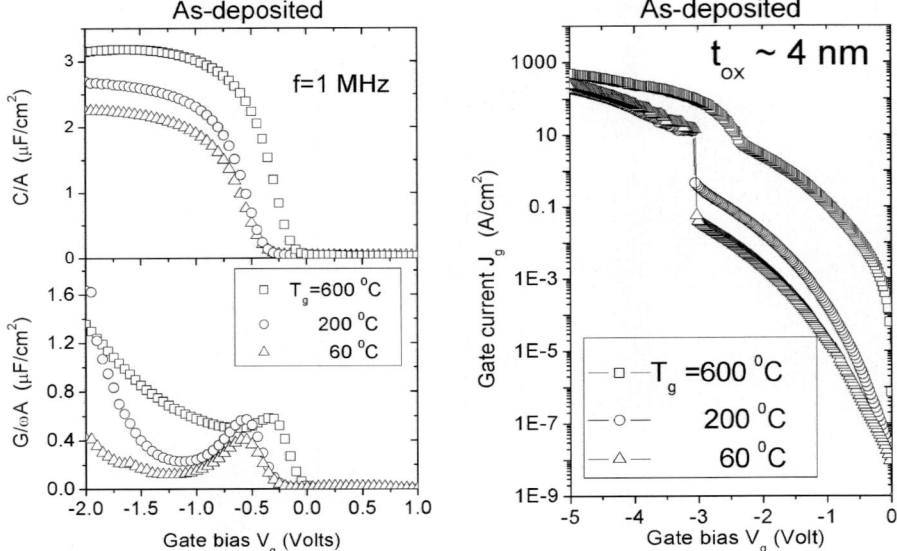

**Fig. 3.** (a) High-frequency capacitance and conductance as a function of gate bias in as-deposited Mo/a-LHO/p-Si MIS capacitors. The LHO was deposited at three different temperatures $T_g$. (b) Gate current in accumulation for the same samples

in clean interfaces. This increases the accumulation capacitance $C_{acc}$ but also increases $J_g$ at the same time.

After forming gas annealing (FGA) at 420 °C for 20 min, the capacitance for the 600 °C sample drops and all samples converge to the same behaviour. A typical response is shown in Fig. 4, which corresponds to the sample deposited at 600 °C. Using our previously published model [28] which takes into account quantum confinement effects, the EOT and the interface state density $D_{it}$ have been estimated to be 1.05–1.1 nm and $1 \times 10^{-12}$ eV cm$^{-1}$, respectively. To explain this, we assume that after FGA, an SiO$_x$ interfacial layer is formed which lowers $C_{acc}$ (increases EOT) and decreases leakage current. FGA also decreases $D_{it}$ by a factor of 5 with respect to the as-deposited samples.

Using amorphous LHO dielectrics prepared by MBD, short-channel (100 nm) N and P MOSFETs were fabricated [29, 30] on 200 mm Si wafers using a pilot processing line. The N MOSFETs with EOT of ∼ 1.9 nm are all functional, showing very good mobility, remarkably low gate leakage current and negligible threshold voltage instability. The last two are important advantages of LHO compared to HfO$_2$, and make LHO a good gate dielectric candidate for future reliable low-operating power transistors.

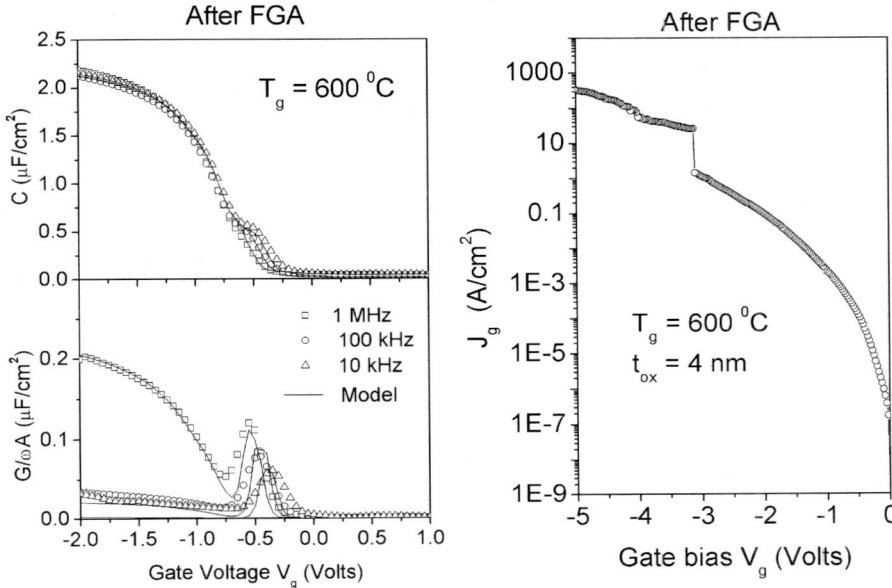

**Fig. 4.** MIS capacitor with a-LHO gate deposited at 600 °C, after FGA treatment. (**a**) High-frequency capacitance and conductance versus gate bias. (**b**) Gate leakage current

## 4 Ceria on Germanium and GaAs Substrates

As stated in the introduction, the interest in Ge transistor devices has been revived recently in order to face the challenges of downscaling to increase the high-frequency performance. Most of the work has been devoted to the $HfO_2$/GeON system, which is a natural choice since it leverages the large amount of work that has been done with $HfO_2$ materials for Si devices. However, the electrical quality of these gate stacks is not adequate [3, 31, 32] and the general consensus is that the $HfO_2$/GeON combination may not be the best solution for Ge MIS capacitors and MOSFET devices. Other oxides including rare earth oxides may be better match to Ge. In the case of GaAs, despite the progress [5] which has been made using $Ga_2O_3-Gd_2O_3$ mixtures, no satisfactory passivating oxide has been found yet. There is recent evidence [33] that binary rare earth oxides ($CeO_2$, $Gd_2O_3$, $Dy_2O_3$) can be deposited directly on Ge, giving good electrical C–V characteristics with reduced $D_{it}$ compared to the $HfO_2$/GeON system. The $CeO_2$ films on Ge are polycrystalline, as could be inferred from X-ray diffraction measurements. Here, we compare side by side $CeO_2$ deposited under the same conditions but on two different substrates: (001)-oriented Ge and GaAs. As seen from Fig. 5, there is a notable difference between the two substrates. $CeO_2$ on Ge gives MIS capacitors with good C–V characteristics (Fig. 5a), showing

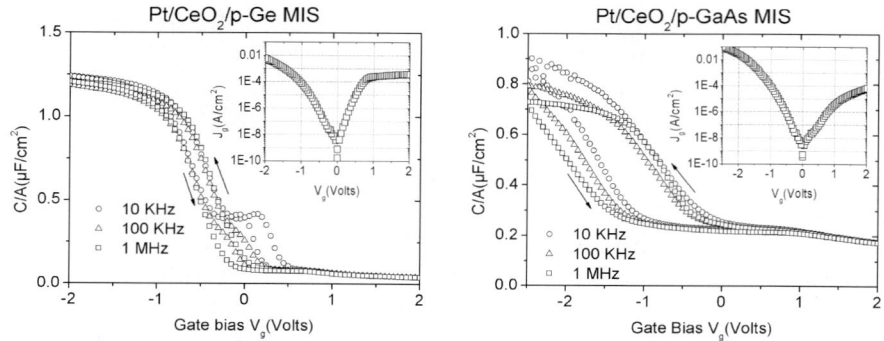

**Fig. 5.** High-frequency capacitance-voltage of MIS capacitors with $CeO_2$ gate dielectric and Pt electrode deposited on (**a**) p-type Ge substrates, (**b**) p-type GaAs substrates. In both samples, the dielectric was deposited at 225 °C and had a nominal thickness of 10 nm. The *insets* show the gate current density

hysteresis less than 100 mV, which is rather small compared to $HfO_2$-based devices [3]. In addition, there is negligible frequency dispersion in accumulation and only small dispersion in depletion, indicating relatively low $D_{it}$. Most important of all is that there is no significant frequency dispersion or low-frequency behaviour in inversion, unlike the case of $HfO_2$/GeON [31,32]. This is an indication that $CeO_2$ provides a better passivation of the surface and a better barrier for metal diffusion from the oxide which could contaminate Ge near the surface. It is proposed here that $CeO_2$ in the +4 oxidation state reduces to $Ce_2O_3$ (+3 oxidation state), acting as a catalyst for the efficient oxidation of the Ge surface. With the help of $CeO_2$, poor-quality Ge suboxides $(GeO_x)$ which may have remained on the surface could turn into good-quality $GeO_2$ which behaves as a good passivating layer.

In the case of $CeO_2$/GaAs (Fig. 5b), the C–V characteristics are not ideal, featuring significant frequency dispersion in accumulation, large hesteresis in excess of 1 V, large stretch-out and sizable threshold voltage shift to negative values, indicating charge trapping into the oxide. Since the $CeO_2$ was prepared under the same conditions, it may be speculated that the charge is trapped in the interfacial layer which is expected to be different in the two types of substrates. The results underlie the importance of interface chemistry which affects the electrical behaviour of the gate stack.

From the insets in Figs. 5a and b, it can be seen that $CeO_2$ is leaky despite the large physical thickness of about 10 nm. The leakage current is even higher in n-type MIS. This may be expected, since $CeO_2$ has a low energy gap of about 3 eV. By studying the temperature variation of the I–V curves in the Ge capacitors, it is found [34] that conduction through the $CeO_2$ is dominated by the Schottky mechanism, with a small barrier height (less than 0.5 eV). This is compatible with a small conduction band offset explaining the high leakage current in the $CeO_2$ MIS. The high leakage of $CeO_2$ implies that this material can be used in a gate stack only in combination with other dielectrics such as $HfO_2$, for example, in order to ensure scalability at low EOT values with low leakage current.

# 5 Summary and Future Outlook

Rare earth oxide thin films and their compounds with other oxides are deposited on Si, Ge and GaAs semiconductors as gate dielectrics for future high-performance transistor devices. Two of these compounds, $La_2Zr_2O_7$ and $La_2Hf_2O_7$ (LHO), are lattice matched with Si and can be deposited epitaxially on Si substrates. LHO in specific can be deposited at high temperature around 770 °C in a local cube-on-cube epitaxial mode on Si(001). However, the film morphology is rough due to the faceting of the Si surface as a result of its interaction with $O_2$ at high temperature prior to oxide growth. At temperatures lower than 650 °C, the films are amorphous and continuous with smooth surface morphology. The electrical and morphological characteristics make them good candidates for future transistors with high-$\kappa$, although the EOT values of about 1.05–1.1 nm are not as low as would be required for high-performance devices. Hf-based oxides and silicates have reached a more mature state and could be used in future device implementations. Rare earth oxides have better chances to play a role in future nanoelectronics if combined with semiconductors other than Si. Especially the rare earth oxides which can be found in more than one oxidation state (as, for example, $CeO_2(Ce_2O_3)$) may be suitable for surface passivation of Ge surfaces, as our data indicate, although they may not be as good for the passivation of GaAs surface; this depends strongly on the surface chemistry between the oxide and the particular semiconductor used. $CeO_2$ itself is leaky and cannot be used in a gate stack unless it is combined with another oxide (i.e., $HfO_2$) which is a better insulator. However, $CeO_2(Ce_2O_3)$ can be used as a model system to study the surface chemistry on Ge surfaces, and then extend the study to other rare oxides having more than one oxidation state. The aim will be to find the most suitable rare earth oxide providing a good passivation layer and being a good insulator, at the same time qualifying as gate dielectric in future high-mobility semiconductor transistor devices.

## Acknowledgements

I would like to thank my students G. Mavrou and A. Sotiropoulos, the postdoctoral fellow Dr. Y. Panayiotatos and my colleagues Dr. A. Travlos of NCSR DEMOKRITOS, Prof. E. Evangelou of the University of Ioannina and Dr. J. Hooker of Philips Research Leuven for their contribution in this work. Also, I acknowledge support from the European Commission in the framework of IST projects INVEST and ET4US.

# References

[1] 2003 edition of the ITRS
    URL http://public.itrs.net/

[2] G. Wilk, R. Wallace, J. Anthony: High-$\kappa$ gate dielectrics: Current status and materials properties considerations, J. Appl. Phys. **89**, 5243 (2001)

[3] A. Dimoulas: Molecular beam deposition of high-$\kappa$ gate dielectrics for advance CMOS, in E. Zschech, C. Whelan, T. Mikolajick (Eds.): *Materials for Information Technologies*, Advances in Thin Film Deposition (Springer, Berlin, Heidelberg 2005) Chap. 1, p. 3

[4] K. C. Saraswat, C. O. Chui, T. Krishnamohan, A. Nayfeh, P. McIntyre: Ge-based high performance nanoscale MOSFETs, Microelectron. Eng. **80**, 15 (2005)

[5] R. Droopad, M. Passlack, N. England, K. Rajagopalan, J. Abrokwah, A. Kummel: Gate dielectrics on compound semiconductors, Microelectron. Eng. **80**, 138 (2005)

[6] W. Tsai, L. Ragnarsson, P. J. Chen, B. Onsia, R. J. Carter, E. Cartier, E. Young, M. Green, M. Caymax, S. D. Gendt, M. Heyns: Comparison of sub 1 nm TiN/HfO$_2$ with poly-Si/HfO$_2$ gate stacks using scaled chemical oxide interface, in *VLSI Technology 2003, Digest of Technical Papers* (Kyoto 2003) p. 21

[7] A. Chin, Y. H. Wu, S. B. Chen, C. C. Liao, W. J. Chen: High quality La$_2$O$_3$ and Al$_2$O$_3$ gate dielectrics with equivalent oxide thickness 5–10 Å, in *VLSI Symp. 2000, Digest of Technical Papers* (Honolulu 2000) p. 16

[8] S. Guha, E. Cartier, M. A. Gribelyuk, N. A. Bojarczuk, M. A. Copel: Atomic beam deposition of lanthanum- and yttrium-based oxide thin films for gate dielectrics, Appl. Phys. Lett. **77**, 2710 (2000)

[9] E. Miranda, J. Molina, Y. Kim, H. Iwai: Effects of high-field electrical stress on the conduction properties of ultrathin La$_2$O$_3$ films, Appl. Phys. Lett. **86**, 232104 (2005)

[10] X. B. Lu, H. B. Lu, Z. H. Chen, X. Zhang, R. Huang, H. W. Zhou, X. P. Wang, B. Y. Nguen, C. Z. Wang, W. F. Xiang, M. He, B. L. Cheng: Field-effect transistors with LaAlO$_3$ and LaAlO$_x$N$_y$ gate dielectrics deposited by laser molecular-beam epitaxy, Appl. Phys. Lett. **85**, 3543 (2004)

[11] G. Vellianitis, G. Apostolopoulos, G. Mavrou, K. Argyropoulos, A. Dimoulas, J. C. Hooker, T. Conard, M. Butcher: MBE lanthanum based high-$\kappa$ gate dielectrics as candidates for SiO$_2$ date oxide replacement, Mater. Sci. Eng. **B**, 85 (2004)

[12] L. Miotti, K. B. Bastos, C. Driemeier, V. Edon, M. C. Hugon, B. Agius, I. J. R. Baumvol: Effects of post-deposition annealing in O$_2$ on the electrical characteristics of LaAlO$_3$ films on Si, Appl. Phys. Lett. **87**, 0222901 (2005)

[13] S. Gaillard, Y. Rozier, C. Merckling, F. Ducroquet, M. Gendry, G. Hollinger: LaAlO$_3$ films prepared by MBE on LaAlO$_3$(001) and Si(001) substrates, Microelectron. Eng. **80**, 146 (2005)

[14] D. S. Yun, C. H. Huang, A. Chin, W. J. Chen, C. X. Zhu, B. J. Cho, M.-F. Li, D. L. Kwong: Al$_2$O$_3$–Ge-on-insulator n- and p-MOSFETs with fully NiSi and NiGe dual gates, IEEE EDL **25**, 138 (2004)

[15] R. A. McKee, F. J. Walker, M. Chisholm: Crystalline oxides on silicon: The first five monolayers, Phys. Rev. Lett. **81**, 3014 (1998)

[16]  C. Zhao, T. Witters, B. Brijs, H. Bender, O. Richard, M. Caymax, T. Heeg, J. Schubert, V. V. Afanas'ev, A. Stesmans, D. G. Schlom: Ternary rare-earth metal oxide high-k layers on silicon oxide, Appl. Phys. Lett. **86**, 132903 (2005)

[17]  T. Heeg, M. Wagner, J. Schubert, C. Buchal, M. Boese, M. Luysberg, E. Cicerrella, J. L. Freeouf: Rare-earth scandate single- and multi-layer thin films as alternative gate oxides for microelectronic applications, Microelectron. Eng. **80**, 150 (2005)

[18]  S. Guha, N. A. Bojarczuk, V. Narayanan: Lattice matched, epitaxial, silicon-insulating lanthanum yttrium oxide eterostructures, Appl. Phys. Lett. **80**, 766 (2002)

[19]  J. W. Seo, J. Fompeyrine, A. Guiller, G. Norga, C. Marchiori, H. Siegwart, J.-P. Locquet: Interface formation and defect structures in epitaxial $La_2Hf_2O_7$ thin films on (111) Si, Appl. Phys. Lett. **83**, 5211 (2003)

[20]  A. Dimoulas, G. Vellianitis, G. Mavrou, G. Apostolopoulos, A. Travlos, C. Wiemer, M. Fanciulli, C. M. Rittersma: $La_2Hf_2O_7$ high-$\kappa$ gate dielectric grown directly on Si (001) by molecular beam epitaxy, Appl. Phys. Lett. **85**, 3205 (2004)

[21]  A. Dimoulas, A. Travlos, G. Vellianitis, N. Boukos, K. Argyropoulos: Direct heteroepitaxy of crystalline $Y_2O_3$ on Si (001) for high-$\kappa$ gate dielectric applications, J. Appl. Phys. **90**, 4224 (2001)

[22]  A. Dimoulas, G. Vellianitis, A. Travlos, V. Ioannou-Sougleridis, A. G. Nassiopoulou: Structural and electrical quality of the high-$\kappa$ dielectric $Y_2O_3$ on Si(001): Dependence on growth parameters, J. Appl. Phys. **92**, 426 (2002)

[23]  J. Kwo, M. Hong, A. R. Kortran, K. L. Queeny, Y. J. Chabal, R. L. Opila, D. A. Muller, S. N. G. Chu: Properties of high-k gate dielectrics $Gd_2O_3$ and $Y_2O_3$ for Si, J. Appl. Phys. **89**, 3920 (2001)

[24]  H. J. Osten, J. P. Liu, E. Bugiel, H. J. Mussing, P. Zaumseil: Growth of crystalline praseodymium oxide on silicon, J. Crystal Growth **235**, 229 (2002)

[25]  A. Fissel, H. J. Osten, E. Bugiel: Towards understanding epitaxial growth of alternative high-$\kappa$ dielectrics on Si(001): Application to praseodymium oxide, J. Vac. Sci. Technol. B **21**, 1765 (2003)

[26]  G. Apostolopoulos, G. Vellianitis, A. Dimoulas, M. Alexe, R. Scholz, M. Fanciulli, D. T. Dekadjevi, C. Wiemer: High epitaxial quality $Y_2O_3$ high-$\kappa$ dielectric on vicinal Si (001), Appl. Phys. Lett. **81**, 3549 (2002)

[27]  E. Bonera, G. Scarel, M. Fanciulli, P. Delugas, V. Fiorentini: Dielectric properties of high-$\kappa$ oxides: Theory and experiment for $Lu_2O_3$, Phys. Rev. Lett. **94**, 027602 (2005)

[28]  G. Apostolopoulos, G. Vellianitis, A. Dimoulas, J. C. Hooker, T. Conard: Complex admittance analysis for $La_2Hf_2O_7/SiO_2$ high-$\kappa$ dielectric stacks, Appl. Phys. Lett. **84**, 260 (2004)

[29]  L. Pantisano, T. Conard, M. Claes, M. Demand, W. Deweerd, S. DeGendt, M. Heyns, M. Houssa, M. Alouaiche, G. Lujan, L. A. Ragnarsson, E. Rohr, T. Schram, J. C. Hooker, Z. M. Rittersma, J. Fompeyrine, J.-P. Locquet, A. Dimoulas: MOSFET with $La_2Hf_2O_7$ and $HfO_2$ high-$\kappa$ dielectrics integrated in a conventional flow, in *Mater. Res. Soc.* (Symposium G, San Francisco 2005)

[30]  Z. M. Rittersma, J. C. Hooker, J.-P. Locquet, C. Marchiori, M. Sousa, J. Fompeyrine, L. Pantisano, T. Schram, M. Rosemeulen, S. DeGendt, G. Vellianitis, A. Dimoulas: Electrical characterization of $La_2Hf_2O_7$ and $HfO_2$ gate dielectric layers deposited by molecular beam epitaxy, J. Appl. Phys. **99**, 24508 (2006)

[31] A. Dimoulas, G. Mavrou, G. Vellianitis, E. K. Evangelou, N. Boukos, M. Houssa, M. Caymax: $HfO_2$ high-$\kappa$ gate dielectrics on Ge (100) by atomic oxygen beam deposition, Appl. Phys. Lett. **86**, 032908 (2005)

[32] A. Dimoulas, G. Vellianitis, G. Mavrou, E. K. Evangelou, A. Sotiropoulos: Intrinsic carrier effects in $HfO_2-Ge$ metal-insulator-semiconductor capacitors, Appl. Phys. Lett. **86**, 223507 (2005)

[33] A. Dimoulas: Electrically active interface and bulk semiconductor in high-$\kappa$/Ge structures, in *Defects in High-k Gate Dielectrics* (Proceedings of the NATO ARW, St. Petersbourg, Russia 2005)

[34] M. Houssa, A. Dimoulas, F. Bellenger, Y. Panayiotatos, A. Sotiropoulos, M. Caymax, M. Heyns: Electrical characterization of $CeO_2$/Ge MIS capacitors, unpublished

# Index

# The Magneto-Electric Properties of RMnO$_3$ Compounds

Thomas T. M. Palstra

Solid State Chemistry Laboratory, Materials Science Centre, University of
Groningen, NL-9752VZ Groningen, The Netherlands
t.t.m.palstra@chem.rug.nl

**Abstract.** The RMnO$_3$ compounds adopt the orthorhombic perovskite crystal
structure for large rare earth ions R, and a hexagonal crystal structure for small
rare earth ions. In both crystal structures, the rare earth ions occupy non-centered
positions which can result in long-range dipolar order at low temperature. Consequently, these materials exhibit coexistence of magnetic and electric ordering. The
crystal packing and chemical bonding of these open-shell systems provides information about mechanisms of magneto-ferroelectric ordering and magneto-electric
coupling.

## 1 Introduction

Multiferroics are materials that simultaneously exhibit more than one type
of ordering. The simultaneous occurrence of magnetic and electric order of
magneto-ferroelectrics is particularly interesting as it combines properties
that could be utilized for information storage, processing and transmission. It
allows both magnetic and electric fields to interact with magnetic and electric
order. However, this property is rare as the existence of partially filled atomic
orbitals, prerequisite for magnetic dipoles or moments, usually precludes the
occurrence of local electric dipoles, which are typically associated with the
presence of either empty $d$-shells and/or an electron lone-pair configuration.

Ferromagnetic materials are commonly used for non-volatile information
storage in tapes, hard drives, etc. They are also used for information processing due to the interaction of electric current and light with magnetic order.
Ferroelectrics find applications due to their large piezoelectric coupling constant, i.e., the coupling between an electric field and strain. This effect is
utilized in devices such as capacitors, microphones and transducers where a
voltage can generate strain and vice versa. Ferroelectric materials not only
exhibit piezoelectric coupling, common for most materials that lack inversion
symmetry, but also possess memory functionality. The electric polarization
remains finite after removing an applied electric field. This property can be
exploited in non-volatile memory devices, where the information stored in
the electric polarization is retained, even after removing the power of the
device. Much interest is being generated by the magneto-ferroelectrics. The
simultaneous magnetic and electric order makes it possible for the magnetic

M. Fanciulli, G. Scarel (Eds.): Rare Earth Oxide Thin Films,
Topics Appl. Physics **106**, 391–400 (2007)
© Springer-Verlag Berlin Heidelberg 2007

polarization to be addressed or switched not only by applying a magnetic field but also by an electric field, or likewise that the electric polarization can be addressed or switched by applying an electric and/or a magnetic field. This property enables completely new device architectures to be designed. Furthermore, it is of interest to materials scientists to understand the associated chemical bonding. This will allow them to circumvent the prevalent mutual exclusion of magnetic and electric dipoles by smart materials engineering in bulk compounds or thin films.

In this manuscript, I focus on two mixed rare earth-transition metal oxides, the orthorhombic perovskites $RMnO_3$, stable for large ionic radius $R$, and the hexagonal compounds $RMnO_3$, stable for small ionic radius $R$. The crystal structure and local coordination numbers are completely different for the two structure types. In the perovskites, the $R^{3+}$ is coordinated by 12 oxygen ions. In the hexagonal $RMnO_3$ structure, the $R^{3+}$ is coordinated by seven oxygen ions, consisting of a trigonal antiprismatic coordination of six plus one additional ion.

# 2 $ABO_3$ Perovskites

This structure class encompasses many ferroelectric materials such as $BaTiO_3$, and can be constructed from corner-sharing $BO_6$ octahedra, and the A-ion is coordinated by eight octahedra (see Fig. 1). A ferroelectric moment can in part be derived from a lone-pair ion on the A-site, such as $Pb^{2+}$ or $Bi^{3+}$, and in part from a small transition metal ion, with $d^0$ electron configuration on the B-site. Both mechanisms can result in long-range dipolar order [1]. Magnetism can be generated by a magnetic rare earth ion on the A-site and/or by a magnetic transition metal ion on the B-site, such as $Mn^{3+}$ or $Fe^{3+}$. However, because transition metal ions generate either magnetic ($d^n$ state) or electric moments ($d^0$ state), magnetic and electric order on the same site is in conflict. A judicious choice of composition can lead to ferroelectromagnetism in compounds such as $Pb(Fe_{0.5}W_{0.5})_3$, and various other combinations. Therefore, the magnetic and electric moments may be dilute, which typically results in ordering at suppressed temperatures. Nevertheless, this approach can be successful, and the overview of *Smolenski* and *Chupis* lists a number of such ferroelectromagnets [2].

Within this class of perovskites, $BiFeO_3$ and $BiMnO_3$ have received considerable attention. They are difficult to prepare as stoichiometric materials and require special synthesis conditions. Both have been grown as thin films by pulsed laser deposition (PLD). $BiFeO_3$ is both ferroelectric and antiferromagnetic at room temperature, which triggers interest for devices. Recent reports of ferromagnetic coupling [3] provide evidence of strong dependence on the oxygen stoichiometry [4]. A non-stoichiometric composition may thus trigger desirable magnetic interactions, but it also increases the electrical conductivity, possibly preventing the application of sufficient electric

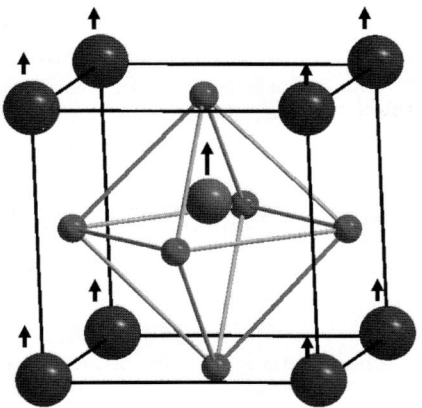

**Fig. 1.** Crystal structure of a perovskite ABO₃. The transition metal B-ions are located near the center of the cube coordinated by an oxygen octahedron. The A-ions are located at the corner of the cube. The *arrows* indicate a tetragonal ferroelectric distortion

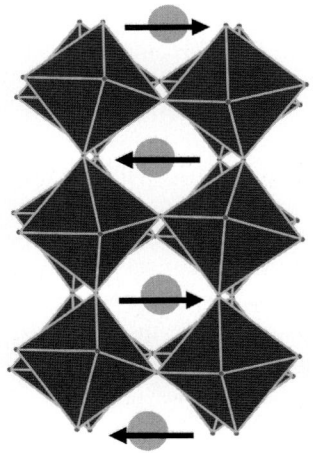

**Fig. 2.** Crystal structure of an orthorhombic perovskite RMnO₃. The Mn-ions are located in the center of the coordinating oxygen octahedra. The R-ions are coordinated by eight octahedra or 12 O-ions, but displaced from the center in an anti-polar fashion, as indicated by *arrows*

fields by electrodes. Ferromagnetism in oxides is often mediated by double exchange interactions, which render a material metallic, screening electric dipoles. Ferromagnetism in ferroelectromagnets should therefore be the result of different interactions, such as single ion anisotropy or Dzyaloshinskii–Moriya coupling, resulting in a canted antiferromagnetic state. Furthermore, the progress in theoretical understanding of the chemical bonding associated with ferroelectricity has impacted greatly on the understanding of magneto-ferroelectrics [5].

The discovery of ferroelectricity resulting from competing magnetic interactions in the perovskites TbMnO₃ and DyMnO₃ [6, 7] has attracted much recent attention and has opened the field to a completely different approach to generating ferroelectricity in oxides. These materials were known as orthorhombic compounds with Pnma (the conventional setting of Pbnm) space group symmetry. This space group exhibits inversion symmetry and thus no ferroelectric order was expected. However, the $Mn^{3+}$-spins in $TbMnO_3$ order in an incommensurate fashion near $40\,K$, which then gives rise to a ferroelectric moment at a lock-in transition, near $30\,K$, below which the incommensurate magnetic wave vector becomes temperature independent. Below this lock-in transition, the electric polarization becomes finite and can be reversed by an electric field. The origin of the emergence of the electric order is not fully understood, but seems to be associated with two competing magnetic orders. The large ionic radius elements on the A-site, such as $LaMnO_3$, give rise to A-type antiferromagnetic order of the Mn-spins, whereas the small ionic radius A-site elements, from $HoMnO_3$ onwards, promote E-type antiferromagnetic order [8]. Small ionic radius A-site cations reduce the $Mn-O-Mn$ bond angles from $180°$ for cubic materials to values less than $145°$. At these bond angles, the perovskite structure is strongly distorted and not only the nearest neighbor magnetic interactions are relevant, but also the next-nearest neighbor interactions. Thus, the magnetic order changes from two antiferromagnetically (AF) and four ferromagnetically (FM) coupled neighbors for A-type order to four AF and two FM coupled neighbors for E-type order. At intermediate bond angles, incommensurate order is stabilized [9], the inversion symmetry is broken and this induces the electric polarization. A strong magneto-electric coupling is evidenced by large magneto-capacitance effects, where the dielectric constant changes by more than a factor 5 on applying a magnetic field of $4\,T$ in $DyMnO_3$ [7]. Furthermore, lattice modulations that are coupled to the magnetic modulation have been detected. The nature of the ferroelectric transition is likely to be improper, where the ferroelectric state is driven by the magnetic ordering. The electric polarization is relatively small, with values lower than $0.2\,\mu C/cm^2$.

# 3 $AMnO_3$ Hexagonal Manganites

Despite the compositional identity with the perovskites, these materials are structurally completely different. They consist of sheets of $MnO_5$ triangular bipyramids connected by $AO_7$ polyhedra. The magnetic ordering is derived from the electronic $d^4$ configuration of $Mn^{3+}$ on the B-site, and from a magnetic rare earth ion on the A-site. Despite conclusions to the contrary in early structure determinations, the $Mn^{3+}$ is located close to the barycenter of the bipyramid [10, 11], and the ferroelectric moment is largely derived from the $A^{3+}$-motion in its coordination sphere.

The structure is stable for small rare earth ions on the A-site, including yttrium, and for $Mn^{3+}$ on the B-site. The $Mn^{3+}$-ion stabilizes the structure by a ligand field effect, because the crystal field splitting of the $d^4$ configuration in a triangular bipyramid results in a non-degenerate electronic state. For all

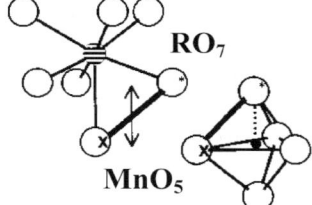

**Fig. 3.** Local coordination of rare earth R- and Mn-ions in hexagonal RMnO$_3$. The R-ion is coordinated by 6 + 1 O-ions and the Mn-ion is coordinated by 5 O-ions forming a trigonal bipyramid (taken from [10, 11])

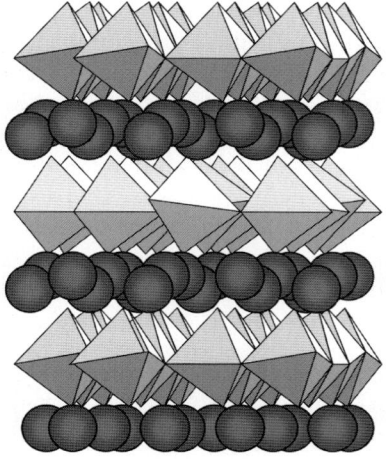

**Fig. 4.** Crystal structure of a hexagonal manganite RMnO$_3$, with R a rare earth element, with ferroelectric distortion. The Mn-ions are located near the barycenter of the trigonal bipyramids of O-ions. The distortion of the Y-layer (*open circles*) is clearly observable (reproduced from [10, 11])

other transition metals on the B-site, and large ionic radius elements on the A-site, the perovskite structure is preferred [12, 13].

These materials were intensely studied in the 1960s, during which time the crystal and magnetic structures were investigated [14]. The ferroelectric ordering sets in above 1000 K, while antiferromagnetic ordering of the Mn-spins occurs near 75 K, with small variations for different R rare earth ions. These ordering temperatures indicate that RMnO$_3$ are proper ferroelectrics, which is in line with a polarization of $5 \mu C/cm^2$ and large displacements inside the unit cell due to the electric ordering. Various corresponding indates RInO$_3$ and gallates RGaO$_3$ are also reported or predicted to be ferroelectric (see review in [12]). The magneto-electric coupling is evidenced by anomalies in the temperature dependence of the dielectric constant at the magnetic ordering [15, 16]. The coupling was convincingly shown by *Fiebig* et al. in non-linear optical experiments, which mapped both electric and antiferromagnetic domain walls [17]. Furthermore, *Lottermoser* et al. showed that in HoMnO$_3$, ferromagnetic ordering of the Ho-spins could reversibly be switched on and off by using an electric field, again indicating the magneto-electric coupling [18]. There have been conflicting results from studies of the nature of the ferroelectric ordering. After the discovery of the ferroelectric properties of this class of materials by *Bertaut* et al. [14], several transition temperatures were reported. Measurements of the pyroelectric current by *Ismailzade* et al. of $YMnO_3$ showed the ferroelectric ordering to be near 930 K [19]. A transition at 1270 K was recorded for $YMnO_3$, associated by *Lukaszewicz* et al. with a tripling of the lattice [20]. *Van Aken* et al. redetermined the crystal structures of several $RMnO_3$ and found, in contrast with earlier reports, that $Mn^{3+}$ is close to the barycenter of the coordinating oxygen bipyramid [10, 11]. They concluded that the ferroelectric moment is mostly associated with the relative motion of the R-ion and its coordinating oxygens [21]. Subsequent neutron diffraction experiments up to 1400 K by *Lonkai* et al. did not reveal any significant coupling to the modes responsible for the ferroelectric transition [22]. They concluded that $RMnO_3$ must be an improper ferroelectric, in agreement with a theoretical analysis of the displacements [23]. *Nénert* et al. argue, based on synchrotron single crystal diffraction and a group theoretical analysis, that these materials are proper ferroelectrics [24], in agreement with the result of *Ismailzade* et al. [19] The situation requires further measurements and analysis.

# 4 Novel Mechanisms

The concept of generating ferroelectricity from geometric magnetic frustration has resulted in theoretical and experimental efforts to formulate new mechanisms. *Van Aken* et al. proposed that the ferroelectric state in $YMnO_3$ is generated by geometric, rather than electronic effects [21]. *Efremov* et al. have proposed that frustration of electronic degrees of freedom other than magnetic interactions may result in ferroelectricity [25]. They predicted that charge ordering in orthorhombic manganites can result in a ferroelectric state. In charge-ordered manganites, the manganese ions can adopt two valence states, $Mn^{3+}$ and $Mn^{4+}$, which order in a regular pattern below the charge ordering temperature. Various charge ordering symmetries have been observed for $(La,Ca)MnO_3$. Such charge order can be site-centered, with a spherical charge distribution around the nucleus, or bond-centered, in which electron density is concentrated between two nuclei. For partial bond-centered and partial site-centered charge ordering, *Efremov* et al. calculated the emergence of a ferroelectric state [25]. It seems natural to expect that future studies of other systems with competing interactions such as orbital orderings will generate novel improper ferroelectrics. In addition, systems with cationic ordering can be expected to show ferroelectric states. Here, the non-centered coordination of $RO_7$, $RO_9$, or $RO_{12}$ clusters can be exploited.

# 5 Conclusions

Ferroelectromagnetic systems, with coexistence of magnetic and electric moments, can be observed in a selected number of compounds. Conventionally, such coexistence is precluded because magnetic systems require unfilled $d$- or $f$-orbitals, whereas electric moments are generated by empty $d$-orbitals or an electron lone-pair configuration. The rare earth ion is special because the magnetism, generated by unfilled $4f$-orbitals, is shielded by filled valence shells. This can lead to magneto-ferroelectric states in both orthorhombic perovskites ABO$_3$ and hexagonal AMnO$_3$. In addition, novel mechanisms including charge ordering may lead to novel magneto-ferroelectric compounds.

**Acknowledgements**

This work is in part the result of the M.Sc. thesis of *Jan-Willem Bos* [13], and the Ph.D. thesis work of *Bas van Aken* [10, 11], Gwilherm Nénert, Umut Adem, Mufti Nandang and the Post Docs Agung Nugroho and Graeme Blake. Fruitful discussions with Gustavo Catalan, Beatriz Noheda, and Rob de Groot are also gratefully acknowledged. The work is in part supported by the MSC$^{plus}$ program, the Stichting FOM (Fundamental Research of Matter), and NWO (Netherlands Organization for Scientific Research).

# References

[1] M. E. Lines, A. M. Glass: *Principles and Applications of Ferroelectrics and Related Materials* (Oxford University Press, New York 2001)

[2] G. A. Smolinskii, I. E. Chupis: Ferroelectromagnets, Sov. Phys. Usp. **25**, 475 (1982)

[3] J. Wang, J. B. Neaton, H. Zheng, V. Nagarajan, S. B. Ogale, B. Liu, D. Viehland, V. Vaithyanathan, D. G. Schlom, U. V. Waghmare, N. A. Spaldin, K. M. Rabe, M. Wuttig, R. Ramesh: Epitaxial BiFeO$_3$ multiferroic thin film heterostructures, Science **299**, 1719 (2003)

[4] W. Eerenstein, F. D. Morrison, J. Dho, M. G. Blamire, J. F. Scott, N. D. Mathur: Comment on epitaxial BiFeO$_3$ multiferroic thin film heterostructures, Science **307**, 1203 (2005)

[5] N. A. Hill: Why are there so few magnetic ferroelectrics?, J. Phys. Chem. B **104**, 6694 (2000)

[6] T. Kimura, T. Goto, H. Shintani, K. Ishizaka, T. Arima, Y. Tokura: Magnetic control of ferroelectric polarization, Nature **426**, 55 (2003)

[7] T. Goto, T. Kimura, G. Lawes, A. P. Ramirez, Y. Tokura: Ferroelectricity and giant magnetocapacitance in perovskite rare-earth manganites, Phys. Rev. Lett. **92**, 257201 (2004)

[8] T. Kimura, S. Ishihara, H. Shintani, T. Arima, K. T. Takahashi, K. Ishizaka, Y. Tokura: Distorted perovskite with e$_\mathrm{g}^1$ configuration as a frustrated spin system, Phys. Rev. B **68**, 060403R (2003)

[9] R. Kajimoto, H. Yoshizawa, H. Shintani, T. Kimura, Y. Tokura: Magnetic structure of TbMnO$_3$ by neutron diffraction, Phys. Rev. B **70**, 012401 (2004)

[10] B. B. Van Aken, A. Meetsma, T. T. M. Palstra: Acta Cryst. C **57**, 230 (2001)

[11] B. B. Van Aken: Ph.D. thesis, Rijksuniversiteit Groningen (2001)
URL http://www.ub.rug.nl/eldoc/dis/science/b.b.van.aken/

[12] S. C. Abrahams: Ferroelectricity and structure in the YMnO$_3$ family, Acta Cryst. B **57**, 485 (2001)

[13] J. W. G. Bos, B. B. van Aken, T. T. M. Palstra: Site disorder induced hexagonal-orthorhombic transition in Y$_{1-x}$Gd$_x$MnO$_3$, Chem. Mat. **13**, 4804 (2001)

[14] F. Bertaut, F. Forrat, P. Fang: Les manganites de terres rares et d'yttrium: Une nouvelle classe de ferroelectriques, C. R. Acad. Sci. **256**, 1958 (1963)

[15] Z. J. Huang, Y. Cao, Y. Y. Sun, Y. Y. Xue, C. W. Chu: Coupling between the ferroelectric and antiferromagnetic orders in YMnO$_3$, Phys. Rev. B **56**, 2623 (1997)

[16] T. Katsufuji, S. Mori, M. Masaki, Y. Moritomo, N. Yamamoto, H. Takagi: Dielectric and magnetic anomalies and spin frustration in hexagonal RMnO$_3$ (R = Y, Yb, and Lu), Phys. Rev. B **64**, 104419 (2001)

[17] M. Fiebig, T. Lottermoser, D. Frohlich, A. V. Goltsev, R. V. Pisarev: Observation of coupled magnetic and electric domains, Nature **419**, 818 (2002)

[18] T. Lottermoser, T. Lonkai, U. Amann, D. Hohlwein, J. Ihringer, M. Fiebig: Magnetic phase control by an electric field, Nature **430**, 541 (2004)

[19] I. G. Ismailzade, S. A. Kizhaev: Determination of the Curie point of the ferroelectrics YMnO$_3$ and YbMnO$_3$, Sov. Phys. Solid State **7**, 236 (1965)

[20] K. Lukaszewicz, J. Karut-Kalicinska: X-ray investigations of the crystal structure and phase transitions of YMnO$_3$, Ferroelectrics **7**, 81 (1974)

[21] B. B. Van Aken, T. T. M. Palstra, A. Filippetti, N. A. Spaldin: The origin of ferroelectricity in magnetoelectric YMnO$_3$, Nature Mater. **3**, 164 (2004)

[22] T. Lonkai, D. G. Tomota, U. Amann, J. Ihringer, R. W. A. Hendrikx, D. M. Többens, J. A. Mydosh: Development of the high-temperature phase of hexagonal manganites, Phys. Rev. B **69**, 134108 (2004)

[23] C. J. Fennie, K. M. Rabe: Ferroelectric transition in YMnO$_3$ from first principles, Phys. Rev. B **72**, R) (2005)

[24] G. Nénert, Y. Ren, H. T. Stokes, T. T. M. Palstra: Symmetry changes at the ferroelectric transition in the multiferroic YMnO$_3$ (2005)
URL cond-mat/0504546

[25] D. V. Efremov, J. van den Brink, D. I. Khomskii: Bond versus site-centred ordering and possible ferroelectricity in manganites, Nature Mater. **3**, 853 (2004)

# Index

# Sesquioxides as Host Materials for Rare-Earth-Doped Bulk Lasers and Active Waveguides

Sebastian Bär, Hanno Scheife, Klaus Petermann, and Günter Huber

Institute of Laser Physics, University of Hamburg, Luruper Chaussee 149,
D-22761 Hamburg, Germany
baer@physnet.uni-hamburg.de

**Abstract.** This Chapter focuses on the optical applications of rare-earth oxides. Due to their thermo-mechanical properties, the sesquioxides of yttrium, lutetium, and scandium are promising host materials for rare-earth-doped solid-state lasers. An Yb-doped $Sc_2O_3$ crystal has been successfully operated in the thin disk laser setup. A maximum output power of $124.5\,W$ with a slope efficiency of nearly $50\,\%$ in the cw mode has been achieved. Additionally, high-quality thin films of these sesquioxides grown by pulsed laser deposition as well as by electron beam evaporation, either on sapphire or on quartz substrates, have been investigated for their use in integrated optics. These films were highly textured along the $\langle 111 \rangle$ direction. In the case of lattice matched systems, e.g., $Lu_{0.92}Sc_{1.08}O_3$ on $\alpha\text{-}Al_2O_3$, epitaxial film growth was achieved. Waveguiding of light at different wavelengths was demonstrated in yttria films with a thickness of $1\,\mu m$ on both types of substrates.

## 1 Introduction

The high-melting sesquioxides of the rare earths (RE) scandium, yttrium, and lutetium (cubic bixbyite structure, space group $Ia3$ or $T_h^7$, see Fig. 1) are interesting laser host materials because of their superior thermo-mechanical properties, strong Stark-splitting, and low phonon energies [1]. Laser operation of sesquioxide crystals doped with various rare earths has been demonstrated. Especially sesquioxides doped with trivalent ytterbium have shown very promising results in first high-power laser experiments, where their high thermal conductivity (nearly twice as high as that of $Y_3Al_5O_{12}$ (YAG)) is of great advantage. Their low phonon energy ensures large energy storage times by minimizing non-radiative relaxation processes, and the large splitting of the lower $Yb^{3+}$ manifold enables efficient laser action in the quasi-for-level operation scheme. Additionally, the broad absorption and emission bandwidths of Yb-doped sesquioxides allow uncritical diode laser pumping and femtosecond pulse generation [1–3].

These materials are also interesting candidates for the thin disk laser concept invented by *Giesen* et al. in 1994 [4], where a thin crystal with thicknesses between $200\,\mu m$ and $400\,\mu m$ is used as active medium. These small thicknesses are possible as the sesquioxides allow very high ytterbium-dopant

M. Fanciulli, G. Scarel (Eds.): Rare Earth Oxide Thin Films,
Topics Appl. Physics **106**, 401–422 (2007)
© Springer-Verlag Berlin Heidelberg 2007

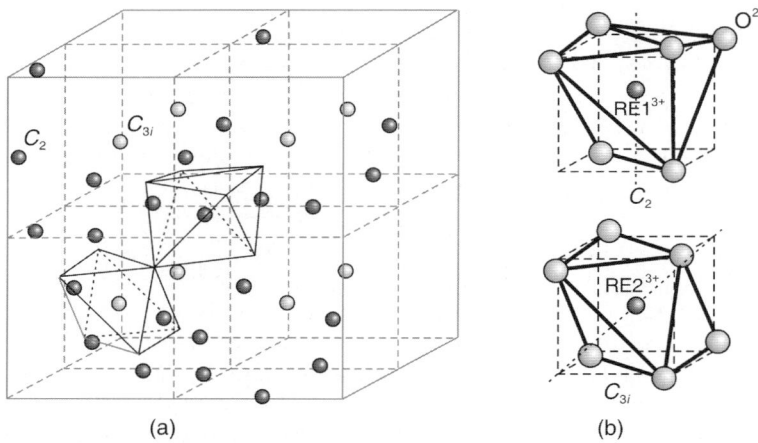

**Fig. 1.** Crystallographic structure of bixbyite: **(a)** unit cell with 24 sites with point group symmetry $C_2$ (non-centrosymmetric) and eight sites with $C_{3i}$ symmetry (centrosymmetric), **(b)** the two different cation sites

concentrations resulting in a high percentage of absorbed pump energy. Experiments with Yb : $Sc_2O_3$ and Yb : $Lu_2O_3$ have shown very promising results, with 124 W of output power at 250 W pump power in the case of scandia and 23 W for 54 W of pump power with lutetia (see Sect. 2) [5].

However, apart from superior properties compared to, e.g., YAG, the production of high-quality bulk samples is complicated by the high melting temperatures around 2500 °C. The heat-exchanger method was adapted for use at the high-melting temperatures to be able to obtain larger single crystals of higher optical quality than with the growth methods used in the past (e.g., Czochralski method). Single crystals of several cubic centimeters in volume can now be grown by this method.

The problems of bulk crystal growth can be overcome if epitaxial layers can be used grown, for example, by the method of pulsed laser deposition (PLD). This is a non-equilibrium process avoiding deviations from stoichiometry by a flash-like evaporation of the target material. As reported by *Grivas* et al., crystalline films with thicknesses of the order of 130 µm can be realized by PLD (Nd : GGG on YAG) [6]. Thus, thin film fabrication by PLD might be an interesting alternative for the fabrication of thin disk laser media as, by this technique, the endfaces of the crystal can be directly equipped with crystalline mirror coatings. The shorter crystal length, which is advantageous for the thermal management of the disk, can then be compensated by higher doping concentrations. A crucial point of this disk fabrication method is the lattice mismatch between substrate and film, which in the worst case can result in disordered polycrystalline growth (see Sect. 3).

Another area of application for rare-earth-doped sesquioxide thin films are passive as well as active waveguide structures, since waveguiding components based on thin film technology are key elements for the development of integrated optics and optoelectronics for optical communications, sensors, interface technologies, data processing, etc. A variety of functional active waveguides have been reported so far [7, 8]. In a crystalline waveguide, the confinement of light generates a large intensity-length product, and thus, the realization of non-linear processes, like upconversion or frequency conversion, which depend strongly on the intensity, is much easier to achieve in crystalline waveguides than in amorphous matrices like glass fibers. Additionally, the guiding of the pump mode as well as the signal mode leads to an excellent overlap of the modes, resulting in lower laser thresholds compared to the corresponding bulk materials [9]. Apart from waveguides of Nd : YAG [10], Nd : GGG [11, 12], and Ti : $Al_2O_3$ [11], also first yttria waveguides have been fabricated, for example, Er : $Y_2O_3$ [13–15].

Up to now, the materials used for integrated optics are mainly compounds of semiconductors, as they are the most promising materials for monolithic integration of the components. Still, in certain cases, the hybrid integration technology might be favorable. Advantages of these dielectric oxides are their wide transparency range (UV to mid-IR) and their lower refractive indices compared to semiconductors, resulting in considerably reduced Fresnel reflections at the endfaces designed for coupling light in and out of the device.

The growing interest in the fabrication and characterization of rare-earth oxide films over the last few years is related to the diverse range of potential applications that can be envisioned apart from waveguide devices, e.g., phosphor materials or high-temperature corrosion protection. Due to the large band gap ($\sim 6\,\mathrm{eV}$) and the high dielectric constant (14–18), these materials can be used in semiconductor devices, e.g., metal-insulator-semiconductor (MIS) diodes, transistor gates, metal-oxide-semiconductor (MOS) capacitors, and dynamic random access memory (DRAM) [16–18].

## 2 Rare-Earth-Doped Sesquioxide Lasers

As discussed above, sesquioxides are promising candidates as host materials for high-power lasers. For these applications, the $Yb^{3+}$ ion is most suitable among the rare-earth elements. This can be explained by the unique properties of trivalent ytterbium. Due to the fact that $Yb^{3+}$ has only two manifolds separated by $10\,000\,\mathrm{cm}^{-1}$, no effective loss processes (excited state absorption, upconversion, and cross relaxation) occur (see Fig. 2a). Compared to neodymium, the ytterbium ion has a very low quantum defect between the absorption (940 nm to 970 nm) and emission (1010 nm to 1100 nm) because of the quasi-four-level scheme, resulting in reduced thermal-induced effects. Additionally, the broad absorption lines around 940 nm and 970 nm are well suited for efficient laser diode pumping (see Fig. 2b).

Up to now, several laser experiments with single crystals of Yb-doped $Y_2O_3$, $Lu_2O_3$, and $Sc_2O_3$ have been performed in different configurations.

404    Sebastian Bär et al.

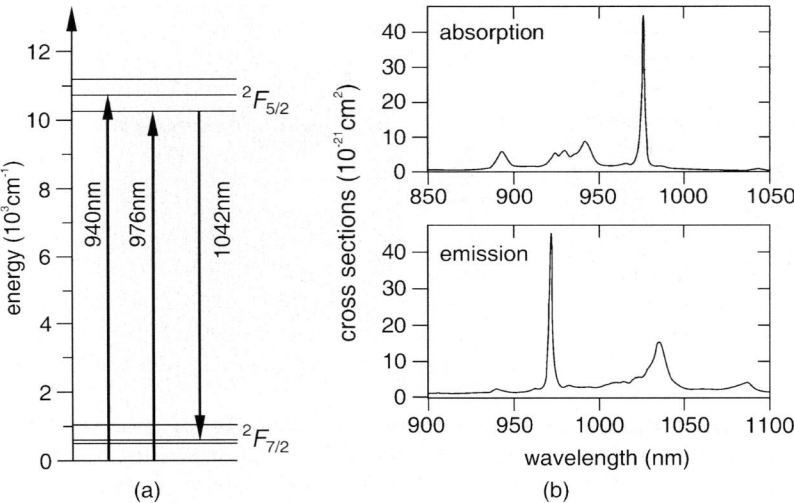

**Fig. 2.** $Yb^{3+} : Sc_2O_3$: (**a**) energy level scheme, (**b**) room temperature absorption and emission spectra

However, most of the lasers are based on the scandia matrix as host for the ytterbium ions, since it is the most efficient system. This can be explained by the largest ground-state splitting, the highest emission cross section, and the very high heat conductivity ($\sim 12\,\mathrm{W/mK}$) at low doping levels [3]. In a simple two-mirror longitudinally pumped cavity, the slope efficiencies range from 37 % to 58 %, depending on the output coupler [1,19]. Using the thin disk laser setup, slope efficiencies of nearly 50 % could be achieved (see below) [4]. Aside of these laser crystal-based systems, also $Yb : Sc_2O_3$ ceramic lasers have been invented. Under diode pumping, a slope efficiency of 9 % was achieved [20].

The thin disk laser concept designed for high-power lasers was invented in 1994 by *Giesen* et al. [4]. The geometry of this setup is depicted in Fig. 3 and will be explained in the following. The laser crystal with thicknesses between 200 µm and 400 µm is provided on the backside with a high-reflection (HR) coating for the pump wavelength as well as for the laser wavelength, and the frontside is anti-reflection (AR) coated also for both wavelengths. The laser resonator is formed by the thin disk mounted on a heat sink and an external output coupler. The pump light of the fiber-coupled laser diodes is collimated and then focused onto the crystal by the parabolic mirror. Because of the small crystal thickness, only a small fraction of the pump power will be absorbed at one pass through the crystal. For that reason, a more complicated pump geometry is used for multi-pass pumping. In this optical system, consisting of the parabolic mirror and three folding prisms, the non-absorbed power will be imaged several times on the crystal to optimize the absorption efficiency. The path of the pump beam for 16 passes is shown is Fig. 3, but

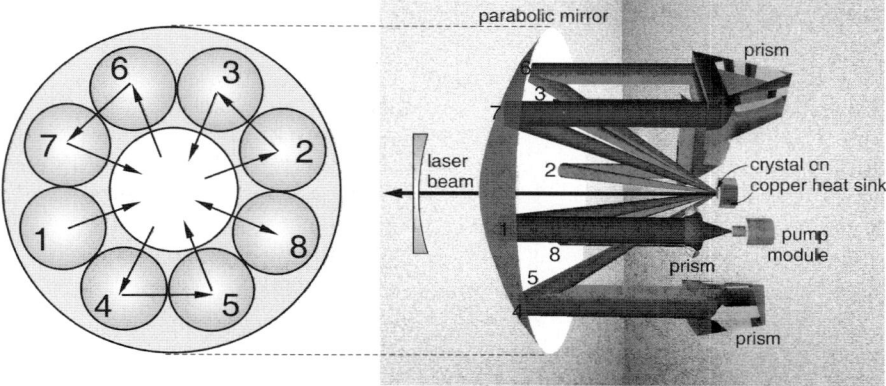

**Fig. 3.** Thin disk laser setup: pump geometry is shown on the *right*. On the *left*, a detailed view of the parabolic mirror is depicted, indicating the multiple reflections for multi-pass pumping; the passes $2 \rightarrow 3$, $4 \rightarrow 5$, and $6 \rightarrow 7$ are realized by folding prisms

currently up to a total of 32 pump light passes are used. The advantage of this geometry is the nearly one-dimensional thermal profile within the crystal, minimizing the formation of thermal lenses at high powers, and thus, extending the applicable power limit to the destruction limit of the crystal. Additionally, a high beam quality even at high laser powers is achievable.

After first experiments, scandia seems the most promising host material in this setup. Using an $Yb(2.8\%) : Sc_2O_3$ crystal, experiments under different pump conditions have been performed. The results are shown in Fig. 4. The maximum output power of 124.5 W at a laser wavelength of 1041 nm was achieved in the continuous wave (cw) mode at an incident power of 254 W at 940 nm, resulting in a slope efficiency of 49 % (see single circular point in Fig. 4). The output coupler had a transmission of 1.7 %. Other experiments with quasi-cw pumping at several pumping ratios (in Fig. 4, the triangles indicate the characteristics at a pumping ratio of 20/100 ms) or somewhat defocused excitation (diamond symbols) showed slightly lower results. Up to the maximum power, no roll-over can be seen, which otherwise would indicate overheating of the crystal.

However, improvement of the system can be achieved by pumping at the zero-phonon line due to the strong absorption. Furthermore, thinner crystals can be used, which reduces the thermal load of the crystal and, therefore, the risk of mechanical damage at higher pump powers.

## 3 Thin Film Preparation

As described in the introduction, the use of thin film production methods is also attractive for the preparation of thin disk laser media. In addition,

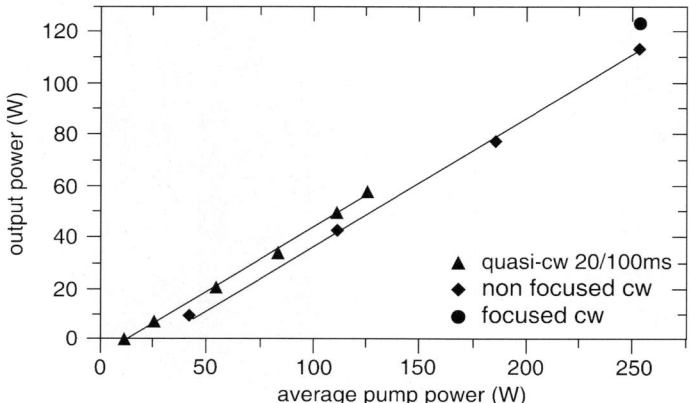

**Fig. 4.** Input-output characteristics of $Yb_0(2.8\%) : Sc_2O_3$ in the thin disk laser setup

the sesquioxides are interesting candidates for integrated optic devices due to their physical and chemical properties. The basis for fabrication of these devices are planar structures which can be fabricated by a variety of physical and chemical deposition techniques.

Here, high-quality crystalline thin films of various rare-earth-doped sesquioxides are presented, which have been grown either by pulsed laser deposition (PLD) or electron beam evaporation (EBV) both on single crystal (0001) sapphire substrates and on single crystal (0001) quartz substrates. The hexagonal structures of sapphire and quartz exhibit lattice constants of $a_{Al_2O_3} = 4.759\,\text{Å}$ and $a_{SiO_2} = 4.91\,\text{Å}$, respectively. As can be seen from Fig. 5, lattice matching is achieved by using different crystallographic orientations for the film and the substrate, because the cubic lattice of the sesquioxides provides a hexagonal arrangement of the ions in the $\{111\}$ planes of the cubic unit cell. Thus, the relation for lattice matching is $3 \times a_{\text{substrate}} \approx \sqrt{2}a_{RE_2O_3}$. The lattice mismatch for $Y_2O_3$ on sapphire (quartz) is $4.8\%$ ($1.8\%$). In the case of the sesquioxides $Lu_2O_3$ and $Sc_2O_3$, the mismatch is only $2.8\%$ ($0.2\%$) and $-2.5\%$ ($-5.8\%$), respectively.

To obtain "perfect" lattice matching, the lattice constant can be customized by mixing the sesquioxides, e.g., $Lu_xSc_{1-x})_2O_3$. These mixed crystals can be described as solid solutions, where the substituents are distributed statistically over the lattice sites in the matrix. As the sesquioxides behave chemically and physically very similarly, most of their properties are not changed in a mixed system. In binary solutions, like $(Lu_xSc_{1-x})_2O_3$, the lattice constant is a linear function of $x$, and thus the new lattice constant $a'$ can be calculated easily by the following equation, termed Vegard's law, $a' = xa_1 + (1-x)a_2$, where $a_1$ and $a_2$ are the lattice constants of the pure single crystals [21]. Applying this equation to sapphire substrates, lattice

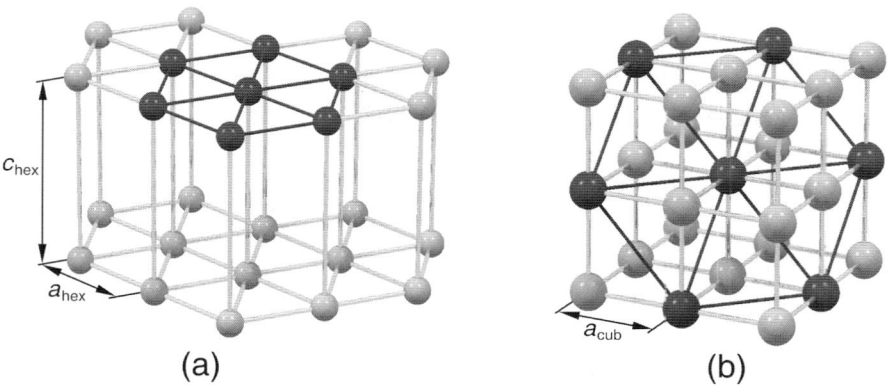

**Fig. 5.** Hexagonal structure (**a**) and cubic structure (**b**). In order to visualize characteristic planes, the corresponding atoms are displayed in a different color

matching is achieved using the compositions of either $Lu_{0.918}Sc_{1.082}O_3$ or $Y_{0.662}Sc_{1.338}O_3$. With the $LuScO_3$ system, perfect lattice matching on sapphire was achieved, proved by X-ray diffraction and surface X-ray diffraction (see Sect. 4.2). For silica substrates, one can use $Y_{0.28}Lu_{1.72}O_3$ to obtain the desired lattice matching.

### 3.1 Pulsed Laser Deposition

In recent years, the method of pulsed laser deposition (PLD) has proven to be a very attractive and practical method for the growth not only of high-quality doped [22, 23] and undoped $Y_2O_3$ films [24], but also of a variety of other crystalline materials and even polymers ($YBa_2Cu_3O_4$ [25], $Gd_3Ga_5O_{12}$, $Y_3Al_5O_{12}$ [26], PMMA [27]), polyacrylonitrile [28].

The targets used for ablation were either pressed and sintered sesquioxide powders or sesquioxide single crystals. The preparation of the sintered targets was as follows: europium-doped $RE_2O_3$ powders were cold-pressed into a pellet of one inch in diameter. To increase the density, the pellets were subsequently sintered for 72 h in air at a temperature of 1700 °C and then slowly cooled to room temperature (0.8 °C/min).

All targets were cleaned prior to each deposition by ablation with 300−500 pulses under oxygen deposition atmosphere conditions to ensure a homogeneous surface morphology. During the deposition process, the laser spot was raster-scanned over the target surface to reduce the possibility of macroscopic damages on the target surface and to avoid stoichiometric changes in the target material. The sapphire substrates as well as the $SiO_2$ subtrates (each type has the dimensions 10 mm × 10 mm × 0.5 mm, RMS roughness < 0.4 nm) were cleaned in acetone and alcohol prior to deposition. Before ablation, the chamber was evacuated to a base pressure of $10^{-8}$ mbar and then filled with oxygen with a partial pressure of $10^{-2}$ mbar. The substrates were placed at a distance of 9 cm in front of the target, and were heated electrically by a boron nitride heating element up to 750 °C before starting the deposition process. During the whole process, the temperature was computer controlled by a pyrometer.

Using an KrF excimer laser (Lambda Physics LPX 305i) operating at a wavelength of 248 nm, a pulse duration of 25 ns and a repetition rate of 10 Hz, $RE_2O_3$ targets were irradiated with a laser fluence of 4.25 J/cm². The film thickness was determined using in situ reflectometry. The growth rate was of the order of a few tenths of an Å per pulse.

## 3.2 Electron Beam Evaporation

Additionally, thin films were also prepared by electron beam evaporation (EBV). This technique is commonly used for insulating and resistor films for electronic components as well as optical anti-reflection coatings, mirrors, and filters. In this technology, the coating material is placed in a water-cooled crucible and the energy to melt, and finally evaporate the material is provided by a high-energy electron beam. To avoid the formation of deep craters in the melt, which can alter the evaporation characteristics, the electron beam is raster-scanned over the crucible. The main advantage of this method is the homogeneous coating of large substrate areas. Still, in contrast to PLD, which is a non-equilibrium process with particle energies up to 100 eV, the thermal equilibrium in the EBV process provides particle energies of only a few 100 meV. As a consequence, the adhesion of the condensated atoms on the substrate surface is lower than that of PLD-fabricated films, resulting in films with lower mechanical stability and a decrease in the packing density $p$ ($p_{PLD} \approx 1$, $0.75 < p_{EBV} < 1$) [29].

A stoichiometric powder mixture of rare-earth-doped sesquioxides was pressed into a molybdenum crucible, which was then placed into an EBV coating system (Balzers Co.). The powder in the crucible was melted using an electron beam, resulting in a black and glassy substance. This procedure was repeated until the crucible was completely filled. The sapphire substrates with the same dimensions as in the PLD experiments were placed into a concentric rotating substrate holder. Before evaporation, the chamber was evacuated to a base pressure of $10^{-7}$ mbar and then refilled with oxygen to a pressure of $3.5 \times 10^{-5}$ mbar. The substrates were heated electrically to 400 °C and the surface was cleaned using an argon discharge lamp. The growth rate of 3.7 Å/s was determined by a quartz crystal microbalance.

Both techniques have their advantages and drawbacks. The thermodynamical non-equilibrium PLD process allows the growth of films with a complex chemical composition, as the stoichiometry is conserved during the material transfer from the target to the substrate. Using EBV, different vapor pressures of the target material elements can lead to a deviation in film composition. For $(Lu/Sc)_2O_3$ films, this was observed by *Rabisch* et al. [30].

Another advantage of PLD are the high energies (up to 100 eV) of the particles hitting the substrate surface. In combination with a heated substrate, the atoms arriving at the surface have enough mobility to form dense, crystalline structures, while EBV films often show a porous, columnar growth behavior. In addition, for PLD the film thickness can be determined more precisely by the number of laser pulses than is the case in EBV. The accuracy in film thickness is of the order of atomic layers. A major drawback of pulsed laser deposition is to achieve homogeneous film growth over large areas, due to the directed plasma expansion in the process. As this problem does not occur in the EBV process, this method can be used for the coating of many substrates at the same time, making this fabrication method very cheap.

# 4 Analytical Techniques and Thin Film Characterization

To characterize the quality of the films with thicknesses between 3 nm and 1 μm, different techniques (AFM, XRD, SXRD, RBS, optical spectroscopy) have been applied. These techniques and their results will be discussed in the following.

## 4.1 Atomic Force Microscopy

The surface morphology of the thin films has been studied using atomic force microscopy (AFM) in the contact mode, which allows a very sensitive regulation of the tip–surface distance.

While the amorphous films are characterized by a randomly distributed grain structure with an in-plane size of the grains of 40 nm to 60 nm and an average height of 5 nm (see Fig. 6a), the crystalline films show an ordered triangular surface morphology (Fig. 6c). This ordered structure is more distinct in PLD-fabricated films than in EBV films, which can be explained by a higher degree of crystallinity. Both films, amorphous and crystalline, have a mean thickness of 500 nm, and the grade of crystallinity was determined by X-ray diffraction measurements. The triangular (2D) or pyramidal (3D) crystallite structure observed in the crystalline films represents the crystallographic cubic-fluorite structure of bulk $Y_2O_3$ and can be assigned to the $\langle 111 \rangle$ growth direction. The in-plane grain size is about 120 nm to 160 nm and is of the same order as the out-of-plane grain size determined by XRD measurements.

The RMS roughness of the crystalline film was determined to be 2.4 nm, a slight increase compared to the 1.7 nm RMS roughness of the amorphous films. A similar surface morphology with pyramidally shaped grains was also observed in [31] and [32]. Interestingly, the same structure is observed at the {111} cleavage plane of an yttria bulk crystal. Figure 6d shows a picture of such a cleavage taken by an optical light microscope. Together with the results

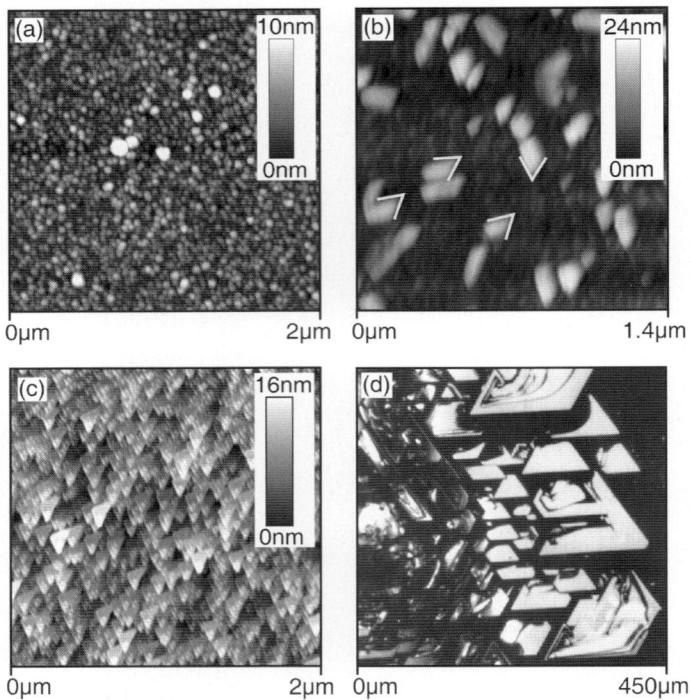

**Fig. 6.** Surface morphology of $Y_2O_3$ films: **(a)** 500 nm amorphous, **(b)** 5 nm crystalline, **(c)** 500 nm crystalline, **(d)** $\langle 111 \rangle$ cleavage of bulk crystal (optical light microscope) [22]

of X-ray diffraction and Rutherford backscattering, it can be concluded that the film surface presented in Fig. 6c shows a $\{111\}$ plane of yttria.

In general, the roughness of thick films, amorphous as well as crystalline, is comparably small (RMS values around 2 nm), which clearly shows that these films can have optical quality and thus, seem to be suitable for waveguide devices or optical coatings.

Thin films with a mean thickness of 5 nm have an uncompletely covered surface and show 3D island growth, as can be seen in Fig. 6b. Also in this case, the shape of the single crystallites with in-plane dimensions between 50 nm and 200 nm and heights ranging from 15 nm to 25 nm is triangular or quadrangular with at least one angle of 60° or 120°. As the "perfect" triangular structure of the $\{111\}$ planes also shows angles of 60°, it can be concluded that the films prefer the $\langle 111 \rangle$ growth direction from the beginning of film formation onwards.

## 4.2 X-Ray Diffraction Studies

The crystal structure of the different sesquioxide films was analyzed by X-ray diffraction (XRD) and surface X-ray diffraction (SXRD) methods. The peak positions of the resulting diffractograms provide information about the lattice structure and growth direction. The peak widths can give information about the size of the crystallites forming the film.

The XRD measurements were performed using either a Philips X'PERT system or a D8-ADVANCE diffractometer from Bruker axs Co. Both systems use the Cu–K$_{\alpha 1}$ radiation with a wavelength of $\lambda_{\mathrm{XRD}} = 1.506$ Å. The measurements were taken in the range of $20° \leq 2\theta \leq 75°$ with an increment of $\Delta 2\theta = 0.02°$. In addition, in-plane measurements were carried out by SXRD. These experiments were performed with synchrotron radiation at beamline BW2 at HASYLAB (DESY, Hamburg, Germany) [33]. The angle of incidence with respect to the surface was varied between 0.2° and 0.3° depending on the thicknesses of the films. The wavelength of the X-rays was 1.24 Å, corresponding to an energy of 10 keV. To facilitate the comparison with the XRD measurements, the $\theta$–$2\theta$ scans were converted to the Cu–K$_{\alpha 1}$ wavelength by

$$\sin \theta_{\mathrm{XRD}} = \frac{\lambda_{\mathrm{XRD}}}{\lambda_{\mathrm{SXRD}}} \sin \theta_{\mathrm{SXRD}} . \tag{1}$$

In addition to the $\theta$–$2\theta$ measurements, $\omega$-scans were taken. In the $\omega$-scans the $\theta$ and $2\theta$ angles were kept constant and the sample was rotated by 360° around the $\omega$-axis perpendicular to the surface of the substrate. These scans show the angular distribution of a distinct atomic plane selected by the $\theta$ and $2\theta$ angles.

### 4.2.1 X-Ray Diffraction

To investigate the grade of crystallinity of PLD-fabricated films for different growth conditions, XRD measurements were performed on Y$_2$O$_3$ films grown at different substrate temperatures and oxygen pressures, because these parameters have been found to be the most important ones. The results of a substrate temperature-dependent series of 500 nm thick films is depicted in Fig. 7. All films were grown at an O$_2$ partial pressure of $5 \times 10^{-2}$ mbar. For the film grown at room temperature, no Y$_2$O$_3$ diffraction peaks are visible, which means that the film is completely amorphous (see Fig. 7a). At a substrate temperature of 300 °C, the yttria {222} and {444} reflection peaks appear at 29.1° and 60.26°, respectively. Additionally, several other reflection peaks are visible but with less intensity than for the {222} and {444} peaks, indicating a preferred growth in the ⟨111⟩ direction (see Fig. 7b). This effect becomes stronger at a substrate temperature of 700 °C (see Fig. 7c). In this case, the {222} peak at 29.19° (FWHM = 0.163°) dominates the spectrum and is even stronger than the $\alpha$-Al$_2$O$_3$ peak (FWHM = 0.1°). The {400}

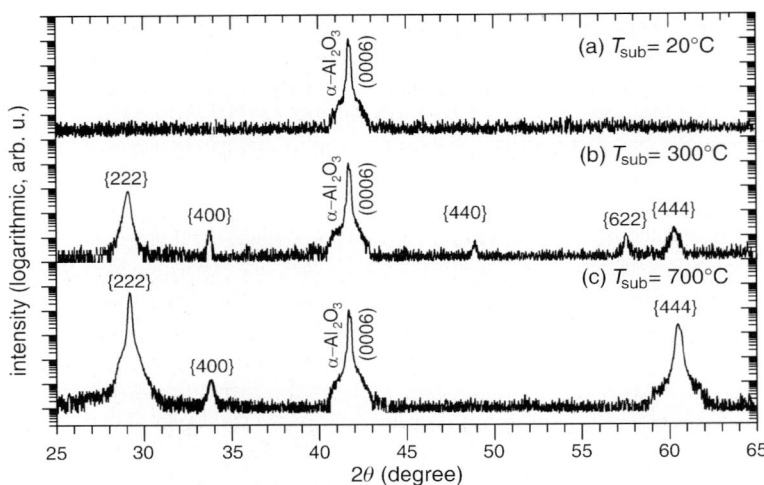

**Fig. 7.** X-ray diffraction patterns of 500 nm thick Eu : $Y_2O_3$ films on $\alpha$-$Al_2O_3$ ($\theta$–$2\theta$-scans) grown at different substrate temperatures $T_{\text{sub}}$ ($p(O_2) = 0.05$ mbar)

peak is also visible, but nearly four orders of magnitude smaller. From these measurements it can be seen that the 500 nm thick films grown at substrate temperatures $> 300\,^{\circ}C$ are uniaxially textured along the $\langle 111 \rangle$ direction of the cubic crystalline phase of $Y_2O_3$, because the X-ray diffraction patterns reveal mainly the {222} and {444} diffraction lines.

The same effect (preferred growth in $\langle 111 \rangle$ direction) is observed when the oxygen partial pressure is varied at constant temperature ($T_s = 700\,^{\circ}C$). If the pressure goes beyond $4 \times 10^{-3}$ mbar, the FWHM of the {222} reflection peak decreases until it reaches a minimum of $0.163°$ at $5 \times 10^{-2}$ mbar. A further increase of the pressure results in a strong increase of the FWHM. Additionally, it was observed that the {400} peak becomes stronger at higher oxygen content. The finite size of the crystallites $B$ ($B \gg \lambda$) can be estimated by the Scherrer equation

$$B = \frac{0.89\lambda}{(\Delta 2\vartheta)\cos\theta}\,. \tag{2}$$

Here, the expression $\Delta 2\vartheta = \Delta 2\theta - \Delta 2\theta_{\text{ref}}$ is the corrected linewidth, where $\Delta 2\theta$ is the observed linewidth and $\Delta 2\theta_{\text{ref}}$ is a reference linewidth of an "infinite" extended single crystal. In this case, the reference crystal was the sapphire substrate and the (0006) peak of $\alpha$-$Al_2O_3$ with a FWHM of $0.1°$ was used as reference linewidth. The broadening of the single crystal diffraction peaks is then limited only by the resolution of the apparatus. Using this equation, the FWHM of $0.163°$ of the {222} peak presented in Fig. 7c gives an out-of-plane diffracting domain size of $B_{\perp} = 160$ nm, which is nearly a third of the film thickness. In general, the fraction of crystallite size to mean film thickness becomes larger the thinner the films are. For example, in a 20 nm thick $Y_2O_3$ film fabricated by PLD, the crystallite size was estimated to be 18 nm, which is nearly the film thickness. The same observation was made for EBV-deposited films. This behavior indicates a maximum crystallite size. If the mean film thickness exceeds this crystallite size, a new layer of crystallites is formed.

In the XRD patterns of EBV-deposited yttria on sapphire substrates and additionally to the peaks of the $\langle 111 \rangle$ growth direction, peaks of $\beta$-$Y_2O_3$ can be identified (cf. Fig. 2 in [30]). This admixing of a different phase can be avoided if a better lattice matched system is used. Figures 8a and b show the diffraction patterns of two $Lu_{0.38}Sc_{1.62}O_3$ films grown on sapphire (thicknesses 24 nm and 3.6 nm), where only the {222} and {444} peaks of the mixed system are visible. The lattice mismatch for this mixture is only $-1.5\%$ instead of $+4.7\%$ for $Y_2O_3$, resulting in crystallites with dimensions of the film thickness.

### 4.2.2 Surface X-Ray Diffraction

To study the geometrical structure of surface and near-surface regions in more detail, the technique of surface X-ray diffraction (SXRD) has been applied. In contrast to XRD, where only information about planes parallel to the surface is obtained, the method of SXRD results in diffraction patterns of planes which are perpendicular to the surface.

Figure 8 summarizes the X-ray studies on two very thin EBV-deposited $(Lu/Sc)_2O_3$ films. Figures 8a and b show the XRD diffraction patterns of two $Lu_{0.38}Sc_{1.62}O_3$ films grown on sapphire (thicknesses 24 nm and 3.6 nm), where only the {222} and {444} peaks of the mixed system are visible. Additionally, surface X-ray diffraction (SXRD) experiments have been performed on these films. The results obtained using the 3.6 nm thick film are presented in Figs. 8c and d. It can be seen that the film is highly textured along the $\langle 111 \rangle$ direction, since only planes perpendicular to the {111} plane are observed in Fig. 8c. A rotation around the surface normal at constant $2\theta = 51.75°$ ($\omega$-scan) reveals six peaks of nearly the same intensity, separated by 60° each. The FWHM of the peaks is $\Delta\omega = 14.7°$. Again, this result indicates the preferred growth along the $\langle 111 \rangle$ direction, as the symmetry of this growth mode predicts these six diffraction peaks.

However, PLD-fabricated films tend to grow with a higher degree of crystallinity, as indicated by XRD and SXRD measurements. The structure of the spectra is very similar but the peak widths are considerably smaller. Regarding the $\omega$-SXRD measurements, the FWHM of the PLD films was in the best case $\Delta\omega = 7.8°$, showing that the grains are well aligned around the surface normal following the hexagonal structure of the $Al_2O_3$ substrate (cf. Fig. 3 in [30]).

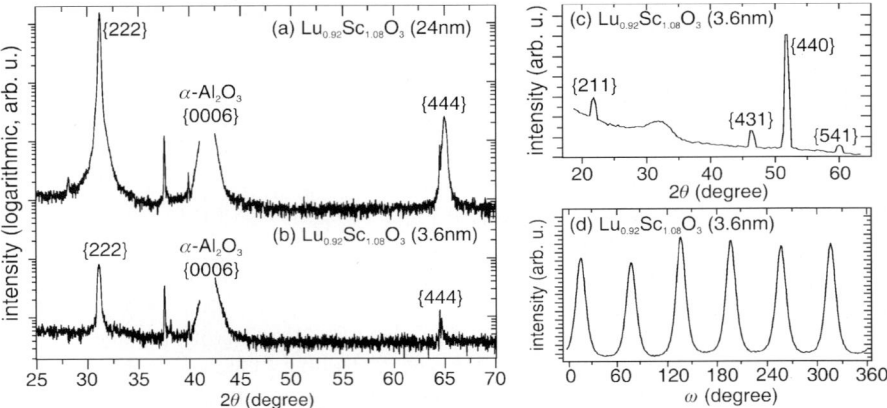

**Fig. 8.** *Left*: XRD pattern of (**a**) a 24 nm thick LuScO₃ film and (**b**) a 3.6 nm thick LuScO₃ film. *Right*: (**c**) SXRD $\theta$–$2\theta$ scan and (**d**) SXRD $\omega$-scan of a 3.6 nm thick LuScO₃ film. Substrate temperature was 400 °C

## 4.3 Rutherford Backscattering

The elemental composition of the films was examinded by Rutherford back-scattering analysis (RBS). Using a 20 MeV $^4\text{He}^+$ beam with a spot size of 1 mm² to probe the samples, the correct stoichiometry of the films could be proven and no impurities have been found. These results are valid for all PLD- and EBV-deposited pure sesquioxide films [10]. Figure 9 (lines a and b) shows the results of the measurement and the simulation of a Eu(1.5 %) : $Y_2O_3$ film with a thickness of 500 nm fabricated by PLD. For PLD films, epitaxial growth of the films along the $\langle 111 \rangle$ direction on the (0001) sapphire was evidenced by the observation of channeling in the RBS experiments (see Fig. 9, line c). A similar effect was described by *Cho* et al. [34], who observed a channeling effect in the system $Y_2O_3(111)$ / Si(111).

## 4.4 Optical Spectroscopy

More information about the quality of the films can be obtained by optical spectroscopy. Therefore, room temperature excitation and emission measurements have been performed in the near-UV and visible spectral range (300 nm to 750 nm) using a modular fluorescence spectrometer (Jobin Yvon Fluorolog). Additionally, excitation measurements have been carried out in the VUV part of the electromagnetic spectrum from 58 nm to 350 nm at the SuperLumi setup at HASYLAB (DESY, Hamburg, Germany) [35].

For first experiments, all films were doped with europium (1.5 % at.). The $Eu^{3+}$-ion was chosen because of its high emission quantum yield, its relatively simple energy level scheme, and the possibility to use it as a structural probe for the sites occupied by the $RE^{3+}$-ions in a given material.

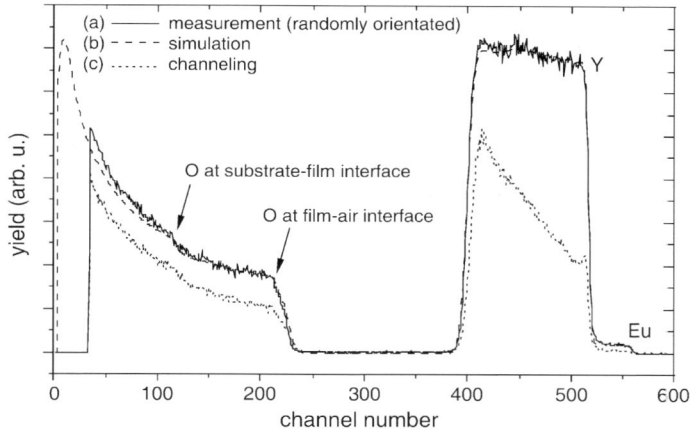

**Fig. 9.** RBS spectrum of a $500\,\mathrm{nm}$ thick $\mathrm{Eu}(1.5\,\%) : \mathrm{Y_2O_3}$ film on alumina

All emission measurements in the visible have been carried out by excitation of the $\mathrm{Eu}^{3+}$-ion into the maximum of charge transfer band around $240\,\mathrm{nm}$. This band originates from interaction between $\mathrm{Eu}^{3+}$-ions and $\mathrm{O}^{2-}$-ions and ranges from $200\,\mathrm{nm}$ to $265\,\mathrm{nm}$. This excitation results in the promotion of the electrons into the $^5D_J$ manifold due to non-radiative processes. The characteristic emission results from the $4f \rightarrow 4f$ electronic relaxation to the $^7F_J$ ground-state manifold. For excitation measurements, the emission wavelength was kept constant at the dominant transition peak at $611\,\mathrm{nm}$. The results are shown in Fig. 10.

The reference spectrum of a $\mathrm{Lu_2O_3}$ bulk crystal in Fig. 10a is characterized by an intense, narrow peak at $611\,\mathrm{nm}$, arising from the hypersensitive $^5D_0 \rightarrow {}^7F_2$ transition of $\mathrm{Eu}^{3+}$ in the $C_2$ site, whereas the less intense features at $586\,\mathrm{nm}$, $592\,\mathrm{nm}$, and $599\,\mathrm{nm}$ belong to the $^5D_0 \rightarrow {}^7F_1$ transition of $\mathrm{Eu}^{3+}$-ions occupying the $C_2$ as well as the $C_{3i}$ site. Down to a thickness of $20\,\mathrm{nm}$, the emission and excitation spectra of the Eu-doped films look similar to those of the corresponding crystalline bulk material, whereas films with a thickness $\leq 5\,\mathrm{nm}$ show a broadened emission behavior. This broadening can be explained by inhomogeneous and amorphous film growth in the first atomic layers. In Figs. 10b–d, the fluorescence spectra of $\mathrm{Eu}(1.5\,\%) : \mathrm{Y_{0.28}Lu_{1.72}O_3}$ films grown on $\mathrm{SiO_2}$ with different thicknesses are shown. If the thickness of the active layer is reduced to $5\,\mathrm{nm}$, but embedded between two undoped layers, the crystalline behavior of the fluorescence is partly reconstructed (see Fig. 10e). Thus, it can be concluded that the luminescence behavior of very thin films is influenced mainly by surface effects (island growth leads to a large surface-to-volume ratio of the crystallites), and partly by interface effects (subplantation of film components into the substrate matrix), depending on the growth parameters. If the growth conditions are optimized, the interface between substrate and film plays a minor role. The main changes in

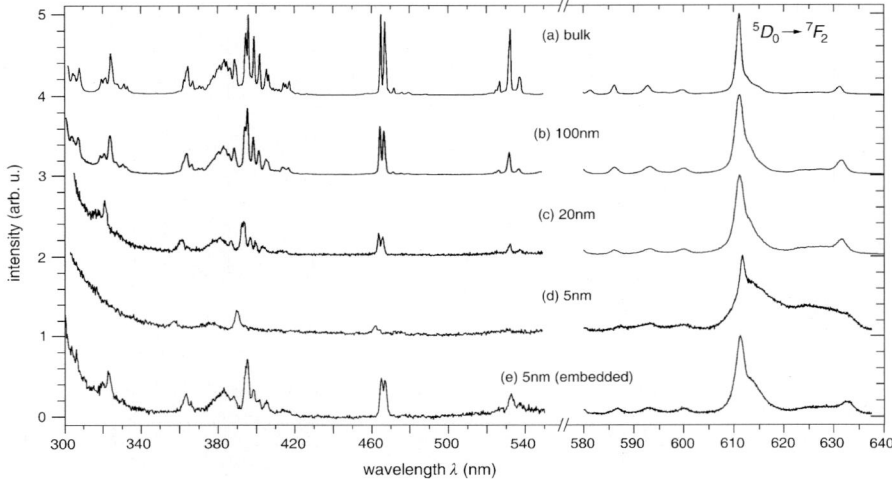

**Fig. 10.** Excitation ($\lambda_{\mathrm{em}} = 611\,\mathrm{nm}$) and florescence ($\lambda_{\mathrm{exc}} = 255\,\mathrm{nm}$) spectra of
(a) Eu : $Lu_2O_3$ bulk, (b) $100\,\mathrm{nm}$ Eu : $YLuO_3$, (c) $20\,\mathrm{nm}$ Eu : $YLuO_3$, (d) $5\,\mathrm{nm}$
Eu : $YLuO_3$, and (e) Eu : $YScO_3$ embedded between undoped layers. The substrate
is $SiO_2$ in all cases

luminescence behavior result from the film–air interface. This was proven by
a 5 nm thick crystalline Eu-doped $Y_2O_3$ film grown on sapphire and covered
with an amorphous $Al_2O_3$ layer which showed the bulk-like luminescence behavior. In the case of EBV-deposited films which are covered by an additional
layer, not only are the spectral characteristics affected but also the quantum
efficiency of the $Eu^{3+}$ fluorescence is increased compared to non-covered thin
films.

To obtain more information about the spectroscopic behavior of thin Eu-
doped films, additional excitation measurements have been performed in the
VUV spectral range. The results of Eu : $Y_2O_3$ films with thicknesses of 40 nm
and 100 nm are shown in Fig. 11. Here, it can be seen that the VUV excitation spectra of the films, which were measured at an emission wavelength of
611 nm, are almost identical to the reference bulk spectrum (see Fig. 11a).

The excitation spectrum of the Eu : $Y_2O_3$ bulk crystal is dominated
by the broad charge transfer excitation state (CTS). The peak at 204 nm
(6.08 eV) corresponds to the well-known excitonic state (ES). Subsequent
to the ES, the interband transitions of $Y_2O_3$ are visible. Excitation of the
excitonic state as well as the CTS lead to the well-known emission spectrum
of europium-doped yttria due to an energy transfer from the ES to the $^5D_J$
states of the $Eu^{3+}$-ion. The same spectroscopic behavior is observed in the
thin films, as can be seen in Figs. 11b and c. The difference in intensity
between the ES and the CTS is not due to the film characteristics but rather
to the thickness-dependent properties of these two excitation mechanisms.

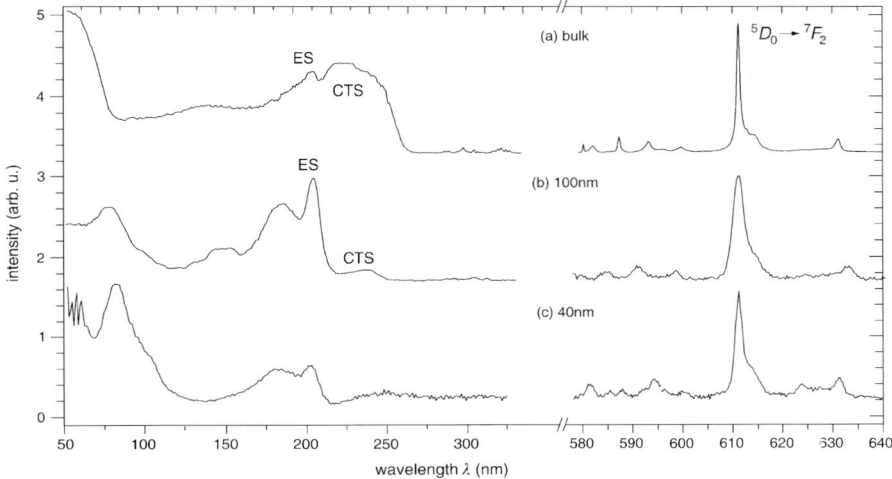

**Fig. 11.** VUV excitation ($\lambda_{\mathrm{em}} = 611\,\mathrm{nm}$) and fluorescence ($\lambda_{\mathrm{exc}} = 204\,\mathrm{nm}$ or $180\,\mathrm{nm}$) of **(a)** Eu : $Y_2O_3$ bulk, **(b)** $100\,\mathrm{nm}$ Eu : $Y_2O_3$, and **(c)** $40\,\mathrm{nm}$ Eu : $Y_2O_3$. ES – excitonic state, CTS – charge transfer state

In general, the excitation of an exciton occurs only in a surface layer with a thickness $\leq 100\,\mathrm{nm}$, whereas the CT process can occur to depths of $5\,\mu\mathrm{m}$. Thus, in thin films the excitonic transitions are expected to be more efficient, i.e., they have higher cross sections, than the CT process.

### 4.4.1 Temporal Luminescence Characteristics

The emission characteristics of the rare-earth-doped PLD-fabricated films remain unchanged over a period of several years. This result is valid for pure sesquioxide films as well as for mixed, lattice matched systems. In contrast, films of pure yttria on sapphire deposited by EBV show a change in the emission characteristics and a decrease in luminescence efficiency with time. As observed by *Rabisch* et al. [30], the quantum efficiency of Eu-doped $Y_2O_3$ films deposited by EBV decreases dramatically after a period of 6 months. A decrease of luminescence intensity of gas-phase-condensed Eu : $Y_2O_3$ nanocrystals is reported in [36]. Responsible for this process seems to be the formation of an amorphous hydroxide phase due to reactions with atmospheric water vapor and $CO_2$.

However, if lattice matched films are fabricated by EBV, this altering process can be avoided. For example, the $(Lu/Sc)_2O_3$ films presented in Sect. 4.2 show, 1 year after fabrication, the same spectral emission behavior and quantum efficiency as directly after deposition. An explanation for this different behavior is the crystallite size and the density of the films. The lattice matched systems have larger crystallites and a higher density than the pure sesquioxide films of the same thickness. Consequently, a higher effective surface area for chemical reactions is provided in the pure $RE_2O_3$ films, leading to the considerable modifications observed.

## 5 Waveguides

The importance of the production of planar waveguides as key elements in integrated optics was already discussed in Sect. 1.

Two yttria waveguides were prepared on $\alpha$-$Al_2O_3$ and $\alpha$-$SiO_2$. The difference in the refractive indices of the film and the substrate materials ($\Delta n \approx 0.39$ ($SiO_2$) or $0.18$ ($Al_2O_3$)) leads to guiding of the light in the sesquioxide layer. The Eu-doped $Y_2O_3$ waveguide on sapphire with a film thickness of $1\,\mu$m was grown by electron beam evaporation (EBV) and then covered with an amorphous alumina film with a thickness of $1\,\mu$m in order to reduce scattering at the film surface and to render the waveguide symmetrical. Since the film growth occurred at room temperature, the yttria film is also amorphous. The second waveguide consists of a crystalline, $1.2\,\mu$m thick Eu : $Y_2O_3$ film prepared by PLD at $700\,^\circ$C.

The dimension of the waveguides, i.e., a film thickness $d \approx 1\,\mu$m, was chosen in order to find a compromise between efficient incoupling of the light and single mode guiding. For a film thickness of $1\,\mu$m, the mode guidance condition predicts only small contributions of the first-order mode.

Waveguide experiments were performed by focusing the light from a blue semiconductor laser ($\lambda = 398$ nm), an Ar-ion laser ($\lambda = 514$ nm) or a helium neon laser ($\lambda = 632$ nm) onto the polished endface of the waveguide using a high NA triplet ($f = 6.5$ mm, NA $= 0.61$). The output endface is imaged onto the CCD chip of a camera using a microscope objective. The camera picture obtained from a $1\,\mu$m thick Eu-doped yttria film on sapphire is shown in Fig. 12a. It can be seen that the output is nearly uniform along the $y$-direction. The measured intensity profile in the $x$-direction can be interpreted as the fundamental mode of the waveguide. A comparison between the experimental result and the numerical simulation is given in Fig. 12b. When the light from the semiconductor laser is coupled into the waveguide, a bright red $Eu^{3+}$ fluorescence can be seen from the side.

The results of the crystalline PLD-fabricated waveguide show less clear intensity distribution. In fact, more scattering in this waveguide was observed. This behavior can be explained by the crystalline structure, which is not perfect single crystalline. As shown in Sect. 4.2, the particle dimension in the thick PLD-fabricated films is of the order of 200 nm, which is a factor of 0.3 to 0.5 smaller than the wavelengths used. This leads to pronounced Mie-scattering. In the amorphous film, the particle size is much smaller and scattering processes play a minor role in loss processes.

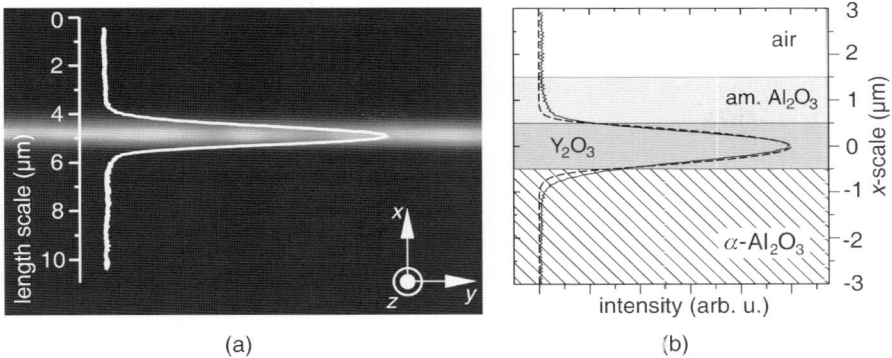

(a)                                        (b)

**Fig. 12.** (a) Waveguide intensity distribution ($\lambda = 514\,\mathrm{nm}$), (b) experimental mode intensity distribution (*solid line*) compared with the theoretical prediction of the fundamental mode (*dashed line*)

## 6 Summary

In this Chapter, the optical applications of rare-earth-doped sesquioxides have been presented. Due to their physical and chemical properties, the rare-earth oxides of yttrium, lutetium, and scandium are well suited as host materials for high-power lasers. Ytterbium-doped sesquioxide crystals grown by the heat-exchange method have been successfully operated in the thin disk laser setup. A maximum output power of 124.5 W with a slope efficiency of nearly 50 % in the cw mode has been achieved.

Additionally, rare-earth oxides seem to be interesting candidates for devices in integrated optics. Thus, thin films of rare-earth-doped sesquioxide films have been fabricated either by pulsed laser deposition or by electron beam evaporation on different substrates. As a result, good-quality crystalline films with thicknesses between 1 nm and 1 μm have been grown by PLD. In general, the quality of the EBV films is inferior to that of the PLD films. Only in the case of latticed matched systems, e.g., $(\mathrm{Lu/Sc})_2\mathrm{O}_3$ on $\alpha\text{-Al}_2\mathrm{O}_3$, do the structural and optical behavior of these EBV films become comparable to those of the PLD films. X-ray diffraction analysis (XRD/SXRD) as well as the results of atomic force microscopy indicate a highly textured film growth of $\mathrm{RE}_2\mathrm{O}_3$ along the $\langle 111 \rangle$ direction. The documented luminescence characteristics of the films (excitation and emission) are almost identical to those of the bulk material. The change in the spectral behavior in thin films is attributed mainly to surface effects, as optically active films buried under non-active layers show the bulk-like emission behavior.

A first step for application in integrated optics is the development of planar waveguiding structures. Thus, first planar yttria waveguides have been fabricated, and guiding of light at different wavelengths has been demonstrated. In the future, these waveguides will be optimized with respect to their optical characteristics, i.e., lattice matched systems with crystalline growth to minimize scattering losses, and use different rare-earth ions (Nd, Er) in order to obtain laser active waveguides.

# References

[1] L. Fornasiero, E. Mix, V. Peters, K. Petermann, G. Huber: New oxide crystals for solid state lasers, Cryst. Res. Technol. **34**, 255–260 (1999)

[2] L. Laversenne, Y. Guyot, C. Goutaudier, M. T. Cohen-Adad, G. Boulon: Optimization of spectroscopic properties of $Yb^{3+}$-doped refractory sesquioxides: Cubic $Y_2O_3$, $Lu_2O_3$ and monoclinic $Gd_2O_3$, Opt. Mater. **16**, 475–483 (2001)

[3] P. Klopp, V. Petrov, U. Griebner, K. Petermann, V. Peters, G. Erbert: Highly efficient mode-locked $Yb : Sc_2O_3$ laser, Opt. Lett. **29**, 391–393 (2004)

[4] A. Giesen, H. Hügel, A. Voss, K. Wittig, U. Brauch, H. Opower: Scalable concept for diode-pumped high-power solid-state lasers, Appl. Phys. B **58**, 365 (1994)

[5] V. Peters, K. Petermann, G. Huber, M. Larionov, J. Speiser, A. Giesen: Growth of sesquioxides for high power thin-disk-laser applications, in *Advanced Solid-State Lasers Conference*, vol. 68, Technical Digest, OSA TOPS (Quebec City, Canada 2002) pp. 150–152

[6] C. Grivas, T. C. May-Smith, D. P. Shepherd, R. W. Eason: On the growth and lasing characteristics of thick Nd : GGG waveguiding films fabricated by pulsed laser deposition, Appl. Phys. A **79**, 1203–1206 (2004)

[7] J. K. Jones, J. P. de Sandro, M. Hempstead, D. P. Shepherd, A. C. Large, A. C. Tropper, J. S. Wilkinson: Channel waveguide laser at 1 micron in Yb-indiffused $LiNbO_3$, Opt. Lett. **20**, 1477–1479 (1995)

[8] E. Flores-Romero, G. V. Vazquez, H. Marquez, R. Rangel-Rojo, J. Rickards, R. Trejo-Luna: Planar waveguide lasers by proton implantation in Nd : YAG crystals, Optics Express **12**, 2264 (2004)

[9] D. C. Hanna, A. C. Large, D. P. Shephard, A. C. Trooper, I. Charier, B. Ferrand, D. Pelenc: Low threshold quasi-three-level 946 nm laser operation of an epitaxially grown $Nd : Y_3Al_5O_{12}$ waveguide, Appl. Phys. Lett. **63**, 7–9 (1993)

[10] C. L. Bonner: *Multi-Watt, Diode-Pumped Planar Waveguide Lasers*, Ph.D. thesis, Faculty of Science, Department of Physics, University Southampton, Southampton (2000)

[11] A. A. Anderson: *Crystalline Planar Waveguide Lasers Fabricated by Pulsed Laser Deposition*, Ph.D. thesis, Faculty of Science, Department of Physics, University of Southampton, Southampton (1998)

[12] S. J. Barrington: *Planar Waveguide Devices Fabricated by Pulsed Laser Deposition*, Ph.D. thesis, Faculty of Science, Department of Physics, University of Southampton, Southampton (2001)

[13] T. H. Hoekstra, L. T. H. Hilderink, P. V. Lambeck, T. J. A. Popma: Photoluminescence and attenuation of spray-pyrolysis-deposited erbium-doped $Y_2O_3$ planar optical waveguides, Opt. Lett. **17**, 1506–1508 (1992)

[14] M. B. Korzenski, P. Lecoeur, B. Mercey, P. Camy, J. L. Doualan: Low propagation losses of an $Er : Y_2O_3$ planar waveguide grown by alternate-target pulsed laser deposition, Appl. Phys. Lett. **78**, 1210–1212 (2001)

[15] P. Lecoeur, M. B. Korzenski, A. Ambrosini, B. Mercey, P. Camy, J. L. Doualan: Growth of Er : $Y_2O_3$ thin films by pulsed laser ablation from metallic targets, Appl. Surf. Sci. **186**, 403–407 (2002)

[16] S. L. Jones, D. Kumar, R. K. Singh, P. H. Holloway: Luminescence of pulsed laser deposited Eu yttrium oxide films, Appl. Phys. Lett. **71**, 404–406 (1997)

[17] A. C. Rastogi, R. N. Sharma: Structural and electrical characteristics of metal-insulator-semiconductor diodes based on $Y_2O_3$ dielectric thin films on silicon, J. Appl. Phys. **17**, 5041–5052 (1992)

[18] S. Zhang, R. Xiao: Yttrium oxide films prepared by pulsed laser deposition, J. Appl. Phys. **83**, 3842–3848 (1998)

[19] K. Petermann, L. Fornasiero, E. Mix, V. Peters: High melting sesquioxides: Crystal growth, spectroscopy, and laser experiments, Opt. Mat. **19**, 67–71 (2002)

[20] J. Lu, J. F. Bisson, K. Takaichi, T. Uematsu, A. Shirakawa, M. Musha, K. Ueda, H. Yagi, T. Yanagitani, A. A. Kaminskii: $Yb^{3+}$ : $Sc_2O_3$ ceramic laser, Appl. Phys. Lett. **83**, 1101–1103 (2003)

[21] L. Vegard: Die Konstitution der Mischkristalle und die Raumfüllung der Atome, Zeitschrift für Physik **5**, 17 (1921)

[22] S. Bär, G. Huber, J. Gonzalo, A. Perea, M. Munz: Pulsed laser deposition of Eu : $Y_2O_3$ thin films on (0001) $\alpha$-$Al_2O_3$, Appl. Phys. A **80**, 209 (2005)

[23] K. G. Cho, D. Kumar, P. H. Holloway, R. K. Singh: Luminescence behavior of pulsed laser deposited Eu : $Y_2O_3$ thin film phosphors on sapphire substrates, Appl. Phys. Lett. **73**, 3058 (1998)

[24] S. Zhang, R. Xiao: Yttrium oxide films prepared by pulsed laser deposition, J. Appl. Phys. **83**, 3842 (1998)

[25] D. Dijkamp, T. Venkatesan, X. D. Wu, S. A. Shaheen, N. Jisrawi, Y. H. Min-Lee, W. L. McLean, M. Croft: Preparation of Y−Ba−Cu oxide superconductor thin films using pulsed laser evaporation from high $T_c$ bulk material, Appl. Phys. Lett. **51**, 619 (1987)

[26] S. J. Barrington, T. Bhutta, D. P. Shepherd, R. W. Eason: The effect of particulate density on performance of Nd : $Gd_3Ga_5O_{12}$ waveguide lasers grown by pulsed laser deposition, Opt. Commun. **185**, 145 (2000)

[27] S. G. Hansen, T. E. Robitaille: Formation of polymer films by pulsed laser evaporation, Appl. Phys. Lett. **52**, 81 (1988)

[28] S. Nishio, T. Chiba, A. Matsuzaki, H. Sato: Control of structures of deposited polymer films by ablation laser wavelength: Polyacrylonitrile at 308, 248, and 193 nm, J. Appl. Phys. **79**, 7198 (1996)

[29] F. Flory, L. Escoubas: Optical properties of nanostructured thin films, Prog. in Quant. Electr. **28**, 89–112 (2004)

[30] L. Rabisch, S. Bär, H. Scheife: Eu doped $(Lu/Sc)_2O_3$ thin films grown by thermal evaporation, Opt. Lett. accepted for publication

[31] K. G. Cho, D. Kumar, D. G. Lee, S. L. Jones, P. H. Holloway, R. K. Singh: Improved luminescence properties of pulsed laser deposited Eu : $Y_2O_3$ thin films on diamond coated silicon substrates, Appl. Phys. Lett. **71**, 3335–3337 (1997)

[32] M. B. Korzenski, P. Lecoeur, B. Mercey, D. Chippaux, B. Raveau, R. Desfeux: PLD-grown $Y_2O_3$ thin films from Y metal: An advantageous alternative to films deposited from yttria, Chem. Mat. **12**, 3139 (2000)

[33] H. Schulte-Schrepping, J. Heuer, B. Hukelmann: Adaptive indirectly cooled monochromator crystals at HASYLAB, J. Sync. Rad. **5**, 682 (1998)
[34] M. H. Cho, D. H. Ko, K. Jeong, I. W. Lyo, S. W. Whangbo, H. B. Kim, S. Choi, J. H. Song, S. Cho, C. N. Whang: Temperature dependence of the properties of heteroepitaxial $Y_2O_3$ films grown on Si by ion assisted evaporation, J. Appl. Phys. **86**, 198–204 (1999)
[35] T. Möller, P. Gürtler, E. Roick, G. Zimmerer: The experimental station superlumi: A unique setup for time- and spectrally resolved luminescence under state selective excitation with synchrotron radiation, Nucl. Instrum. Met. A **246**, 461 (1986)
[36] B. M. Tissue: Synthesis and luminescence of lanthanide ions in nanoscale insulating hosts, Chem. Mater. **10**, 2837–2845 (1998)

# Index

# Index

# Topics in Applied Physics

Printing: Krips bv, Meppel
Binding: Stürtz, Würzburg